Advances in
Cryogenic Engineering
Materials

VOLUME 38, PART A

A Continuation Order Plan is available for this series. A continuation order will bring delivery of each new volume immediately upon publication. Volumes are billed only upon actual shipment. For further information please contact the publisher.

An International Cryogenic Materials Conference Publication

Advances in Cryogenic Engineering
Materials

VOLUME 38, PART A

Edited by
F. R. Fickett
National Institute of Standards and Technology
U.S. Department of Commerce
Boulder, Colorado

and
R. P. Reed
Cryogenic Materials, Inc.
Boulder, Colorado

PLENUM PRESS · NEW YORK and LONDON

The Library of Congress cataloged the first volume of this title as follows:

Advances in cryogenic engineering. v. 1–
  New York, Cryogenic Engineering Conference; distributed
by Plenum Press, 1960–
    v. illus., diagrs. 26 cm.
  Vols. 1–   are reprints of the Proceedings of the Cryogenic Engineering
Conference, 1954–
  Editor: 1960–   K. D. Timmerhaus.

  1. Low temperature engineering—Congresses.   I. Timmerhaus, K. D., ed. II
Cryogenic Engineering Conference.
TP490.A3   660.29368   57-35598

Proceedings of the Ninth Intenational Cryogenic Materials
Conference (ICMC), held June 11–14, 1991, in Huntsville, Alabama

ISBN 0-306-44183-7

© 1992 Plenum Press, New York
A Division of Plenum Publishing Corporation
233 Spring Street, New York, N.Y. 10013

All rights reserved

No part of this book may be reproduced, stored in a retrieval system, or transmitted
in any form or by any means, electronic, mechanical, photocopying, microfilming,
recording, or otherwise, without written permission from the Publisher

Printed in the United States of America

FOREWORD

The ninth International Cryogenic Materials Conference (ICMC) was held on the campus of the University of Alabama at Huntsville (UAH) in collaboration with the Cryogenic Engineering Conference (CEC) on June 11-14, 1991. The continuing bond between these two major conferences in the field of cryogenics is indicative of the extreme interdependence of their subject matter. The major purpose of the conference is sharing of the latest advances in low temperature materials science and technology. However, the many side benefits which accrue when this many experts gather, such as identification of new research areas, formation of new collaborations which often cross the boundaries of both scientific discipline and politics, and a chance for those new to the field to meet the old-timers, may override the stated purpose.

This 1991 ICMC was chaired by F.R. Fickett of the National Institute of Standards and Technology. K.T. Hartwig, of Texas A&M served as Program Chairman with the assistance of eleven other Program Committee members.

We especially appreciate the contributions of the CEC board and its Conference Chairman, J. Hendricks of Alabama Cryogenic Engineering, to the organization of this joint conference. UAH hosted the conference. The local arrangements and management, under the watchful eye of Ann Yelle and Mary Beth Magathan of the UAH conference staff, were excellent.

Participation in the CEC/ICMC continues to exceed expectations with 650 registrants for the combined conference. The ICMC attracted more than 200 contributed and invited papers covering all aspects of cryogenic materials research. The two ICMC plenary papers discussed superconductivity at high and low temperatures from an applications viewpoint. The first, by David Larbalestier of the University of Wisconsin, described the rapid advances occurring in the development of ceramic superconductors and the second, by Carl Henning of the Lawrence Livermore Laboratory brought us up to date on the International Thermonuclear Experimental Reactor (ITER). Papers describing advances in high temperature superconductivity (HTS), especially in YBCO and BSCCO conductors, accounted for about half of the superconductivity papers. The prospects for commercially viable HTS conductors for a range of applications appears excellent. Development of the more conventional low temperature superconductors (LTS) continues under the encouragement of large projects such as SSC, ITER, LHC, and others. The use of very fine filaments in LTS wire and cable for ac applications has led to a renewed interest in magnetization and loss measurements as well as new and clever processing techniques. Pure metal cryoconductors were not as much in evidence this time, although a successful generator using hydrogen-cooled aluminum was described and data were presented on a number of alloys commonly used in construction of low temperature apparatus.

The combined conferences are held biennially. The next CEC/ICMC will be hosted by Los Alamos National Laboratory (LANL) and held July 12-16, 1993 at the Albuquerque Convention Center. K.T. Hartwig of Texas A&M will be the ICMC Chairman and W. Stewart of LANL will chair the CEC. The ICMC Board sponsors conferences on special topics in other years. In 1992 a conference on cryogenic materials will be held in Kiev, Ukraine in collaboration with the International Cryogenic Engineering Conference (ICEC).

This volume is the second largest ever published by ICMC. From before the start of the conference to the day when the final copy was sent to the publisher, the task of keeping this volume on track has fallen to Anne Shuri. The editors are grateful for her accurate and diligent performance of the many tasks associated with its production.

F.R. Fickett
R.P. Reed

BEST PAPER AWARDS

Awards for the best papers of the 1989 ICMC proceedings, <u>Advances in Cryogenic Engineering - Materials</u>, Volume 36, were presented at the 1991 conference. Selection is made by the awards committee from nominations of the editors in the categories of superconductors, structural materials, and student paper. It was a pleasure to present these three awards. We thank the authors for their exemplary contributions.

PROXIMITY EFFECT OF FLUX PINNING STRENGTH IN SUPERCONDUCTING Nb-Ti WITH
THIN α-Ti RIBBONS

T. Matsushita, S. Otabe, and T. Matsuno

Kyushu University

ELECTRICAL INSULATION SYSTEM FOR SUPERCONDUCTING MAGNETS ACCORDING
TO THE WIND AND REACT TECHNIQUE

P. Bruzzone, K. Nylund, and W.J. Muster

ETH High Voltage Engineering Group
ABB - IFE
EMPA, Swiss Laboratory for Materials Research

CREEP OF PURE ALUMINUM AT CRYOGENIC TEMPERATURES

L.C. McDonald and K.T. Hartwig
Texas A&M University

1991 INTERNATIONAL CRYOGENIC MATERIALS CONFERENCE

BOARD OF DIRECTORS

A.F. Clark .... National Institute of Standards & Technology
Gaithersburg, Maryland, USA

E.W. Collings .... Battelle Memorial Institute
Columbus, Ohio, USA

D. Evans .... Rutherford Appleton Laboratory
Chilton, Oxon, England

F.R. Fickett .... National Institute of Standards & Technology
Boulder, Colorado, USA

H.C. Freyhardt .... University of Göettingen
Göettingen, Germany

G. Hartwig .... Kernforschungszentrum Karlsruhe
Karlsruhe, Germany

K.T. Hartwig .... Texas A&M University
College Station, Texas, USA

T.S. Kreilick .... Hudson International Conductors
Ossining, New York, USA

Y.Y. Li .... Institute of Metal Research
Shenyang, Peoples' Republic of China

H. Maeda .... National Research Institute for Metals
Tsukuba, Ibaraki, Japan

T. Okada .... The Institute of Scientific and Industrial Research, Osaka University
Ibaraki, Osaka, Japan

V.M. Pan .... Institute of Metal Physics
Kiev, Ukraine

R.P. Reed .... National Institute of Standards & Technology
Boulder, Colorado, USA

D.B. Smathers .... Teledyne-Wah Chang Albany
Albany, Oregon, USA

L.T. Summers .... Lawrence Livermore National Laboratory
Livermore, California, USA

K. Tachikawa .... Tokai University
Hiratsuka, Kanagawa, Japan

G.R. Wagner .... Westinghouse Electric Corporation
Pittsburgh, Pennsylvania, USA

K.A. Yushchenko .... E. O. Paton Welding Institute
Kiev, Ukraine

L. Zhou .... Northwest Institute for Nonferrous Metal Research
Baoji, Shaanxi, Peoples' Republic of China

CONTENTS

Foreword v

Best Paper Awards vii

1991 International Cryogenic Materials Conference Board of Directors ix

PART A – STRUCTURAL MATERIALS

Structural Alloys

MECHANICAL PROPERTIES OF INCOLOY 908 – AN UPDATE 1
    I.S. Hwang, R.G. Ballinger, M.M. Morra, and M.M. Steeves

THE EFFECT OF INDIUM ADDITIONS ON THE CRYOGENIC TENSILE
PROPERTIES OF SUPERPLASTICALLY DEFORMED Al–Cu–Li–Zr
ALLOY 2090 11
    E.L. Bradley III, D. Chu, and J.W. Morris, Jr.

DEFORMATION AND FRACTURE OF Al-Li ALLOYS IN
MECHANICAL-IMPACT TESTS 19
    R.P. Reed, C.N. McCowan, N.J. Simon, and J.D. McColskey

MICROSTRUCTURE AND CRYOGENIC TENSILE FRACTURE
BEHAVIOR OF AN Al–Li–Zn–Zr ALLOY 29
    X.J. Jiang and Y.Y. Li

MICROSTRUCTURAL INFLUENCE ON THE WORK HARDENING OF ALUMINUM–
LITHIUM ALLOY 2090 AT CRYOGENIC TEMPERATURES 37
    D. Chu, C. Tseng, and J.W. Morris, Jr.

EFFECTS OF Nb_3Sn HEAT TREATMENT ON THE STRENGTH AND TOUGHNESS
OF 316LN ALLOYS WITH DIFFERENT CARBON CONTENTS 45
    R.P. Reed, R.P. Walsh, and C.N. McCowan

METASTABLE AUSTENITES IN CRYOGENIC HIGH
MAGNETIC FIELD ENVIRONMENTS 55
    J.W. Chan, D. Chu, A.J. Sunwoo, and J.W. Morris, Jr.

PROPERTY EVALUATION OF Ni BASE ALLOY FOR SUPERCONDUCTING
GENERATORS AND ITS APPLICATION TO SEAL WELDED JOINTS 61
    N. Suzuki, T. Murakami, K. Suzuki, S. Asai,
    M. Tanaka, and Y. Kobayashi

VAMAS SECOND ROUND ROBIN TEST OF STRUCTURAL
MATERIALS AT LIQUID HELIUM TEMPERATURE 69
    T. Ogata, K. Nagai, K. Ishikawa, K. Shibata, and S. Murase

Structural Alloys

EFFECT OF HYDROGEN CHARGING ON AMBIENT AND CRYOGENIC MECHANICAL
PROPERTIES OF A PRECIPITATE-STRENGTHENED AUSTENITIC STEEL 77
  Luming Ma, Guojun Liang, and Y.Y. Li

DEFORMATION MEASUREMENTS OF MATERIALS AT LOW TEMPERATURES
USING LASER SPECKLE PHOTOGRAPHY METHOD 85
  S. Nakahara, Y. Maeda, K. Matsumura, S. Hisada, T. Fujita, and K. Sugihara

HYDROGEN EFFECT ON THE MECHANICAL PROPERTIES OF INCOLOY 907
FROM AMBIENT TO CRYOGENIC TEMPERATURE 93
  K. Yang, X. Zhao, Y. Xie, and Y.Y. Li

Structural Alloys – Weldments

THICK-SECTION WELDMENTS IN 21-6-9 AND 316LN STAINLESS
STEEL FOR FUSION ENERGY APPLICATIONS 101
  D.J. Alexander and G.M. Goodwin

JOINING OF AUSTENITIC STAINLESS STEELS FOR CRYOGENIC APPLICATIONS 109
  T.A. Siewert and C.N. McCowan

Structural Alloys – Creep

CREEP OF INDIUM AT LOW TEMPERATURES 117
  R.P. Reed and R.P. Walsh

1100 HOUR CREEP TEST RESULTS FOR OFHC COPPER: VALIDATION OF
PREVIOUSLY PUBLISHED RESULTS 127
  L.C. McDonald and K.T. Hartwig

FATIGUE AND FATIGUE CRACK GROWTH PROPERTIES OF 316LN AND
INCOLOY 908 BELOW 10 K 133
  A. Nyilas, Jinbo Zhang, B. Obst, and A. Ulbricht

DEFORMATION STRUCTURES IN HIGH-CYCLE FATIGUE OF
0.1N-32Mn-7Cr STEEL AT CRYOGENIC TEMPERATURES 141
  O. Umezawa and K. Ishikawa

NEAR-THRESHOLD FATIGUE CRACK GROWTH OF AUSTENITIC
STAINLESS STEELS AT LIQUID HELIUM TEMPERATURE 149
  K. Suzuki, J. Fukakura, and H. Kashiwaya

LONG-CRACK FATIGUE THRESHOLDS AND SHORT CRACK
SIMULATION AT LIQUID HELIUM TEMPERATURE 159
  R.L. Tobler, J.R. Berger, and A. Bussiba

SUPERCONDUCTOR CONDUITS: FATIGUE CRACK GROWTH RATE AND
NEAR-THRESHOLD BEHAVIOR OF THREE ALLOYS 167
  A. Bussiba, R.L. Tobler, and J.R. Berger

HIGH-CYCLE FATIGUE PROPERTIES OF TITANIUM ALLOYS
AT CRYOGENIC TEMPERATURES 175
  O. Umezawa, K. Nagai, T. Yuri, T. Ogata, and K. Ishikawa

COLD THERMAL FATIGUE OF AUSTENITIC STAINLESS STEEL 183
  A. Nishimura and Y. Mukai

Structural Alloys – Toughness

INFLUENCE OF AGING ON THE FRACTURE TOUGHNESS OF
CRYOGENIC AUSTENITIC MATERIALS, EVALUATED BY
A SIMPLE TEST METHOD 191
    J. Kübler, H.-J. Schindler, and W.J. Muster

EFFECTS OF BORON ON INCREASING TOUGHNESS OF HIGH STRENGTH
HIGH MANGANESE NON-MAGNETIC STEELS 199
    H. Tanaka, K. Fujita, and K. Shibata

THE CHARPY IMPACT TEST AS AN EVALUATION OF 4 K
FRACTURE TOUGHNESS 207
    H. Nakajima, K. Yoshida, H. Tsuji, R.L. Tobler,
    I.S. Hwang, M.M. Morra, and R.G. Ballinger

CHARPY SPECIMEN TESTS AT 4 K 217
    R.L. Tobler, A. Bussiba, J.F. Guzzo, and I.S. Hwang

SELECTED RESIDUAL EFFECTS UPON TEMPERATURE TRANSITIONS 225
    Y. Katz, M. Kupiec, and A. Bussiba

Radiation Effects

REACTOR NEUTRON AND GAMMA IRRADIATION OF
VARIOUS COMPOSITE MATERIALS 233
    N.A. Munshi and H.W. Weber

RADIATION DAMAGE OF GLASS-FIBER-REINFORCED
COMPOSITE MATERIALS AT LOW TEMPERATURES 241
    T. Okada, S. Nishijima, T. Nishiura, K. Miyata, Y. Yamaoka, and S. Namba

EFFECTS OF FABRIC TYPE, SPECIMEN SIZE, AND
IRRADIATION ATMOSPHERE ON THE RADIATION
RESISTANCE OF POLYMER COMPOSITES AT 77 K 247
    S. Egusa, M. Sugimoto, H. Nakajima, K. Yoshida, and H. Tsuji

A RADIATION-RESISTANT EPOXY RESIN SYSTEM FOR TOROIDAL
FIELD AND OTHER SUPERCONDUCTING COIL FABRICATION 255
    N.A. Munshi

EFFECTS OF RADIATION ON INSULATION MATERIALS 261
    R. Pöhlchen

RADIATION EFFECTS ON HIGH CURRENT DIODES AT CRYOGENIC
TEMPERATURES IN AN ACCELERATOR ENVIRONMENT 271
    D. Hagedorn and W. Nägele

Special-Use Materials

ALUMINA DISPERSION-STRENGTHENED COPPER ALLOY MATRIX
Ti ADDED Nb_3Sn WIRE BY THE TUBE PROCESS 279
    S. Nakayama, S. Murase, K. Shimamura, N. Aoki, and N. Shiga

ELECTRICAL RESISTIVITY OF NANOCRYSTALLINE Ni-P ALLOYS 285
    K. Lu, Y.Z. Wang, W.D. Wei, and Y.Y. Li

MAGNETIC PROPERTY OF GADOLINIUM HYDRIDES 293
    H. Yayama and A. Tomokiyo

Special-Use Materials

QUENCH PROTECTION DIODES FOR THE LARGE
    HADRON COLLIDER (LHC) AT CERN 299
    D. Hagedorn and W. Nägele

COEFFICIENT OF FRICTION MEASUREMENTS OF SOLID
    FILM LUBRICANTS AT CRYOGENIC TEMPERATURES 307
    L.O. El-Marazki

MAGNETO-TRANSPORT PROPERTIES OF FILAMENTARY ALUMINUM
    CONDUCTORS IN MAGNETIC FIELDS, 12-30 K 315
    W.N. Lawless, C.F. Clark, and R.W. Arenz

TRIBOLOGICAL BEHAVIOR OF 440C MARTENSITIC STAINLESS
    STEEL FROM –184°C TO 750°C 323
    A.J. Slifka, R. Compos, T.J. Morgan, J.D. Siegwarth, and D.K. Chaudhuri

Resins – Mechanical Properties

SCALING TESTS ON SMOOTH AND NOTCHED SPECIMENS OF
    POLYIMIDE (SINTIMID) AT CRYOGENIC TEMPERATURES 331
    K. Humer, H.W. Weber, and E.K. Tschegg

SEVERAL PROPERTIES OF IMPREGNATING EPOXY
    RESINS USED FOR SUPERCONDUCTING COILS 339
    H. Moriyama, Y. Inoue, H. Mitsui, Y. Sanada, and Y. Kobayashi

Composites – Mechanical Properties

SHEAR BEHAVIOR OF GLASS-REINFORCED SYSTEMS
    AT LOW TEMPERATURES 347
    P.E. Fabian, C.S. Hazelton, and R.P. Reed

SHEAR FRACTURE TESTS (MODE II) ON FIBER REINFORCED
    PLASTICS AT ROOM AND CRYOGENIC TEMPERATURES 355
    E.K. Tschegg, K. Humer, H.W. Weber

COMPRESSION AND SHEAR TESTS OF VACUUM-IMPREGNATED
    COMPOSITES AT CRYOGENIC TEMPERATURES 363
    N.J. Simon, R.P. Reed, and R.P. Walsh

CRYOGENIC FATIGUE TESTING OF GLASS REINFORCED EPOXY TUBES 371
    M.K. Abdelsalam

GLASS–FILM–GLASS HYBRID ORGANIC COMPOSITES
    FOR FORCED-FLOW FUSION MAGNETS 379
    S. Ueno, S. Nishijima, T. Okada, and M. Maruyama

FRACTURE-MECHANICAL CHARACTERIZATION OF FIBER REINFORCED
    PLASTICS IN THE CRACK-OPENING-MODE (MODE I) 387
    E.K. Tschegg, K. Humer, and H.W. Weber

FRACTO-EMISSION FROM CRYSTALLINE AND NON-CRYSTALLINE
    MATERIALS AT CRYOGENIC TEMPERATURES 397
    S. Owaki, T. Okada, S. Nakahara, and K. Sugihara

Composites – Mechanical Properties

FRICTION AND WEAR OF RADIATION RESISTANT COMPOSITES, COATINGS
   AND CERAMICS IN VACUUM AND LOW TEMPERATURE ENVIRONMENT 405
   A. Lipski and M. Ruschman

AN INNOVATIVE PROCESS FOR THE IMPREGNATION
   OF MAGNET COILS AND OTHER STRUCTURES 413
   D. Evans and J.T. Morgan

LARGE SCALE TESTS OF COMPOSITE SUPPORT STRUTS FOR
   SUPERCONDUCTING MAGNETIC ENERGY STORAGE RINGS 421
   R.P. Walsh, R.P. Reed, J.D. McColskey, M. Tupper, and E. Johnson

FRICTION AND WEAR OF A THREE-DIMENSIONAL
   FABRIC-REINFORCED PLASTIC AT ROOM TEMPERATURE
   AND LIQUID NITROGEN TEMPERATURE 429
   S. Nishijima, T. Okada, P.C. Michael, and Y. Iwasa

MODIFICATION OF THE ASTM D 3039 TENSILE
   SPECIMEN FOR CRYOGENIC APPLICATIONS 437
   T.J. Eisenreich and D.S. Cox

THERMAL INSULATING SUPPORT SYSTEMS FOR RADIATION ENVIRONMENTS 445
   Y. Ohtani, S. Nishijima, T. Okada, and K. Asano

FATIGUE BEHAVIOR OF UD-CARBON-FIBRE
   COMPOSITES CRYOGENIC TEMPERATURES 453
   G. Hartwig and K. Pannkoke

PART B – SUPERCONDUCTORS

Superconductivity Reviews

RESULTS OF THE FIRST VAMAS INTERCOMPARISON
   OF AC LOSS MEASUREMENTS 459
   K. Itoh, H. Wada, and K. Tachikawa

RECENT PROGRESS ON HIGH-T_c SUPERCONDUCTING MATERIAL
   RESEARCH IN THE PEOPLE'S REPUBLIC OF CHINA 469
   L. Zhou

IMPLICATIONS OF HIGH TEMPERATURE SUPERCONDUCTORS
   FOR POWER GENERATION 479
   C.E. Oberly, G. Kozlowski, and R. T. Fingers

General Superconductor Theory, Measurement, and Processing

PROGRESS AND PROBLEMS IN UNDERSTANDING HYSTERESIS LOSS 491
   W.J. Carr, Jr.

AC LOSSES IN ULTRA-FINE FILAMENTARY SUPERCONDUCTORS 501
   J.R. Cave and A.M. Campbell

NORMAL ZONE SOLITON IN LARGE COMPOSITE SUPERCONDUCTORS 509
   R. Kupferman, R.G. Mints, and E. Ben-Jacob

General Superconductor Theory, Measurement, and Processing

THE DOMAIN MOVEMENT IN COMPOSITE SUPERCONDUCTOR
WITH CONTACT RESISTANCE 517
   A.A. Akhmetov

AN AC TECHNIQUE TO SEPARATE INTRA GRAIN
AND INTER GRAIN CURRENT CONTRIBUTIONS
TO THE MAGNETIZATION OF HIGH-T_c RINGS 523
   A.R. Kruper, H. Hemmes, and L.J.M. van de Klundert

THERMAL CONDUCTIVITY LIMITED RECOVERY CURRENTS IN HTS 531
   O. Christianson

CRITICAL NUCLEI AND SOME TEMPERATURE PULSE
PECULIARITIES IN HTS CURRENT CARRYING COMPOSITES 539
   A.A. Akhmetov, V.A. Altov, O.V. Filatova, V.V. Sytchev,
   E.A. Trukhacheva, and I.V. Yakovlev

THE EFFECT OF A NON-UNIFORM CRITICAL CURRENT DENSITY ON THE
RESISTIVE TRANSITION IN COMPOSITE SUPERCONDUCTORS 545
   D. ter Avest, E.M.J. Niessen, and L.J.M. van der Klundert

TRANSPORT AND INDUCED CURRENTS DISTRIBUTION IN
SUPERCONDUCTING TRANSPOSED CABLES 553
   V.E. Sytnikov, I.B. Peshkov, G.G. Svalov, and Yu.V. Prismakov

COMPARISON OF TRANSPORT CRITICAL CURRENT MEASUREMENT METHODS 559
   L.F. Goodrich and A.N. Srivastava

A COMPARISON OF SELF FIELD CORRECTED TO NON-CORRECTED
CRITICAL CURRENT DATA FOR SUPERCONDUCTING WIRES 567
   J.C. McKinnell, M.B. Siddell, and D.B. Smathers

Low-Temperature Superconductors – Nb_3Sn

CRITICAL CURRENT DENSITIES IN MULTIFILAMENTARY Nb_3Sn
COMPOSITE CONDUCTORS FOR AC USE 573
   K. Yasohama, Y. Kubota, H. Kobayashi, T. Ogasawara,
   S. Akita, S. Torii, and K. Ueda

IMPROVEMENT OF THE STRUCTURE AND PROPERTIES OF
INTERNAL TIN Nb_3Sn CONDUCTORS 579
   E. Gregory, G.M. Ozeryansky, and B.A. Zeitlin

DEVELOPMENT OF MULTIFILAMENTARY IN-SITU
PROCESSED Nb_3Sn SUPERCONDUCTING WIRES 587
   M. Sugimoto, M. Tange, K. Gotoh, N. Sadakata, T. Saitoh,
   O. Kohno, and Y. Ikeno

MAGNETIC HYSTERESIS LOSSES AND OTHER PROPERTIES OF
SUBMICRON-FILAMENT BRONZE-PROCESSED Nb_3Sn WIRES 595
   S. Sakai, K. Miyashita, K. Kamata, K. Tachikawa,
   T. Taniguchi, K. Endoh, and H. Hatakeyama

DEPENDENCE OF THE CRITICAL CURRENT DENSITIES IN Nb_3Sn COMPOSITE
CONDUCTORS ON THE PROCEDURE OF APPLYING THE MAGNETIC FIELD 603
   Y. Kubota, K. Yasohama, N. Takeuchi, S. Ban, M. Chiba,
   N. Miyazawa, K. Uno, and T. Ogasawara

Low-Temperature Superconductors – Nb_3Sn

COMPARISONS OF AC LOSSES OF Nb_3Sn SINGLE
STRANDS AND US-DPC CONDUCTOR 611
   C.Y. Gung, M. Takayasu, M.M. Steeves, T. Painter,
   B. Oliver, D. Reisner, and M.O. Hoenig

CRITICAL CURRENTS OF Nb_3Sn WIRES OF THE US-DPC COIL 619
   M. Takayasu, M.M. Steeves, T.A. Painter, C.Y. Gung, and M.O. Hoenig

THE INFLUENCE OF STRAIN INTRODUCED BY DIFFERENTIAL
THERMAL CONTRACTION ON THE CRITICAL CURRENT
OF Nb_3Sn SUPERCONDUCTORS 627
   G. Pasztor and B. Jakob

THERMAL CONDUCTION IN FULLY IMPREGNATED Nb_3Sn WINDINGS
FOR LHC TYPE OF DIPOLE MAGNETS 635
   A. den Ouden, J.M. van Oort, L. Burnod,
   H.H.J. ten Kate, and L.J.M. van de Klundert

CRITICAL-CURRENT DEGRADATION IN Nb_3Sn COMPOSITE WIRES
DUE TO LOCALLY CONCENTRATED TRANSVERSE STRESS 643
   S.L. Bray and J.W. Ekin

Low-Temperature Superconductors – Nb–Ti

QUANTITATIVE ANALYSIS OF SAUSAGING IN Nb BARRIER CLAD
FILAMENTS OF Nb-46.5wt% Ti AS A FUNCTION OF
FILAMENT DIAMETER AND HEAT TREATMENT 647
   Y.E. High, P.J. Lee, J.C. McKinnell, and D.C. Larbalestier

FINE FILAMENT NbTi AND Nb_3Sn CONDUCTORS 653
   K. Heine, H. Krauth, A. Szulczyk, and M. Thöner

DEVELOPMENT AND MANUFACTURE OF FINE FILAMENT
NbTi SUPERCONDUCTIVE WIRES AT AISA 661
   Hoang Gia Ky, P. Sulten, G. Grunblatt

METALLURGICAL, PHYSICAL, AND SUPERCONDUCTING
PROPERTIES OF SERIES OF NbTi AND NbTiMn ALLOYS 667
   D.S. Pyun and E.W. Collings

FURTHER DEVELOPMENTS IN NbTi SUPERCONDUCTORS
WITH ARTIFICIAL PINNING CENTERS 675
   H.C. Kanithi, P. Valaris, L.R. Motowidlo, B.A. Zeitlin, and R.M. Scanlan

CHARACTERIZATION OF Nb-Ti SUPERCONDUCTORS
WITH ARTIFICAL PINNING STRUCTURES 685
   D.R. Dietderich, S. Eylon, and R.M. Scanlan

IMPROVED RESPONSE TO HEAT TREATMENT OF Nb-Ti
ALLOYS BY THE ADDITION OF Zr 691
   P.J. Lee, J.C. McKinnell, and D.C. Larbalestier

COMPOSITE SUPERCONDUCTORS WITH
COPPER–ALUMINUM STABILIZING MATRIX 699
   V.E. Keilin, O.P. Anashkin, A.V. Krivikh, I.V. Kiriya,
   I.A. Kovalev, P.I. Dolgosheev, A.V. Rychagov, and V.E. Sytnikov

Low-Temperature Superconductors – Nb–Ti

ALUMINUM STABILIZED SUPERCONDUCTING CABLES 703
    V.E. Sytnikov, A.V. Rychagov, Ju.P. Ipatov, A.M. Jetymov,
    P.I. Dolgosheev, B.V. Marjanchik, V.E. Keilin, and A.V. Krivyh

SHAPE MEASUREMENTS OF THE RESISTIVE TRANSITION IN SSC STRAND 709
    W.H. Warnes and W. Dai

MAGNETIZATION DECAY OF SSC-TYPE STRANDS IN VARIOUS
    SHORT TEST SAMPLE CONFIGURATIONS 715
    K.R. Marken, M.D. Sumption, E.W. Collings, and R.M. Scanlan

DIFFUSIONAL REACTION RATES THROUGH THE Nb WRAP IN SSC AND
    OTHER ADVANCED MULTIFILAMENTARY Nb 46.5wt%Ti COMPOSITES 723
    K.J. Faase, P.J. Lee, J.C. McKinnell, and D.C. Larbalestier

TRANSPORT CURRENT EFFECTS ON FLUX CREEP AND MAGNETIZATION
    IN Nb-Ti MULTIFILAMENTARY CABLE STRANDS 731
    R.W. Cross

MEASUREMENTS OF THE RESISTIVE TRANSITION IN
    VERY LONG SAMPLE SUPERCONDUCTORS 737
    D. ter Avest, G. Schoenmaker, H.G. Knoopers, and L.J.M. van de Klundert

REDUCTION OF AC LOSSES IN SUPERCONDUCTING WIRE WITH VERY
    FINE FILAMENTS BY WEAKENING FLUX-PINNING STRENGTH 745
    T. Matsushita and E.S. Otabe

TEMPERATURE AND FIELD DEPENDENCE OF SHORT TERM DECAY
    AND LOSS IN MULTIFILAMENTARY SUPERCONDUCTORS –
    A PROXIMITY EFFECT INTERPRETATION 751
    M.D. Sumption and E.W. Collings

DEVELOPMENT OF THE PRODUCTION TECHNOLOGY FOR FINE FILAMENT
    SUPERCONDUCTING CABLES USED IN ACCELERATOR MAGNETS 759
    H. Ii, S. Meguro, T. Suzuki, K. Ogawa, and M. Ikeda

kA-CLASS A.C. SUPERCONDUCTING CABLE DEVELOPMENT 767
    E.S. Yoneda, D. Ito, I. Takano, S. Akita,
    S. Torii, T. Kumano, and E. Suzuki

CABLE DEGRADATION OF SSC STRAND 775
    W.H. Warnes, W. Dai, J. Seuntjens, and D.W. Capone II

EFFECT OF TWIST PITCH, SAMPLE LENGTH, AND FIELD ORIENTATION
    ON THE PROXIMITY EFFECT ENHANCED MAGNETIZATION OF FINE
    FILAMENTARY MULTIFILAMENTARY STRANDS 783
    M.D. Sumption and E.W. Collings

CRITICAL CURRENT CAPACITY OF SUPERCONDUCTORS AT
    DIFFERENT AC FREQUENCIES 791
    V.R. Karasik, V.S. Vysotsky, S.G. Derjagin, V.N. Tsikhon,
    Zhou Lian, Wu Xiaozu, Zhou Guixiang, Yang Xiaodong, and Teng Xinkang

ROOM TEMPERATURE MECHANICAL PROPERTIES OF Nb-Ti/Cu
    SUPERCONDUCTING COMPOSITES 797
    He Liu and W.H. Warnes

Low-Temperature Superconductors – New Materials

AC SUSCEPTIBILITY AND MAGNETIZATION OF Nb-TUBE PROCESSED
Nb$_3$Al COMPOSITE WIRES 805
   K. Itoh, M. Yuyama, T. Kuroda, and H. Wada

AC LOSSES IN A MULTIFILAMENTARY Nb$_3$Al/Cu COMPOSITE
SUPERCONDUCTOR MADE BY A JELLY ROLL PROCESS 813
   T. Ando, Y. Takahashi, M. Nishi, Y. Yamada, K. Ohmatsu, and M. Nagata

HYSTERESIS LOSS IN MULTIFILAMENTARY V$_3$Ga WIRES 821
   K. Tachikawa, T. Ajioka, M. Endo, and Y. Tanaka

STRAIN EFFECTS IN Nb$_3$Al MULTIFILAMENTARY WIRES 827
   K. Katagiri, M. Ohgami, T. Okada, T. Kuroda, H. Wada,
   K. Inoue, K. Noto, M. Ishii, K. Watanabe, and H. Kodaka

SYNTHESIS OF A15 Nb$_3$(Al,Ge) THROUGH THE σ PHASE 835
   K. Tachikawa, J. Ninomiya, and M. Terada

STRAIN EFFECTS IN PbMo$_6$S$_8$ WIRES 843
   K. Katagiri, K. Seo, T. Okada, Y. Kubo, F. Fujiwara,
   K. Noto, Y. Morii, M. Ishii, and K. Watanabe

ELECTRON TRANSPORT IN NIOBIUM-SILICON MULTILAYERS 851
   G.K. Sherrill, E.J. Cukauskas, and L.H. Allen

High-Temperature Superconductors – YBCO

MELT-PROCESSED YBCO/Ag COMPOSITE TAPES WITH NICKEL CLADDING 859
   G. Kozlowski, C.E. Oberly, R.E. Leese, and J.C. Ho

EVALUATION OF Y-Ba-Cu-O TUBES PREPARED BY TAPE CASTING
AND SUBSEQUENT RATE-CONTROLLED SINTERING 867
   N. Kenny, T.R. Shrout, F. Rodriguez, C.E. Oberly,
   G. Kozlowski, I. Maartense, R. Spyker, and J.C. Ho

FRACTURE BEHAVIOUR AND CRITICAL CURRENT DENSITY OF
SILVER SHEATHED HIGH T_c SUPERCONDUCTING TAPES 875
   K. Osamura, S. Ochiai, and K. Hayashi

AC LOSS AND DYNAMIC RESISTANCE OF A HIGH-T_c STRAND CARRYING
A DIRECT CURRENT IN A TRANSVERSE AC MAGNETIC FIELD 883
   E.W. Collings, K.R. Marken, M.D. Sumption, J.R. Clem,
   S.A. Boggs, and M.V. Parish

MAGNETIC PROPERTIES OF HIGH T_c OXIDE SUPERCONDUCTING
FILAMENT PRODUCED BY SUSPENSION SPINNING METHOD 893
   T. Goto, Y. Kino, and C. Yamaoka

NEUTRON IRRADIATION EFFECTS ON YBCO SINGLE CRYSTALS 901
   F.M. Sauerzopf, H.P. Wiesinger, H.W. Weber, and G.W. Crabtree

MICROSTRUCTURE AND ELECTRIC PROPERTIES
OF MELT PROCESSED Y-Ba-Cu-O 907
   S. Gauss and S. Elschner

EFFECTS OF SILVER ADDITION AND ZIRCONIUM SUBSTITUTION
ON YBa$_2$Cu$_3$O$_{7-x}$ PREPARED BY A CITRIC ACID SALT PROCESS 915
   T. Asaka, Y. Okazawa, Y. Shiomi, and K. Tachikawa

High-Temperature Superconductors – YBCO

OBSERVATION ON MICROSTRUCTURES OF YBCO SUPERCONDUCTOR BY
POWDER MELT PROCESS 923
    Wang Keguang, Zhou Lian, Zhang Pingxiang, Wang Shuqian,
    Wang Jinrong, and Ji Ping

FABRICATION, PROPERTIES AND MICROSTRUCTURE OF THE YBCO
SUPERCONDUCTOR PREPARED BY THE POWDER MELTING PROCESS 929
    Zhou Lian, Zhang Pingxiang, Wang Keguang,
    Wang Jingrong, Du Zehua, and Ji Ping

MAGNETIC AND PINNING FORCE STUDIES OF
POWDER-MELT-PROCESSED YBCO SUPERCONDUCTOR 935
    Zhou Lian, Wang Jingrong, Zhang Pingxiang,
    M.D. Sumption, K.R. Marken, and E.W. Collings

THE EFFECT OF ATMOSPHERE ON SYNTHESIS OF YBCO
SUPERCONDUCTOR BY POWDER MELTING PROCESS 943
    Zhang Pingxiang, Zhou Lian, Ji Ping, Du Shejun,
    Wang Jingrong, Wang Keguang, and Wu Xiaozu

T_c, H_{c2}, AND OXYGEN ORDERING IN QUENCHED
OXYGEN DEFICIENT $YBa_2Cu_3O_{7-\delta}$ 949
    M. Däumling, L.E. Levine, and T.M. Shaw

LOW FIELD AC LOSS AND LOWER CRITICAL FIELD
IN PMP HIGH-T_c SUPERCONDUCTORS 957
    E.W. Collings, K.R. Marken, Jr., M.D. Sumption, Zhou Lian,
    Wang Jingrong, Zhang Pingxiang, and W.J. Carr, Jr.

DYNAMIC RESISTANCE OF SUPERCONDUCTING $YBa_2Cu_3O_x$ SINTERED POWDER
AT 81 K: LIQUID VERSUS VAPOR NITROGEN ENVIRONMENT 965
    J. Moreland, W.P. Dubè, and L.F. Goodrich

STUDY OF MAGNETIC FLUX TRAPPED IN $YBa_2Cu_3O_{7-x}$ + 5wt%Ag
SUPERCONDUCTING COMPOSITE 973
    R.L. Spyker, G. Kozlowski, C.E. Oberly, and J.C. Ho

SYSTEMATIC STUDY OF Ni-DOPING EFFECT ON
MELT-PROCESSED YBCO SUPERCONDUCTOR 977
    G. Kozlowski, D. Hansley, C.E. Oberly, J.C. Ho,
    X.W. Cao, R.L. Spyker, and R.E. Leese

STRUCTURE OF SINGLE CRYSTALLINE YBaCuO THIN
FILMS WITH HIGH CRITICAL CURRENT DENSITY
PREPARED BY DC-MAGNETRON SPUTTERING 983
    B.C. Yang, X.P. Wang, C.Q. Wang, R.K. Wang, C.G. Cui, and S.L. Li

POLYCRYSTALLINE $Y_1Ba_2Cu_3O_{7-\delta}$ FILMS PREPARED BY CVD METHOD 991
    K. Watanabe, S. Awaji, H. Yamane, H. Kurosawa,
    T. Hirai, N. Kobayashi, and Y. Muto

AC SUSCEPTIBILITY OF SUPERCONDUCTIVE THIN FILMS 997
    H.R. Kerchner, J.O. Thomson, and R. Feenstra

A NEW GEOMETRY FOR LASER ABLATION FOR THE
PRODUCTION OF SMOOTH THIN SINGLE LAYER
$YBa_2Cu_3O_{7-x}/PrBa_2Cu_3O_{7-x}$ MULTILAYER FILMS 1005
    R.J. Kennedy

High-Temperature Superconductors – YBCO

IN-SITU DEPOSITION OF YBaCuO FILMS ON BOTH SIDES OF
TWO-INCH DIAMETER WAFERS BY OFF-AXIS SPUTTERING 1013
    T.T. Braggins, J.R. Gavaler, and J. Talvacchio

SPUTTERED THIN FILMS OF $Y_1Ba_2Cu_3O_{7-x}$ ON MgO SUBSTRATES 1019
    L.H. Allen, E.J. Cukauskas, G.K. Sherrill, R.T. Holm, and P.K. Van Damme

INFRARED RADIATION DETECTION WITH A YBCO MICROBRIDGE 1027
    H. Imokawa, T. Yotsuya, Y. Suzuki, and S. Ogawa

High-Temperature Superconductors – Bi-Based

PROCESSING OF HIGH-T_c SUPERCONDUCTOR WIRES
FOR MAGNET APPLICATION 1033
    K. Heine, J. Tenbrink, H. Krauth, and M. Wilhelm

IMPROVED UNIAXIAL STRAIN TOLERANCE OF
THE CRITICAL CURRENT MEASURED IN Ag-SHEATHED
Bi_2-Sr_2-Ca_1-Cu_2-O_{8+x} SUPERCONDUCTORS 1041
    J.W. Ekin, S.L. Bray, T.A. Miller, D.K. Finnemore, and J. Tenbrink

EFFECTS OF STRAIN ON CRITICAL CURRENTS IN
Ag-SHEATHED BiSrCaCuO TAPES 1045
    T. Kuroda, M. Yuyama, K. Itoh, and H. Wada

DEGRADATION AND RECOVERY IN THE SUPERCONDUCTING PHASES
OF THE Ag-SHEATHED BPSCCO SUPERCONDUCTING WIRES 1053
    Li Chengren, Li Yalu and Zhou Lian

EFFECTS OF PROCESSING, HEAT-TREATMENT TEMPERATURE,
AND STARTING POWDERS ON THE CRITICAL CURRENT
DENSITIES OF Ag-SHEATHED Bi-Pb-Sr-Ca-Cu-O WIRES AND TAPES 1057
    Xi Zhengping, Zhou Lian, Ji Chunlin, Wu Suihua, and Cheng Zhenguo

THE EFFECT OF DOUBLE STAGE COLD ROLLING TREATMENT ON THE
MICROSTRUCTURE AND THE CRITICAL CURRENT OF SILVER CLAD
$(Bi,Pb)_2Sr_2Ca_2Cu_3O_x$ SUPERCONDUCTING TAPES 1065
    Wai Lo and B.A. Glowacki

THE INFLUENCE OF CARBON IMPURITIES ON J_c IN Ag/Bi(2223) TAPES 1073
    R. Flükiger, A. Jeremie, B. Hensel, E. Seibt, J.Q. Xu, and Y. Yamada

FABRICATION OF BiSrCaCuO/Ag COMPOSITE SUPERCONDUCTORS
BY DIP-COATING PROCESS 1081
    K. Togano, H. Kumakura, K. Kadowaki, H. Kitaguchi,
    H. Maeda, J. Kase, J. Shimoyama, and K. Nomura

INVESTIGATION OF HEAT TREATMENT CONDITIONS FOR
SILVER SHEATHED Bi2212 SUPERCONDUCTING WIRES 1087
    K. Shibutani, T. Egi, S. Hayashi, R. Ogawa, and Y. Kawate

TRANSPORT CRITICAL CURRENT DENSITY OF
$Bi_2Sr_2CaCu_2O_x$ SINGLE CRYSTAL 1097
    K. Shibutani, S. Hayashi, T. Egi, I. Shigaki,
    R. Ogawa, Y. Kawate, K. Kitahama, and S. Kawai

High-Temperature Superconductors – Bi-Based

REVERSIBLE MAGNETIZATION OF $Bi_2Sr_2CaCu_2O_8$ SINGLE CRYSTALS 1103
   H. W. Weber, W. Kritscha, F.M. Sauerzopf,
   G.W. Crabtree, Y.C. Chang, and P.Z. Jiang

IRREVERSIBILITY LINE IN BULK SUPERCONDUCTING Bi–Pb–Sr–Ca–Cu–O 1111
   T. Matsushita, E.S. Otabe, T. Nakatani, B. Ni, K. Yamafuji,
   T. Umemura, K. Egawa, M. Wakata, and S. Utsunomiya

MICROSTRUCTURE OF Bi–Sr–Ca–Cu–O PREPARED BY MELT PROCESS 1119
   K. Egawa, T. Umemura, S. Kinouchi, M. Wakata, and S. Utsunomiya

HIGHLY ORIENTED Bi-SYSTEM BULK SAMPLE PREPARED
   BY A DECOMPOSITION-CRYSTALLIZATION PROCESS 1127
   Xi Zhengping, Zhou Lian, and Ji Chunlin

THE MECHANISM OF HIGH-T_c PHASE FORMATION
   IN Bi-BASED SUPERCONDUCTORS 1133
   Qiu Xinliang and Li Chengren

ELECTRICAL RESISTIVITY AND DIFFERENTIAL THERMAL ANALYSES
   OF Ca_2PbO_4-DOPED Bi(Pb)–Sr–Ca–Cu–O SUPERCONDUCTORS 1139
   J.C. Ho, C.Y. Wu, G. Kozlowski, C.E. Oberly, and R.L. Spyker

STUDY OF RELAXATION EFFECTS IN THE
   LEAD-DOPED Bi–Sr–Ca–Cu–O SYSTEM 1147
   G. Kozlowski, A.K. Sarkar, C.E. Oberly, R. Spyker,
   I. Maartense, T.L. Peterson, and J.C. Ho

THE PROPERTIES OF METAL CONTACTS ON BSCCO SUPERCONDUCTORS 1155
   T. Kusaka, Y. Suzuki, T. Yotsuya, S. Ogawa, S. Miyake, and T. Aoyama

DEGRADATION OF $Pb_{0.3}Bi_{1.7}Sr_2Ca_2Cu_3O_x$ SUPERCONDUCTORS 1161
   G.L. Larkins Jr., Q. Lu, D. Albaijes,
   C. Levay, R. Laurence, and W.K. Jones

Cryoconductors, Capacitors, and Magnetic Materials

RECENT DEVELOPMENTS WITH THE EDDY CURRENT DECAY
   METHOD FOR RESISTIVITY MEASUREMENTS 1169
   K.T. Hartwig, L.C. McDonald, and H. Zou

ELECTRICAL RESISTIVITY OF COPPER ALLOYS BETWEEN 76 K AND 300 K 1177
   C.A. Thompson and F.R. Fickett

ENERGY STORAGE AT 77 K IN MULTILAYER CERAMIC CAPACITORS 1183
   W.N. Lawless and C.F. Clark

LOW TEMPERATURE MAGNETIC BEHAVIOR OF "NONMAGNETIC" MATERIALS ... 1191
   F.R. Fickett

Indexes

AUTHOR INDEX 1199

MATERIALS INDEX 1203

SUBJECT INDEX 1205

MECHANICAL PROPERTIES OF INCOLOY 908 - AN UPDATE*

I. S. Hwang, R. G. Ballinger, M.M. Morra and M. M. Steeves

Massachusetts Institute of Technology
Department of Materials Science and Engineering
Plasma Fusion Center
Cambridge, Massachusetts 02139

ABSTRACT

Incoloy 908 is a nickel-iron base superalloy with its coefficient of thermal expansion and mechanical properties optimized for use in Nb_3Sn superconducting magnets. Thermo-elastic, tensile, fatigue crack growth, fracture toughness,and Charpy impact properties, and the results of conduit pressurization tests are summarized for base and weld metal. A limited number of stress rupture tests were also performed in air. The average yield strength (0.2% offset) for the solution annealed and aged base metal is 1200 MPa at 4.2 K. The fracture toughness, K_{IC}, is greater than 230 MPa$\sqrt{m}$ at 4.2 K. The fatigue behavior at 4.2K is comparable to austenitic stainless steels. Fatigue crack growth rates are a factor of three lower at 4.2 K than 298 K and are independent of heat treatment. At 4.2 K, the 20% cold-work-then-aged material has a 20% higher yield strength and a 10% higher ultimate tensile strength. Gas tungsten arc weld (GTAW) metal with or without Incoloy 908 filler metal exhibited comparable yield and about 10% lower tensile strength when compared with that of the base metal after a 200 hour age at 650°C. Fracture toughness, tensile elongation and Charpy absorbed energy were about 40% of those of the base metal. Leak-before-break behavior was observed in an internal pressurization test at room temperature for a geometry identical to that of the US-Demonstration Poloidal Coil conduit. The stress rupture performance is better than other low COE alloys of a similar type to that of Incoloy 908.

INTRODUCTION

Incoloy 908, a nickel-iron base superalloy developed for use in Nb_3Sn superconducting magnets, has a chemical composition that has been optimized for coefficient of thermal expansion, good structural properties, phase stability, workability and weldability. Strengthening is achieved by precipitation-hardening which occurs concurrent with the superconductor reaction heat treatment. The alloy precipitates $Ni_3(Al,Ti,Nb)$, γ', as the strengthening phase. In this paper the results of mechanical property measurements for solution-annealed-then-aged and cold-worked-then-aged base and weld material are summarized. The 20% cold-worked-then-aged condition was chosen to represent the as-fabricated cable-in-conduit conductor (CICC) condition. The results summarized in this paper are: (1) tensile, (2) fatigue, (3) fracture toughness , (4) Charpy absorbed energy, and (5) leak-before-break behavior.

---

* Supported by the U.S.-DOE Grant DE-FG02-91ER-54110.

MATERIALS

The Incoloy 908 test materials were produced by vacuum induction melting by Inco Alloys International, Huntington, West Virginia. Ingots were homogenized at 1191°C for 16 hours and fast cooled. A series of forging and reheating steps resulted in 27 mm, 16 mm and 2.4 mm thick plates which were then solution-annealed at 980 C for 1 hour and air cooled. The 26 mm plate was further cold worked by 20%. Gas tungsten arc weld (GTAW) metal was produced for the 27 mm plate using Incoloy 908 filler metal by butt-welding double-grooved plates in the rolling direction using a flowing argon gas shield. GTAW without filler metal (autogenous welding) was made for the 2.4 mm plate. The chemical compositions of the materials are shown in Table 1.

Specimens for tensile, fracture toughness and fatigue testing were cut from each plate. The tensile axis for base metal specimens was oriented in the rolling direction. This orientation is representative of the hoop direction in a solenoid magnet winding. The specimen axis for weld metals was oriented perpendicular to the welding direction to represent material behavior with flaws in the through thickness direction. Specimens were then subjected to three heat treatments in vacuum representing typical Nb_3Sn reaction treatments: (1) 650°C for 200 hours, (2) 700°C for 100 hours, and (3) 750°C for 50 hours. The US-Demonstration Poloidal Coil (US-DPC) conduit material experienced a more complex process [1]. Starting with the solution-annealed 2.4 mm x 80 mm wide sheet, the final conduit is formed into a tube and seam welded by autogenous GTAW welding The weld is quenched by 77 K nitrogen vapor. The diameter is reduced by 20% and the final square (with rounded corners) shape formed using a Turk's head die. The entire assembly is then aged at 650°C for 200 hours.

The coefficient of thermal expansion of Incoloy 908 was measured at the National Institute of Standards and Technology (NIST) by dilatometric method [2,3]. The thermal strain due to cooling from the Nb_3Sn reaction temperature (650 C) to 4.2 K was determined to be about 0.4% greater than Nb_3Sn but 0.5% smaller than that of austenitic stainless steels. The elastic moduli (volume averaged) of the solution-annealed and the 20%-cold-worked Incoloy 908 after ageing at 650 C for 200 hours were measured by acoustic techniques at NIST [4]. The results of this analysis are shown in Table 2. The elastic moduli increased in a linear fashion from 295 K to 70 K but changed very little from 70 K to 5 K. The solution-annealed-

TABLE 1. Chemical Composition of Incoloy 908 for Solution-Annealed (SA) Plate, 16 mm 20% Cold-Worked (CW) Plate, and 2.4 mm US-DPC Conduit Plate. All values in (wt%).

	SA (27 mm) Plate		20% CW (16 mm) Plate	US-DPC CONDUIT	
ELEMENT	Base Metal	Weld Metal	Base Metal	Base Metal	Weld Metal
Iron	40.7	40.8	40.8	40.8	40.6
Chromium	3.98	3.99	4.12	4.06	4.03
Cobalt	0.013	0.014	---	0.020	0.020
Manganese	0.041	0.041	0.09	0.041	0.039
Silicon	0.13	0.13	0.17	0.15	0.15
Niobium	2.94	2.88	3.04	2.91	2.94
Aluminum	0.93	0.94	1.10	0.94	0.93
Titanium	1.74	1.69	1.54	1.75	1.74
Carbon	0.011	0.014	0.013	0.011	0.012
Oxygen	0.0013	0.0022	---	0.0014	0.0011
Nitrogen	0.0020	0.0027	---	0.0021	0.0017
Phosphorus	0.003	0.002	0.002	0.002	0.001
Sulfur	0.002	0.002	0.001	<0.002	<0.002
Boron	0.006	0.006	0.004	0.005	0.004
Nickel	Balance	Balance	48.7	Balance	Balance

TABLE 2. Elastic Properties of Incoloy 908 for Solution-Annealed (SA)-Then-Aged Material and Cold-Worked (CW)-Then-Aged (650C for 200 Hours) Material [4].

Temperature T (K)	Elastic Modulus E (GPa)		Shear Modulus G (Gpa)		Bulk Modulus B (GPa)		Poisson's Ratio v	
	SA	CW	SA	CW	SA	CW	SA	CW
5	182.3	184.2	69.96	70.93	154.2	152.6	0.3029	0.2987
70	182.0	184.0	69.86	70.86	153.5	151.8	0.3025	0.2980
295	179.0	180.9	68.99	69.97	147.0	145.6	0.2971	0.2929

then-aged material showed isotropic behavior whereas the 20% cold-worked-then-aged material showed a 1.5% higher elastic modulus in the rolling direction.

EXPERIMENTAL PROCEDURES

Tensile, fatigue and fracture toughness tests were performed using a cryogenic test facility at M.I.T.[5]. Specimen strain or crack opening displacement were measured using extensometers designed for use at 4K. The extensometers were calibrated at room temperature on a weekly-basis to a precision of 25 micrometer. Tensile tests were performed in accordance with the Proposed ASTM Method for Tension Testing at Liquid Helium Temperature [6]. Flat tensile test specimens with a 2.3 mm thickness, a 3.2 mm width and a 16 mm long reduced section were fabricated by wire electrodischarge machining (EDM). The specimen design was proportionally scaled down from a sheet metal specimen design in accordance with ASTM E8-88. All tests were conducted at a nominal strain rate of 2×10^{-4} sec^{-1} in displacement-control. Strain measurements were recorded to a maximum of 2%.

Fatigue crack growth rate test specimens of the compact tension type with a 12.7 mm thickness were fabricated in accordance with ASTM E647-88. Specimens were oriented in the LT direction. Crack length was determined using an unloading compliance technique after the method of Saxena and Hudak [7] using an extensometer mounted on the specimen front surface. Tests were performed at constant ΔK with a 10Hz sine wave and stress ratio (R) of 0.1.

Fracture toughness (J based) was determined using 25.4 mm thick compact tension specimens fabricated in accordance with ASTM E813-88. Specimens for the US-DPC conduit material were scaled down to a half size with the thickness further reduced to 1.8 mm. The minimum thickness requirement for ASTM E813-88 was not met in the latter specimens. Therefore only K_Q, unqualified toughness values, could be determined. Specimens were precracked by fatigue loading at room temperature with a final maximum stress intensity factor of 33 MPa$\sqrt{m}$. The ratio of initial crack length to specimen width ranged from 0.60-0.70. An extensometer, mounted at load-line, was used for measurement of load-line displacement Crack length was determined from the unloading compliance at the load-line. Specimen rotation and moving crack effects were taken into account in the crack length determination. Testing was conducted under computer control in accordance with ASTM E813-88 in COD-control at a typical displacement rate of 0.1 mm/min. Data analysis was in accordance with ASTM 1152-87.

Charpy specimens were full size (10 x 10 x 55 mm) with a 2 mm deep notch in strict accordance with ASTM Method E 23-88. Specimens from the US-DPC conduit were fabricated by stacking five 2 mm thick laminates, which were then by electron beam welded alongside the edges without welding in the notch region. The majority of tests was performed using a standard C-type pendulum machine at the Superconducting Magnet Laboratory of the Japan Atomic Energy Research Institute (JAERI) and the Materials Reliability Division of the U.S. National Institute of Standards and Technology (NIST). Some 77 K and 298 K tests were made using a U-shaped pendulum machine at the Mechanical Behavior Laboratory of MIT. With the exception of the MIT machine all testing machines meet the requirements of

ASTM Method E 23-88 with a maximum energy of 300 J. Overlap in test conditions among the three laboratories was insured allowing a cross check of data to be made. The variation of test data between laboratories was about 10%.[8] There is no standard test procedure for Charpy testing at 4.2 K and tests were performed using the flow method [9].

Internal pressurization tests were made using a meter-long section of the US-DPC conduit after 650°C for 200 hour heat treatment. A hydraulic fitting was attached at the center of the specimen using a reinforcement weld design that was developed for the US-DPC helium inlet joint [1]. Both ends of the specimen were plugged by GTA welding. One specimen containing a rewelded through wall crack (by GTA) 50% of the wall thickness using Incoloy 908 filler metal and tested in order to simulate unaged repair weld behavior. Using a hydraulic

TABLE 3. Tensile Properties of Incoloy 908 for Solution-Annealed (SA) Material (27 mm Plate) and Cold-Worked (CW) Material (12.7 mm Plate)

T (K)	Heat Treatment	σ_y (MPa)	σ_{UTS} (MPa)	Elongation (%)
4.2	SA (27 mm Plate)	662	1130	36.9
	SA + 20% CW	1254	1613	20.6
	SA + 650°C/200h	1227±14	1892±31	28.5±1.4
	SA + 700°C/100h	1258±31	1883±0	26.0±2.3
	SA + 750°C/50h	1199±14	1778±21	25.8±0.5
	20%CW + 650°C/200h	1489±28	1903±14	24.0±0.5
	20%CW + 700°C/100h	1434±6.9	1882±14	27.0±4
	20%CW + 750°C/50h	1320±3	1799±0	26.5±1.5
	20%CW + SA + 650°C/200h	1096±21	1779±3	24±0.3
	20%CW + SA + 700°C/100h	1151±13.8	1744±21	24±0
	20%CW + SA + 750°C/50h	979±13.8	1579±41	23.5±5
	SA + Filler Weld + 650°C/200h	1279±44	1648±34	8.0±3.7
77	SA (27 mm Plate)	662	1082	59.4
	SA + 20%CW	1199	1454	19.1
	SA + 650°C/200h	1189±17	1664±38	21.7±0.3
	SA + 700°C/100h	1192±0	1682±0	24.2±0.2
	SA + 750°C/50h	1117±7	1603±18	25.9±0.5
	20%CW + SA + 650°C/200h	1070	1680	18
	20%CW + SA + 700°C/100h	1130	1660	24±0
	20%CW + SA + 750°C/50h	900	1510	25
	SA + Filler Weld + 650°C/200h	1137±41	1413±76	10.0±5.1
298	SA (27 mm Plate)	389	717	**
	SA + 20%CW	1025	1135	**
	SA + 650°C/200h	1075±41	1433±0	16.5±0.8
	SA + 700°C/100h	1103±0	1396±4	15.2±0.5
	SA + 750°C/50h	1041±7	1344±21	16.1±0.1
	20%CW + 650°C/200h	1279±10	1499±17	19.0±1.1
	20%CW + 700°C/100h	1241±14	1451±3	21.0±1
	20%CW + 750°C/50h	1248	1413	17
	20%CW + SA + 650°C/200h	951±7	1358±34	14.8±0.8
	20%CW + SA + 700°C/100h	986±0	1365±0	14.8±0
	20%CW + SA + 750°C/50h	821±14	1227±21	17.0±2.0
	SA + Filler Weld + 650°C/200h	1062±14	1316±7	7.8±2.2
	US-DPC Weld + 650°C/200h	1182	1347	5.0

* Two tests were made except for those data without error values.
** Not determined

compressor system the specimen internal pressure was increased at 500 atm./sec rate at room temperature to test pressure. The test pressure was gradually increased until failure was detected by a pressure drop.

Stress rupture tests were performed on notched, round bar specimens fabricated and tested in accordance with ASTM E-139-83.

RESULTS

Tensile Properties

Tensile properties of 20% cold-worked (CW) and solution-annealed (SA) materials are summarized in Table 3 for unaged and aged conditions. The 20% cold-work resulted in about 90% increase in yield strength and about 45% increase in ultimate tensile strength from the solution-annealed material at 4.2 K. An even greater effect of cold-work on the yield and ultimate tensile strength is observed at room temperature. Solution-anneal-then-aged materials staring from the 20% cold-work showed about 10% lower yield and tensile strength than those of the solution annealed-then-aged material.

The lowest strength at 4.2 K are obtained after the 750°C/50 hours age due to the onset of γ' coarsening. The difference in tensile properties between room temperature and 4.2 K is only moderate for the heat treatments studied. As temperature is decreased from room temperature to 4.2 K both yield strength and elongation increase by about 20%. This moderate temperature dependence is in contrast to a factor of three increase in yield strength from room temperature to 4.2 K observed with Type 316LN stainless steel and a high Cr-Ni stainless steel [5]. Weld metal aged at 650C for 200 hours showed comparable yield strength but about 10% lower ultimate tensile strength than the base metal at all temperatures. Elongation of weld metals is only about 50% of base metal elongation. Tensile properties of aged weld metal remained constant over the temperature range studied. However an increased scatter in elongation between specimens is observed at lower temperature.

Fatigue Crack Growth Rate

Fatigue crack growth rates were determined for the solution-annealed-then-aged material and 20% cold-worked-then-aged material both at room temperature and 4.2 K. Results are shown in Figures 1 and 2. All three heat treatment conditions exhinit similar crack growth rates, within a factor of two. About a factor of three decrease in the crack growth rate is observed for the solution-annealed-then-aged material when the temperature is decreased from room temperature to 4.2 K. Weld metal showed lower fatigue crack growth rate than base metal.

Fatigue growth rate data for the solution-annealed-then-aged material and the 20% cold-worked-then-aged material were analyzed by fitting to a Paris' Law. The results of this analysis are shown in Table 4. All three heats of material show very similar crack growth rate behavior to that of stainless steel Type 316 at 4.2 K. However the weak temperature dependence of fatigue behavior makes the slope of the da/dN vs. ΔK curve much lower than that of stainless steel. This weak temperature dependence is attributed to precipitation hardening as the primary strengthening mechanism.

Fracture Toughness, Charpy Absorbed Energy and Pressurization Test Results

The J-integral fracture toughness data were obtained in accordance with ASTM E 813-88. Due to the complexity of the J-integral technique especially at 4.2 K, our test system including software was verified using materials with known fracture toughness values.. Excellent agreement with the round-robin data and data reproducibility was observed [5].

The results of these tests are summarized in Table 5. The data summarized include the yield strength and fracture toughness of Incoloy 908 for the solution-annealed-then-aged materials and the welded-then-aged material. Base metal exhibits the high and uniform fracture toughness and yield strength at 4.2K. However, the fracture toughness of GTA weld metal is

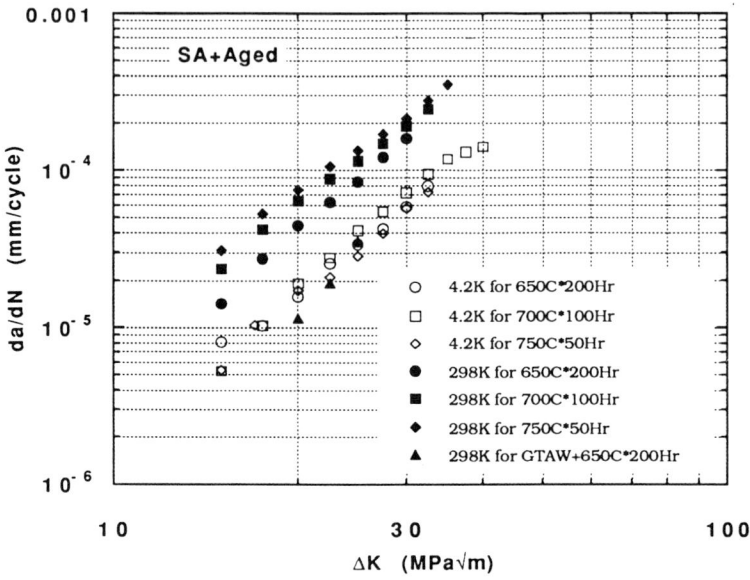

Figure 1. Fatigue Crack Growth Rates of Solution-Annealed-Then-aged Incoloy 908.

less than 50% of the base metal values. The fracture surfaces of base metal showed dimpled rupture. The weld metal fracture surface also showed dimpled rupture with increased dimple size and coarse particles at the base of most dimples. These particles were rich in Nb and appeared to have promoted void nucleation and lowered fracture toughness.

Charpy absorbed energy data is also summarized in Table 5. Only data at 77 K and 298 K are considered meaningful due to adiabatic heating at lower temperatures[8,9]. The effect of 20% cold work is to increase yield strength and decrease Charpy absorbed energy. GTA welding results in a significant reduction in Charpy absorbed energy.

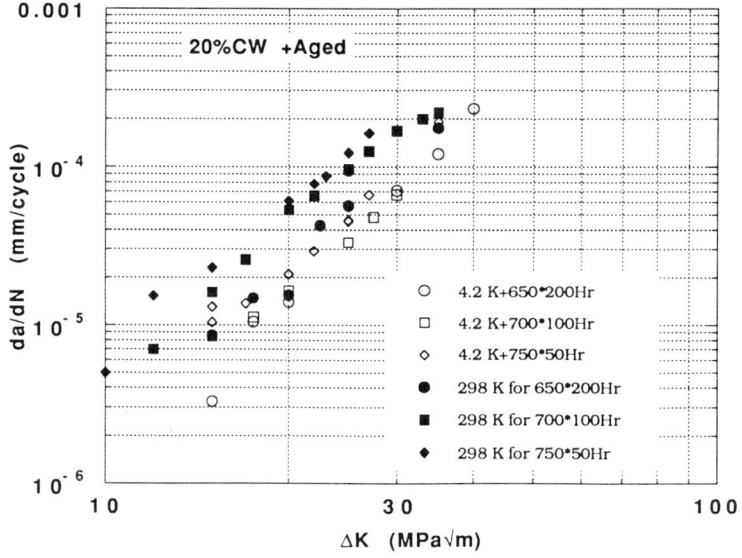

Figure 2. Fatigue Crack Growth Rates of 20% Cold-Worked-Then-Aged Incoloy 908.

TABLE 4. Paris' Law Constants of Incoloy 908 for Fatigue Crack Growth Rate Data $\left(\frac{da}{dN} = C(\Delta K)^m\right)$

MATERIAL	$C \times 10^{-11}$ (mm/cycle) 4.2 K	$C \times 10^{-11}$ (mm/cycle) 298 K	m 4.2 K	m 298 K
SA + 650°C for 200 hours	150	197	3.41	3.03
20%CW + 650°C for 200 hours	7.03	27.3	4.06	3.79
SA + 700°C for 100 hours	69.5	872	3.38	2.95
20%CW + 700°C for 100 hours	199	258	3.04	3.25
SA + 750°C for 50 hours	111	1850	3.18	2.76
20%CW + 750°C for 50 hours	77.2	349	3.45	3.25

TABLE 5. Yield Strength, Fracture Toughness, and Charpy Absorbed Energy of Solution-Annealed-Then-Aged Incoloy 908

MATERIAL	298K σ_y (MPa)	298K K_{IC} (MPa√m)	298K CVN (J)	77K σ_y (MPa)	77K K_{IC} (MPa√m)	77K CVN (J)	4K σ_y (MPa)	4K K_{IC} (MPa√m)	4K CVN (J)
SA + 650°C * 200hr	1075±41	196±5	73±1	1189±17	243±6	62±2	1227±14	235±5	71±1
SA + 700°C * 100hr	1103±0	176±5	---	1192±0	219±5	---	1258±31	220±2	---
SA + 750°C * 50hr	1041±7	160±0	---	1117±7	211±1	---	1199±14	240±4	---
SA + Filler GTAW *	1061±14	106±8	---	1137±41	---	29±1	1278±45	105±1	---
20% CW *	---	---	35±1	---	---	---	---	---	---
SA 2.4mm + GTAW *	---	---	44±1	---	30±1	---	---	---	33±1
SA 2.4mm + GTAW+SA*	---	---	75±1	---	---	---	---	---	---
US-DPC (Round)*	---	---	---	---	---	---	---	---	16±2
US-DPC (Square) *	1182	91**	---	---	---	17±5	---	---	---

* 650C for 200 hours final heat treatment was given.
** K_O Value due to inadequate specimen thickness.

The internal pressurization test of the one-meter section of US-DPC conduit specimen after 650 C for 200 hours of heat treatment resulted in a leak and pressure drop at 1,200 atm. Failure occurred in the filler weld region used for the high pressure fitting attachment. The crack was about 4 mm long in a direction perpendicular to the conduit length. The crack was repaired by welding to produce 50% wall penetration and re-tested. The repair weld leaked with a pressure drop at 1,450 atm. after a significant plastic deformation. Assuming the initial crack size equaled 1.3 mm the crack tip stress intensity at 1450 atm. was estimated to be about 35 MPa$\sqrt{m}$. This is a conservative lower bound for the fracture toughness of the unaged repair weld.

Stress Rupture Properties

The results of the stress rupture tests are shown in Figure 3. The data is sparse but preliminary results indicate that the performance is at or better that other low COE alloys.

DISCUSSION

Mechanical properties of Incoloy 908 and a Cr-Nb austenitic (JK1) steel [10] at 4.2 K are compared on a yield strength-fracture toughness diagram, shown in Figure 4. The solution-annealed-then-aged Incoloy 908 meets the Japanese target (the so-called JAERI box) for the magnet sheath alloy development program characterized by a 1,200 MPa yield and 200 MPa$\sqrt{m}$ fracture toughness. However fracture toughness of the JK1 is decreased significantly due to grain boundary sensitization after ageing [11]. The 20% cold worked-then-aged Incoloy 908 is expected to have a fracture toughness lower than 200 MPa due to its high yield strength. The decrease in Charpy absorbed energy and elongation of the 20% cold-worked-then-aged material supports this expectation. Additional toughness testing for 20% cold-work-then-aged material is in progress to determine the strength-toughness relationship.

The toughness of welded-then-aged material is only 40% of the base metal. Nb-rich coarse precipitates in the weld are responsible for the reduced toughness. A program to improve the weld properties in in progress.Preliminary results are very promising. The internal pressurization test showed that the GTA weld metal of Incoloy 908, nevertheless, has a fracture toughness that is adequate to insure the leak-before-break behavior in the US-DPC design.

Aged base and weld metal after shows a weak temperature dependence with respect to mechanical properties down to 4.2 K. Similar temperature insensitivity is observed for other precipitation-hardened superalloys including alloy 718 [12] and JBK-75 [13]. The temperature

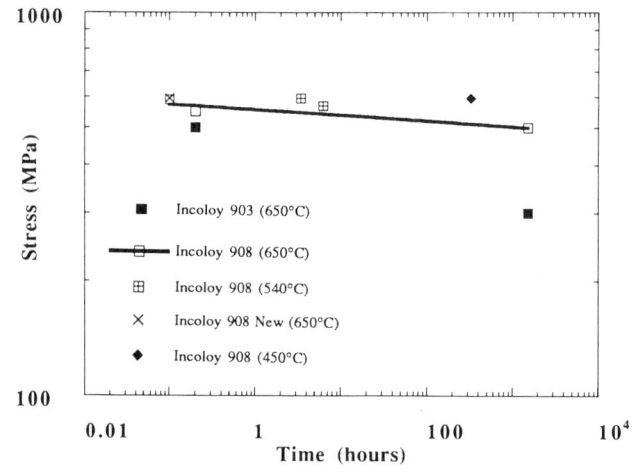

Figure 3. Stress Rupture Properties of Incoloy 908.

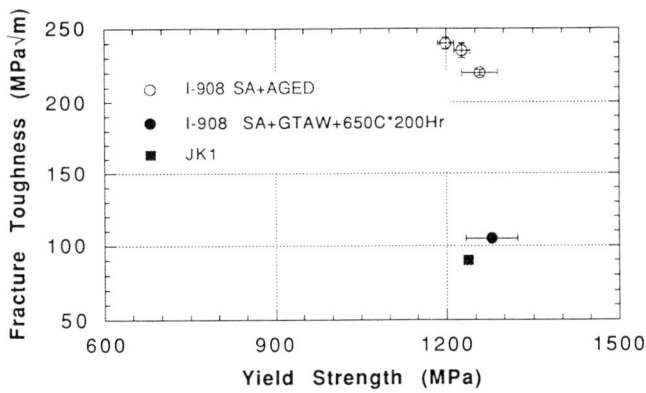

Figure 4. Yield Strength-Fracture Toughness Diagram for Incoloy 908 and JK1 (Heat Treated at 650°C for 180 Hours).

dependence of mechanical properties of the precipitation hardened alloys may be rationalized based on the fact that the primary source of strengthening, precipitates, are immobile in this temperature range and dislocation behavior has been shown to be temperature insensitive [14]. The weak temperature dependence provides several significant advantages. Structural integrity can be maintained during rapid temperature transients with sustained active loading which may be possible during off-normal conditions during the operation of large scale magnets. This behavior also allows for high compressive preloading at room temperature without yielding which can be employed to minimize peak fatigue loading during operation. Sophisticated mechanical tests that are not feasible at 4.2 K due to the requirements of expensive and complicated apparatus can be performed at room temperature for temperature-insensitive materials. The result can be a meaningful simulation of actual behavior at 4.2 K.

CONCLUSIONS

1. Incoloy 908, has been shown to exhibit adequate yield strength and high fracture toughness over the temperature range between 4.2 K and 298 K. The excellent mechanical properties combined with its low thermal expansion coefficient compatible with Nb_3Sn, makes Incoloy 908 an attractive material for Nb_3Sn magnets.

2. The 20% cold-worked-then-aged Incoloy 908 has 20% higher yield strength than the solution-annealed-then-aged material. Fracture toughness of the cold-worked-then-aged material is expected to decrease from 235 MPa$\sqrt{m}$ obtained with solution annealed then aged material at 4.2 K. Optimum combination of solution-anneal and cold-work will allow optimization of mechanical properties for specific design requirements.

3. Tensile properties, fatigue crack growth rate and fracture toughness show weak dependences on temperature compared with those of austenitic stainless steel due to primary strengthening by immobile precipitates.

ACKNOWLEDGEMENTS

We would like to acknowledge that Charpy impact tests of Incoloy 908 have been performed by Ralph Tobler of the NIST, Boulder, Colorado and Hideo Nakajima of the Naka Fusion Research Establishment, JAERI, Japan.

REFERENCES

1. M.M. Steeves, et al., "Test Results, US-DPC Program," CEC-ICMC Conf., Huntsville, AL, (June 1991).

2. J.W. Ekin, Report SR-724-34-84, U.S NBS, Boulder, Colorado (1984).
3. J.W. Ekin, Report SR-724-28-86, U.S NBS, Boulder, Colorado (1986).
4. H. Ledbetter, private communication with M.M. Morra, U.S. NIST, Boulder, Colorado (1990).
5. M.M. Morra, et al., *Proc.of MT-11*, Tsukuba, Japan, 831-838 (August 28-September 1 1989).
6. R.L. Tobler, ASTM Symp. on Elastic-Plastic Frac. Mechanics, Bueno Vista, FL., (November 1989).
7. A. Saxena, S.J. Hudak, Jr., *Int'l J. Frac.,* 14:5 (1978).
8. I.S. Hwang, et al., submitted to the *Journal of Testing and Evaluation*.
9. R.L. Tobler, et al., *J. of Testing and Evaluation*, 19:1:34-40 (1991).
10. R.L. Tobler, Report NISTIR 3944-13, U.S. NIST, Boulder Colorado, (September, 1990).
11. R.L. Tobler, private communication with I.S. Hwang, (May 1991).
12. R.L. Tobler, *Advances in Cryogenic Engineering,* 16:669-674 (1976).
13. P.K. Liaw, et al., *Austenitic Steels at Low Temperatures*, Plenum, NY, 171-185 (1983).
14. J.P. Hirth, J. Lothe, "Theory of .Dislocations," John Wiley and Sons, NY, 681 (1982).

THE EFFECT OF INDIUM ADDITIONS ON THE CRYOGENIC TENSILE PROPERTIES OF SUPERPLASTICALLY DEFORMED Al-Cu-Li-Zr ALLOY 2090

E. L. Bradley, III, D. Chu, and J. W. Morris, Jr.

Center for Advanced Materials, Lawrence Berkeley Laboratory, and
Department of Materials Science and Mineral Engineering
University of California, Berkeley 94720

Superplastic aluminum-lithium alloys are promising aerospace materials based on their high stiffness and low density, as compared to other superplastic aluminum alloys, and their ability to be formed to complex shapes that minimize weight. However, superplastic forming precludes the use of a pre-aging stretch which promotes homogeneous nucleation of T_1 (Al_2CuLi) and is a partial source of the excellent properties of 2090. Indium was added to superplastic 2090 in order to improve the post-formed mechanical properties. Tensile tests performed at 300 and 77 K show that the strengths are better than those demonstrated for regular 2090 after superplastic deformation. Failure occurs by transgranular shear and intergranular separation at 300K in both the formed and unformed material. At 77K failure occurs intergranularly prior to the geometric tensile instability, which is consistent with brittle fracture. It is believed that grain boundary intermetallics are promoting this fracture mode change at low temperatures.

INTRODUCTION

Aluminum-lithium alloys are well known as having higher stiffnesses and lower densities than other structural aluminum alloys[1]. The alloy 2090-T8 (stretch + age) has demonstrated increasing strength and toughness with decreasing temperatures[2] down to 4K with a microstructure of unrecrystallized, pancake-shaped grains and δ' (Al_3Li) and T_1 (Al_2CuLi) precipitates. While δ' forms homogeneously through a spinodal decomposition[3] upon quenching, T_1 nucleates heterogeneously at grain boundaries and dislocations[4]. As a result, 2090 is stretched prior to aging in order to provide more homogeneous nucleation of T_1 by increasing the number of intragranular dislocations.

There is interest in the aerospace industry in superplastic forming (SPF) since complex parts can be made lighter than is possible with conventional fabrication routes[5]. Small, equiaxed grains are required for superplasticity, and the parts are formed to net shape; thus, superplastic 2090 cannot duplicate the microstructure of the T8 condition, and no stretch can be applied to the post-formed part to promote homogeneous T_1 nucleation. For this reason, indium (In) was added to the superplastic modification of 2090 in order to improve the mechanical properties of the alloy. It has been demonstrated that In increases the yield strength of Al-Li-Cu alloys in general[6] and 2090 in the T6 condition (no stretch + age)[7]. This increase in strength in Al-Cu-Li-In alloys is the result of the precipitation of the metastable strengthening phase θ' (Al_2Cu)[6]. Blackburn and Starke[8] have shown that indium also tends to promote homgeneous nucleation of T_1 in 2090-type alloy compositions.

The post-SPF tensile properties of 2090+In at 300 and 77K were tested and compared to those of unformed material and a superplastic modification of 2090 without In in order to determine the effectiveness of the indium addition in improving the properties of post-formed 2090.

EXPERIMENTAL PROCEDURE

The alloy tested was supplied by Reynolds Aluminum in the form of 0.32 cm (0.125 in.) sheet of the composition (in weight %) Al-2.54Cu-2.30Li-0.18Zr-0.21In-0.05Fe-0.03Si. The sheets were then superplastically deformed into pans (Figure 1) by Northrop at 502°C (935°F) at a forming pressure of 100-150 psi and a back pressure of 200 psi to prevent cavitation.

The pans were cut into tensile blanks and solution heat treated at 550°C (1022°F) for 30 minutes. This temperature was chosen based on the presence of a dissolution reaction (on alloy heating) around 530°C (986°F) as measured by differential scanning calorimetry. The blanks were then aged at 160°C (320°F) for 72 hours to near peak strength. This aging regimen was determined by hardness measurements taken from both SPF deformed and undeformed areas of the pan aged at 160°C for up to 96 hours.

Tensile tests were performed using flat, subsize tensile specimens with a 2.54 cm (1 in) gage length and an orientation with respect to the SPF pan as shown in Figure 2. The true superplastic thickness strain in the deformed section was 0.9 while those from the flange section were undeformed. The SPF tensile samples were much thinner than the samples from the non-deformed area since the starting sheet was 0.32 cm (0.125 in.) thick while the pan bottom was between 0.13-0.15 cm (0.052-0.058 in.) thick. The specimens were pulled at a constant stroke rate of 0.0038 cm/min (0.0015 in/min) and the strain was measured using an extensometer.

Optical microscopy was performed on samples polished to 0.05 µm and etched with Keller's reagent (95.5% H_2O, 2.5% HNO_3, 1.5% HCl, and 0.5% HF). The fracture surfaces were studied using a scanning electron microscope at 20 kV. The precipitate distribution within the grains was studied using a Phillips 301 transmission electron microscope (TEM) at 100kV. The TEM samples were jet electropolished using a mixture of 20% HNO_3 and 80% Methanol at -30°C at 18V.

Figure 1. Superplastically formed 2090+In pan.

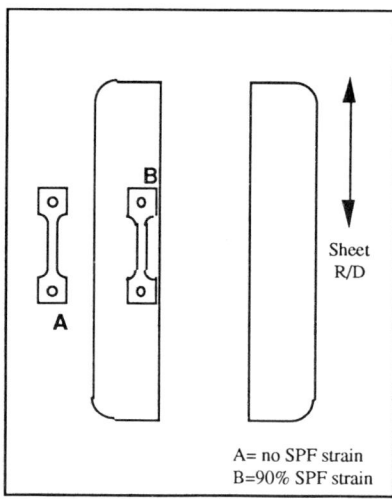

Figure 2. Orientation of tensile specimens in the pan with respect to the sheet rolling direction

RESULTS AND DISCUSSION

The as-received 2090+In from both the formed and unformed regions is shown in Figure 3, and one can see that the microstructure is totally recrystallized. Grain growth occurred during SPF, resulting in larger grains; however, the subsequent solution heat treatment caused the grains in the unformed material to coarsen to a size approaching that in the formed material. Note that the dark spots are not voids resulting from cavitation but are the consequence of etching. The aged microstructure was observed with the TEM and Figure 4 shows that δ', $\beta'(Al_3Zr)$, T_1, and θ' precipitates were found in the post-formed material. There were no differences observed in the precipitation distribution and morphology between the post-formed and unformed material.

The data for the tests is shown in Table 1 and it is apparent that the SPF deformation did little to impair the properties of the material. The yield strength increases with decreasing temperature for both the unformed and post-formed material. However, the elongations do

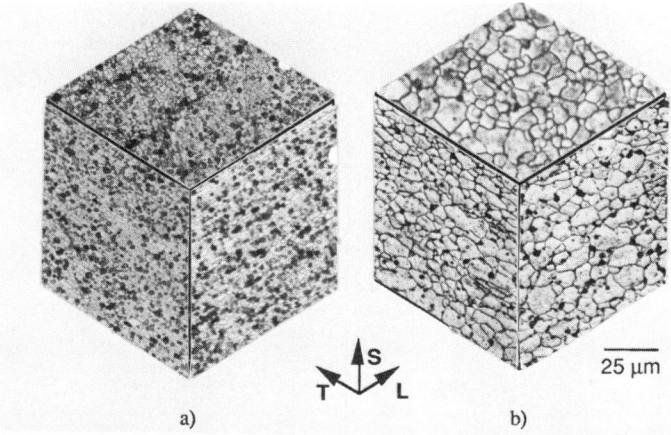

Figure 3. Microstructure of the as-received 2090+In in the a) unformed and b) superplastically formed conditions.

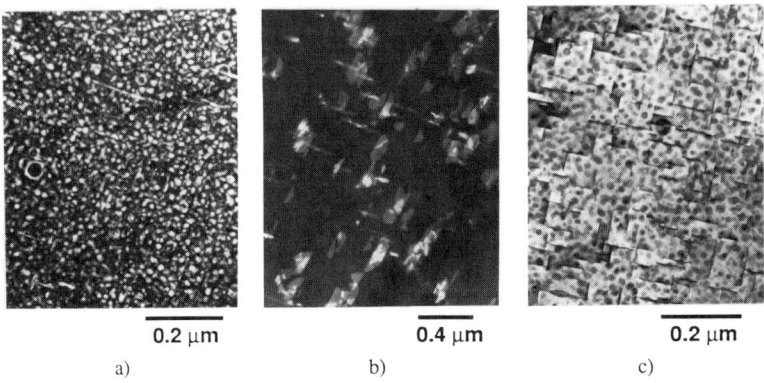

Figure 4. TEM micrographs of the aged microstructure in the superplastically formed 2090+In showing a) δ' and β'(centered dark field), b) T_1 (CDF), and c) θ'(bright field).

not improve, as is common in Al-Li alloys. The similarity in properties between the formed and unformed material is due to the fact that (i) the solution heat treatment caused grain coarsening in the unformed material so that the grain sizes were equivalent, (ii) the precipitate distributions were very similar, and (iii) the cavitation in the formed material was held to a minimum by the application of back pressure during SPF.

The effect of the indium is apparent in Figure 5, where the properties of superplastic 2090 are compared with the 2090+In alloy. In all cases the yield strengths are greater for the indium alloy than those of the 2090 without indium tested by Glazer et. al.[9] and is the result of θ' which is not present in any appreciable quantities in regular 2090. The 2090+In exhibits greater elongations at room temperature while the regular 2090 shows increasing elongations with lower temperatures comparable to those of the T8 condition[9]. The yield and ultimate strengths of the 2090+In at 300K are equal to or greater than those of post-formed 8091 (ys = 391 MPa, uts = 495 MPa)[10].

Since 2090 typically shows an increase in elongation with decreasing temperature[2], the observed decrease in elongation in this alloy for both the formed and unformed material indicates that a change in fracture mode may be occurring at lower temperatures. Measurements of work hardening rates and true stress vs. strain at failure reveal that the Considere criterion ($\partial\sigma/\partial\varepsilon < \sigma$) was not satisfied for either the formed or unformed material tested at 77K, although it is just satisfied for the both forms at 300K. This indicates that the mode of failure was brittle. Scanning electron microscopy of the fracture surfaces reveals large amounts of intergranular fracture. Figure 6 shows that at 77K the unformed material

Table 1. Tensile properties of 2090+In sheet before and after superplastic forming

True SPF Strain (%)	Test Temp (K)	Yield Strength (ksi)	Tensile Strength (ksi)	Total elongation (%)
0	300	383 (56)	519 (75)	10.4
	77	428 (62)	593 (86)	10.0
90	300	386 (56)	540 (78)	9.7
	77	424 (62)	557 (81)	4.8

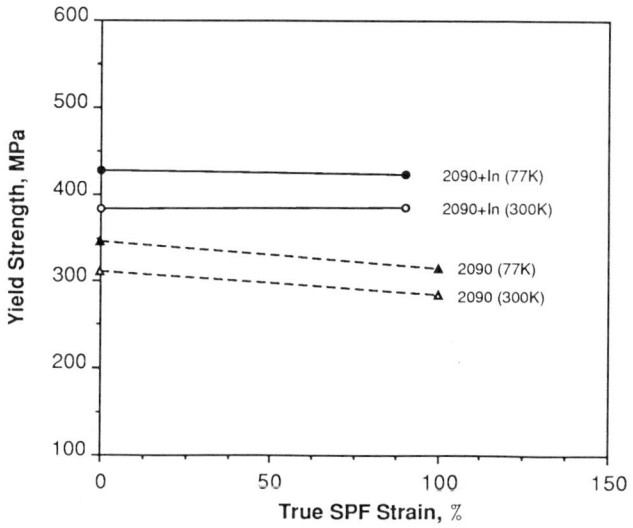

Figure 5. Comparison of strengths of 2090 and 2090+In with superplastic strain.

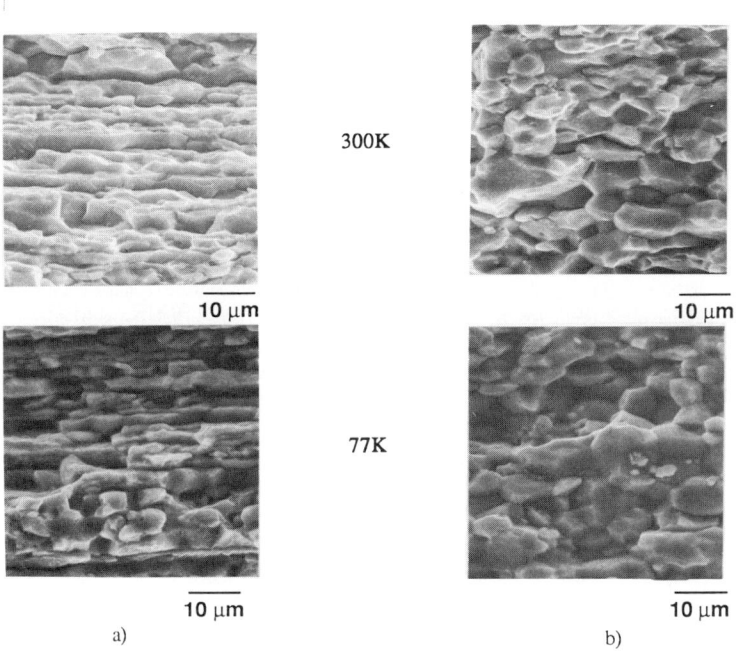

Figure 6. SEM micrographs of the fracture surfaces of a) unformed and b) formed 2090+In at 300 and 77K.

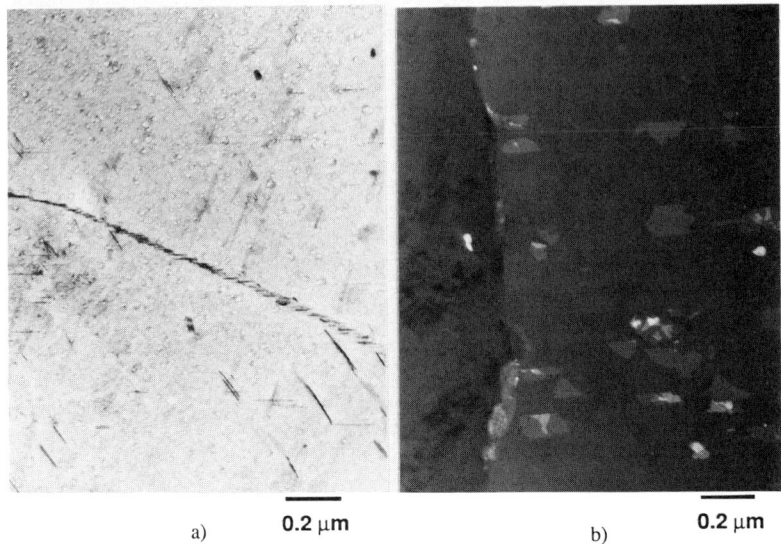

Figure 7. TEM micrographs showing T_1 precipitation on a) subgrain (BF) and b) grain boundaries (DF).

does exhibit some transgranular shear, but the post-formed material is purely intergranular. There are also precipitates present on the intergranularly fractured regions which may have played a role in embrittling the boundaries.

Further evidence explaining the brittle fracture of the 2090+In alloys at low temperatures is given in Figure 7 which shows that T_1 is present on both grain and subgrain boundaries in the material. This is also believed to have contributed to the relatively low elongations observed at 77K in both the formed and unformed materials. The boundaries are not able to accommodate as much deformation due to the presence of the T_1 and other intermetallics, and this results in the intergranular fracture seen in Figure 6. Improvements in the grain boundary microstructure should result in better low temperature properties.

CONCLUSIONS

The addition of indium to 2090 raises the yield strength of the alloy in both the unformed and formed conditions to levels equal to 8091. However, the elongations of the alloy do not increase with decreasing temperature due to the intrusion of a brittle grain boundary fracture mode at 77K. This fracture mode is believed to result from the presence of T_1 and coarse intermetallics at the grain boundaries.

ACKNOWLEDGEMENTS

The authors gratefully acknowledge Hiro Hayashigatani and Russell Cinque for their assistance with the microscopy. This research was supported by the Director, Office of Energy Research, Office of Basic Energy Science, Material Sciences Division, of the U. S. Department of Energy under Contract No. DE-AC03-76SF00098.

REFERENCES

1. Aluminum-Lithium Alloys, T. H. Sanders, Jr. and E. A. Starke, Jr., eds., The Metallurgical Society, Warrendale, PA, (1981).

2. J. Glazer, S. L. Verzasconi, R. R. Sawtell, and J. W. Morris, Jr., Mechanical Behavior of Aluminum-Lithium Alloys at Cryogenic Temperatures, Metall. Trans. A, 18A:1695, (1987).
3. A. G. Khachaturyan, T. F. Lindsey, and J. W. Morris, Jr., Theoretical Investigation of the Precipitation of δ' in Al-Li, Metall. Trans. A, 19A:249, (1988).
4. B. Noble and G. E. Thompson, T_1 (Al_2CuLi) Precipitation in Aluminum-Copper-Lithium Alloys, Metal Sci J., 6:167, (1972).
5. A. J. Barnes, Advances in Superplastic Aluminum Forming, in: "Superplasticity in Aerospace," H. C. Heikkenen and T. R. McNelley, eds., The Metallurgical Society, Warrendale, PA, (1988).
6. J. M. Silcock, T. J. Heal, and H. K. Hardy, The Structural Aging Characteristics of Ternary Aluminum-Copper Alloys with Cadmium, Indium,and Tin, J. Inst. Metals, 84:23, (1955-56).
7. S. Verzasconi and J. W. Morris, Jr., Cryogenic Mechanical Properties of Low Density Superplastically Formable Al-Li-Alloys, in: "Proc. 5th Intl. Conf. Al-Li Alloys," T. H. Sanders, Jr. and E. A. Starke, Jr., eds, Materials and Component Engineering Publications, Birmingham, U.K. (1989).
8. L. Blackburn and E. A. Starke, Jr., Effect of In Additions on Microstructure-Mechanical Property Relationships for an Al-Cu-Li Alloy, in: "Proc. 5th Intl. Conf. Al-Li Alloys," T. H. Sanders, Jr. and E. A. Starke, Jr., eds, Materials and Component Engineering Publications, Birmingham, U.K. (1989).
9. J. Glazer, T. G. Nieh, and J. W. Morris, Jr., Tensile Behavior of Superplastic Al-Cu-Li-Zr Alloy 2090 at Cryogenic Temperatures, (1987).
10. W. S. Miller and J. White, The Development of Superplastic 8090 and 8091 Sheet, in: "Superplasticity in Aerospace," H. C. Heikkenen and T. R. McNelley, eds., The Metallurgical Society, Warrendale, PA, (1988).

DEFORMATION AND FRACTURE OF Al-Li ALLOYS IN MECHANICAL-IMPACT TESTS

R. P. Reed, C. N. McCowan, N. J. Simon, and J. D. McColskey

Materials Reliability Division
National Institute of Standards and Technology
Boulder, Colorado, U.S.A.

ABSTRACT

Mechanical-impact tests in liquid and gaseous oxygen atmospheres aid in the assessment of Al-Li alloys for their use in cryogenic tankage for Advanced Launch Systems. In these tests, a falling plummet-striker pin assembly impacts a small disc specimen. Compressive, radial shear, and tensile (hoop) stresses are imparted to the disc specimens. The combination of these stresses and the high strain rate results in unique deformation and fracture characteristics. These characteristics include adiabatic shear bands, microcracking, specimen splitting, specimen heating (local melting), radial deformation in preferred orientations, and shear lips at the specimen-striker pin interface. These deformation and fracture modes are described and compared for various tempers of Al-Li alloys 8090, 2090, 2095, and alloy 2219. Distinctions between the Al-Li alloys and alloy 2219 are discussed.

INTRODUCTION

At the National Institute of Standards and Technology (NIST), this study is part of a broader program to assess the suitability of new, high-strength Al-Li alloys for cryogenic tankage of the Advanced Launch System by determining their relative compatibility with liquid and gaseous oxygen. Program objectives are (1) to provide well-characterized specimens of state-of-the-art commercial alloys for liquid-oxygen (LOX) and gaseous-oxygen (GOX) measurements at two qualified NASA test laboratories, Marshall Space Flight Center (MSFC) and White Sands Test Facility (WSTF); (2) to study the roles of intrinsic material variables (fracture properties, chemical composition, and second-phase particles) on ignition characteristics; and (3) to assess the extrinsic test variables (impact energy, pressure, temperature; specimen size, orientation, and location within the LOX or GOX specimen cup) on reaction tendencies.

In this program, seven alloys were tested for compatibility with liquid oxygen by using the open-cup and pressurized mechanical-impact facilities at MSFC and the open-cup facilities at WSTF. The results of these tests have been previously reported.[1,2] Mechanical-property tests on these alloys have also been conducted at low temperatures,[3] and the mechanical and physical properties on Al-Li alloys and alloy 2219 have been compiled.[4]

The sudden collision of the striker pin with the specimen during a mechanical-impact test produces unique deformation and fracture characteristics in Al-Li alloys and alloy 2219. The general features of these deformation and fracture characteristics are discussed in this paper.

MATERIALS

The alloys studied were 8090-T8771, 2090-T81, 2095 (T851 and T651), and 2219 (T87, T851, and T37). Their chemical compositions and suppliers, as well as the quantity of material and the dates that they were received, are given in previous reports.[1-3] All compositions fall within the Aluminum Association specifications for each alloy. Specimens were machined as discs, 12.7 mm in diameter and 1.6 or 3.2 mm thick.

MECHANICAL-IMPACT TEST MACHINE

The ASTM standard for determining the impact sensitivity of materials in LOX requires the use of a mechanical-impact tester (Fig. 1). Its structural frame consists of three vertical guides that maintain the alignment of the plummet assembly as it falls onto the striker pin. The plummet is a 9.09-kg mass that falls from a maximum height of 1.1 m. The frame is mounted on a 100-mm thick stainless steel base plate, which in turn is mounted on a 0.6-m cube of concrete. The plummet, when released from the height corresponding to the desired impact energy, falls freely with minimal friction from the alignment guides. The plummet strikes the top of a hardened 17-4 stainless steel striker pin that rests on the specimen. The plummet then rebounds off the top of the striker pin. The steel base plate is precooled to the desired temperature with liquid nitrogen (LN_2). The specimen, an Al-alloy cup, a stainless steel cup insert, and the striker pin are precooled in the desired test environment (LOX or LN_2) and are kept submerged in a cryogen tray until ready for testing. These parts are then removed from the cryogen tray and installed in the impact machine.

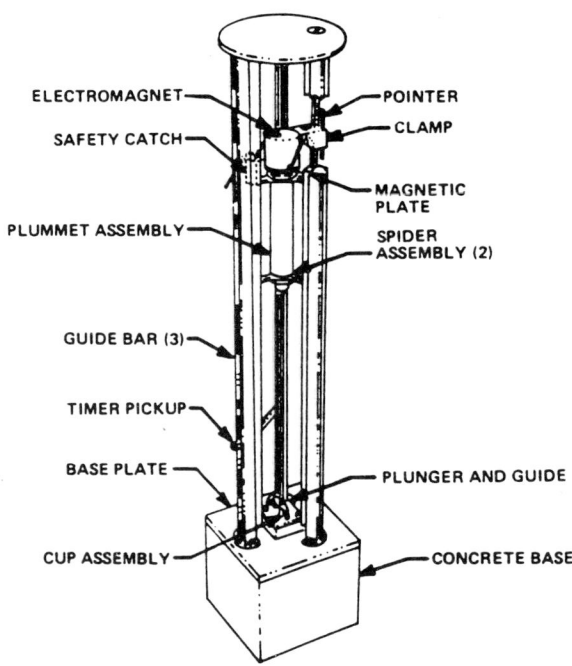

Fig. 1. ABMA mechanical-impact tester.

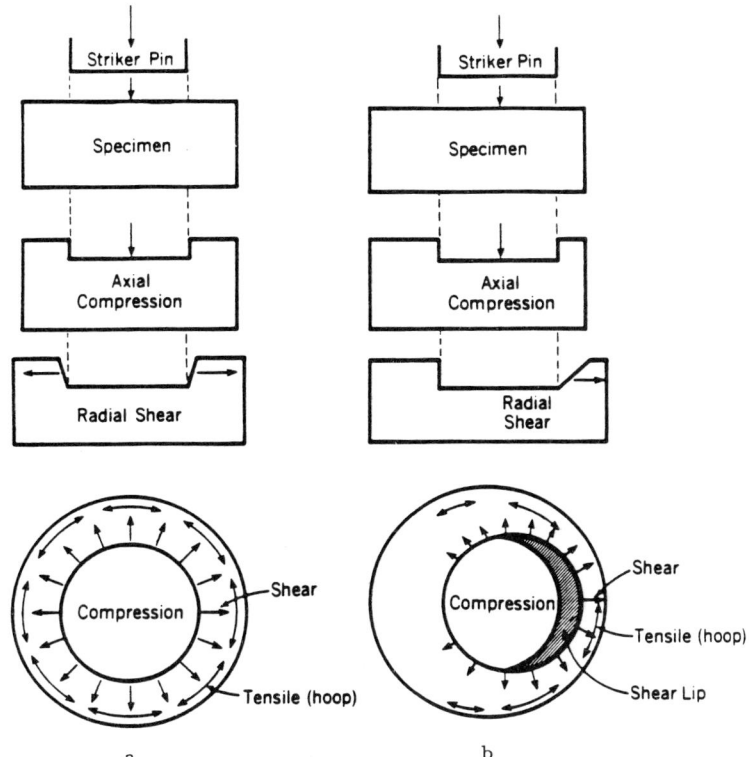

Fig. 2. Deformation modes in a mechanical-impact test specimen: (a) concentric strike, (b) eccentric strike.

In the mechanical-impact test, the indentation of the striker pin produces compressive forces in the specimen under the pin, radial shear forces toward the specimen perimeter outside the indentation of the striker pin, and tensile (hoop) forces around the outer perimeter of the specimen. The tensile forces are produced from the radial specimen plastic deformation that results in lateral specimen expansion. The strain rates are high (~3 × 10^3 s^{-1}); therefore, nearly adiabatic heating conditions exist.

ANISOTROPIC RADIAL SHEAR DEFORMATION

Following impact, the specimen experiences axial compression from the top and is constrained at the bottom by the support cup. The yield stress is exceeded, and the axial compression forces produce an indentation from the striker pin in the specimen from 0.2 to 0.7 mm for a plummet-impact potential-energy parameter of 8 to 10 kg·m. The radial shear forces result from accommodation of the axial plastic deformation and the constrained lower specimen surface. The forces for isotropic deformation produced under these impact conditions are shown in Fig. 2. The radial plastic deformation of the specimen is anisotropic and is larger in the rolling (L) direction of the alloy (see Fig. 3, for example). Eccentricity of the striker-pin impact indentation with respect to the center of the specimen also results in anisotropic radial shear; more extensive deformation occurs in the specimen area that is smallest (between the indentation and circumference) and, therefore, provides less constraint. The anisotropy of the radial shear forces from eccentricity of the striker-pin impact is illustrated in Fig. 2(b).

Since the flow strength in the transverse (T) orientation is less (typically about 20 MPa less) than in the L orientation, we cannot associate the

21

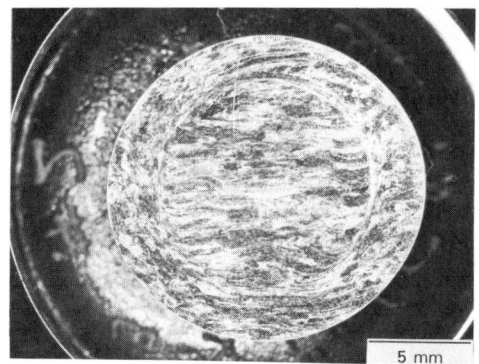

Fig. 3. Long axis of a deformed specimen parallel to rolling direction. Etched 2095-T851, tested in open cup at 6 kg·m.

more extensive plastic deformation in the L orientation (in the radial specimen direction) with reduced flow strength in that direction. Instead, the more extensive deformation in the radial L orientation must be associated with less specimen constraint from the circumferential hoop forces. The relatively low circumferential flow strength in the T specimen direction results in more radial deformation in the L orientation.

SPLITS

When the plummet-impact potential-energy parameter is large (~6 to 10 kg·m), the energy absorbed by the specimen is sufficient to impose tensile stresses on the specimen perimeter in excess of the tensile strength of the Al alloys. A crack initiates along the specimen perimeter, in the region of large, radial shear deformation (close to the shear lip), and grows inward. A shear lip is formed by the radial shear forces; the radial shear strain acts to expand the original impact area to produce a tapered impression in the specimen. This effect is pointed out in later figures. The crack tends to propagate parallel to the rolling direction (L) and usually grows into the region of high compressive forces under the striker pin. In this paper, these cracks are called *splits* because they tend to divide the specimen into two regions; splits are illustrated in Figs. 4 and 5. All alloys exhibited the tendency to crack in the L direction. This is understandable: the fracture toughness in the TL orientation is lower than in the LT orientation for these alloys.

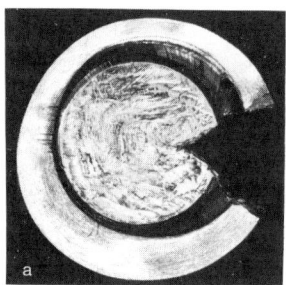 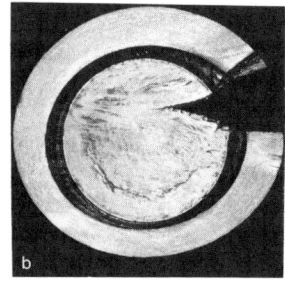 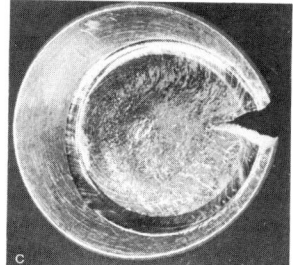

Fig. 4. Typical splits extending under the striker-pin indentation for (a) alloy 2090-T81, open-cup LOX, 6 kg·m; (b) 2095-T851, open-cup LOX, 10 kg·m; (c) 2219-T87, open-cup LOX, 10 kg·m.

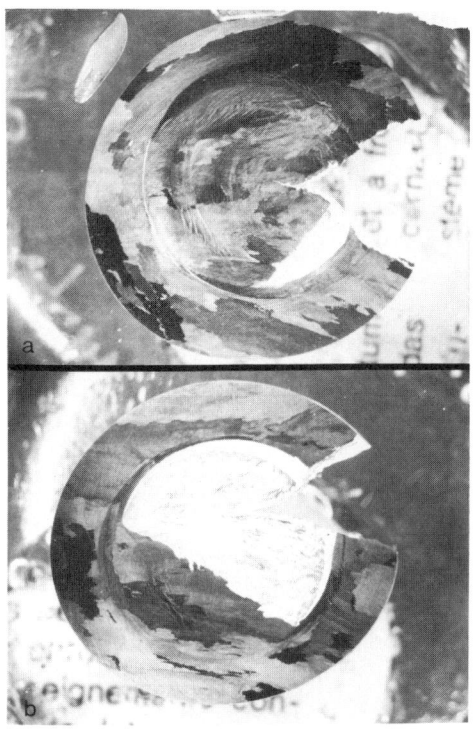

Fig. 5. Crack growth along L orientation in 2090-T81 at
(a) 10 kg·m and (b) 8 kg·m, open-cup LOX.

At 300 K (GOX) and 10, 8, and 6 kg·m, specimens of 2090-T81 split, but specimens of 2095-T851 and 2219-T87 did not. At 76 K (LOX) and 6 kg·m and above, all 2090 specimens split. At 76 K and 6 kg·m, some specimens of 2095, 8090, and 2219-T851 split. At 76 K, more specimens split at higher energies.

At low temperatures, the micromechanism of split fracture in all 2095 and 2219 tempers was microvoid coalescence. Limited delaminations in the plane of the specimen were observed in 2095-T851. At both temperatures, the fracture micromechanism for splitting in 2090-T81 was a combination of intergranular cracking and transgranular shear. The SEM photographs of these fracture mechanisms (Fig. 6) show alternate layers of shear followed by intergranular cracking, which look like a delaminated crack front. At higher magnification, the transgranular shear faces reveal microvoid coalescence on a very fine scale. For alloy 2090-T81, intergranular cracking was more prevalent in specimens tested in LOX (90 K) than in those tested in GOX (300 K). This increase of intergranular cracks at low temperatures corresponds to the reduction of fracture toughness in the SL and ST orientations at 76 K from that at 300 K.[3] Intergranular cracking associated with a split in 2090-T81 (tested in LOX at 10 kg·m) is illustrated in Fig. 7. The crack obviously tends to grow in the long-grain direction and within long-grain boundaries.

MICROCRACKS

Microcracks were found on the outside diameter of the specimens where the tensile stress is high, in the area of radial shear plus tensile deformation between the shear lip and the specimen perimeter, or within the impacted region of high compressive stress. Microcracks were observed in all alloys and were more prevalent in specimens that were impacted at low temperature

 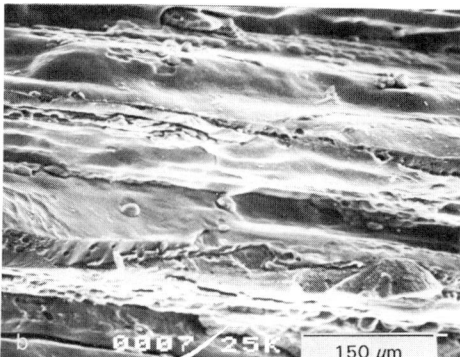

Fig. 6. The fracture surface of a split in an open-cup (LOX) of 2090-T81: (a) low magnification, (b) local deformation features.

(LOX). Both inter- and transgranular cracking were observed. Microcracks in 2090-T81 are shown in Fig. 8. In this alloy, as discussed earlier, all microcracks tended to follow grain boundaries. Microcracks are not unexpected in these tests, owing to the large biaxial stress fields (compression-shear or shear-tensile) in many regions of the specimens.

ADIABATIC SHEAR BANDS

Large adiabatic shear bands were observed in the Al-alloy specimens immediately under the striker-pin indentation and under the shear lip. The scale of the adiabatic shearing varied with grain size. In Fig. 9, shear bands adjacent to the shear lip are shown for alloys 2090-T81 and 2095-T851. The shear bands in 2090-T81 are very large and correspond to the length of the grains (1 to 5 mm). In 2095-T851 and 2219-T87, the shear-band size was restricted by the grain size.

Shear bands are associated with both localized deformation and heating. We estimate that the time of specimen deformation is between 7×10^{-5} and 1.4

Fig. 7. Intergranular microcracking associated with a split in 2090-T81 (10 kg·m, pressurized LOX).

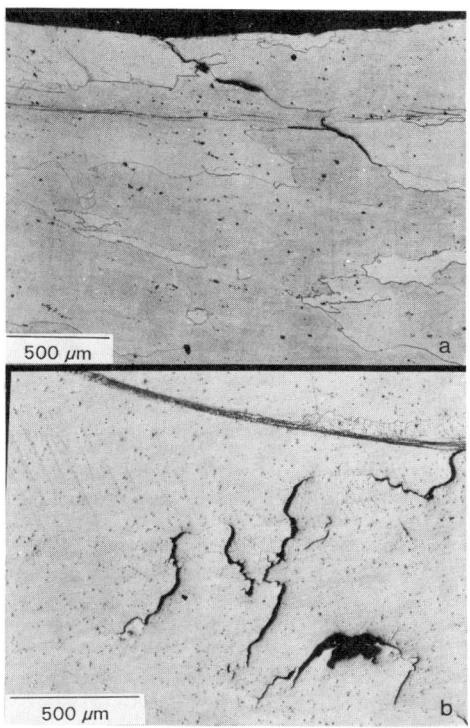

Fig. 8. Microcracks in 2090-T81: (a) adjacent to the shear lip (8 kg·m, GOX) and (b) in a region between bottom of the shear lip (indicated by curved "line" on the top) and the specimen perimeter (10 kg·m, GOX).

$\times$ 10^{-4} s, and that the time for propagation of a thermal transient is 2.8 $\times$ 10^{-3} s.[1,2] Thus, the establishment of thermal equilibrium lags mechanical equilibrium, leading to local hot spots. This is precisely the condition that produces adiabatic shearing, the process where thermal softening promotes continued local deformation that, in turn, promotes more local heating. The process may be continuous and, thus, lead to local melting. The adiabatic shear bands also lead to localization of excessive deformation; evidence of microcracking has been observed to be associated with these bands, particularly near grain boundaries. We find evidence for microcracking where the shear bands terminate at the grain boundaries.

An argument can be made that Al-Li alloys are expected to have more adiabatic shearing than alloy 2219. In Table 1, typical values of specific heat (C) and thermal conductivity (K) at 90 and 300 K are listed for each alloy. Also included in Table 1 are the ratios $C \cdot \rho / K$, which enable estimations of the time to propagate a thermal transient (t_c):

$$t_c = \frac{x_t^2 \cdot C \cdot \rho}{K} ,$$

where x_t is the length of thermal path (assumed to be constant for both alloys for this discussion) and ρ is the density (alloy 2219 is about 10% more dense than alloy 2090). Table 1 shows that the time to propagate a thermal transient for alloy 2090 is twice that of alloy 2219. Therefore, thermal stability should not be achieved as quickly in alloy 2090, and nearly adiabatic conditions should occur more frequently in this alloy.

Fig. 9. Adiabatic shear bands adjacent to the shear lip in alloys 2090-T81 (a) and 2095-T851 (b).

Table 1. Thermal Transient Parameters

Alloy	Specific Heat, C, J/(kg·K)		Thermal Conductivity, K, W/(m·K)		Transient Ratio, $C \cdot \rho / K$, s/m^2 × 10^4	
	90 K	300 K	90 K	300 K	90 K	300 K
2219	520	890	70	125	2.1	2.0
2090	540*	930	32	72	4.4	3.3

*estimate

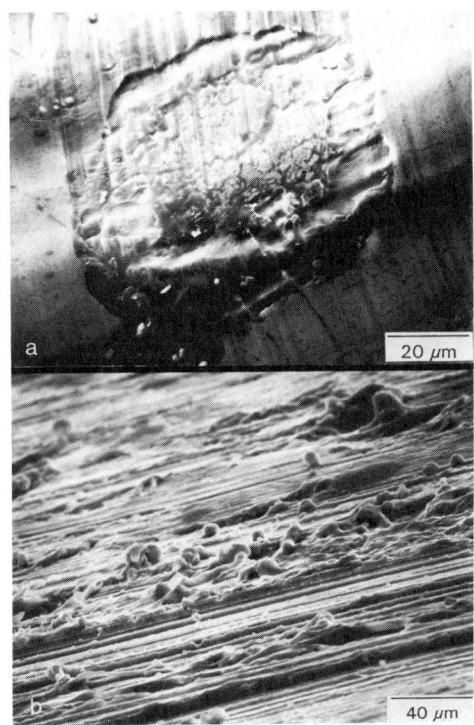

Fig. 10. Regions of localized melting on the shear lip (a) and on the compression surface (b) of 2090-T87 (10 kg·m, GOX).

LOCALIZED MELTING

We estimated that an increase in the local absorbed energy of only a factor of 8 is required to induce local melting, assuming that mechanical equilibrium precedes thermal equilibrium.[1,2] We argued that testing irregularities, such as concavity of the striker-pin surface and nicks along the striker-pin perimeter could easily result in local increases of absorbed energy of this magnitude.

There is good evidence of localized melting in Al-Li alloys. Localized melting may be identified by two microstructural features: (1) a smooth surface on which no machining nor indications of deformation are present, and (2) a smooth, curved interface contour, as distinct from the abrupt, sometimes linear interfaces from deformation shear.

In Fig. 10, a region of local melting on the shear lip is shown; notice evidence of solidification cracking on its surface. In Fig. 11, evidence for melting at the base of the shear lip is shown for 2219-T851. Here the metal has melted, flowed onto the base of the indentation, and in the process, obliterated the specimen machining marks. The irregularities on the shear lip probably represent inclusion-striker-pin reactions in which the inclusions have been partially separated from the matrix.

SUMMARY

The mechanical-impact test imparts a large energy into specimens in a short time. The energy produces large compression, shear, and tensile forces in the disc-shaped specimen; anisotropic deformation of the specimen results.

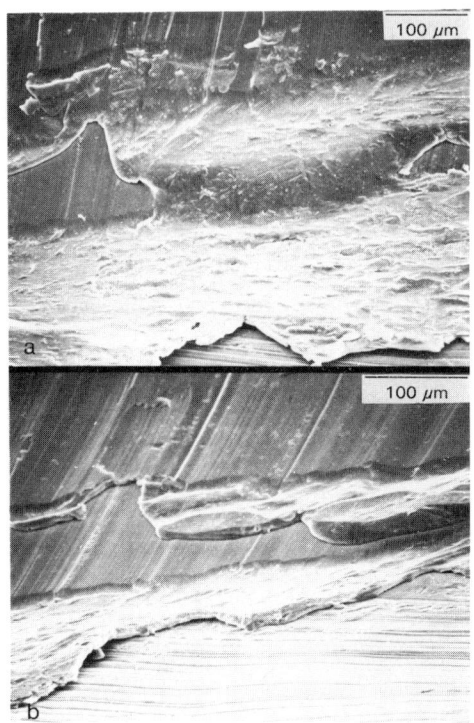

Fig. 11. Localized melting in 2219-T851 (10 kg·m, GOX) at the root of the shear lip. In both (a) and (b), the shear lip is on the top, and the indentation (with specimen machining markings still showing) is on the bottom.

The large forces produce local microcracking, specimen splitting, adiabatic shear bands, and local melting.

ACKNOWLEDGMENTS

Our research was supported by the Advanced Launch System, Edwards Air Force Base, Bao Nguyen, Program Manager.

REFERENCES

1. R. P. Reed, N. J. Simon, J. D. McColskey, J. R. Berger, C. N. McCowan, J. W. Bransford, E. S. Drexler, and R. P. Walsh, "Aluminum Alloys for Cryogenic Tanks: Oxygen Compatibility Vol. 1," AL-TR-90-063, Astronautics Lab (AFSC), Edwards AFB, California (September, 1990).
2. R. P. Reed, N. J. Simon, J. D. McColskey, J. R. Berger, C. N. McCowan, J. W. Bransford, E. S. Drexler, and R. P. Walsh, "Aluminum Alloys for Cryogenic Tanks: Oxygen Compatibility Final Report: Vol. 2," Al-TR-90-063, Astronautics Lab (AFSC), Edwards AFB, California (September, 1990).
3. R. P. Reed, P. T. Purtscher, N. J. Simon, J. D. McColskey, R. P. Walsh, J. R. Berger, E. S. Drexler, and R. Santoyo, National Institute of Standards and Technology, Boulder, Colorado (1991), to be published as an Edwards AFB report.
4. N. J. Simon, E. S. Drexler, and R. P. Reed, "Review of Mechanical and Physical Properties of Al-Li Alloys and Alloy 2219," Al-TR-90-064, Astronautics Lab (AFSC), Edwards AFB, California (September, 1990).

MICROSTRUCTURE AND CRYOGENIC TENSILE FRACTURE

BEHAVIOR OF AN Al–Li–Zn–Zr ALLOY

X. J. Jiang and Y. Y. Li

Institute of Metal Research, Academia Sinica
Shenyang 110015, China

ABSTRACT

The microstructure and cryogenic tensile properties of an Al–Li–Zn–Zr alloy in different aging conditions have been studied. The temperature dependence of the tensile properties of the alloy were evaluated from ambient to liquid nitrogen temperature in the long transverse direction. The results show that the tensile strength and elongation of the alloy are much increased with the decrease of test temperature. The variation in tensile fracture mode and deformation behavior are proved to depend on the aging conditions and test temperature closely. The effect of microstructure on tensile deformation and fracture behavior of the alloy at room and cryogenic temperatures have been discussed. δ' particles are the major fractor to the improvement of ductility for the alloy at low temperatures.

INTRODUCTION

Aluminum–lithium alloys have attracted much attention in recent years because of their distinct advantages over conventional aluminum alloys, such as high strength, high modulus and low density[1]. Moreover, many researches on the mechanical behaviour of Al–Li alloys indicate that the properties of the alloys in L-T orientation improve significantly with decreasing temperature[2-7]. So the alloys have been considered for aircraft and space vehicle applications, including cryogenic tankages.

Now several possible mechanisms have already been proposed to explain these curious properties of Al–Li alloys at low temperatures. One is the low melting point phases exist at grain boundaries, they solidify at low temperatures and remain liquid at room temperature[2]. Another is a large number of crack delaminations perpendicular to the fracture surface in an L–T oriented specimen and in plane crack deflections at cryogenic temperatures and thus increases the low temperature toughness[3]. The increase in tensile elongation and fracture toughness have also been attributed to the increase in strain–hardening capability[4]. At the same time, the deformation mode changes from planar slip at room temperature to a fairly homogeneous slip distribution at 77K is suggested for the explanation of improvement of tensile elongation at 77K[5]. Although the low temperature behaviors of Al–Li alloys have been described, the exact mechanism of such phenomena has not been established yet.

The purpose of the present investigation is to examine the low temperature tensile behaviour of an Al–Li–Zn–Zr alloy in different aging conditions. The effect of microstructure on tensile deformation and fracture mode of the alloy at room and cryogenic temperatures have also been discussed.

EXPERIMENTAL PROCEDURES

The ingot, which was melted in vacuum induction furnace and cast in argon, was homogenized at 763K for 18h and 798K for 2h. It was held at 723K for 2h and extruded to a 28mm diameter bar, then hot rolled to a sheet of 5mm thickness after keeping at 723K for 1h. The composition of the alloy is(in wt.%): Li 2.29, Zn 3.65, Zr 0.08, Fe+Si<0.3, and balance Al.

The specimens were solution treated at 798K for 1h, quenched in cold water, then artificially aged at 403, 433 and 493K for 16h, corresponding to under–, peak– and over–aged conditions, respectively.

The tensile specimens were machined from the sheet in the L–T direction. Tensile tests were carried out at a strain rate of $2 \times 10^{-3} s^{-1}$, at temperatures from 77 to 293K. For temperatures of 150 and 220K the specimens were immersed in liquid nitrogen fllowed by warming in gaseous nitrogen; 77K was attained by immersion in liquid nitrogen. Tests at 293K were performed in air. Each value of the tensile properties in this paper is the mean of two experimental results.

To understand the precipitates of the alloy in different aging conditions and the temperature dependence of the deformation behaviour, thin foils for TEM observations taken from aged sheet and fractured tensile samples were prepared by mechanical grinding and electrolytic polishing using twin–jet polishing in a 25 vol.% HNO_3–methanol electrolyte solution below 248K. Scanning electron microscopy was used to identify the microscopic fracture mechanisms of the fracture surfaces of the failed specimens.

RESULTS

Microstructure

Optical metallographic observation revealed that the grains in samples subject to three tempers were unrecrystallized with the shape of pancake extending along the rolling direction. This is attributed to the presence of zirconium in the alloy. Zirconium atoms combine with aluminum atoms to form Al_3Zr dispersoids to impede the recrystallization.

A detailed TEM examination was performed to study the microstructure of the alloy in different aging conditions. The alloy in the under– and peak–aged conditions were strengthened by fine homogeneous matrix distributions of coherent, spherical, order $\delta'(Al_3Li)$ precipitates and $\beta'(Al_3Zr)$ dispersoids(Fig.1a,b). No evidence of other precipitate phases was observed in samples of these two tempers. During aging, with the coarsening of δ' particles within grains, δ' phase near grain boundaries disappeared continously, resulting in δ'–precipitate–free–zone(PFZ), which were ~ 15nm in under–aged and ~ 35nm in peak–aged specimens.

With aging to over–aged condition, the δ' particles further coarsen (Fig.1c), accompanied by the precipitation of many large lath– and block–like phases in matrix and at grain boundaries, respectively(Fig.1d). Equlibrium δ–phases also precipitate at grain boundaries. Lath–like phase is considered as phase B and block–like phase is designated as phase A. Their growth depends on the disappear of δ' particles, and PFZ was correspondingly wider (~ 500nm). Energy dispersive X–ray analysis facility(EDAX) equipped on TEM was employed for detection of elements in phases A and B. Suppose the composition of phase A and B is $Al_x Li_y Zn_{10-(x+y)}$, the average results of EDAX analyses showed: $x+y \approx 8$ for A and $x+y \approx 9$ for B. Because lithium is undetectable using this manner, the exact composition of these two phases cannot be decided.

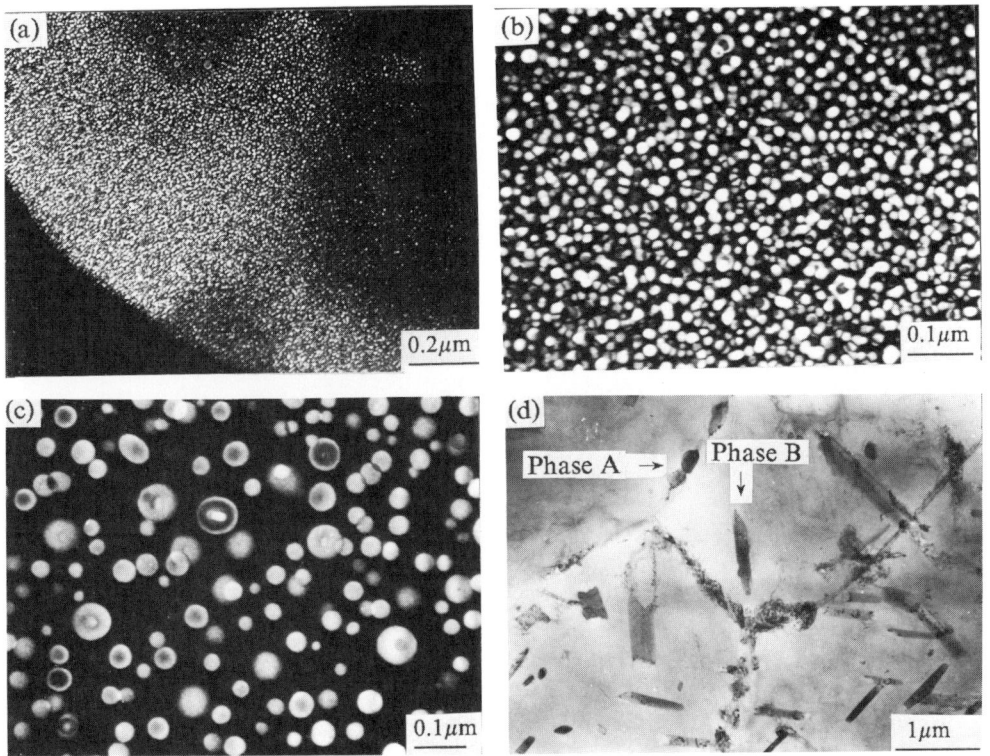

Fig.1 Transmission electron micrographs showing the prominent microstructural features of the alloy in under-aged (a), peak-aged (b) and over-aged conditions (c,d).

Tensile Properties

The tensile properties as a function of test temperature obtained from the material in different aging conditions are given in Table 1. These data clearly show an marked increase in ultimate tensile strength at 77K compared to those tested at room temperature. But the temperature dependence of the elongation relates closely to the heat treatment conditions. For the under- and peak-aged specimens, the elongation improve significantly as the temperature drops, however, in over-aged condition, it shows a little reduction at 77K. Whereas the yield strength only exhibits a slight increase with decreasing test temperature for all tempers.

Fractography

The fracture surfaces are inclined approximately 45° to the tensile axis for all tensile specimens of different heat treatment when tested at ambient temperature. As the test temperature drops, the macroscopic fracture paths for peak- and over-aged specimens change gradually, and become perpendicular to the tensile axis as tested at 77K.

Fig.2 shows the scanning electron micrographs of the fracture surfaces of tensile specimens in different aging conditions broken at 293 and 77K. The under-aged specimen shows a mixture of intergranular and transgranular fracture at 77K instead of a transgranular mode at 293K. When tested at 293K, transgranular shear fracture with facets appears, some dimples formed on the facets (Fig. 2a). At 77K the tensile fracture

Table 1 Tensile Properties of the Alloy at Various Test Temperatures

Aging Condition	Test Temp. (K)	Y.S. (MPa)	U.T.S. (MPa)	Elong. (%)
Under–aged	293	192	313	15.0
	220	194	307	17.4
	150	205	326	18.8
	77	217	397	23.5
Peak–aged	293	251	368	12.0
	220	250	360	11.8
	150	249	363	13.0
	77	254	420	16.5
Over–aged	293	145	249	14.7
	220	141	263	15.8
	150	152	285	16.6
	77	178	364	14.0

surface shows portions of intergranular cracking plus transgranular mode with small dimples(Fig. 2b). Transganular shear fracture is a predominant mode on the fracture surfaces of peak– and over–aged specimens at 293K(Fig. 2c,e). There is little evidence of intergranular fracture along the high angle grain boundaries of the over–aged sample(Fig. 2e). However, they become fully intergranular as tested at 77K, and the fine dimples presented on the intergranular facets(Fig. 2d,f). Therefore, it is obvious that the variation in tensile fracture mode occurs with decreasing temperature for the specimens subjected to different aging conditions.

Moreover, delamination along grain boundaries can also be found at 77K, but there is no intensified delamination appears in all samples broken at low temperatures compared to ambient temperature.

Deformation Behavior

The temperature dependence of the deformation behavior in the alloy during tension was studied by TEM analysis of thin foils obtained from regions adjacent to the cracking surface of the tensile specimens, as show in Fig. 3. Planar slip is the predominant deformation mode for under– and peak–aged samples tested at room temperature(Fig.3a,c). These slip bands are most often extended across the grains. At 77K, the dislocation distribution is fairly homogeneous throughout the tensile test in under–aged specimen(Fig.3b) with only some evidence of weak slip bands. When the peak–aged specimens were tested at 77K, the deformation was characterized by a predominantly relatively homogeneous slip(Fig.3d) compared with the well–defined intense slip bands at 293K. That is to say, the homogeneity of deformation increases with decreasing test temperature. Whereas the specimens in over–aged condition show the same deformation behavior when tested at room and cryogenic temperatures, and no slip bands are detected.

DISCUSSION

The present investigation confirms the prior observation of increasing mechanical properties with decreasing temperature in Al–Li alloys[2-7], except the elongation of the

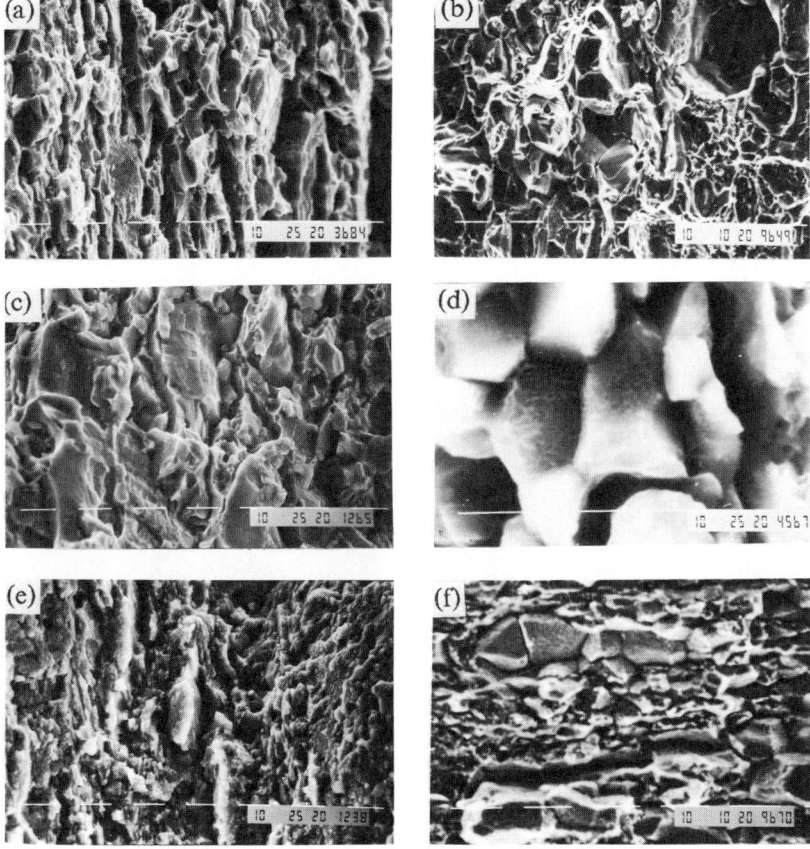

Fig.2 SEM micrographs of the fracture surfaces of the alloy tested at 293K (a,c,e) and 77K (b,d,f). (a,b–under–aged, c,d–peak–aged and e,f–over–aged condition).

over–aged specimen. The data in Table 1 indicate that the temperature dependence of the tensile properties relates to the heat treatment conditions of the alloy. That is, the type, size and distribution of precipitates of the alloy have important effects on the cryogenic properties.

The temperature dependence of ultimate tensile strength is closely associated with grain structure of the alloy. The pancake–like grain structure has higher strain hardening capacity at low temperatures. While the ultimate tensile strength relates to the strain hardening behavior. So it is proposed that the improvement in the ultimate tensile strength for the alloy at cryogenic temperatures may be attributed to the increase in strain hardening capacity[4].

TEM observation of general microstructural features shows that the alloy is hardened by ordered dispersive $\delta'(Al_3Li)$ particles in under– and peak–aged conditions (Fig. 1a,b). The coherent δ' particles can be easily sheared by the moving dislocations since their slip systems are coincident with those of the matrix. Once the δ' precipitates have been sheared, the flow stress is reduced on the slip plane, and further slip is encouraged on that plane. Therefore, planar slip is the predominant deformation mode for the alloy tested at ambient temperature (Fig. 3a,c). The specimens in peak–aged condition have the δ' particles with greater diameter and higher

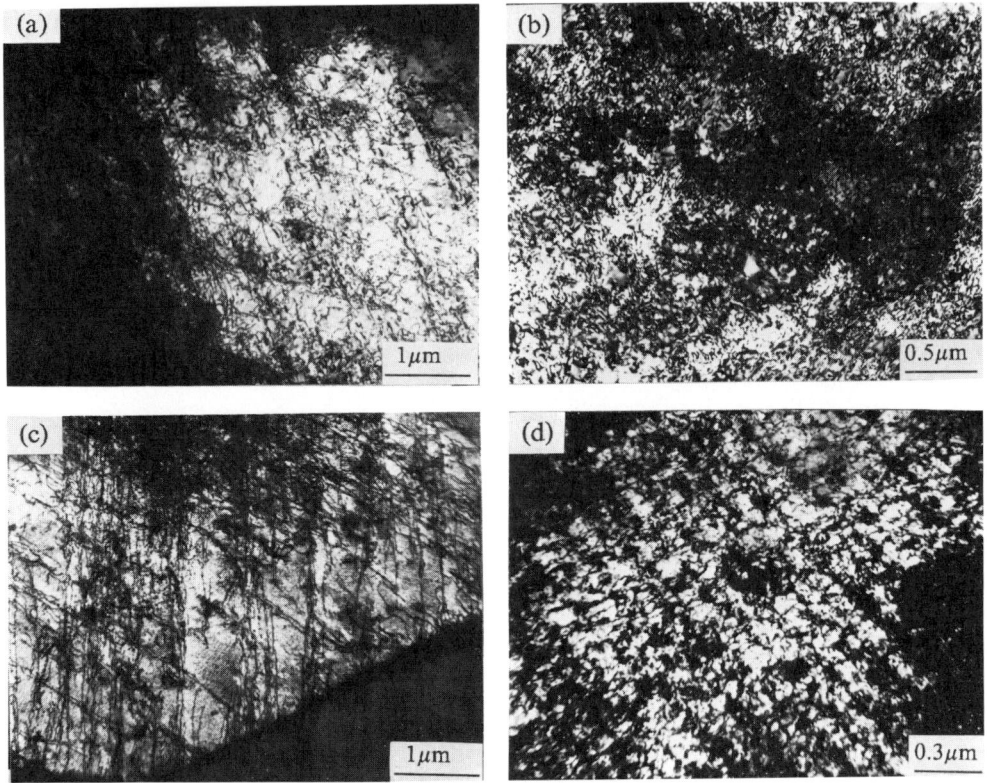

Fig.3 TEM micrographs illustrating the dislocation structure and slip distribution near the fracture surface of the alloy in under–(a,b) and peak–aged (c,d) conditions tested at 293K (a,c) and 77K (b,d).

volume fraction. These δ' should give rise to planar slip during plastic deformation. Moreover, it also can be observed by TEM that the slip bands extend across grain boundaries. However, intense planar slip would make the fracture process dominated by slip localization in slip bands and cause slip bands decohesion, most tend to be of the slip band decohesion type of failure[6]. Therefore, the transgranular cracking occurs along slip bands for under– and peak–aged samples (Fig. 2a,c).

As the temperature decreases, the movement of dislocations slows down, and the antiphase boundary energy of the δ' particle increases, the cutting of δ' particles by dislocation becomes more difficult. This could partly disperse extensive slip localization and account for a more uniform deformation mode[5] instead of extensive planar slip. Result in the transition of deformation from well–defined intense planar slip at 293K to homogeneous mode for under– and relatively homogeneous slip for peak–aged specimens at 77K, respectively. Usually, a planar deformation will lead to a lower ductility. At low temperatures, a more homogeneous mode in the alloy as well as dispersed strain localization will make the fracture of specimen need more plasticity and cause a higher ductility. However, weakening tendency for extensive planar slip lead to decreasing tendency for the domination of slip band decohesion. At cryogenic temperatures, the strength of matrix increases due to the increased work hardening ability of the matrix with decreasing temperature, but the strength of grain boundary remains approximately constant, so the grain boundary strength is less than that of the matrix at 77K[7]. When slip bands encounter grain boundaries, large amount of dislocations will pile up there because intense planar slip cannot form at low temperatures. It needs more energy for dislocations pass through the grain boundary. Therefore, the slip bands will interact with grain boundaries and give intergranular failure. Thus the intergranular mode seems to be more important for fracture than the decohesion of slip bands in the peak–aged alloy at 77K (Fig. 2d). For under–aged specimen, few weak slip bands exist at 77K result in some intergranular zones (Fig. 2b).

For the over–aged condition, with the coarsening and reduced volume fraction of δ'–phase and precipitation of phase B, the planar deformation during tension has been suppressed, so no slip band can be observed by TEM at ambient and low temperatures. The precipitation of grain boundary phases δ and A and wide PFZ weaken the grain boundary. Therefore, early microcracks appear near the coarse grain boundary precipitates and in PFZ during tensile test, this results in a portion of intergranular failure at 293K (Fig. 2e) and fully intergranular fracture at 77K (Fig. 2f), and lead to lower ductility.

Furthermore, from the experimental results, it can be concluded that the nature of δ' particles, i.e., the transition of deformation mode associated with δ' particles, are responsible for the improvement of ductility of the alloy at cryogenic temperatures.

CONCLUSION

1. The tensile properties of an Al–Li–Zn–Zr alloy are enhanced as test temperature dorps, but the elongation of the alloy in over–aged condition goes down slightly at 77K.

2. In over–aged condition, many large phases A and B precipitate in matrix and at grain boundaries of the alloy, respectively. Their compositions are $Al_x Li_y Zn_{10-(x+y)}$, and $x+y \approx 8$ for A and $x+y \approx 9$ for B.

3. The improvement of ductility for the alloy in under– and peak–aged conditions at cryogenic temperatures may be attributed to the increase of homogeneity of deformation.

4. The δ' particles are the major factor to the improvement of ductility for the alloy at low temperatures.

REFERENCES

1. T.H. Sanders, Jr. and E.A. Starke, Jr. eds. "Aluminum–Lithium Alloys V ", MCEP, Birmingham, U.K.(1989).
2. D. Webster, The Effect of Low Melting Point Impurities on the Properties of Aluminum–Lithium Alloys, Metall. Trans. A, 18A:2181(1987).
3. R. C. Dorward, CRYOGENIC TOUGHNESS OF Al–Cu–Li ALLOY AA 2090, Scr. Metall., 20:1379(1986).
4. J. Glazer, S. L. Verzasconi, R. R. Sawtell and J. W. Morris, Jr., Mechanical Behavior of Aluminum–Lithium Alloys at Cryogenic Temperatures, Metall. Trans. A, 18A:1695(1987).
5. K. Welpann, Y. T. Lee and M. Peters, LOW TEMPERATURE DEFORMATION BEHAVIOUR OF 8090, in ref.1, p.1513.
6. H. J. Roven, E. A. Starke, Jr., Φ. Søpdahl and J. Hjelen, EFFECTS OF TEXTURE ON DELAMINATION BEHAVIOR OF A 8090–TYPE Al–Li ALLOY AT CRYOGENIC AND ROOM TEMPERATURE, Scr. Metall., 24:421(1990).
7. D. Dew–Hughes, E. Creed and W. S. Miller, Grain boundary failure in an Al–Li alloy, Mater. Sci. Technol., 4:106(1988).

MICROSTRUCTURAL INFLUENCE ON THE WORK HARDENING OF ALUMINUM-LITHIUM ALLOY 2090 AT CRYOGENIC TEMPERATURES

D. Chu, C. Tseng, and J.W. Morris, Jr.

Center for Advanced Materials, Lawrence Berkeley Laboratory, and
Department of Materials Science and Mineral Engineering
University of California, Berkeley

ABSTRACT

Previous studies indicate that the work hardening characteristics of a Vintage III 2090-T81 12.7-mm (0.5-in.) plate is highly dependent on the through-thickness position.[1,2] This dependency has been linked to two distinctly different microstructures existing as laminates within the plate.[3] An investigation of the two microstructures by both optical and transmission electron microscopy reveal a difference in the distribution of grains and subgrains. It is believed that these factors affect the work hardening characteristics by governing the manner in which slip is transmitted from grain to grain. The findings are positive as they suggest a practical means by which improvements in work hardening can be obtained through modifications of the microstructure at the polygranular level.

INTRODUCTION

Due to their potential as a structural material with lower density, higher stiffness and strength relative to current high strength aluminum alloys,[4,5] aluminum-lithium alloys have become candidate materials for aerospace applications.[6,7] One particular alloy which has received a great deal of attention in recent years is the Al-Cu-Li-Zr alloy 2090. Previous studies have revealed an excellent strength-toughness combination for 2090 at cryogenic temperatures.[6,8-13] Subsequently, much of the work on 2090 has concentrated on understanding the source of the excellent mechanical properties observed at cryogenic temperatures.[1-3,6,8-14] Although the determining mechanism remains an issue of debate, it is generally agreed upon that the deformation behavior, the ability of a material to work harden, plays an important part in determining the subsequent fracture toughness.[6,8,10,12]

The majority of work on 2090 and other like alloys associates increases in work hardening with greater slip homogeneity.[15-17] Jata and Starke[16] relate this greater homogeneity with a decrease in slip planarity. The planarity of slip is, in turn, controlled by the ability of dislocations to cross-slip. Since cross-slip is a thermally activated process, it is sensitive to temperature and to solute atoms that interact with dislocations. When cross-slip is difficult, as it is at cryogenic temperatures, work hardening is expected to increase due to the decreased ability of a dislocation to bypass an impenetrable obstacle. Following this line of reasoning, the improved work hardening observed in 2090 and other aluminum-lithium alloys has been regarded as predominantly a temperature-induced phenomenon.[6,8,10,16]

General theories on work hardening attribute work hardening to microstructural features at the dislocation level. Evidence of this can be found in studies of low carbon steel.[18,19] Results suggest that for a certain temperature range, the flow stress in low carbon steels is controlled predominantly by the motion of screw dislocations and not by other larger microstructural features.[20] The implications of this conclusion is not promising since it suggests that the work hardening, as it is controlled by dislocation motion, will remain unchanged despite further modifications of the polygranular structure.

Although the inability to influence the deformation behavior through manipulation of the polygranular structure may be true in some materials such as low carbon steel, this does not appear to be the case for aluminum-lithium alloys. Studies on a 12.7-mm (0.5-in.) Vintage III 2090-T81 plate material[1-3] reveal two distinctly different microstructures existing as individual laminates within the same plate. Tensile tests conducted on the two microstructures reveal a significant difference in the work hardening temperature response.[2,3] This finding is both promising and potentially important as it implies that, at least for this alloy system, the work hardening is controlled to some extent by microstructural features at the polygranular level.

Ongoing research has continued with transmission electron microscopy toward understanding the microstructural source of the observed differences in work hardening. This paper reviews our current understanding and the results from preliminary microscopy studies.

EXPERIMENTAL PROCEDURE

The material investigated in this study was a Vintage III 2090 plate that was provided by Alcoa in the form of a 12.7-mm (0.5-in.) plate in the -T81 condition (solution heat treated, stretched and peak-aged). This temper is also referred to as -T8E41 in earlier publications. The composition was determined by atomic absorption spectroscopy to be Al-3.05Cu-2.16Li-0.11Zr. The process is a proprietary one that includes a thermomechanical reduction into plate form, a solution heat treatment, a subsequent 6 to 8 pct stretch, and a final peak age at 163 °C for approximately 24 hours.

Mechanical tests were done at one of four temperatures: 300 K (room temperature), 200 K (dry ice and alcohol), 77 K (liquid nitrogen), or 4 K (liquid helium). Tests were performed in servohydraulic testing machines equipped for cryogenic testing. Details of the data collection and analysis are documented in earlier papers.[1-3]

The as-received microstructure was examined by both optical microscopy and transmission electron microscopy (TEM). The areas examined were selected from the regions corresponding to mid-thickness and quarter-thickness of the plate. Optical samples were polished to 0.05 µm and etched using Keller's reagent (2.5 pct HNO_3, 1.5 pct HCl, 0.5 pct HF (40 pct), balance H_2O) for approximately 30 seconds. Corresponding TEM samples were prepared by mechanically grinding to 0.25 mm (0.010 in.) and punching out 3-mm disks. The disks were electropolished with the double jet technique at -30 °C in a 4:1 mixture of methanol and nitric acid. The polishing condition was set at 18 to 20 V.

RESULTS AND DISCUSSION

A complete tabulation of the measured mechanical properties from both tensile tests and fracture toughness tests can be found in an earlier publication.[2] Those pertinent to the present discussion are highlighted here.

Figure 1 plots the yield strength as a function of test temperature and distance from the centerline of the Vintage III 2090-T81 plate material. This profile is an improvement over the yield strength profile exhibited by its predecessor, Vintage I 2090, which decreases monotonically from mid-thickness to both surfaces.[21-24] Earlier work has related the parabolic yield strength profile associated with Vintage I 2090 to a parallel variation in the amount of hot-rolled texture components.[22,23] Despite the variation in texture found in Vintage I 2090, the temperature response of the work hardening remains independent of the through-thickness position. Although the absolute values differ, work hardening curves obtained from tensile specimens at differing through-thickness positions exhibit a steady and monotonic increase with decreasing temperature, as depicted schematically in Figure 2.

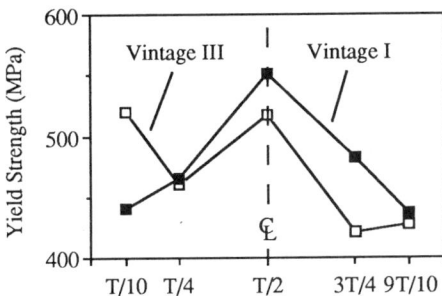

Figure 1. Comparison of the longitudinal yield strength as a function of the through-thickness position for Vintage I and Vintage III 2090-T81 at 300 K.

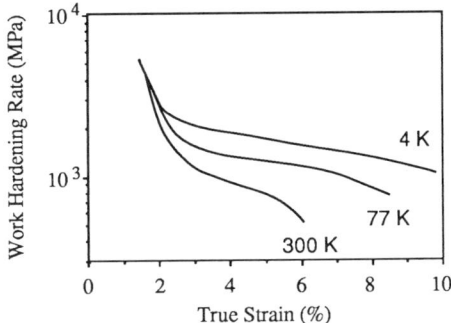

Figure 2. Schematic showing the improvement in work hardening with decreasing temperature in Vintage I 2090.

The most striking difference observed resides in the temperature response of the work hardening at mid-thickness and quarter-thickness in the Vintage III material. The improvement in the work hardening with decreasing temperature at mid-thickness is similar to that observed in the Vintage I material, as depicted in Figure 3a. In contrast, work hardening plots obtained from quarter-thickness exhibit little to no improvement with decreasing temperature until 4 K (Figure 3b). Since both regions are extracted from the same plate, they receive the same heat treatment and contain the same composition and intragranular microstructure. Subsequently, differences in the work hardening characteristic must lie at the polygranular level.

An optical study reveals that such a polygranular difference does exist. Figure 4 displays the microstructure at mid-thickness (Figure 4a) and quarter-thickness (Figure 4b). Although varying grain sizes are observed at mid-thickness, the average grain size is still significantly smaller than that within the quarter-thickness region. A difference in subgrain development is also revealed in Figure 4; the mid-thickness region exhibits a strong distribution of subgrains compared to quarter-thickness.

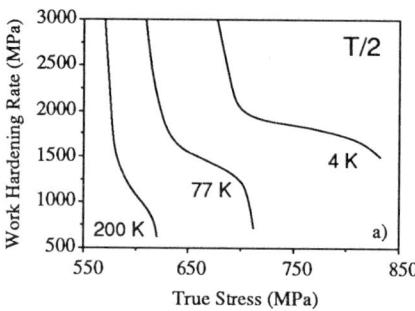 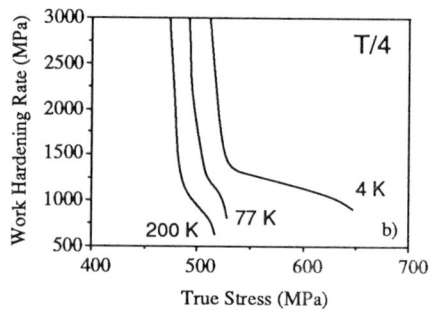

Figure 3. Temperature variation of the work hardening rate as a function of true stress for 2090-T81 at a) mid-thickness and b) quarter-thickness.

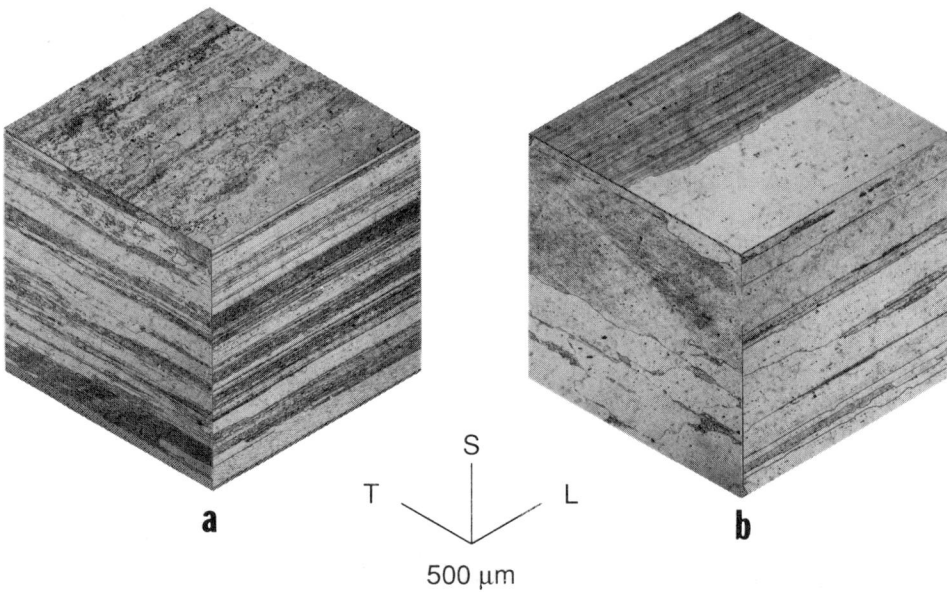

Figure 4. Optical micrograph of the microstructure at a) mid-thickness and b) quarter-thickness. (XBB 915-3801)

Results from preliminary TEM studies lend further support to the observations made by optical microscopy. Samples in the longitudinal plane of the as-received material show that the two regions differ in subgrain distribution and grain morphology. Extensive subgrain formation is observed in the mid-thickness region as shown in Figure 5. The subgrains are homogeneously distributed, despite a relatively large range of grain thicknesses exhibited at this region; this grain thickness variation has been mentioned above (Figure 4). Figure 5 illustrates the subgrain formation at mid-thickness within one of the thicker grains (Figure 5a) and within a collection of thinner grains (Figure 5b). At quarter-thickness, however, subgrain formation is limited and inhomogeneously distributed (Figure 6); isolated grains composed of subgrains are found between large grains that do not contain subgrains.

Figure 5. TEM micrograph of the grain structure at mid-thickness. Extensive subgrain formation is seen homogeneously distributed throughout a) one grain, and b) several grains (grain boundaries denoted by arrows).

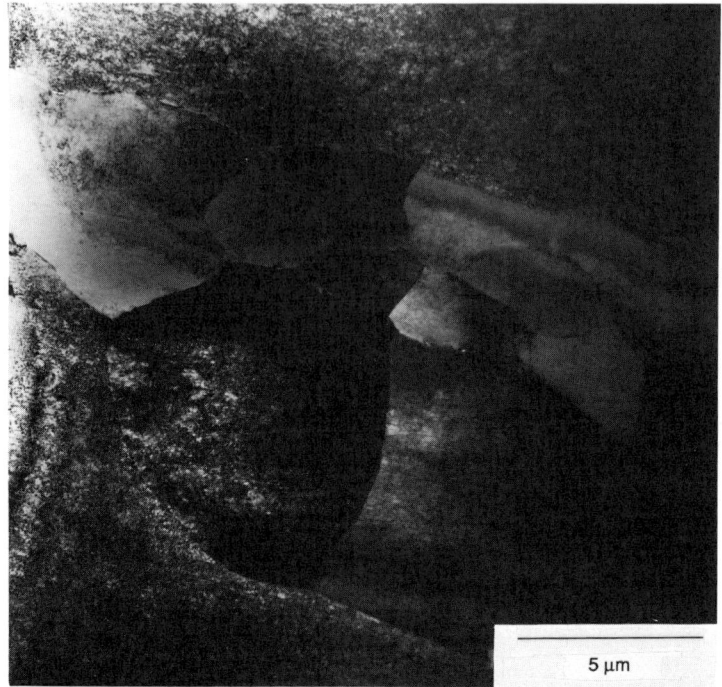

Figure 6. TEM micrograph of the grain structure at quarter-thickness. Note the inhomogeneous distribution of subgrains. Isolated grains composed of subgrains are found between large grains that do not contain subgrains. (XBB 915-4150)

The grain size/shape distributions in the two regions are also dissimilar. Flat and elongated grains are seen at mid-thickness (Figures 5a and 5b). As previously discussed, these grains also vary in thickness. In contrast, the quarter-thickness grains appear more equiaxed and possess a greater variation in grain size (Figures 6 and 7) than those in Figure 5b. Smaller grains such as R and S are typically surrounded by large grains (C and D) in this region (Figure 7). Many of these smaller grains are situated between large grains (Figure 7).

The microstructural differences discussed above will undoubtedly affect the volume fraction of grain boundaries. In addition, the degree of grain and subgrain boundary misorientation should also be affected. It appears that these features play a strong role in determining the work hardening behavior; all other intragranular features are ruled out since the two microstructures contain the same constituents and receive the same aging treatment. It is speculated that the grain and subgrain boundaries dictate the work hardening behavior by governing the manner by which slip is transmitted from grain to grain. Evidence of such an interplay has been observed through an optical slip-relief study where greater elongations are found to be coupled with a greater tendency for out-of-plane rotation of subgrains.[14]

The results accumulated and discussed above are promising as they imply that a practical means may exist by which improvements in the work hardening behavior can be

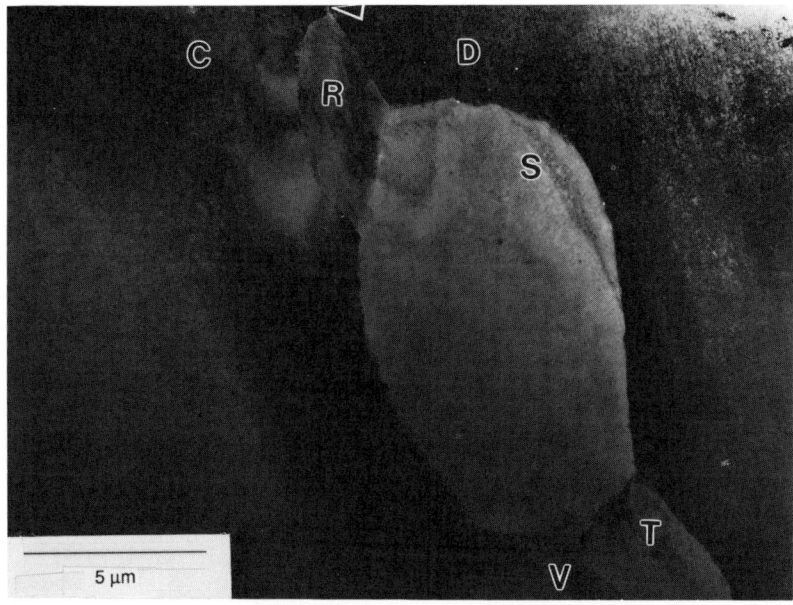

Figure 7. TEM micrograph at quarter-thickness showing the formation of small grains (R, S, T, V) between large grains (C, D). The original grain boundary is denoted by an arrow. (XBB 915-4151)

obtained through selected modifications of the polygranular microstructure. Continuing TEM studies on both as-received and prestrained specimens will attempt to verify and establish that aspect of the microstructure responsible for controlling the work hardening behavior.

CONCLUSION

Preliminary microscopy studies on the two microstructures evidenced in the alloy Vintage III 2090-T81 reveal a difference in the size and distribution of grains and subgrains. These differences in the polygranular structure are believed to be partially responsible for the varying temperature responses observed in the work hardening characteristic. It is speculated that the degree of grain boundary misorientation may also be important. The microstructural influence on the work hardening at the polygranular level implies the existence of an achievable means by which the work hardening and, to some extent, the subsequent strength and toughness can be controlled by modifications of the grain structure.

ACKNOWLEDGEMENTS

The authors thank the Alcoa Technical Center for providing the materials for this research. This study was supported by the Director, Office of Energy Research, Office of Basic Energy Science, Material Sciences Division of the U.S. Department of Energy under Contract No. DE-AC03-76SF00098. D. Chu was supported under a National Science Foundation Graduate Fellowship for the entire period of this work.

REFERENCES

1. D. Chu and J.W. Morris, Jr., <u>Metall. Trans. A</u>, to be published Aug 1991.
2. D. Chu: Master's Thesis, LBL 30170, University of California, Berkeley, CA, Dec. 1990.
3. D. Chu and J.W. Morris, Jr.: in "Light Weight Alloys for Aerospace Applications", E.W. Lee, ed., TMS, Warrendale, PA, 1991.
4. E.S. Balmuth and R. Schmidt: in <u>Aluminum-Lithium Alloys</u>, T.H. Sanders, Jr. and E.A. Starke, Jr. eds., TMS-AIME, New York, NY, 1981, pp. 69-88.
5. G.H. Narayanan, B.L. Wilson, W.E. Quist, and A.L. Wingert: Technical Report No. D6-51411, The Boeing Company, Seattle, WA, Aug. 1982.
6. J. Glazer: Ph.D. Thesis, LBL-27607, University of California, Berkeley, CA, July 1989.
7. W.E. Quist and G.H. Narayanan: in <u>Treatise on Materials Science and Technology: Aluminum Alloys - Contemporary Research and Applications</u>, A.K. Vasudevan and R.D. Doherty, eds., Academic Press, Inc., San Diego, CA, 1989, vol. 31, p. 219-54.
8. J. Glazer, S.L. Verzasconi, E.N. Dalder, W. Yu, R.A. Emigh, R.O. Ritchie, and J.W. Morris, Jr.: <u>Adv. Cryog. Eng.</u>, 1986, vol. 32, pp. 397-404.
9. R.C. Dorward: <u>Scripta Metall.</u>, 1986, vol. 20, pp. 1379-83.
10. J. Glazer, S.L. Verzasconi, R.R. Sawtell, and J.W. Morris, Jr.: <u>Metall. Trans. A</u>, 1987, vol. 18A, pp. 1695-701.
11. D. Webster: <u>Metall. Trans. A</u>, 1987, vol.18A, pp. 2181-93.
12. K.T.V. Rao, W. Yu, and R.O. Ritchie: <u>Metall. Trans. A</u>, 1989, vol. 20A, pp. 485-97.
13. K.T.V. Rao and R.O. Ritchie: <u>Acta Metall. Mater.</u>, 1990, vol. 38, pp. 2309-26.
14. D. Yao, D. Chu, and J.W. Morris, Jr.: in <u>Light Weight Alloys for Aerospace Applications</u>, E.W. Lee, ed., TMS, Warrendale, PA, 1991.
15. A.H. Cottrell and R.J. Stokes: <u>Proc. Roy. Soc. A</u>, 1955, vol. 233, pp. 17-34.
16. K.V. Jata and E.A. Starke, Jr.: <u>Scripta Metall.</u>, 1988, vol. 22, pp. 1553-56.
17. K. Welpmann, Y.T. Lee, and M. Peters: <u>Mater Sci. and Eng. A</u>, 1990, vol. 129, pp. 21-34.
18. S.A. Vincent: Master's Thesis, LBL 24443, University of California, Berkeley, CA, Dec. 1987.
19. P.E. Johnson, J.H. Schmitt, S.A. Vincent, and J.W. Morris, Jr.: <u>Scripta Metall.</u>, 1990, vol. 24, pp. 1447-52.
20. P.E. Johnson: PhD Thesis, University of California, Berkeley, CA, June 1991.
21. R.J. Bucci, R.C. Malcolm, E.L. Colvin, S.J. Murtha, and R.S. James: Final Report No. NSWC TR 89-106, Aluminum Company of America, Alcoa Center, PA, Sept. 1989.
22. A.K. Vasudèvan, W.G. Fricke, Jr., R.C. Malcolm, R.J. Bucci, M.A. Przystupa, and F. Barlat: <u>Metall. Trans. A</u>, 1988, vol. 19A, pp. 731-32.
23. A.K. Vasudèvan, W.G. Fricke, Jr., M.A. Przystupa, and S. Panchanadeeswaran: in <u>8th Int. Conf. on Textures of Materials</u>, J.S. Kallend and G. Gottstein, eds., TMS, Warrendale, PA, 1988, pp. 1071-77.
24. R.J. Rioja, B.A. Cheney, R.S. James, J.T. Staley, and J.A. Bowers: Structure-Property Relationships in Al-Li Alloys: Plate and Sheet Products, Aluminum Corporation of America, Alcoa Center, PA, March 1989.

EFFECTS OF Nb$_3$Sn HEAT TREATMENT ON THE STRENGTH AND TOUGHNESS OF 316LN ALLOYS WITH DIFFERENT CARBON CONTENTS

R. P. Reed, R. P. Walsh, and C. N. McCowan

Materials Reliability Division
National Institute of Standards and Technology
Boulder, Colorado, U.S.A.

ABSTRACT

A series of six 316LN-type alloys with different carbon contents were evaluated for suitability as conduit-sheath materials for the Nb$_3$Sn superconductors that will be used in high-field magnets. Impact energies (at 76 to 295 K) and tensile strengths (at 4, 76, and 295 K) of annealed and "sensitized" alloys were compared. Embrittlement by aging at 700°C for 100 h (the Nb$_3$Sn reaction conditions) was determined to be a function of carbon content; alloys with very low carbon content retained adequate toughness.

INTRODUCTION

The increasing use of Nb$_3$Sn as the conductor for high-field magnets is placing special emphasis on the development of adequate structural support for conduit sheath, which will carry substantial Lorentz-force loads. The assembly sequence requires the sheath to endure the Nb$_3$Sn reaction heat treatment of approximately 100 h at 700°C. Following exposure to the Nb$_3$Sn reaction conditions, the sheath alloy must retain acceptable strength, toughness, and fatigue resistance at 4 K. Poloidal-field coil pulses will impose fatigue loads on the thin-walled (about 5 mm) sheath. Calculations of fatigue-induced crack growth, originating from a minimum detectable crack (5% of wall thickness) and growing through the sheath wall, suggest that the magnet pulses must be restricted to less than 40 000 operational cycles.

Candidate sheath alloys include the U.S. alloys Incoloy 908 and A-286 (JBK-75), as well as U.S.-, Japanese-, and European-modified 316LN alloys. This paper addresses the control of carbon content in 316LN-type alloys to prevent the deleterious loss of toughness at low temperatures from the extensive heat treatment at 700°C.

At the Nb$_3$Sn reaction temperature of 700°C, most grades of austenitic stainless steel experience extensive carbide precipitation at grain boundaries. Heat treatment of nitrogen-strengthened austenitic steels at 700°C leads to nitride and carbide formation. (Both nitrides and carbides form more quickly at about 700°C than at higher or lower temperatures.) Nitrogen readily reacts with chromium to form Cr$_2$N. Small particles or platelets of Cr$_2$N form within the austenite grains in minutes.[1] Carbon reacts with solid-solution elements, especially chromium, to form M$_{23}$C$_6$ particles. Initially, after about 1 h of heating, these particles tend to precipitate at grain boundaries; later, after about 10 h, carbides begin to form at incoherent twin boundaries.[2]

The formation of carbides at interfaces, such as grain boundaries, within the austenitic structure creates a nearly continuous path of lower toughness for fracture, and indeed, leads to severe embrittlement of austenitic steels at low temperatures. On the other hand, nitride precipitation within austenite grains does not generate a path for cracking and should not change the fracture mechanism.

The effects of nitride and carbide formation on the low-temperature yield strength have not been studied. Depletion of nitrogen and carbon from interstitial solid solution leads to reduced solid-solution strengthening, but the addition of precipitates contributes to particle strengthening at low temperatures. The effects of solid-solution strengthening of 316LN alloys through carbon and nitrogen additions have been previously studied by Reed and Simon.[3]

The effects of nitride and carbide precipitation on toughness have been studied by Muster et al.[4] In a series of 316-type alloys that had been heat-treated at 700°C for 50 h, they correlated the Charpy impact energy (C_V) at 77 K and the tensile elongation (Elong.) at 4 K with the parameter [C] + 0.1 [N], where [C] and [N] are the carbon and nitrogen contents in weight percent. At [C] + 0.1 [N] $\geq$ 0.05 wt.%, the toughness and ductility values were very low (C_V < 40 J and Elong. < 10%). At lower contents (0.035 wt.%), the values rose abruptly to C_V > 60 J and Elong. $\simeq$ 40% at [C] + 0.1 [N]. This transition may be attributed to insufficient carbon to form continuous $M_{23}C_6$ intergranular precipitation for easy, intergranular crack propagation.

Shimada[5] studied the effects of phosphorus and boron on the tensile properties and fracture toughness of 316LN alloys following extensive aging at 700°C. His alloys contained less than 0.01 wt.% carbon. The addition of phosphorus in amounts greater than 0.01 wt.% decreased the fracture toughness at 4 K of aged alloys, but the addition of boron in amounts greater than 0.003 wt.% actually increased the low-temperature toughness. The deleterious effect of phosphorus addition on toughness was attributed to enhancement of diffusion rates by the presence of phosphorus at grain boundaries. Aging temperatures between 650 and 725°C for 75 h did not affect the tensile yield strength of these alloys at 4 K.

This paper extends the work of Muster et al.[4] to other alloys and heat treatments. We added tensile strength data to suggest that nitride precipitation contributes to the low-temperature yield strength of austenitic steels.

MATERIALS

All materials were 316LN-base alloys. Their chemical compositions are listed in Table 1 in order of increasing carbon content. Alloys A and B had low phosphorus and chromium contents, and alloy F had low molybdenum content. All the carbon and nitrogen contents were determined by the same commercial source. The average grain diameters are also listed in Table 1 for the conditions of annealed (1200°C, 1 h) and aged (700°C, 100 h following annealing). Aging was performed in an argon atmosphere, and specimens were water-quenched following both annealing and aging treatments. Average variability of the grain size was ±50%. Hardness values for all alloys were very similar; they ranged from 86 to 89 R_B.

Table 1. Chemical Composition and Grain Size of Alloys

Alloy	Grain Size (μm) Annealed	Aged	C	N	P	O	Ni, Cr, Mo, Mn
A	91	80	0.0042	0.158	0.002	0.005	11, 18, 2.5, 1.4
B	70	80	0.0042	0.160	0.002	0.007	11, 18, 2.5, 1.4
C	50	130	0.0132	0.164	0.015	—	13, 17, 2.6, 1.5
D	80	80	0.0209	0.170	0.016	—	14, 17, 2.7, 1.3
E	242	210	0.0215	0.181	0.019	—	14, 18, 2.7, 1.0
F	103	160	0.0243	0.099	0.022	—	14, 17, 2.2, 1.8

Carbides formed during aging. The extent of carbide precipitation depended on carbon content. In alloys A and B, only discrete carbide particles were observed at the grain boundaries. In alloys C through F, semicontinuous carbide particles formed along high-angle grain boundaries. Following cold work, carbides were observed even along coherent twin boundaries after aging at 700°C. (Alloy C was also aged after about 15% cold rolling.)

EXPERIMENTAL PROCEDURES

All materials were received as plate. Tensile specimens were machined parallel to the rolling direction (L); the notch in the impact specimens was positioned with the crack normal to the rolling direction in the plane of the plate (LT). All tensile specimens were 5.08 mm in diameter with a 38.1-mm reduced section, except for alloy C specimens, which were 3-mm-thick, flat specimens with a 25.4-mm gage length.

Tensile tests were conducted at a constant crosshead rate of 1.27 mm per min. Strain-gage extensometers were used to sense strain, and a commercial load cell was used for load measurement. Tests were conducted in boiling liquid helium (4 K), boiling liquid nitrogen (76 K), and at room temperature. Intermediate temperatures were obtained by carefully balancing the cold nitrogen gas and the heating elements connected to both specimen grips. The standard 0.2%-offset technique was used to measure yield strength. Elongation was determined by measuring the final gage length and comparing it to the initial 25.4-mm-scribed gage length.

Impact tests were conducted with standardized machines. The standard thickness (10 mm) from the tip of the notch to the opposite side of the specimen was used for all alloys except alloy C. Owing to the size of the original alloy-C product form, a thickness of only 5.08 mm could be obtained. Impact energies from these specimens were adjusted by considering the ratio of their thickness to those of the other alloys. Intermediate temperatures were obtained by warming the specimens from 76 K. The capacity of the impact machine was about 340 J; thus, tests at higher temperatures were restricted.

EXPERIMENTAL RESULTS

The tensile data are listed in Table 2. The aged alloys had higher tensile yield strengths (σ_y) but lower tensile ultimate strengths (σ_u) at low temperatures than the annealed alloys. However, at temperatures above 76 K, the σ_u values became approximately equivalent. Ductilities (elongation, reduction of area) were always lower for aged alloys; they were significantly lower for the aged alloys at 4 K.

Table 2. Tensile Data

Alloy	C (wt.%) [N (wt.%)]	Temp. (K)	σ_y (MPa) Ann.	σ_y (MPa) Aged	σ_u (MPa) Ann.	σ_u (MPa) Aged	R.A. (%) Ann.	R.A. (%) Aged	Elong. (%) Ann.	Elong. (%) Aged
A	0.0042 [0.158]	4	924		1560		61		48	
B	0.0042 [0.160]	4	942		1585		56		41	
C	0.0132 [0.164]	4	877	927	1475	1394	38	23	57	24
D	0.0209 [0.170]	4	1037		1570		55		—	
E	0.0215 [0.181]	4	990	1093	1450	1330	50	7	38	6
		76	729	769	1220	1040	73	13	61	13
		125		592		904		29		30
		175	467	504	841	851	78	45	50	46
		295	265	286	614	630	81	60	59	51
F	0.0243 [0.099]	4	780	815	1290	1210	57	31	49	28
		76	597	610	1090	1020	78	33	67	37
		125		474		859		50		39
		175	351	376	755	739	75	64	59	50
		295	212	223	528	536	80	75	70	64

Alloy E was heat-treated at temperatures ranging from 550 to 1000°C for 1 h following annealing at 1200°C for 0.5 h. The annealing at a higher temperature precluded subsequent grain growth during heat treatment at lower temperatures. The tensile data at 4 K for alloys with these heat treatments are given in Table 3. For reference, the tensile data for the 700°C, 100-h heat treatment are also included. Clearly, an increase in yield strength is associated with the lower temperature heat treatments.

The impact data for annealed and aged alloys are listed in Table 4. The impact energies of the aged alloys were always lower than the energies of the annealed alloys. Alloy C was received in a 10% cold-worked condition, then aged (700°C for 100 h), annealed, and then aged again. Comparison of the two sets of data after aging suggests that cold working substantially adds to the sensitization of the structure.

The impact data at 76 K for both the annealed and aged conditions are plotted versus carbon content in Fig. 1. All impact energies exceeded 185 J for annealed alloys. For aged alloys, a transition occurs at approximately 0.010 wt.% [C] from high-impact energies (at low [C]) to low-impact energies (at high [C]).

Table 3. Effects of Heat Treatment Temperature on Tensile Properties at 4 K (Alloy E)

Heat Treatment Conditions	σ_y (MPa)	σ_u (MPa)	R.A. (%)	Elong. (%)
550°C, 1 h	1030	1460	47	43
700°C, 1 h	1030	1460	50	38
700°C, 100 h	1093	1330	7	6
850°C, 1 h	997	1420	52	34
1000°C, 1 h	987	1420	52	40

Table 4. Impact Data

Alloy	C (wt.%)	Condition[†]	Impact Energy (J) at Test Temperature (K)						
			76	125	152	173	182	230	295
A	0.0042	Annealed	210	252					
			222	274					
		Aged	142	167			264		
			143	167					
B	0.0042	Annealed	222	239			338		
			222	248					
		Aged	147	171	196		262		
			156	180	204		295		
C	0.0132	10% CW	159	171			184	212	228
			162						
		Annealed	222						316
			230						332
		10% CW, Aged	17	24			43	57	96
			19						
		Annealed, Aged	48						147
			51						150
D	0.0209	Annealed	239	274					347
			242	294					
		Aged	55	81			133	194	254
			60	83			140	224	295
								228	
E	0.0215	Annealed	315						
			325						
		Aged	27	51			92	155	164
			39	56				156	166
F	0.0243	Annealed	186	210		248			
			206	230		268			
		Aged	57	81		103		174	186
			58	93		117		179	201

*1 J = 1.335 ft·lbf
[†]CW = cold worked

DISCUSSION

The addition of nitrogen and carbon affects both the strength and toughness of austenitic stainless steels:

Strength

The contributions of [N] and grain size to the tensile yield strength of 316LN alloys at 4 K have been analyzed by Reed and Simon.[3] They found that tensile yield strength (σ_y in MPa) can be expressed by

$$\sigma_y = 429 + 2500[N] + 740 d^{-1/2}, \tag{1}$$

where d is the average grain diameter in μm. From this relationship, the σ_y values of the annealed alloys in this program were predicted. These values are plotted versus the measured values (Table 2) in Fig. 2; there is a reasonable correlation. For this comparison, the σ_y of alloy F was increased by 25 MPa to account for its low molybdenum content, following the analyses of

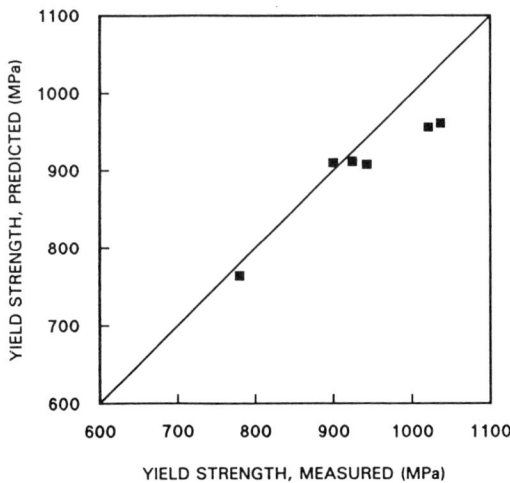

Fig. 1. Tensile yield strengths measured at 4 K versus tensile yield strengths predicted from Eq. 1 for 316LN alloys.

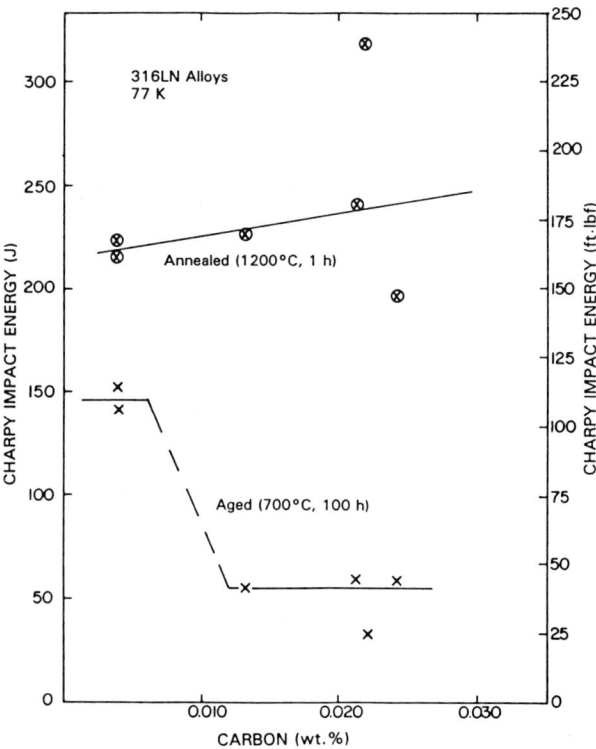

Fig. 2. Impact energy at 76 K as a function of carbon content for annealed and aged 316LN alloys.

Reed and Simon.[3] Possible explanations for the variability include inconsistencies in grain size and the amount of nitrogen or other chemical elements along the gage length of the specimens.

The yield strengths of the aged alloys were higher than the yield strengths of the annealed alloys. Both nitrogen and carbon were in interstitial solid solution in the annealed alloys. Aging at 700°C results in CrN_2 and $M_{23}C_6$ precipitation, and as a result, both elements were likely to be severely depleted in solid solution. From Eq. 1, the maximum reduction of σ_y at 4 K is about 400 MPa when no nitrogen exists in solid solution and about 25 MPa when no carbon exists in solid solution (from Reed and Simon[6]). Under these conditions, CrN_2 and $M_{23}C_6$ precipitates would contribute about 475 MPa to σ_y at 4 K, since the difference between the σ_y values of annealed and aged conditions was about 50 MPa. The difference was substantial, and precipitate strengthening should be explored further.

The σ_y data at 4 K of alloy E following heat treatment at various temperatures (Table 3) suggest that the increased strengthening from aging can be attributed to CrN_2. The 1-h holding time is insufficient to permit substantial carbide formation, but it is well within the time requirements for nitride formation.[1,2,4] Another indication of the lack of carbide precipitation is that specimens aged for 1 h retained good ductility at low temperatures.

The probable role of carbide particles in strengthening is supported by the recent work of Shimada,[5] who found no strengthening effect after aging for 75 h at 700°C despite the [N] of 0.2 wt.% in his alloys. The [C] was low, below 0.01 wt.%. Therefore, less $M_{23}C_6$ particles were produced in his alloys. However, the other distinction between Shimada's alloys and alloys E and F is the [P]. The [P] of his alloys ranged from 0.003 to 0.007 wt.%; alloys E and F had 0.019 and 0.022, respectively. Therefore, if Shimada's premise is correct about the role of phosphorus to enhance nitride and carbide precipitation, the presence of more phosphorus in our alloys also suggests enhanced precipitation. Unfortunately, we did not have sufficient material of alloys A and B (low [P]) to study this premise.

Toughness

The fracture surfaces of impact specimens of annealed and aged alloys tested at 76 K are shown in Fig. 3. In annealed specimens, microvoid coalescence, as depicted by a ductile, dimpled type of fracture surface, is the dominant fracture mechanism. The fracture surface of the aged, low-carbon alloys retained a dominant ductile, dimpled appearance (Fig. 3a), but the fracture mechanism of the aged, high-carbon alloys changed to an intergranular mode (Fig. 3b). At temperatures above 76 K, more characteristics of microvoid coalescence were observed on the impact-specimen fracture surfaces.

Dissimilarities of fracture appearance between annealed and aged tensile specimens, tested at 4 K, were also noted (Fig. 4). In annealed specimens of alloys C through F, very wide and shallow dimples, typical of shear failure, were found. In aged specimens of these alloys, intergranular fracture dominated, and clear outlines of the grain structure were observed on the fracture surfaces.

With scanning electron microscopy, we observed microvoids near the carbide precipitates at the grain boundaries close to the fracture-surface of specimens (Fig. 5). Presumably, the voids formed in solute-free zones that had been depleted of alloying elements that contribute to strengthening. In such a depleted zone, local deformation and subsequent microvoid formation would occur prematurely. Note that the microvoids in Fig. 5 are always

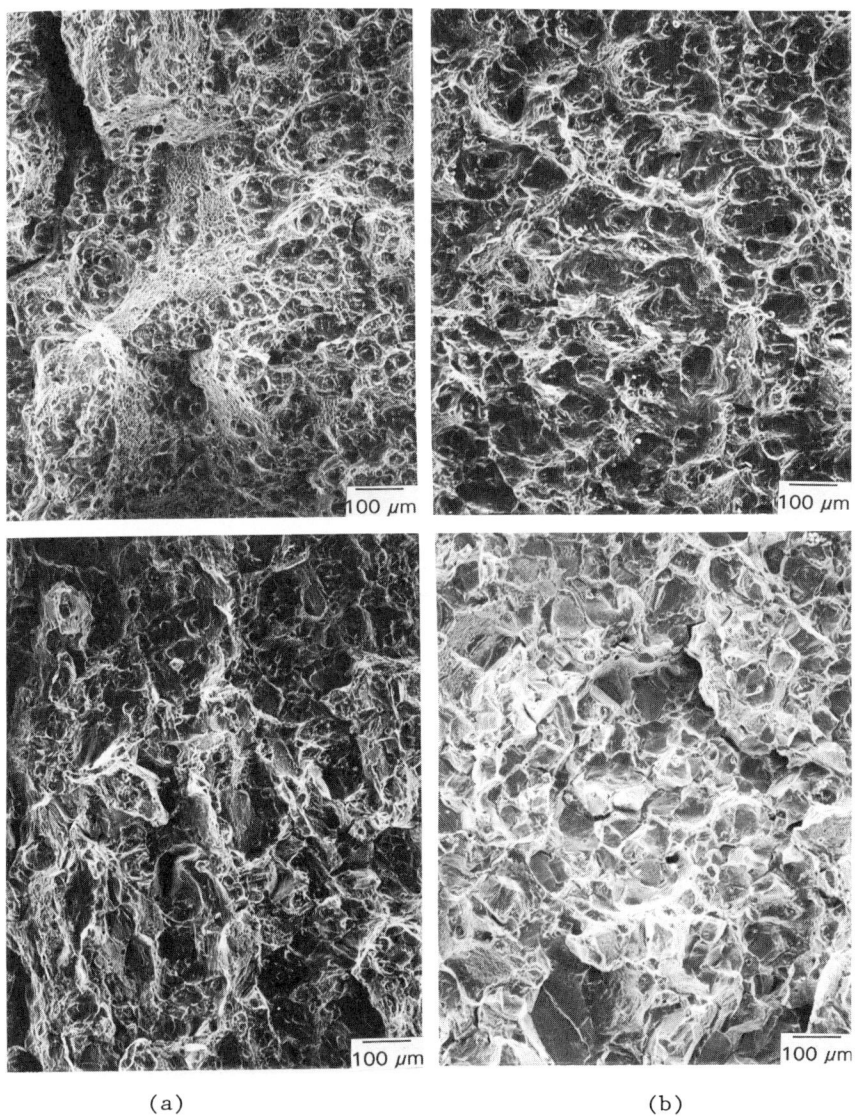

Fig. 3. Fracture surfaces of impact specimens tested at 76 K; (a) alloy A and (b) alloy D in annealed (top) and aged (bottom) conditions.

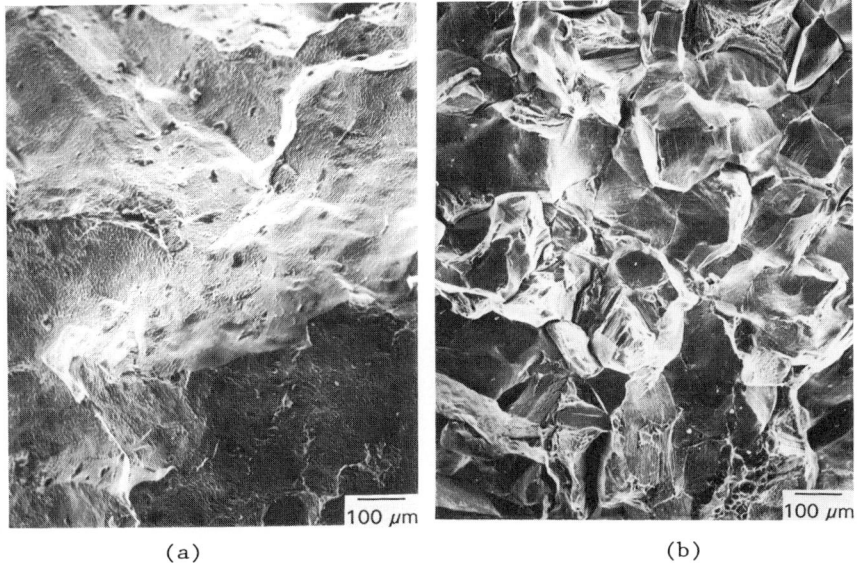

Fig. 4. Fracture surfaces of tensile specimens, tested at 4 K, of alloy E in annealed (a) and aged (b) conditions.

associated with microcracking in the carbide particles. This type of void and microcracking formation has been observed in other alloy classes.

SUMMARY

Aging at 700°C increased the tensile yield strength at 4 K of 316LN alloys with N > 0.10, C < 0.02, and P > 0.016. For short aging times (1 h), the increased strengthening was attributed to CrN_2 precipitates. For long aging times (100 h), both CrN_2 and $M_{23}C_6$ precipitates may have contributed.

Following extensive aging at 700°C, the toughness (Charpy-impact energy) and tensile ductility at low temperatures were strongly influenced by carbon contents. For alloys with carbon contents of less than 0.01 wt.%, toughness at 76 K and tensile ductility at 4 K improved significantly following aging

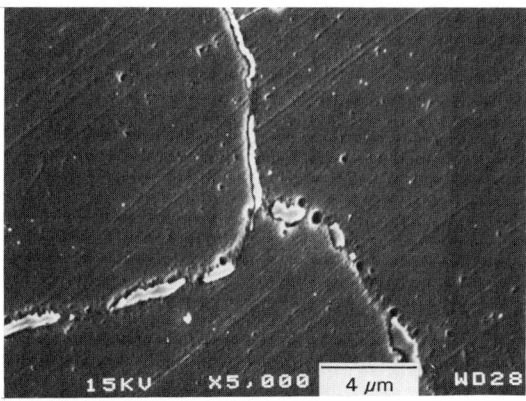

Fig. 5. Microvoids formed near carbide particles in the vicinity of grain boundaries not far from the fracture surface of the impact specimen of alloy C, tested at 76 K.

53

at 700°C. This abrupt change in fracture energy has been demonstrated to reflect a change in the mechanism of impact and tensile fracture: at low carbon contents (<0.01 wt.%), the mechanism was microvoid coalescence, at higher carbon contents, the fracture mechanism was intergranular cracking through carbide precipitates.

REFERENCES

1. M. O. Spiedel and P. J. Uggowitzer, *Moderne Stähle*, ETH, Zurich (1987).
2. B. Weiss and R. Stickler, *Metall. Trans.* 3:851-860 (1972).
3. R. P. Reed and N. J. Simon, in: *Advances in Cryogenic Engineering—Materials*, vol. 34, Plenum Press, New York (1988), pp. 165-172.
4. W. J. Muster, J. Kübler, Ch. Hochhaus, *Swiss Mater.* 2: 25-31 (1989).
5. M. Shimada, *ISIJ Int.* 30:579-586 (1990).
6. R. P. Reed and N.J. Simon, in: *Advances in Cryogenic Engineering—Materials*, Vol. 30, Plenum Press, New York (1984), pp. 127-136.

METASTABLE AUSTENITES IN CRYOGENIC HIGH MAGNETIC FIELD ENVIRONMENTS

J. W. Chan, D. Chu, A. J. Sunwoo, and J. W. Morris, Jr.

Center for Advanced Materials
Lawrence Berkeley Laboratory
Berkeley, CA 94720

ABSTRACT

The fracture behavior of austenitic stainless steels of differing stability, AISI 310S, 304, and 304L, in a 4.2 K, 8 T magnetic field environment are examined. 304L specimens with different amounts of work at different rolling temperatures are also examined. The different rolling conditions are used to produce stability differences independent of those inherent to chemistry differences. The application of an 8 T magnetic field at 4.2 K leads to measured fracture toughness above and below that without an applied field in the metastable alloys, with the amount and direction of change a function of the stability of the alloy. Stable 310S alloy does not exhibit significant fracture toughness changes. This difference in fracture behavior is attributed to the enhancement of martensitic transformation about the crack tip during the fracture process in a magnetic field.

INTRODUCTION

Austenitic stainless steels have been used as structural alloys in high field superconducting magnets. Some of the candidate structural alloys for the next generation of magnetic confinement fusion reactors are of this type. In this application the alloys sustain high stresses in high strength magnetic fields at 4.2 K. It is known that plastic deformation at low temperatures induces martensitic transformation in some of these alloys and that the presence of a strong magnetic field enhances the transformation. If the structural alloys selected for this application undergo martensitic transformation under service conditions, there may be unanticipated effects, such as changes in the tensile, fatigue, and fracture properties that can potentially degrade the performance of the device. Even if the base alloys remain austenitic under service conditions, there remains the possibility that welds, which commonly suffer from chemical inhomogeneities because of segregation during solidification, can in some regions be metastable. This necessitates complete characterization of the mechanical behavior: tensile, fatigue, and fracture toughness properties of these metastable alloys at the anticipated operating conditions to better understand the controlling mechanisms. Previous work demonstrates that the presence of a strong magnetic field can influence both tensile[1-3] and fracture properties[4,5] of AISI 304 stainless steels at 4.2 K. The change in 4.2 K mechanical properties of these metastable alloys with the application of a magnetic field is associated with a martensitic transformation during deformation. Both increases[5,6] and decreases[4] in 4.2 K fracture toughness were observed in an 8T magnetic field. Some of the mechanisms affecting fracture toughness that arises because of the in situ transformation were identified in previous work[5]. In this work, AISI310S, an alloy stable with respect to deformation at 4.2 K, and AISI304L, a metastable alloy, with various prior processing conditions are examined to help further clarify the role of alloy stability on the direction and magnitude of fracture toughness change in an applied magnetic field.

EXPERIMENTAL

The amount of transformation that can occur during the J_{IC} test is varied both through changes in chemistry and through prior deformation and transformation. Matched sets of specimens were tested at 4.2 K with 0 T and 8 T applied fields.

Chemistry-related stability changes were accomplished by using 304L and 310S. AISI 304L is metastable while 310S is a stable alloy at 4.2 K. The compositions of the 304L and 310S plates used in this work are listed in Table 1.

The amount of transformation available during crack initiation and propagation was controlled independently of chemistry by using 304L specimens with different prior deformation. Heavy prior deformation reduces the amount of in situ transformation by both causing a prior transformation and by stabilizing the austenite. 304L plates were given a 20% reduction (31.8mm to 25.4mm) at 300 K and at 993 K. The 993 K treatment is well above the reversion temperature for 304L, so the plate remains austenitic. Deformation at that temperature will, however, stabilize the alloy against subsequent transformation.[7] In addition, 304L plates were given a 13% (31.8mm to 27.6mm) rolling reduction at 77 K, reducing its stability. Thus the order of the alloys in terms of increasing stability is 304L processed at 77 K, 304L, 304L processed at room temperature, 304L processed at 993 K, and 310S. Compact tension (CT) and tensile specimens were then machined from these plates.

J_{IC} tests were performed in an 8 T magnetic field at 4.2 K in the bore of a NbTi superconducting solenoid and without an applied magnetic field at 4.2 K. A single specimen compliance technique, in which load line displacement measurements are made using a clip gage, was used. The specimens were precracked at room temperature to a nominal a/w of 0.6. J_{IC} tests were conducted according to ASTM 813-88.

The 304L specimens that were cold-worked at 77 K contained significant amounts of martensite and were thus subject to brittle fracture. Thus valid J_{IC} measurements were not obtainable. Instead, matched sets of specimens with this processing were made into CT specimens suitable for use in K_{IC} tests in both 0 T and 8 T magnetic fields at 4.2 K. The specimens were precracked at room temperature to an a/w of 0.5. The tests were conducted according to ASTM399-83.

Table 1. Alloy compositions in wt%.

	Cr	Ni	Mn	Si	C	N	O	P	S	Mo
304L	18.90	8.29	1.84	0.33	0.019	0.087	0.011	0.024	0.015	0.43
310S	24.66	19.11	1.84	0.52	0.025	0.062	0.010	0.024	0.012	0.41

Table 2. 4.2 K tensile properties.

	σ_y [MPa]	σ_u [MPa]	total elongation[%]
304L	555	1328	37
993 K-rolled 304L	536	1677	50
293 K-rolled 304L	994	1826	45
77 K-rolled 304L	2144	---	8
310S	772	1220	68

RESULTS

The 4.2 K tensile properties for the alloys processing conditions examined are listed in Table 2. The 77 K-rolled 304L tensile specimens failed in a brittle mode after only 8% elongation. The 4.2 K K_{IC} for the 77 K-rolled specimens and the $K_{IC}(J)$ for the others in both 0T and 8T magnetic fields are shown in Figure 1. All of the metastable specimens tested failed in a quasi-cleavage micromode while the stable 310S failed through void coalescence.

Both increases and decreases in fracture toughness are detected with an applied magnetic field, depending on the alloy stability. The fracture toughness values decrease in an 8T magnetic field for specimens of low stability, increase for specimens of higher stability, and do not change appreciably for fully stable 310S specimens.

DISCUSSION

Previous work[4-6] on 304 stainless steel has shown that the 4.2 K fracture toughness can change with the application of a magnetic field. AISI304 stainless steels are metastable, tending to transform from an fcc structure to a more stable bcc martensite at low temperatures. The transformation is driven by the free energy difference between the γ and

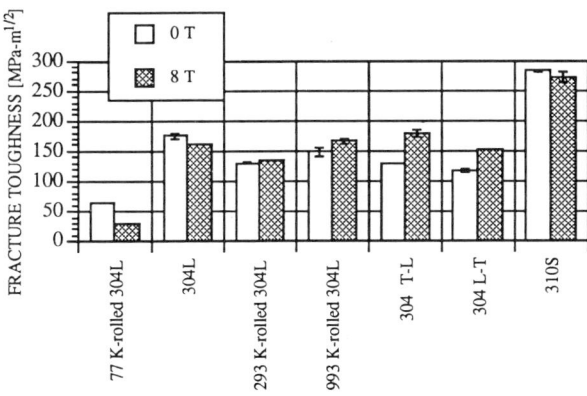

Figure 1. Fracture toughness of alloys tested at 0T and 8T magnetic fields. The values are $K_{IC}(J)$ except for the 77 K-rolled 304L which are from K_{IC} tests. Error bars denote spread in values. The values for 77K-rolled 304L are from one specimen in each condition only while the others are from at least two tests in each condition. The values for 304 are from ref. 5.

the α' phases and is opposed by the increase in strain and surface energy due to the transformation. Increasing the free energy difference between the two phases at a constant temperature increases the driving force for the transformation at that temperature. Low temperature, mechanical stress, and the presence of a magnetic field will increase this free energy difference. Small amounts of prior plastic deformation below the M_d temperature also may facilitate transformation by promoting the nucleation of the martensite phase.[8,9] Thus the presence of these conditions will tend to shift the transformation temperature upward and enhance the extent of the transformation to α'. The change in transformation behavior in the presence of a strong magnetic field was proposed as the cause of the observed changes in fracture toughness.[5,6]

The current data support the proposed explaination that the observed fracture toughness changes are a result of the transformation behavior change in the magnetic field. The fracture behavior of 304L plates that have been subjected to different prior processing conditions indicates that in situ martensitic transformation is the proximate cause of the observed changes in an 8T magnetic field. Rolling of the 304L plates at different temperatures produces differences in the amount of transformation that occurs during deformation and fracture. This is seen more clearly in Figure 2, where the data in Figure 1 is replotted as a percentage change in 4.2 K fracture toughness with the application of an 8T magnetic field. The least stable alloy, the 77 K-rolled 304L, has the largest percentage reduction while the fully stable alloy has not changed significantly. The alloys with intermediate stabilities progress from a slight decrease for annealed 304L to negligible change for 293 K-rolled 304L to increases in the 993 K-rolled 304L and in the 304 studied previously.[5] Such behavior suggests that the location where significant transformation first occurs with respect to the crack tip, as well as the amount of the transformation is important in determining the fracture toughness. A material that transforms easily produces a brittle zone ahead of the advancing crack, reducing its measured fracture toughness. The reduction in fracture toughness of martensite containing 304L is indicated by the low K_{IC} value for the specimens partially transformed by rolling at 77 K. A material that delays significant transformation until the higher strains present closer to the crack tip will avoid exposure of the brittle martensite regions to the tensile stress peak ahead of the crack, reducing locally brittle fracture and allowing the fracture toughness enhancing mechanisms identified in previous work[5] to dominate. The presence of a magnetic field during fracture will enhance this behavior. Completely stable specimens will not show an effect in the magnetic field.

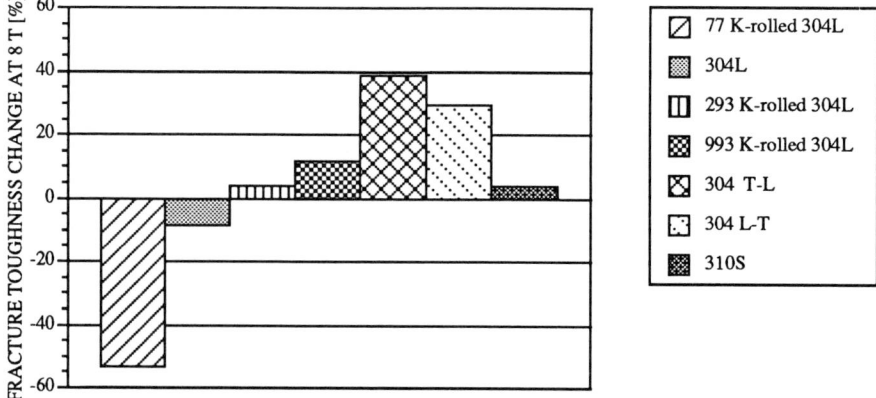

Figure 2. Percentage change in fracture toughness in an 8T magnetic field plotted approximately in order of alloy stability.

CONCLUSIONS

The magnitude and the direction of change in 4.2 K fracture toughness with the application of an 8T magnetic field is a function of the stability of the alloy.

The least stable alloy shows a large reduction in 4.2 K fracture toughness with an applied magnetic field. The amount of fracture toughness reduction with an applied magnetic field decreases as the stability of the specimens increase. At intermediate stability, the direction of fracture toughness change reverses and improvements in fracture toughness are observed. For the fully stable condition, no significant changes in fracture toughness are observed.

ACKNOWLEDGEMENTS

The authors would like to thank C. Tseng for help in preparing this manuscript. This work is supported by the Director, Office of Energy Research, Office of Development and Technology, Office of Magnetic Systems, Base Technology Division of the U. S. Department of Energy under Contract No. DE-AC03-76SF00098.

REFERENCES

1. Fultz, B. and J. W. Morris, Jr., Acta Metall., 34, no. 3:379, 1986.

2. Fultz, B., G. O. Fior, G. M. Chang, R. Kopa, and J. W. Morris, Jr., Adv. Cry. Eng. Mat., 32:377, 1986.

3. Fultz, B., G. M.Chang, R. Kopa, and J. W. Morris, Jr., Adv. Cry. Eng. Mat., 30:253, 1983.

4. Fukushima, E., S. Kobatake, M. Tanaka, and H. Ogiwara, Adv. Cry. Eng. Mat., 34:367, 1988.

5. J. W. Chan, J. Glazer, Z. Mei, J. W. Morris, Jr., Adv. Cry. Eng. Mat., 36:1299, 1989.

6. Fukushima, E., S. Kobatake, M. Tanaka, and H. Ogiwara, presented at the 11th Conference on Magnet Technology, Tsukuba, Japan, Sept 1989.

7. Reed, R. P., Acta Metall., 10:865, 1962.

8. Reed, R. P., Martensitic Phase Transformations, Materials at Low Temperatures, R. P. Reed and A. F. Clark, eds., American Society for Metals, Metals Park, Ohio, 1983, pp. 295.

9. Strife, J. R., M. J. Carr, and G. S. Ansell, Met. Trans. A, 8A:1471, 1977.

PROPERTY EVALUATION OF Ni BASE ALLOY FOR SUPERCONDUCTING GENERATORS AND ITS APPLICATION TO SEAL WELDED JOINTS

Nobuhisa SUZUKI, Toshiaki MURAKAMI, Kenichi SUZUKI, Satoru ASAI, Minoru TANAKA

Heavy Apparatus Engineering Laboratory, Toshiba Corporation
2-4, Suehiro-cho, Tsurumi-ku, Yokohama 230, Japan

Yoshio KOBAYASHI

Super-GM
5-14-10, Nishitenma, Kita-ku, Osaka 530, Japan

INTRODUCTION

Property requirements demanded on structural materials for rotors of superconducting generators are quite strict while there are many themes in the technical development of such materials. Components used in the cryogenic temperature are of course required to be nonmagnetic material. Additional requirements demanded on cryogenic structural materials for high response excitation superconducting generators can be summarized as follows.

- 0.2% yield strength > 800 MPa (RT)   (1)
- Charpy absorbed energy > 25 J   (4K)   (2)
- Electrical resistivity > 90 $\mu\Omega\cdot$cm (4K) (3)

As a candidate material which satisfy the above property requirements, we can cite precipitation-hardened Ni base alloys, for example, Inconel718. However, production of Inconel718 is limited to about 2 tons due to problems related to segregation etc., and therefore fabrication of large-sized ingots for manufacturing generator components may be difficult. Also, there may be problems regarding weldability, especially from the viewpoint of microcracking in the heat affected zone, among strength-supporting parts or the parts requiring functional capability such as seal welded joints. Under such background circumstances, we conducted studies on chemical composition by considering producibility of large-sized ingots and weldability of the alloy as well as property evaluations of materials incorporating such requirements. Further, included in this paper are the application of the alloy to seal welded joints and its unified fatigue strength evaluation method.

STUDY ON CHEMICAL COMPOSITION

Within the ranges that satisfy the property requirements expressed by Equations (1) through (3), we conducted a study on chemical composition based on the following policies.

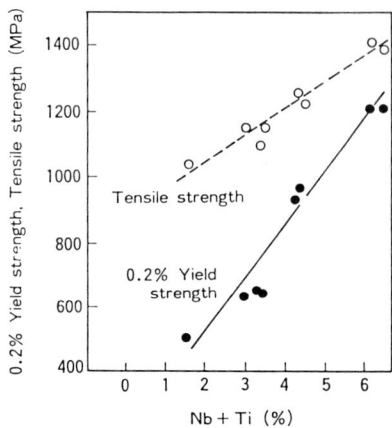 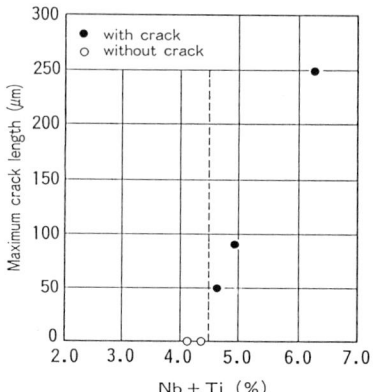

Fig. 1. Effects of (Nb+Ti) content on strengths and maximum crack lengths

(i) To avoid segregation which poses problems in producing large-sized ingots, the percentage of the precipitation-hardening element Nb+Ti is to be reduced as much as possible within the range that satisfies strength requirements.
(ii) To reduce amounts of C, Si, Mn, P, S, and B, which are judged to have apparently adverse effects on weldability, and to determine percentage range of Nb+Ti that does not cause weld cracking.

What are mentioned under (i) above are equally advantageous, as explained in the next section, from the standpoint of toughness improvement. Therefore, using 20mm thick plates fabricated from a 20kg ingot obtained by vacuum induction melting, we studied the relations among Nb+Ti percentage, 0.2% yield strength, tensile strength and weld crack lengths. As is evident from the test results shown in Fig. 1, when Nb+Ti percentage are reduced, 0.2% yield strength and tensile strength are decreased in approximately linear form. From this Fig. 1, what has become clear is that to satisfy the required property value of 0.2% yield strength shown in Eq. (1), more than about 4.0% of Nb+Ti is necessary. Further, from the relations between weld crack lengths and Nb+Ti percentage, we have come to know that weldability is improved as a result of decrease of Nb+Ti percentage and that weld cracking is not caused when the percentage is less than about 4.5%. Based on such findings, we have defined the range shown below as Nb+Ti percentage of improved Ni base alloy.

$$4.0\% \leq Nb+Ti \leq 4.5\% \qquad (4)$$

PROPERTY EVALUATIONS OF IMPROVED Ni BASE ALLOYS

Test rings were fabricated by incorporating the above-mentioned study results, and property evaluation tests were conducted over the range from 4K to room temperature. These tests were actually conducted around the items indicated in Equations (1) through (3). Dimensions of the test rings are 650mm in outside diameter, 590mm in inside diameter, and 1000mm in axial length, and circumferential welding lines are provided at axial center sections. Chemical composition of the test material of the rings and filler metal are shown in Table 1. Test rings were forged, and crystal grain sizes were made smaller by using a ring roll mill. After fabricating two rings from the same ingot by these processes, we applied solution heat treatment, welding, re-solution heat treatment, and aging heat treatment. Heat treatment conditions are as follows.

· Solution heat treatment   1010°C X 3H   (oil quench)
· Aging heat treatment   720°C X 8H + 620°C X 8H (air cool)

Welding was conducted by gas tungsten arc welding (abreviated as GTAW) with welding current of 220A. and microstructures of weld metal and heat affected zone were confirmed to be free from microcracks. Using the test pieces cut off from the test rings, we tested the mechanical and physical properties from 4K to room temperature.

Tensile Properties

Figure 2 shows the tensile test results on base metals and weld metals. From these test results, we are able to realize that both 0.2% yield strength (YS) and tensile strength (TS) increased along with temperature decrease but variations of ductility involving elongation (EL), reduction of area (RA), etc. were relatively limited. Also, from the test results on the base metals, we note that YS and TS of this alloy are slightly lower than those of the reference data[1,2] of Inconel718. This state was created because the percentage of Nb+Ti, was reduced on this alloy. Further, from the test results on the weld metals, the values of TS were found to be slightly lower than those on the base metals, but both EL and RA marked the levels approximately equal to those of the base metals.

Charpy Absorbed Energy and Fracture Toughness Values

The results of Charpy impact tests and fracture toughness tests on the base metals and the weld metals are shown in Fig. 3. The values of Charpy absorbed energy of the base metals and the weld metals of improved Ni base alloy are slightly greater than those of Inconel718 in reference data[1]. This is because Nb+Ti percentage was reduced in this Ni base alloy. The same state also exists on the fracture toughness values. The absorbed energy and fracture toughness values of weld metals revealed levels higher than those of the base metals, and this state corresponds to the fact that tensile strength of the weld metals is a little less than that of the base metals. Temperature dependence characteristics of toughness values of these base metals and weld metals are rather limited.

Electrical Resistivities

Shown in Fig. 4 are the temperature dependence characteristics of electrical resistivities on the base metals and the weld metals as well as their magnetic field dependence characteristics at 4K. While the electrical resistivities decrease as temperature drops, levels of such decrease are not great. The electrical resistivities of the weld metals are about equal to those of Inconel718[2], or about 100 $\mu\Omega\cdot$cm at 4K. On the other hand, the electrical resistivities of the base metals are slightly smaller when compared with those of the weld metals, or about 90 $\mu\Omega\cdot$cm at 4K. This difference of values can be proved from the fact that the more the chemical composition of Ni percentage is reduced, the more the electrical resistivity decreases. In other words, Ni composition percentage (51.06%) of the weld metals tested this time is approximately same as that of Inconel718 (50-55%) and Ni composition percentage of the base metals is lower than this level, namely, being 39.57%. Also, it was discovered that the electrical resistivities in magnetic fields at 4K remained constant throughout the range of 0-7T.

All of these mechanical and physical property values satisfy the requirement values of Eqations (1) through (3) for whole temperature range of 4K to room temperature. Explained hereinafter are the results of applying this alloy to those joints with small geometries like seal welded joints, especially pertaining to the applicability to joints consisting of dissimilar materials. At the same time, a study was made on the fatigue strength evaluation method of such joints.

Table 1. Chemical Compositions (wt%)

	C	Si	Mn	P	S	Ni	Cr	Cu	Aℓ	Ti	Nb	Co	B	Fe
Base metal	.015	.14	.14	.003	.001	39.57	14.71	.01	.33	1.75	2.71	.007	.0006	Bal
Filler metal	.018	.02	.16	.002	.003	51.06	14.08	—	.28	1.73	2.68	—	.0002	—

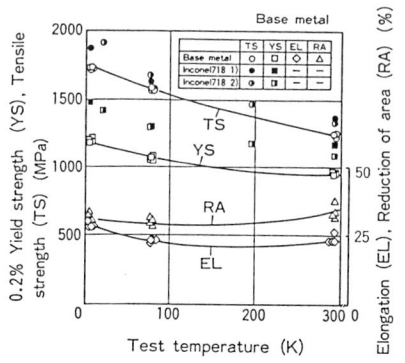
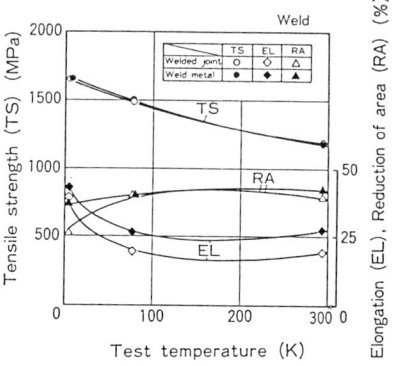

Fig. 2. Tensile test results

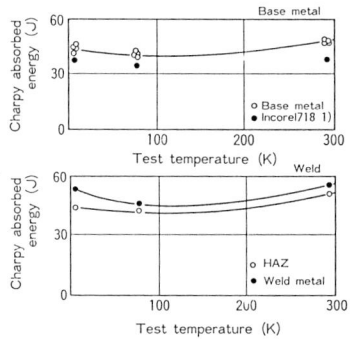
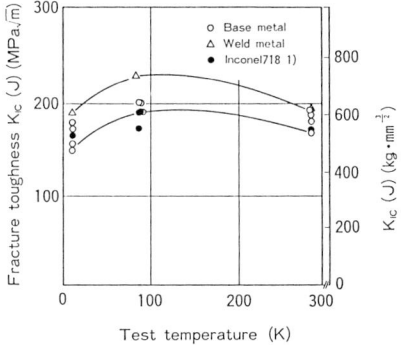

Fig. 3. Charpy impact and fracture toughness test results

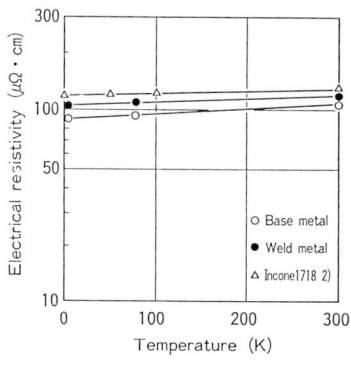
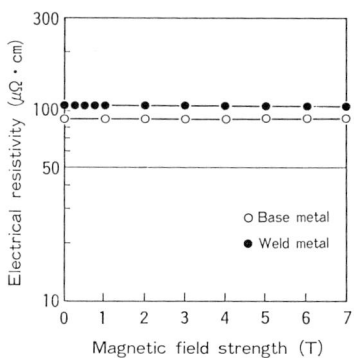

Fig. 4. Electrical resistivity test results

APPLICATION OF IMPROVED Ni BASE ALLOY TO SEAL WELDED JOINTS

Weldability of Seal Welded Joints

Many seal welded joints are provided on the rotors of superconducting generators and these joints are designed to possess the sealing function so that vacuum insulation effects can be provided among the multiple cylinders. Besides sealing effects, another property requirement demanded from the seal welded joints is flexibility to permit an opening at certain levels against loads by centrifugal force etc. Therefore, we first studied the geometries of seal welded joints and followed with examinations of weldability. The lip geometry joint as shown in Fig. 5 is judged to satisfy the property requirements on the above-mentioned seal welded joints. Therefore, by studying width "a" from the standpoint of weldability and length "b" from the viewpoint of flexibility, we selected two pairs of joint geometries, namely, one being lip geometry 1 with width of 2mm and length of 7mm and the other lip geometry 2 with width of 1mm and length of 5mm. On these two types of lip geometries, the welding tests were conducted. Shown in Fig. 5 are the geometries of fatigue specimens explained in the subsequent section, but actual welding tests were conducted using the plates with thickness t of 150mm. Using the previously mentioned improved Ni base alloy (hereinafter abbreviated as IN) and SUS304L (hereinafter abbreviated as SUS), the welding tests were conducted on the combinations of IN-IN (same-material joint) and SUS-IN (dissimilar-material joint). Hereinafter, identifications of the seal welded joints will be made by combining materials and lip geometries, for example, SUS-IN lip1.

Welding wires were selected based on combinations of materials and lip geometries. On the welding tests, Inconel718 was used for IN-IN lip1, Non-filler for IN-IN lip2, and Inconel 82 for SUS-IN lip1. Welding method of GTAW, the same as for test rings explained in the preceding section, was employed. All these weldings were conducted by one weld pass with welding currents of 30-60A.

To confirm the integrity of the welded joints, we conducted macrostructural and microstructural observations of the cross-sectional areas. As an example of this cross-sectional macrostructure, the welded joint of SUS-IN lip1 is shown in Fig. 5. On any combinations of materials and lip geometries, no macroscopic nor microscopic cracks were observed. Based on the seal welded joints whose integrity was confirmed as mentioned above, we then

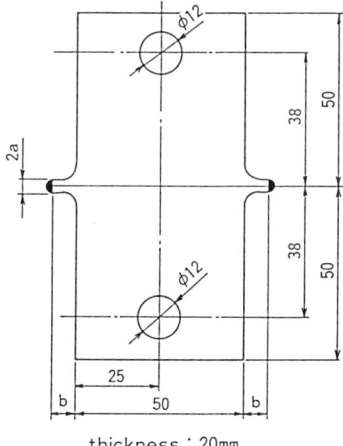
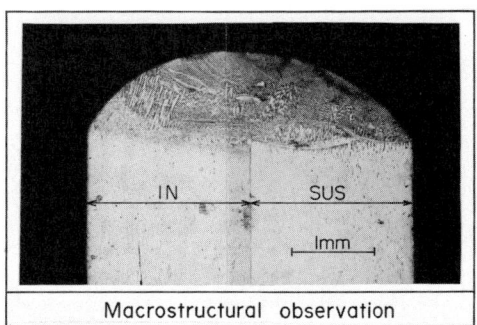

Fig. 5. Lip geometry and macrophotograph of seal welded joints

conducted the following fatigue strength evaluation tests. These tests were conducted using as-welded fatigue specimens.

FATIGUE STRENGTH EVALUATION METHODS

Test Methods

While various loads including centrifugal force and thermal stress are applied to the seal welded joints of superconducting generators, we conducted fatigue tests by simulating the deformation of the most strict mode I. As shown in Fig. 5, the lip joints are arranged symmetrically on the fatigue specimens. This arrangement was selected so that similarly to actual generators, conditions can be satisfied to cause the distortion angle to become zero at crack tip. The tests were conducted by using an electric hydraulic servo type fatigue testing machine and applying pulsating load control. Further, two temperature levels, namely, room temperature and liquid nitrogen temperature (77K), were selected to evaluate the temperature dependence characteristics of fatigue strength, and clip gauges were used to measure displacements of center-section openings on the top and bottom of the test specimens at proper number of cycles. Three types of the test specimens explained in the preceding section were used.

Test Results and Observations

Figure 6 shows the fatigue test results at room temperature and 77K in the form of relations between nominal stress ranges and number of fracture cycles. Since the tests were conducted with pulsating load, nominal stress range ΔS is approximately equal to the value obtained by dividing the maximum load with the test specimen cross-sectional area (50 X 20 mm^2). From this Fig. 6, it is evident that under any material and lip geometry combinations, the fatigue strength at 77K is greater than that at room temperature. Also, when compared under the same test temperature basis, we are able to note that the fatigue strength is higher on lip geometry 1 with larger dimensions than on lip geometry 2 with smaller dimensions. Further, even among the specimens with same lip geometry 1, those with IN-IN material combination show higher fatigue strength than those with SUS-IN combination. Consequently, after adjusting the test results by using nominal stress range ΔS, we are able to state that the fatigue strength differs depending on material combinations, lip geometries, and test temperature.

Shown in Fig. 7 are the relations between the number of cycles and the lip opening displacements and crack lengths of weld metals obtained from the room temperature test results of SUS-IN lip1. In this paper, lip opening displacement is defined as opening displacement between lip at center section of test specimen. The crack lengths appearing in this Fig. 7 are the values measured by using a reading microscope with resolution of 1/100 mm, and these lengths represent the mean values of four points on top, bottom, right, and left sides of the test specimen. From this figure, we are able to know that the crack lengths and the lip opening displacements display a similar trend of increase. Also evident is that the number of cycles, where the line diagram of lip opening displacements - number of cycles starts deviating from the straight line determined by initial inclination, is approximately coincident with the number of cycles where the crack lengths begin increasing. Therefore, we can realize that the number of cycles corresponds to the crack initiation life. The minimum value of lip opening displacements indicates the residual plastic deformation value. Based on the above-mentioned method, we clarified the crack initiation life from the lip opening displacements on all test results. To estimate this crack initiation life, used were the mean values of number of cycles respectively determined from maximum and minimum values of lip opening displacements.

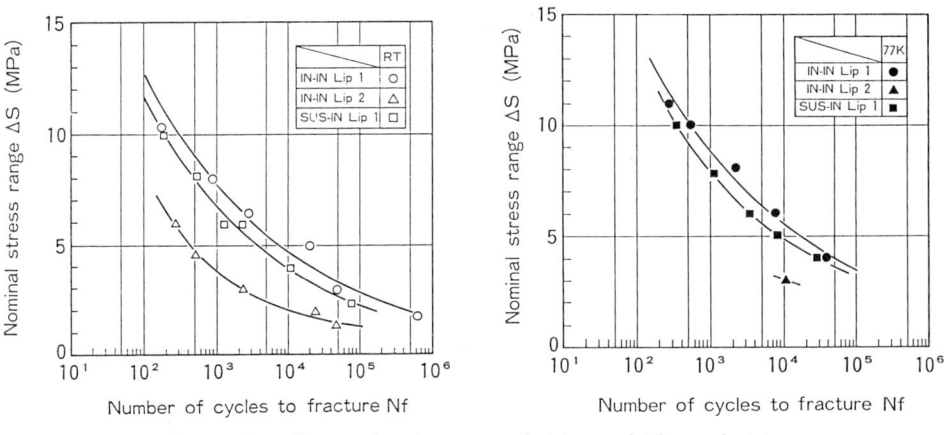

Fig. 6. Nominal stress - fatigue life relations

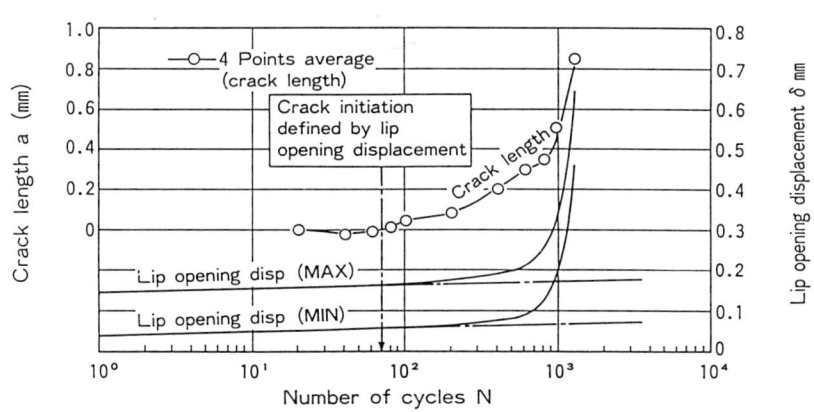

Fig. 7. Variations of crack lengths and lip opening displacements with number of cycles

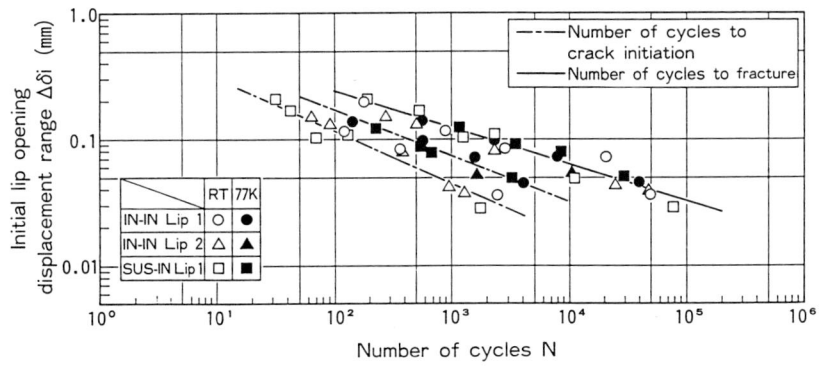

Fig. 8. Relations between initial lip opening displacement range and number of cycles

Shown in Fig. 8 are the relations among lip opening displacements $\Delta\delta i$, crack initiation life, and number of cycles to fracture. Here, $\Delta\delta i$ represents the relative values obtained, as calculated as follows, by subtracting the minimum values from the maximum values of lip opening displacements at initial number of cycles.

$$\Delta\delta i = \delta\max - \delta\min \quad \text{(at initial cycles)} \tag{5}$$

When the data is adjusted by using $\Delta\delta i$, it becomes clear that the crack initiation life is longer at 77K than at room temperature, and that the relations between these two factors are linear on both logarithms. This fact corresponds to that the yielding point rises more at low temperature than at room temperature and that, in the relation of nominal stress range ΔS - number of fracture cycles, the fatigue strength is greater at low temperature than at room temperature. Also, under the same test temperature, we note that the crack initiation life can be collectively evaluated by $\Delta\delta i$ regardless of differences on material combinations or lip geometries. On the other hand, as for the relations between $\Delta\delta i$ and number of fracture cycles, all test results show linear relation on both logarithms irrespective of differences on material combinations, lip geometries, and test temperature levels. Consequently, the crack propagation life can be judged long at room temperature and short at 77K by the number of cycles equivalent to the difference between the crack initiation life at room temperature and that at 77K. However, regarding the fatigue at seal welded joints, since the ratio occupied in the fracture life by the crack propagation life is extremely large, this crack propagation life may possibly involve much limited variations by temperature difference. We are presently applying further studies on this point.

CONCLUSIONS

(1) From the results of material strength tests and welding tests on plates by varying Nb+Ti contents, range of 4.0% $\leq$ Nb+Ti $\leq$ 4.5% is obtained. The material (test rings) which satisfy the above chemical composition are confirmed to meet the property requirement values of Equation (1) through (3) within the full temperature range between room temperature and 4K.

(2) As a result of applying Ni base alloy to seal welded joints, satisfactory weldability has been confirmed. From the fatigue tests of this joints, crack initiation life has been found to be longer at 77K than at room temperature and unified relation between lip opening displacements and number of fracture cycles has been obtained regardless of material combinations, lip geometries, or test temperature levels.

ACKNOWLEDGMENT

This work was performed as a part of "R&D on Superconducting Technology for Electric Power Apparatuses" as a subject of Super-GM under the Moonlight Project of Agency of Industrial Science and Technology, MITI, being consigned by New Energy and Industrial Technology Development Organization (NEDO).

REFERENCES

1. M. Kohno, T. Moriyama, M. Shimada, and A. Suzuki, Mechanical Properties of Alloy A286 at Cryogenic Temperatures, ISIJ 71:1956 (1985).
2. "Handbook on Materials for Superconducting Machinery," No. MCIC-HB-04, Metals and Ceramics Information Center, Battelle, Columbus, Columbus Laboratories, Columbus, Ohio, (1977).

VAMAS SECOND ROUND ROBIN TEST OF STRUCTURAL MATERIALS

AT LIQUID HELIUM TEMPERATURE[1]

T.Ogata, K.Nagai, K.Ishikawa, K.Shibata*, and S.Murase**

National Research Institute for Metals, Tsukuba Labs.
Tsukuba, Ibaraki 305, Japan
*The University of Tokyo, Bunkyo, Tokyo, Japan
**Toshiba Corp., Kawasaki, Kanagawa, Japan

ABSTRACT

A second international round robin(RR) tensile and fracture toughness test program at liquid helium temperature has been coordinated under the Versailles Project on Advanced Materials and Standards (VAMAS). This RR-test was carried out in order to refine procedures of the tests and clarify remaining problems, so that some testing conditions were limited. The results were considerably improved compared to the those in first RR-test; average yield strength was 1065 MPa with a standard deviation of 15 MPa (1.4% to the average) and fracture toughness was 263 MPa$\sqrt{m}$ with a standard deviation of 15.3 MPa$\sqrt{m}$(5.8%).

INTRODUCTION

The VAMAS has more than ten research areas aiming at the development in the evaluation of advanced materials and its standardization through international collaboration. Cryogenic structural materials subgroup in the area 06 has conducted interlaboratory comparisons of mechanical properties, tensile properties and fracture toughness, in order to develop the understanding on mechanical property determination at liquid helium temperature and establish the unified method.

The first intercomparison programs(previous RR-tests)[1,2] aimed at identifying problems and errors which might occur in the testings, so test conditions and procedures were more or less in accordance with room temperature testing standards and were not intentionally unified among the institutes. And some interesting results were discussed; the necessity of a basic approach to microstrain measurement at liquid helium temperature and load-cell calibration for tensile test and the effect of the testing variables for fracture toughness test.

---

[1]This work was sponsored by the Science and Technology Agency, Japan.

Table 1. List of the laboratories and personnel taking part in the second interlaboratory testing program

Institute	Nation	Key Person	Tensile test	Fracture toughness
NIST	U.S.A.	R. P. Reed	O	O
KfK	F.R.G.	A. Nyilas	O	O
EMPA	Switzerland	W. Muster	O	O
Techn. University of Vienna	Austria	E. Tschegg	O	
sep	Frans	J.M.De Monicault	O	
The University of Tokyo	Japan	K. Shibata	O	O
Kobe Steel, Ltd.	Japan	M. Shimada	O	O
Toshiba Corp.	Japan	S. Murase		O
Mitsubishi Heavy Industry,Ltd.	Japan	M. Satoh	O	O
Mechanical Engineering Lab.	Japan	K.Hirano		O
Nat'l Res. Inst. for Metals	Japan	K. Ishikawa	O	O

Based on the results of the first programs, the second programs were designed as to specify the testing conditions in accordance with the latest draft of testing standards. Main purpose of this RR-test is refining procedures of tensile tests and draft document-1988 of fracture toughness test through clarifying remaining problems under the testing conditions.

Eleven research laboratories from six nations have participated and conducted tensile and/or fracture toughness tests on SUS 316LN. Table 1 lists the laboratories and personnel, and the test carried out. This paper presents the detail program and the experimental results of the second RR-test.

MATERIAL AND SPECIMENS

The test material is SUS 316LN, which was the same heat of the material used in previous RR-tests, and is commercially available and has a high yield strength and good fracture toughness favorable for cryogenic structural materials. From the results of previous RR-tests, SUS 316LN was selected as a most appropriate material for the purpose of this RR-test.

Chemical Composition

The steel was provided by a Japanese maker as the form of hot-rolled plates with a thickness of 30 mm. Each material was obtained from the same heat. The chemical composition of SUS 316LN is given in Table 2.

Specimens

Tensile and fracture toughness specimens were cut from center of the plate and were machined at NRIM according to geometries given by participants. Tensile specimens were machined in the L orientation (load axis is parallel to the hot-rolled direction. Most fracture toughness specimens were CT specimens with a 25 mm thickness and a 50 mm wide, a 20 mm

Table 2. Chemical composition of SUS316LN.(wt.%)

C	Si	Mn	P	N	S	Ni	Cr	Mo
0.019	0.50	0.84	0.025	0.18	0.001	11.16	17.88	2.62

Table 3   The testing conditions

	1st RRT	2nd RRT
No. of Participants	18 (6 country)	11 (6country)
Materials	SUS316LN YUS170	SUS316LN Ti-5Al-2.5Sn ELI(in progress)
Control mode	free	strain- and stroke-control (exclusive of load-control)
Strain rate	free	< 1x $10^{-3} s^{-1}$ (tensile test) < 1mm/min (fracture toughness test)
Report of calibration	none	calibration data and chart of load-cell and extensometer
Reference standard	ASTM E813-81 (fracture toughness test)	latest draft of ASTM or JIS (tensile test) ASTM E813-87 (fracture toughness test)

thickness proportional to the CT specimen and a 15 mm thickness specimens were also used. The notch orientations were LT. Most specimens were precracked at room temperature.

TEST PROCEDURE

The differnce in testing conditions between the first RR-test and the second RR-test is summarized in Table 3.

Tensile Tests

Tests were carried out with constant crosshead control and nominal strain rates less than $1 \times 10^{-3} s^{-1}$ according to existing testing standards.

The calibration of load cell and extensometer was performed according to own standards of each laboratory and the calibration charts were reported with test results, yield strength(YS), tensile strength(UTS), elongation(El), reduction of area(RA), and Young' modulus(E).

Fracture Toughness Tests

Most tests were performed using computer-controlled unloading compliance method. Test conditions and procedures were in accordance with the revised testing standards ASTM E813-87 and -88 ANNEX in principle. The data obtained according to E813-81 was optionally reported. And more the stroke rate was limited less than 1 mm/min. The calibration of load cell and extensometer was also carried out and reported. In multiple specimen method, crack length was directly observed. Fracture toughness $K_{Ic}(J)$ was estimated from J_{Ic} using the expression $K_{Ic}(J) = (J_{Ic} \cdot E)^{1/2}$. The initial crack length, a_0 was optically measured after testing.

Summary of the procedure of E813-87 and E813-81 are illustrated in Figure 1. Main revised points from 81 to 87 are 1) J-integral is calculated separately such as elastic region and plastic region, 2) power law regression line is determined by least square method, and 3) the intersection of the regression line with the 0.2mm offset line defines J_Q. Fracture

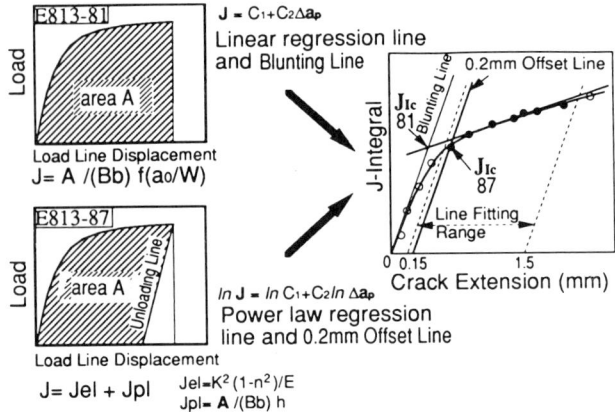

Figure 1 Summary of the procedure of ASTM E813-87 and E813-81.

toughness value of E813-87 can be calculated from the saved data under E813-81 also, from all load-displacement data or from intermediate data set obtained at each unloading procedure such as crack length, area A, load, and compliance.

RESULTS

Tensile Properties

Figure 2 shows the results of tensile tests, YS and UTS is plotted for each participant. Participants are in order of the average ultimate tensile strength obtained in the previous test. Open circles are results of first round robin test and solid triangles are results of thesecond round robin test. Scatter of the data became obviously less in the second RR-test.

Table 4 gives a summary of tensile test results. The results were considerably improved compared to the those in first RR-test except for RA. Average yield strength was 1065 MPa with a standard deviation of 15 MPa (1.4% to the average). Figure 3 shows the YS, UTS, El, and RA as a function of strain rate. There are no measurable effects on the properties for the strain rate less than 10^{-3} s^{-1}.

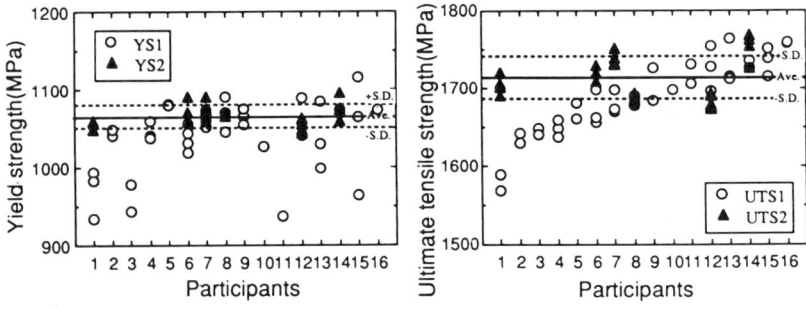

Figure 2 Yield strength and ultimate tensile strength of first RR-test and second RR-test at 4K. (Ave:Average, SD:standard deviation)

Table 4 Testing variables and tensile properties of SUS 316LN.

Code	Machine type	Diameter (mm)	Gauge length (mm)	Strain rate x10-4s-1	Strain measurement	Elongation determ.	Yield strength YS(MPa)	Ultimate strength UTS(MPa)	Elongation El(%)	Reduction of area RA(%)	Young's modulus (GPa)
H	Servo	6.25	40	10	Extenso	Marks	1075	1763	47.4	47.4	185
H	Servo	6.25	40	10	Extenso	Marks	1096	1768	47.8	47.8	193
H	Servo	6.25	40	10	Extenso	Marks		1726	46.0	41.0	
H	Servo	6.25	40	10	Extenso	Marks	1059	1755	46.3	41.0	174
I	Motor	6.25	35	4.8	Extenso	Overall	1048	1674	45.7	42.0	200
I	Motor	6.25	35	4.8	Extenso	Overall	1045	1677	48.0	51.0	195
I	Motor	6.25	35	4.8	Extenso	Overall		1690	45.7	62.0	
I	Motor	6.25	35	4.8	Extenso	Overall	1062	1680	50.0		194
I	Motor	6.25	35	4.8	Extenso	Overall	1054	1680	51.0		199
I	Motor	6.25	35	4.8	Extenso	Overall	1046	1691	49.0		193
I	Motor	6.25	35	4.8	Extenso	Overall	1046	1693	48.0		199
I	Motor	6.25	35	4.8	Extenso	Overall	1043	1680	47.0		197
J	Motor	5	30	2.8	Extenso	Marks	1059	1706	49.6	47.0	201
J	Motor	5	30	2.8	Extenso	Marks	1049	1701	50.0	48.0	204
J	Motor	5	30	2.8	Extenso	Marks	1059	1721	48.5	39.0	203
J	Motor	5	30	2.8	Extenso	Marks	1059	1691	50.0	48.0	199
K	Motor	6	60	4.2	Extenso	Marks	1090	1733	46.3	44.0	195
K	Motor	6	60	4.2	Extenso	Marks	1073	1732	46.7	51.0	201
K	Motor	6	60	4.2	Extenso	Marks	1065	1751	48.5	55.0	189
K	Motor	6	60	4.2	Extenso	Marks	1075	1739	46.2	53.0	191
K	Motor	6	60	4.2	Extenso	Marks	1060	1732	47.3	55.0	197
L	Motor	7	42	2	Gauge	Overall	1071	1685	51.4	58.7	208
L	Motor	7	42	2	Gauge	Overall	1066	1693	51.2	59.6	208
M	Motor	7	48	1.7	Extenso	Marks	1070	1710	46.9	45.0	211
M	Motor	7	48	1.7	Extenso	Marks	1070	1730	46.0	45.3	221
M	Motor	7	48	1.7	Extenso	Marks	1060	1730	51.1	44.8	212
M	Motor	7	50	1.7	Extenso	Marks	1090	1730	51.1	50.1	201
M	Motor	7	50	1.7	Extenso	Marks	1090	1720	49.7	46.2	202
Average							1065	1714	48.3	48.8	199
Standard deviation & (%Ave/SD)							15(1.4)	28(1.6)	1.9(3.9)	6.2(12.7)	9.2(4.6)
Average (1st RRT)							1039	1687	51.1	53.6	207
Standard dev. (1st RRT) (%Ave/SD)							46(4.4)	48(2.8)	2.7(5.3)	5.7(10.7)	13(6.3)

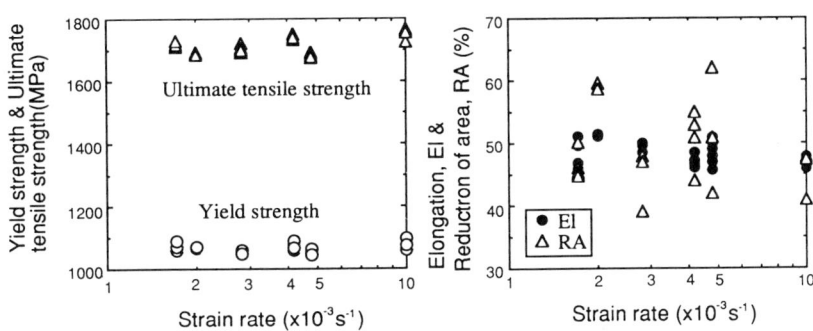

Figure 3 Yield strength, ultimate tensile strength, elongation, and reduction of area measurements at 4K.

Table 5 Testing variables and fracture properties of 316LN.

	B mm	W mm	a0 mm	ΔKmax MPa√m	machine type	control mode	rate mm/min	crack length prediction	Y.R. GPa	Jic87 kJ/m	Kic87 MPa√m	Jic81 kJ/m	Kic81 MPa√m	Kic(J) MPa√m
a	25	50	32	21	servo	strain	0.4	unloading	202	332	259	302	247	259
	25	50	32	21	servo	strain	0.4	unloading	202	277	237	289	242	237
	25	50	32	21	servo	strain	0.4	unloading	202	320	254	285	240	254
d	25	50		29	servo	strain	0.36	unloading	205	344	265.4	317	255	265.4
	25	50		29	servo	strain	0.36	unloading	205	290	243.7	269	234.7	243.7
	25	50		29	servo	strain	0.36	unloading	205	344	265.4	318	255.3	265.4
f	20	40			Motor	strain	0.5	unloading	206			[440]	[300]	[300]
h	15	50		28	Motor	stroke	0.5	Multiple	205				250	250
i	25	50	33	26.7	Motor	strain	0.5	unloading	215	355	274	303	253	274
	25	50	33	26.7	Motor	strain	0.5	unloading	216	377	282	322	261	282
	25	50	33	26.7	Motor	strain	0.5	unloading	216	366	278	312	256	278
	25	50	33	26.7	Motor	strain	0.5	unloading	216	319	259	275	241	259
	25	50	33	26.7	Motor	strain	0.5	unloading	213	387	286	350	272	286
Average										337	264	304	246	263
Standard deviation & (% of Ave/SD)										34.4	15.5(5.9)	23.5	7.8(3.2)	15.3(5.8)
Average 1st RRT													247	
Standard deviation 1st & (% of Ave/SD)													31(12.6)	

Fracture Toughness

The fracture toughness results, J_{Ic}, and $K_{Ic}(J)$ are summarized in Table 5. Some institutes reported the fracture toughness using E813-81. Both $K_{Ic}(J)-87$ and $K_{Ic}(J)-81$ are listed. The average and the deviation value were calculated from the valid data. The average fracture toughness (including both $K_{Ic}(J)-87$ and $K_{Ic}(J)-81$) was 263 MPa√m with a standard deviation of 15.3 MPa√m (5.8 % to the average). The scatter in fracture toughness also decreased in this test.

DISCUSSION

Tensile Test

From the results of the first round robin tests, the calibration error of load-cell was pointed out. So participants have carried out second round robin tests, paying much attention and effort to the calibration. This improved preparation for the testing would reduce the scatter of properties among participants.

The effects of other testing variables such as machine type and strain measurement on the properties were also discussed in the previous test. Table 6 shows the tensile properties for

Table 6 Tensile properties for each testing variable.

Testing variables		YS(MPa)	UTS(MPa)	El(%)	RA(%)	E(GPa)
Servo	Ave.	1077	1753	46.9	44.3	184
n=4	Dev.	19	19	0.9	3.8	9.5
Motor	Ave.	1063	1707	48.5	49.7	201
n=24	Dev.	14	24	1.9	6.2	7.4
Extenso.	Ave.	1064	1716	48.1	47.8	198
n=26	Dev.	16	28	1.8	5.5	9.2
Gauge	Ave.	1069	1689	51.3	59.2	208
n=2	Dev.	4	6	0.1	0.6	0
Marks	Ave.	1071	1730	48.1	47.1	199
n=18	Dev.	14	21	1.8	4.5	10.9
Overall	Ave.	1053	1684	48.7	54.7	199
n=10	Dev.	10	7	2.2	8.2	5.4

the previous test. Table 6 shows the tensile properties for each testing variable. In this test, most tests were carried out using motor-driven type testing machine and extensometer for strain-measurements. So, the effect of these variables is impossible to discuss here. Table 6 also gives the statistics for the different way in extension for elongation determination. There is no significant difference between 'distance between marks on reduced section' and 'overall length' in this test.

These results and other report[4] prove that there would be no problem in cryogenic tensile tests for this kind of materials just except for RA, which shows larger scatter than other properties in this test.

Fracture Toughness

Figure 4 shows the fracture toughness plotted for each participant. Scatter in fracture toughness also decreased in this test. From this we could confirm the small deviation under limited testing condition, however, further study will be required for the standardization.

Figure 5 presents the comparison of the fracture toughness between the servo-hydraulic testing machine and the motor-driven machine. The servo-hydraulic machine gives a little bit lower toughness than the motor-driven testing machine, which could be due to the difference in machine compliance.

The effects of strain rate and specimen thickness on the fracture toughness for SUS 316LN were pointed out in previous test, however, the effects were not clear in this test. The effects of the initial crack length a_0 and the maximum stress intensity factor during the fatigue precracking K_{max} were not clear in this test.

E813-87 and E813-81

Fracture toughness obtained by E813-87 is slightly higher than that obtained by E813-81 in this test. This tendency to result in higher value is seen in other materials which have higher toughness like SUS 316L

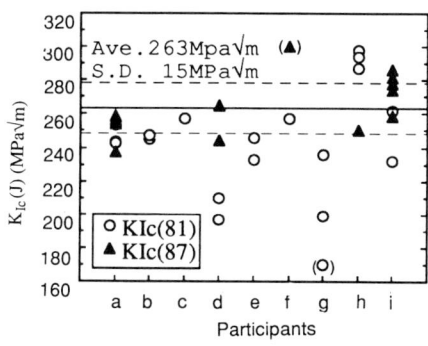

Figure 4   Fracture toughness measurements at 4K.

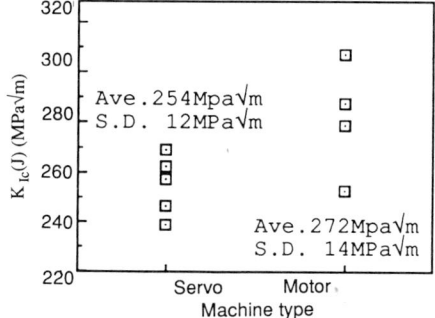

Figure 5   Fracture toughness in the servo-hydraulic and the motor-driven machine.

CONCLUSION

Eleven laboratories from six nations performed the round robin tensile test and fracture toughness test. We agreed the following significant points:

1. Under improved testing condition and surroundings, the tensile results show good agreement, especially the standard deviation of yield strength was 15 MPa and 1.4 % to the average.
2. The standard deviation of fracture toughness was also improved to be 15.3 MP$\sqrt{}$m and 5.8 % to the average under limited condition, there still remains further discussion on the effect of the serration and testing variables; side-grooving, strain rate.
3. We continue this interlaboratory study to develop our understanding on mechanical properties determination and we hope these results also contribute to the standardization or its refinement.

ACKNOWLEDGMENTS

Authors wish to express appreciation to all participants for their collaboration and contribution to these activities.

REFERENCES

1. K. Nagai, T. Ogata, K. Ishikawa, K. Shibata, and E. Fukushima, VAMAS Interlaboratory Tensile Test at Liquid Helium Temperature, Cryogenic Materials '88, 2:893-900(1988)
2. T. Ogata, K. Nagai, K. Ishikawa, K. Shibata, and E. Fukushima, VAMAS Interlaboratory Fracture Toughness Test at Liquid Helium Temperature, Adv. Cryo. Eng. 36:1053- 1060(1990)
3. Standard Test Method for J_{Ic} A Measure of Fracture Toughness, Designation E 813-81, 1986 Annual Book if ASTM Standards, Vol. 03.01, Amer. Soc. Test. Maters., Philadelphia (1986)
4. H. Nakajima, Y. Yoshida, S. Shimamoto, R. L. Tobler, R. P. Reed, R. P. Walsh, and P. T. Purtsher, Interlaboratory Tensionand Fracture Toughnes Test Resultsfor CSUS-JN1 austenituc Stainless Steel at 4K, Adv. Cryo. Eng. 36B:1069-1076 (1990)

EFFECT OF HYDROGEN CHARGING ON AMBIENT AND CRYOGENIC MECHANICAL PROPERTIES OF A PRECIPITATE-STRENGTHENED AUSTENITIC STEEL

Luming Ma, Guojun Liang, and Yiyi Li

Institute of Metal Research
Academia Sinica
Shenyang, China

ABSTRACT

The method of high-pressure hydrogen charging was used to charge hydrogen into smooth and notched tensile specimens of γ' precipitate-strengthened austenitic steel JBK-75. The hydrogen content in the charged specimens was 25.2 ppm (by weight). In the test temperature range 293 to 77 K, hydrogen had no obvious effect on strength, but it caused some decrease in ductility at 223 to 295 K. The steel is not notch sensitive, and hydrogen charging had little effect on notch sensitivity at ambient and low temperatures. With decreasing temperature, both strength and ductility increased, and hydrogen damage greatly decreased. Increasing the aging temperature and time tends to increase the hydrogen damage of the steel, but hydrogen had less effect on aged strength. The steel that had been given an appropriate heat treatment had excellent cryogenic mechanical properties and resistance to hydrogen damage. The steel had very stable microstructure at low temperatures; no phase transition occurred as a result of strain and hydrogen at 293 to 77 K. Fine grain and fine γ' precipitates decreased hydrogen damage; the presence of η phase at grain boundaries increased hydrogen damage.

INTRODUCTION

Hydrogen embrittlement (HE) of precipitate-strengthened austenitic steel (PSAS) has received considerable attention.[1-4] The reason may be the low matrix and weld strength of single-phase austenitic steel; in spite of its resistance to HE, its yield strength is usually less than 400 MPa. There are expected to be a variety of applications in hydrogen environments (including pumps, valves, and special vessels) that will demand higher strength steel. Thus, an austenitic matrix strengthened by precipitation might be selected not only for its high strength but also for its hydrogen compatibility and easy formability and weldability. One such steel is A-286, which has the highest strength of the PSAS. JBK-75 steel has evolved to improve the weldability and hydrogen compatibility of A-286. Some research on the hydrogen compatibility of these two steels has been reported,[1,3] but no attention has been paid to the relation between PSAS aging characteristics and hydrogen damage (HD). There was one report on the occurrence of cryogenic HE in austenitic steel.[5] Whether HE would occur under the interaction of hydrogen, low temperatures, and strain must be considered for cryogenic vessels containing hydrogen.

The purposes of this study were (1) to understand the tendency for HD in one PSAS, JBK-75, at ambient and cryogenic temperatures, (2) to determine how to reduce HD, and (3) to identify the relationship between the microstructure and HD.

EXPERIMENTAL PROCEDURES

The nominal composition of the JBK-75 steel tested was Fe-30Ni-15Cr-1.5Mo-2.0Ti-0.25Al-0.25V-0.002B. Specimens were taken from hot-rolled bar stock, 20 mm in diameter. Smooth, tensile specimens were 5 mm in diameter with a gage length of 25 mm. Notched tensile specimens were 5 mm in diameter at the notch; the notch was 60°, 1 mm deep, with a root radius of 0.1 mm; the stress concentration factor for the notch was 4.55. All specimens were solution-treated at 1253 K for 1 h and then water quenched. Specimens for cryogenic tests were aged 8 h at 1013 K and then air cooled. The average austenite grain size and γ' precipitate sizes were about 33 μm and 6 to 11 nm, respectively.

The aged specimens were hydrogen charged in an autoclave at 573 K and 10 MPa in high-purity hydrogen for 4 days. The hydrogen contents in the specimens before and after charging were 0.9 and 25.2 ppm (by weight), respectively. Results of hydrogen analysis indicated that, for the 5-mm-diameter specimen, hydrogen concentration saturated after hydrogen charging. About 1% of the hydrogen was released after the charged specimens had been exposed to ambient conditions for 7 days.

After hydrogen charging, the specimens were kept at different low temperatures for 15 min, and then tensile tests were conducted at a crosshead rate of 2.5 mm/min. Diffraction analyses of the specimens deformed at cryogenic temperatures were conducted with a Philips PW 1140 X-ray diffractor; the specimen surfaces had been electrolytically polished before they were analyzed. Foil specimens were analyzed by a EM420 TEM operating at 100 kV. The ψ_L and δ_L parameters were used to evaluate the HD of the PSAS, where ψ_L and δ_L are the loss rates of reduction of area (RA) and elongation, respectively, for smooth tensile specimens. The definition of the loss rate for a property is

$$\text{Loss rate for property} = \frac{\text{uncharged property} - \text{charged property}}{\text{uncharged property}}$$

RESULTS AND DISCUSSION

Effect of Hydrogen on Cryogenic Tensile Properties

The effect of hydrogen charging on the ambient and cryogenic tensile properties of JBK-75 steel is shown in Table 1. Hydrogen had no effect on the yield strength ($\sigma_{0.2}$) and ultimate tensile strength (σ_b) at different temperatures. Although hydrogen results in some loss of ductility, the degree of loss depended on the test temperature. As temperature decreased, both strength and ductility increased. Figure 1 shows the HD tendency for the steel at cryogenic temperatures: The steel was not very sensitive to HD at room temperature, and the HD decreased with decreasing temperature. At temperatures less than 223 K, ψ_L was less than 10% and δ_L was less than 5%.

The very low tendency of JBK-75 steel to HD at cryogenic temperatures can be explained by dislocation theory. During deformation, mobile dislocations can carry hydrogen atoms (in the "Cottrell atmosphere"). Near room temperature, hydrogen atoms are moved by dislocations and transferred to obstacles (e.g., grain boundaries, phase boundaries, and inclusions). The

Table 1. Effect of Hydrogen on Tensile Properties

Test Temp., K	Uncharged				Hydrogen Charged				δ_L, %	ψ_L, %
	$\sigma_{0.2}$*	σ_b	δ, %	ψ, %	$\sigma_{0.2}$	σ_b	δ, %	ψ, %		
293	763	1109	28.9	58.1	763	1110	26.1	43.4	9.7	25.3
223	778	1152	30.2	57.7	775	1153	29.6	51.4	2.0	10.9
153	806	1190	31.3	57.3	793	1207	33.1	56.3	0	1.7
77	876	1412	41.6	60.0	868	1417	41.6	59.2	0	1.3

*σ in MPa

result is hydrogen accumulation and dislocation pile-ups, which subsequently form microcracks at these locations that decrease ductility. At cryogenic temperatures, hydrogen atoms cannot move easily, the dislocations do not carry many hydrogen atoms, and therefore the tendency to HD is lower.

The interaction of cryogenic temperatures, hydrogen, and strain on the steel were examined by X-ray-diffraction and magnet-measuring techniques. After maximum strain at 77 K, no strain- or hydrogen-induced martensite transformation (or any other phase transformation) occurred whether or not the steel was hydrogen charged. Figure 2 shows the X-ray diffraction patterns for the steel deformed to 0.25 true plastic strain at 77 K. The diffraction peaks of only γ (including the austenite matrix and γ' precipitates) appeared; no peaks of α' and ϵ martensite appeared. Hydrogen charging decreased the γ_{111} and γ_{200} peaks and shifted them a little toward the lower 2θ angles. It indicates that hydrogen charging slightly expands the lattice. The lattice constants a_0 and a_0^H measured 3.5906 Å before and 3.5929 Å after hydrogen charging. Magnetic measurements of specimens deformed at cryogenic temperatures were made with a ferrite detector. No magnetic α' phase was found in the charged specimens deformed to maximum strain at 77 K. The above results indicate that the microstructure of JBK-75 steel is very stable at cryogenic temperatures. No embrittlement resulted from the interaction of low temperatures, hydrogen, and strain. The increase in strength and slight HD at cryogenic temperatures evidently were independent of the martensite transformation. The increases in both strength and ductility at cryogenic temperatures are possibly related to the fact that cryogenic temperatures aid cross slip and increase the strain-hardening rate.[6]

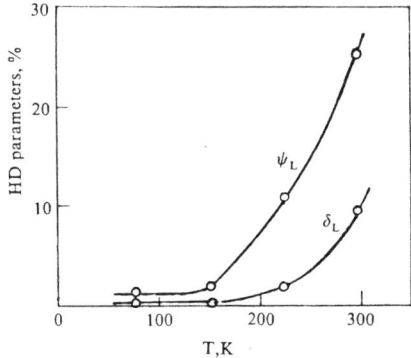

Fig. 1. HD tendency for JBK-75 steel at cryogenic temperatures.

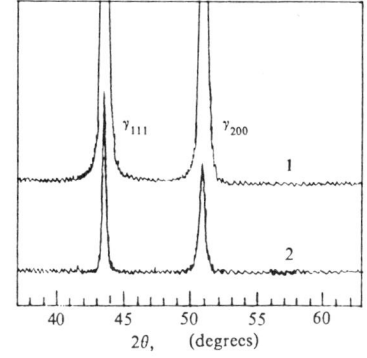

Fig. 2. X-ray diffraction patterns of specimens deformed in tension at 77 K (CuK$_\alpha$, 0.25 strain). 1 - uncharged, 2 - hydrogen charged

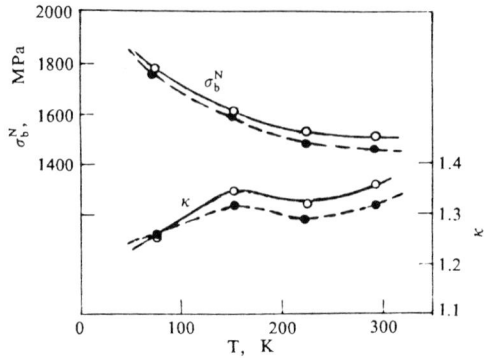
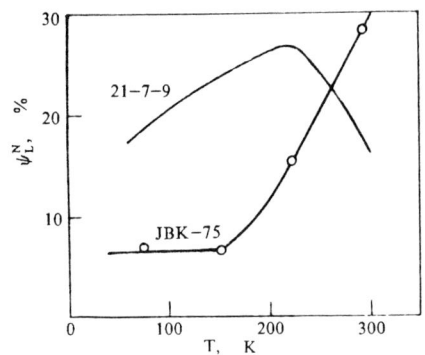

Fig. 3. Effect of hydrogen on σ_b^N and notch sensitivity.
—— uncharged
--- hydrogen charged

Fig. 4. HD tendency for notched specimens at cryogenic temperatures.

Effect of Hydrogen on Notched Tensile Properties and Toughness

The effects of hydrogen charging on notched tensile strength (σ_b^N), notch sensitivity, and ductility for notched specimens are presented in Figs. 3 and 4, where ψ_L^N is the loss rate of RA for a notched tensile specimen, and k is the notch sensitivity factor. The σ_b^N increased as temperature decreased. Hydrogen slightly decreased σ_b^N, and the effect was very small below 200 K. Hydrogen slightly increased the notch sensitivity. At various temperatures, all the notch sensitivity factors were >1; thus, JBK-75 steel is not sensitive to the notch effect. When all notches are identical, ψ_L^N can be used to compare the extent of HD in the specimens;[7] it can be regarded as an auxiliary parameter for evaluating HD. Figure 4 shows HD tendency for notched specimens at cryogenic temperatures; the ψ_L^N values of 21-7-9 steel are also give for comparison.[8] For JBK-75 steel, the ψ_L^N values were less than 10% at temperatures less than 190 K, indicating that JBK-75 has very good resistance to HD at cryogenic temperatures.

The $\sigma_b^N/\sigma_{0.2}$ value for round bar specimens can be used to express the magnitude of plane-strain fracture toughness, K_{Ic}.[9] The effects of hydrogen and temperature on the $\sigma_b^N/\sigma_{0.2}$ value are shown in Fig. 5. The fracture toughness of the steel increased slightly with decreasing temperature. Hydrogen charging had less effect on the toughness. At temperatures higher than 200 K, hydrogen caused a slight decrease in toughness. The toughness value of a high-strength aluminum alloy, 2219-T87, commonly used at low temperatures, is also given in Fig. 5.[10] It can be seen that the fracture toughness of JBK-75 steel is quite good.

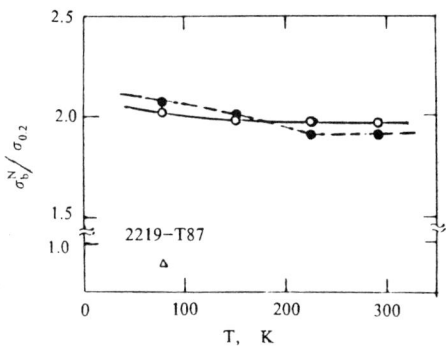

Fig. 5. Effect of hydrogen on toughness of the steel.
—— uncharged
--- hydrogen charged

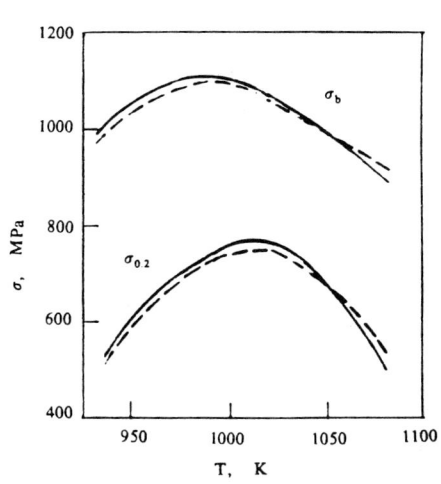 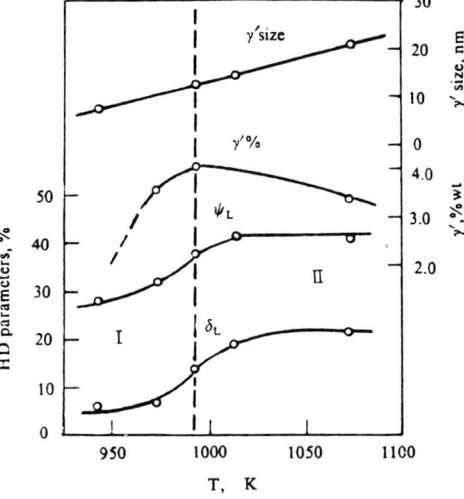

Fig. 6. Effect of T_A and hydrogen on room-temperature strength (aged 16 h). —— uncharged --- hydrogen charged

Fig. 7. Effect of T_A on HD and γ' precipitates at room temperature (aged 16 h).

Aging Characteristics and Hydrogen Damage

We examined the effects of different aging conditions and hydrogen charging on the ambient tensile properties of JBK-75 steel that had been solution-treated at 1253 K for 1 h and water quenched. The effects of aging temperature (T_A) and hydrogen on strength at room temperature are shown in Fig. 6, and curves showing the relationship of T_A with HD and γ' size are plotted in Fig. 7. The T_A had a pronounced effect on strength; maximum strength occurred at 993 to 1013 K. Hydrogen charging had a very small effect on aged strength, but it resulted in an obvious decrease in ductility. The HD increased with increasing T_A; however, the increment of HD increase became small in the high-T range. The curves relating HD to T_A are closely related to the precipitated γ' and η phases. Spherical precipitates of γ' (fcc structure) were observed by TEM at aging temperatures ranging from 943 to 1073 K. For specimens aged 16 h at various temperatures, TEM dark-field photographs were used to measure the γ' sizes, which are shown in Fig. 7. The γ' size increased with increasing T_A, and there is a good linear correlation between γ' size and T_A.

Figure 7 has been divided into two regions: region I, without η phase, and region II, with η phase. The dashed line separating the regions corresponds to the maximum amount of γ' precipitate (about 4 wt.%). In region I, the HD depended on the γ'-phase precipitation. The size of γ' increased with increasing T_A; as a result, HD increased. The η phase had a hcp (Ni_3Ti) structure and was incoherent with the matrix (see Fig. 8). In region II, two kinds of η phase precipitated: intergranular and intragranular. The intergranular η phase usually formed in disc and platelet shapes, as shown in Fig. 9. The intragranular η phase formed mainly in cellular and isolated platelet shapes. The cellular η phase had a lamellae structure (Fig. 8) in which thin η platelets were parallel to each other; the γ matrix was seen between two η platelets. In region II, HD depended mainly on the role of intergranular η (η_{GB}). However, the effect of γ' precipitates on HD became small in region II because the space between γ' particles became larger and the amount of γ' decreased with increasing T_A (even though the size of the γ' particles increased). At T_A from 993 to 1013 K, η phase precipitated mainly

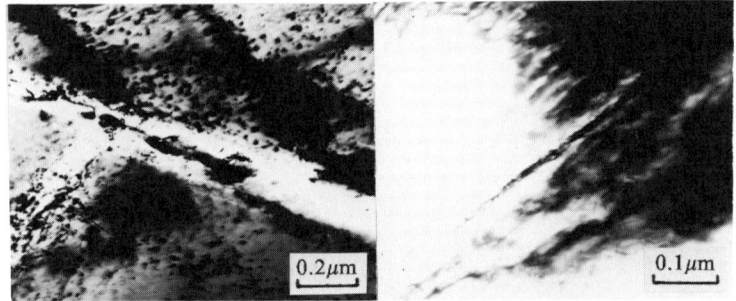

Fig. 8. TEM micrographs of cellular η phase and its diffraction pattern (zone axis: $[2\bar{1}\bar{1}0]_\eta$, $[0\bar{1}\bar{1}]_\gamma$; aging: 1073 K, 16 h).

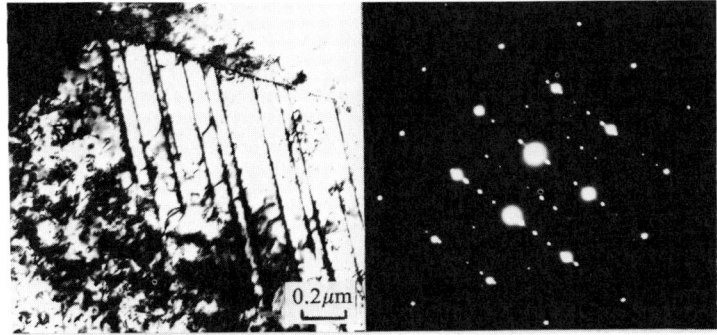

Fig. 9. TEM micrographs of η phase precipitated at grain boundaries.

at grain boundaries, and η_{GB} resulted in a continuously increasing HD. When T_A was higher than 1013 K, more intragranular, cellular η phase precipitated (Fig. 8), with some η_{GB}. The intragranular η phase had very little effect on HD; therefore, the ψ_L-T_A curve reaches a plateau. Although the cellular η had less effect on HD, it decreased the intrinsic ductility and strength of the steel. Experimental results of aging 16 h indicate that the HD of JBK-75 steel depends on γ' precipitates. The HD increased as γ' size increased for γ' < 13 nm; however, for γ' > 13 nm, the HD depended mainly on η_{GB}.

Aging time had an obvious effect on the HD of the steel specimens aged at the same temperature. Generally, HD increased with increasing aging time. A typical example is given in Table 2.

Effect of Grain Size on Hydrogen Damage

Curves showing the relationship between grain size and HD are plotted in Fig. 10 for specimens aged 16 h at 993 K. For the same aging conditions, the finer the grain size, the lower the HD tendency. This relationship is possibly associated with the fact that fine grains lead to uniform plastic deformation, and consequently, to uniform distribution of hydrogen during deformation.

Table 2. Effect of Aging Time on Hydrogen Damage (T_A = 1013 K)

Time, h	ψ_L, %	δ_L, %	γ' size, nm	η_{GB}
16	41.9	19.1	14	a few
8	25.3	9.7	10	none

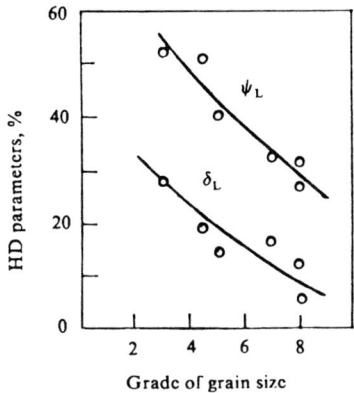

Fig. 10. Effect of grain size on HD.

CONCLUSIONS

Examination of JBK-75 precipitate-strengthened austenitic steel that had been charged with as much as 25.2 ppm hydrogen and tested at 293 to 77 K led to the following conclusions:

1. The steel has superior mechanical properties and a very low tendency to HD at low temperatures; it also has acceptable HD resistance at ambient temperature.

2. Hydrogen has little effect on strength ($\sigma_{0.2}$ and σ_b), σ_b^N, and toughness at temperatures from 293 to 77 K, but it decreases ductility somewhat as a function of temperature. The steel is not sensitive to notch effects, and hydrogen has little effect on the notch sensitivity.

3. The steel has a stable microstructure when subjected to the interaction of cryogenic temperatures, hydrogen charging, and strain; no martensite transformation or embrittlement occurs.

4. Aging conditions have a pronounced effect on HD. Lower T_A and short aging time tend to decrease HD. When no η phase is present, HD increases with γ' particle size. The precipitate of η_{GB} tends to increase HD, whereas intergranular η has little effect on the HD. Fine grain size helps to prevent HD.

ACKNOWLEDGMENTS

We are pleased to acknowledge Prof. C. G. Fan, Mr. D. H. Li, and Mr. X. J. Zhao for their help in the hydrogen charging tests.

REFERENCES

1. A. W. Thompson and J. A. Brooks, Hydrogen performance of precipitation-strengthened stainless steels based on A-286, *Metall. Trans.* 6A:1431 (1975).
2. A. W. Thompson, Hydrogen-induced ductility loss in commercial precipitation-strengthened stainless steels, *Metall. Trans.* 7A:315 (1976).
3. B. C. Odegard, Jr. and A. J. West, The effect of η phase on the hydrogen compatibility of a modified A-286 superalloy: microstructural and mechanical properties observations, in: *Proceedings, International Conference on Hydrogen Effects in Metal*, vol. 597 (1980).

4. P. D. Hicks and C. J. Altstetter, Internal hydrogen effects on tensile properties of iron- and nickel-base superalloys, *Metall. Trans.* 21A:365 (1990).
5. M. Kuribayashi and H. Okabayashi, Influence of heat treatment conditions on mechanical properties of hydrogenated type 304 stainless steel at low temperature, *Trans. Jap. Inst. Met.* 25:623 (1984).
6. J. Glazer, S. L. Verzasconi, R. R. Sawtell, and J. W. Morris, Jr., Mechanical behavior of aluminum-lithium alloys at cryogenic temperatures, *Metall. Trans.* 18A:1659 (1987).
7. Y. Y. Li, L. M. Ma, and G. J. Liang, The effect of high-pressure charging with hydrogen on mechanical properties of an austenitic steel, *Chin. J. Met. Sci. Technol.* 3:74 (1987).
8. L. M. Ma, Y. Y. Li, and G. J. Liang, Effect of hydrogen on 21-7-9 austenitic steel at low temperature, in: *Advances in Cryogenic Engineering—Materials,* vol. 34, Plenum, New York (1988), p. 325.
9. J. W. Kaufman, G. T. Sha, R. F. Kohn, and R. J. Bucci, "Cracks and Fracture," ASTM STP 601, American Society for Testing and Materials, Philadelphia (1976), p. 169.
10. L. M. Ma, J. K. Han, R. L. Tobler, R. P. Walsh, and R. P. Reed, Cryogenic fatigue of high-strength aluminum alloys and correlations with tensile properties, in: *Advances in Cryogenic Engineering—Materials,* vol. 36, Plenum, New York (1990), p. 1143.

DEFORMATION MEASUREMENTS OF MATERIALS AT LOW TEMPERATURES

USING LASER SPECKLE PHOTOGRAPHY METHOD

Sumio Nakahara, Yukihide Maeda*, Kazunori Matsumura*,
Shigeyoshi Hisada, Takeyoshi Fujita, and Kiyoshi Sugihara

Department of Mechanical Engineering, Kansai University
*Graduate Student of Mechanical Engineering, Kansai University
3-3-35 Yamatecho, Suita, Osaka 564, Japan

ABSTRACT

We observed deformations of several materials during cooling down process from room temperature to liquid nitrogen temperature using the laser speckle photography method. The in-plane displacements were measured by the image plane speckle photography and the out-of-plane displacement gradients by the defocused speckle photography. The results of measurements of in-plane displacement are compared with those of FEM analysis. The applicability of laser speckle photography method to cryogenic engineering will also be discussed.

INTRODUCTION

When a device is exposed to the cooling process from room temperature to the cryogenic temperature region, its structural materials and supporting elements will be deformed and subjected to intense stress by thermal contraction. In order to design cryogenic devices, it is necessary to understand their deformation processes at low temperatures.

From this view point, several techniques have been used in order to observe the deformation processes. Conventionally, the deformation of materials has been measured by applying resistance strain gauges. In the cryogenic regions, however, strain gauges are not convenient because the gauge factors are not kept constant for the wide range of temperatures.

Optical methods for deformation measurement, on the other hand, are usually independent of temperatures, and also able to make the measurement non-contact. Holographic and speckle interferometries have been proven to be important tools of the deformation measurements[1]. They are applicable to the objects with optical rough surfaces at sensitivities of the order of wavelength of light. These techniques require the measurement environment to be interferometrically stabile.

---

* Y. Maeda is now with Hitachi Zosen Company Ltd.,
  and * K. Matsumura with Matsushita Electric Works Ltd.

The laser speckle photography (LSP) method[1-10] is another optical method of deformation measurements. The LSP method is less sensitive than the interferometric methods, and hence does not impose severe stability requirements on measuring system. The simplicity of the technique of the speckle photography make it a powerful method of measurement in the field of engineering[8-10]. Speckle photography consists of a two-step process of recording and reading-out processes. In the recording process, doubly exposed photograph is made on a single frame of photographic film for the laser illuminated object before and after deformation. The reading-out is made point-by-point illumination of normal incidence of narrow laser beam on the record, giving parallel equi-spaced fringe pattern (Young's fringe).

Archbold[2], Butters[3] and Ennos[2,4] showed that image-plane speckle patterns is sensitive to in-plane displacement. The spacing of the fringes is proportional to the in-plane displacement of speckles at the observation point in the object image. The direction of the fringes is perpendicular to the direction of the displacement vector. Two-dimensional deformation distribution can thus be obtained by surveying two-dimensionally the double-exposed photographic record.

On the other hand, Tiziani[5], Gregory[6] and Archbold[2] showed that defocused speckle patterns are predominantly sensitive to the gradients of out-of-plane displacement. In this case, the displacement of speckles on a recording film is proportional to the tilt of the object surfaces.

In this paper, we present the measurement of the deformation by thermal contraction on several materials during the cooling down process from room temperature to liquid nitrogen temperature through both LSP and defocused LSP method.

EXPERIMENTAL

In this study, some of the commercially available materials, aluminum (JIS:A1050P), stainless steel (JIS:SUS304 and SUS310S) and high-purity alumina ceramics(99%) plates were examined. Disk shaped specimen is :150 mm

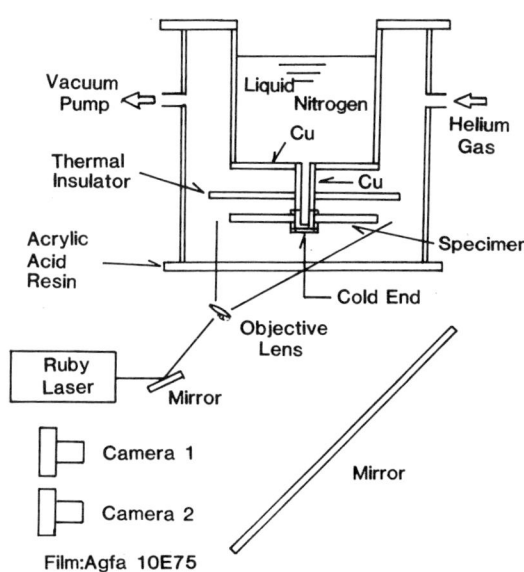

Fig. 1. Optical system and cryostat for LSP method.

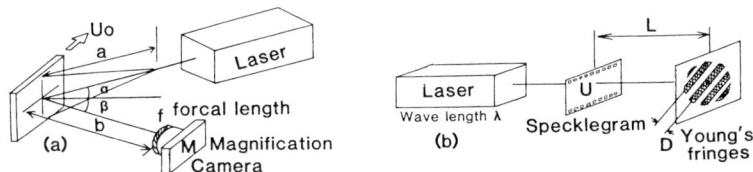

Fig. 2. Optical arrangement for (a) LSP recording and (b) reading-out.

in diameter and 2 mm in thickness. The surfaces of metal specimen were sandpapered and that of alumina ceramics was vapor-deposited with aluminum.

The experimental setup is illustrated in Fig. 1 ; the optical system for the LSP method and the cross sectional view of the cryostat. The cryostat has an observation window (270 mm in diameter) of an acrylic acid resin (PMMA) of 15 mm thickness. The specimen was fixed to the end of a cooling device by a screw. By pouring liquid nitrogen into the inner vessel of the cryostat, the temperature of center of the sample is lowered to about 77K, while the outer portion of the specimen is cooled down by the thermal conduction. In order to prevent condensation of water vapor to the sample surface , the air in the vessel was substituted by helium gas, and then, was evacuated to a fair vacuum. The temperature distribution of the specimen was measured at the points, 25, 40, 50 and 70 mm from the center by using the chromel-alumel thermocouples (K-type) of diameter of 0.05 mm.

The speckle photographs were recorded through a f/2.8 Nikkor lens of 135 mm focal length on the 35 mm photographic film(Agfa Gevaert, 10E75 film, 2800 lines/mm) or a f/5.6 Nikkor lens of 135 mm on the 70 mm film (10E75 film). The specimen to camera separation was adjusted so that the image of the object substantially filled the whole area of photographic film, giving a magnification factor M of about 0.16 on 35 mm film and 0.4 on a 70 mm film, for the specimen to camera lens distance 125 cm and specimen to laser source position 75 cm.

The sample was illuminated through a reflection mirror by a pulsed ruby laser (pulse width 1 msec). A microscope objective lens was used to expand the laser beam over the area of the specimen in the cryostat. Each photographic film was given two exposures, the first one serves as the reference of the deformation, and the second records speckle-pattern-movement due to the deformation of the object. By using two cameras, we have a series of double-exposed speckle photographs(specklegram), each of which records the deformation information during two exposures.

The optical arrangement for LSP is shown in Fig. 2. The displacement caused by the thermal contraction is derived from the Young's fringes (see Fig.2(b)) that are generated by the optical Fourier transformation of the specklegrams. In the case of the LSP method for in-plane displacement, the fringes are perpendicular to the direction of the displacement and have a spacing D inversely proportional to magnitude U of the displacement of speckle pattern on the film surface, which is given by

$$U = \lambda L/D,$$

and the displacement of the object U_o is given by

$$U_o = U/M,$$

where λ is wavelength of the light, L the distance between the speckle photograph and the viewing plane, M is the magnification of the imaging system.

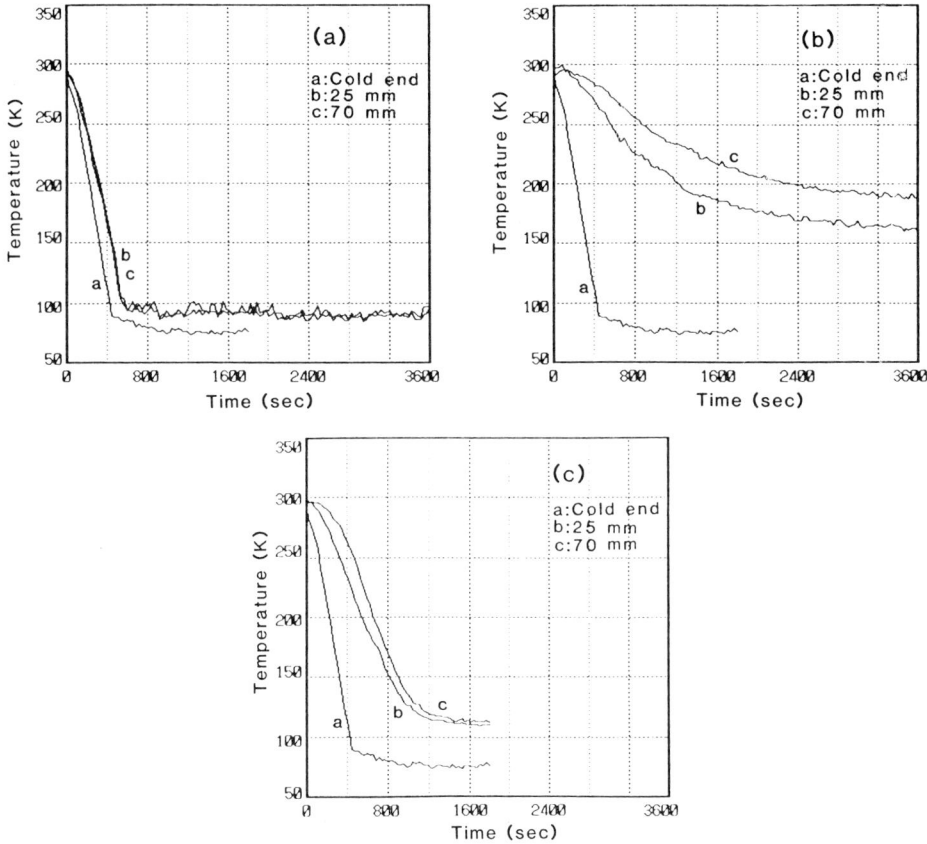

Fig. 3. Temperature history at the cold end, the points 25 and 70 mm from the center for (a) aluminum, (b) stainless steel and (c) alumina.

In the case of the defocused LSP method, the tilt of the specimen is derived from the Young's fringes. The magnitudes of the deflection of the specimen were derived from the tilt of the each points. The tilt angle $\delta\phi$ of the surface of specimen is obtained from the speckle movement U at the defocused plane by the relation

$$\delta\phi = U(a+b-f)/af(1+\cos \alpha /\cos \beta),$$

where α is the beam incidence angle, β the beam scatter angle, a the distance of the light source from the object surface, b the distance of the object surface from the lens and f the focal length of the lens.[6]

RESULTS AND DISCUSSION

Figures 3-a) and -b) show the temperature change of the specimen during the cooling down process at the three points of the specimens. It takes about 10 minutes for the aluminum sample, 70 minutes for the stainless steel, and 25 minutes for alumina ceramics for the temperatures to reach the steady state. The temperature differences between the front and the back surfaces were less than a few degrees for aluminum and stainless steel. It is found that, in the case of stainless steel, the temperature gradient near the center of specimen is larger than that at the outer region, and in the case of aluminum, the temperature gradient was not observed except near by center as shown in Fig. 4.

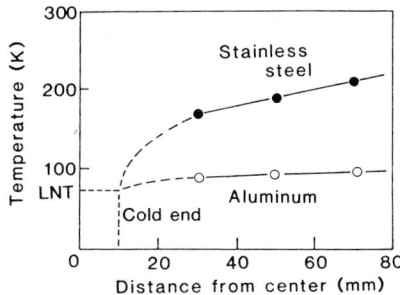

Fig. 4. Temperature distribution on the aluminum and stainless steel disk in steady state.

Experimental results of two-dimensional thermal contraction are shown in Figs. 5-a) to -d), where the magnitude and direction of the displacement are represented by arrows. We assume that Young's fringe on the center region of the sample comes from an uniform displacement of the optical system. The net displacement, hence, induced by the thermal contraction will be obtained by subtracting the uniform displacement from each experimental values. As a result of the processing, the direction of the contraction points almost toward the center of specimen. In Fig. 5-d) the two-dimensional thermal contraction of the alumina ceramics specimen are shown. The distribution is complicated in comparison with metal specimens.

Figures 6-a) and -b) show the radial distribution of the magnitude of displacement due to thermal contraction. The solid lines are calculated

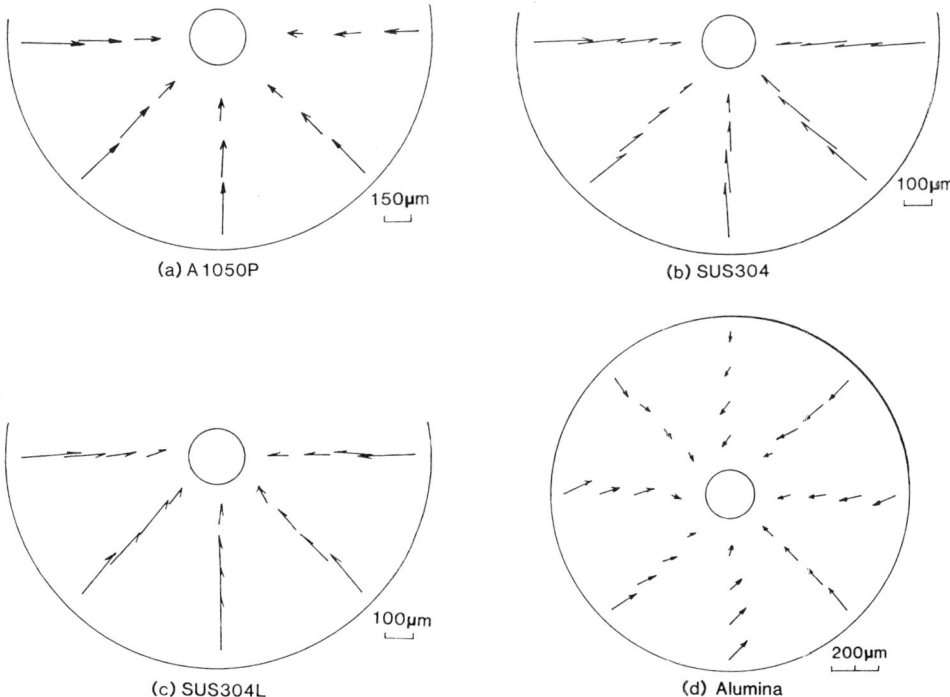

Fig. 5. Typical examples of the two-dimensional thermal contraction on the (a) aluminum, (b) stainless steel, SUS 304, (c) SUS 304L, and (d) alumina.

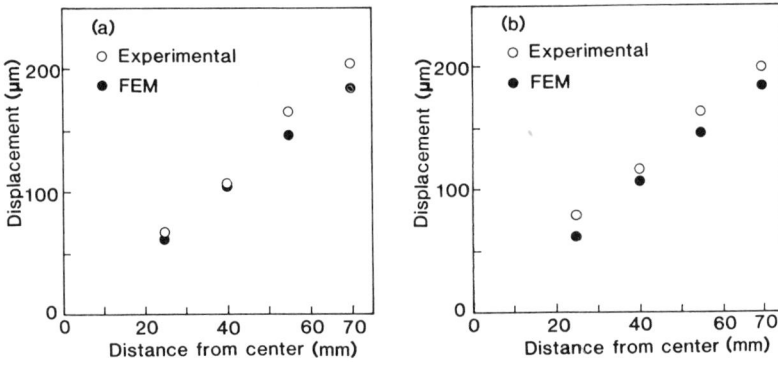

Fig. 6. Radial distribution of In-plane displacement due to thermal contraction and together with calculated value of the thermal contraction by FEM.

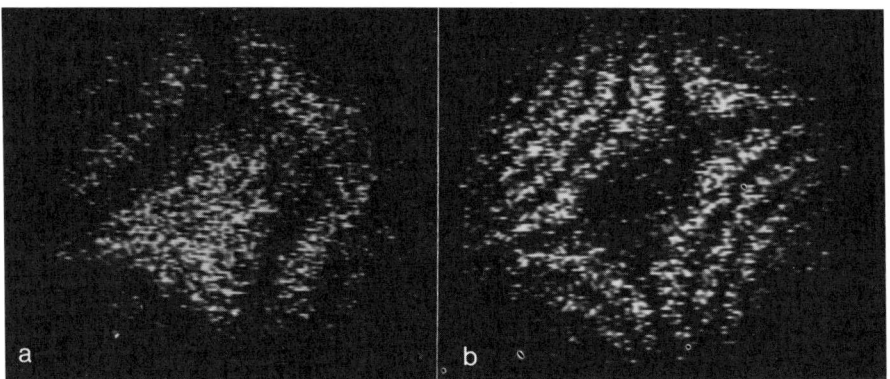

Fig. 7. The out-of-plane displacement distribution in the cooling process, (a) 600 sec and (b) 609 sec after cooling as observed using ESPI.

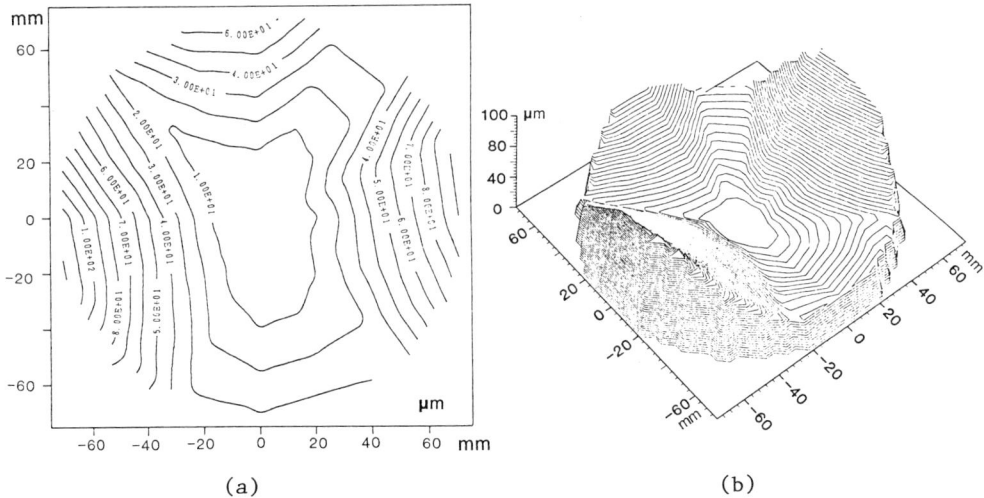

Fig. 8. The (a) contour map and (b) the bird's-eye view of the out-of-plane displacement of stainless steel.

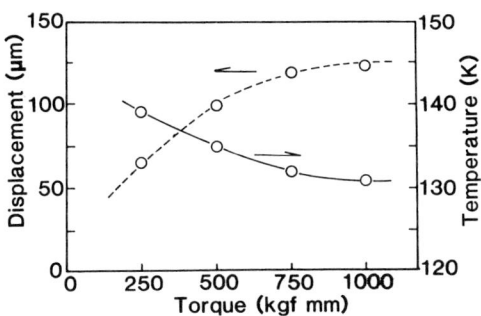

Fig. 9. The maximum out-of-plane displacement and temperature of the steady state on the different torque of a clamp screw.

values of the thermal contraction by FEM. In the case of aluminum and stainless steel, good agreements are observed between the calculated and experimental values.

In the case of out-of-plane deformation, the speckle photography method suffers an intrinsic sign ambiguity in the assignment of the displacement vector field. Therefore, electronic speckle pattern interferometry (ESPI) is used for examining the real time behavior of the deformation[1,4,7]. We can determine the direction of out-of-plane displacement by the fringe-movement of ESPI as shown in Fig. 7.

The magnitude of out-of-plane deformation are shown in Fig. 8-a) for the stainless steel specimen fixed to the cold end with a torque of a clamp screw. The contour map of the out-of-plane displacement are represented by solid lines. The net tilt was obtained in the same way as in the case of in-plane displacements to remove the uniform tilt of the optical system.

Figure 8-b) shows the bird's-eye view of the magnitude of out-of-plane deflection. The temperature of the steady state depends on the magnitude of the torque as shown in Fig. 9. It is, therefore, assumed that the deflection depends on the magnitude of a thermal stress due to the temperature difference before and after cooling.

SUMMARY

The two-dimensional in-plane displacement and out-of-plane deflection due to thermal contraction were measured respectively by using LSP and defocused LSP methods at the cryogenic environments on aluminum, stainless steel and alumina ceramics specimens. The present investigation can be summarized as follows:

1) The magnitude of the in-plane displacement measured by using LSP method were in agreement with the calculation by FEM. In the case of the alumina ceramics, it is found that the distribution of the in-plane displacement is complicated in comparison with metal specimen.

2) All the specimens appear to be deflected by thermal contraction from the results of the defocused LSP. It is found that the degree of deflection is caused by the differences of the temperature distribution.

3) By using LSP method, the behavior of thermal contraction of materials were successfully measured, and the applicability of LSP method to the cryogenic engineering are experimentally confirmed. This method offers a noncontact and temperature-independent method to measure the displace-

ment of materials subjected to large temperature changes, provided that there is no loss of correlation among the speckle patterns.

ACKNOWLEDGMENTS

This work is partly supported by the Science Research Promotion Fund from the Japan Private School Promotion Foundation and also by Grant-in-Aid for Scientific Research on Priority Areas from the Ministry of Education, Science and Culture, No.01647007. One of the authors,(S.N.), is supported by the Kansai University Research Grant.

REFERENCES

1. R. Jones and C. Wykes, "Holographic and Speckle Interferometry", Cambridge University Press, Cambridge (1983).
2. E. Archbold and A. E. Ennos, Displacement Measurement from Double Exposure Laser Photographs, Optica Acta, 19: 253-271 (1972).
3. J. N. Butters and J. A. Leendertz, Journal of Physics E: Scientific Instruments, 4: 277-279 (1971).
4. A. E. Ennos, Speckle Interferometry, in: "Topics in Applied Physics Vol.9 : Laser Speckle and Related Phenomena", J. C. Dainty, ed., 2nd Edition, Springer-Verlag, Berlin (1984) 203-253.
5. H. J. Tiziani, A Study of the Use of Laser Speckle to Measure Small Tilts of Optically Rough Surfaces Accurately, Optics Communications, 5: 271-276 (1972).
6. D. A. Gregory, Basic Physical Principles of Defocused Speckle Photography : A Tilt Topology Inspection Technique, Optics and Laser Technology, Oct.: 201-213 (1976).
7. O. J. Løkberg and G. A. Slettemoen, Basic Electronic Speckle Pattern Interferometry, in: "Applied Optics and Optical Engineering Vol.X", R. R. Shannon and J. C. Wyant, eds., Academic Press, San Diego (1987) 455-504.
8. F. P. Chiang, R. Anastasi, J. Beatty and J. Adachi, Thermal Strain Measurement by One-Beam Laser Speckle Interferometry, Applied Optics, 19: 2701-2704 (1980).
9. E. A. Fuchs and R. E. Rowlands, Photomechanical stress analysis under cryogenic environments, in: "Advances in Cryogenic Engineering--Materials", Vol.30, R. P. Reed and A. F. Clark, eds., Plenum Press, New York (1984) 111-117.
10. S. Nakahara, T. Fujita, K. Sugihara, S. Nishijima, M. Takeno and T. Okada, Two-Dimensional Thermal Contraction of Composites, in: "Advances in Cryogenic Engineering - Materials", Vol.32, R. P. Reed and A. F. Clark, eds., Plenum Press, New York (1986) 209-215.

HYDROGEN EFFECT ON THE MECHANICAL PROPERTIES OF INCOLOY 907 FROM AMBIENT TO CRYOGENIC TEMPERATURE

K. Yang, X. Zhao, Y. Xie, and Y.Y. Li

Institute of Metal Research, Academia Sinica
Shenyang, China

ABSTRACT

The investigation of hydrogen effect on the mechanical properties of Incoloy 907, a Fe-Ni-Co superalloy, was made from ambient to cryogenic temperature (77 K) with and without hydrogen charging under different aging treatments. It was found that there exists an obvious IHE tendency in Inco 907 from ambient to cryogenic temperature, which is affected by the microstructure of the alloy. From ambient to LN temperature an obvious maximum of IHE susceptibility in the alloy was found at the temperature of $-50°C$ for all the aging treatments. The occurance of the hydrogen embrittlement in Inco 907 and its temperature dependence can be well explained by the interaction between the hydrogen and the dislocation in the alloy

INTRODUCTION

The 900-series superalloys represent a relative new class of aerospace materials designed to provide not only enough high strength but also a low coefficient of thermal expansion[1,2]. This unique property combnation has promoted their increased use in gas turbine engine applications over a wide range of temperatures. It was reported[3] that the 900-series, low-expansion superalloys are also highly resistant to high-pressure hydrogen embrittlement, which has led their use in the space shuttle main engine.

The 900-series alloys are iron-based and contain appreciable amounts of Ni and Co. Their microstructures are fully austenitic and strengthened by the addition of Al, Ti and Nb to form hardening precipitates, usually the γ' phase, in the matrice. However, in recent years, some work[4-7] has clearly shown that serious internal hydrogen embrittlement (IHE) exists in precipitated hardening austenitic superalloys, including Inco 903[4,5], which has attracted attention of researchers. Since liquid hydrogen is widely used for the fuel in the space shuttle main engine, the use of the 900-series alloys may get in touch with the hydrogen environment from the cryogenic temperature to the higher temperature. Therefore, further study of hydrogen embrittlement, especially IHE, in 900-series alloys and its temperature dependence should be of both academical and practical importances.

Incoloy 907 alloy[8] is one of the 900-series superalloys, which was developed based on Incoloy 903 and has a chemical composition somewhat different from Inco 903, mainly in strengthening elements Al and Nb. Although some research work has been reported for Inco 903, less work has been published for Inco 907. Similarly, research work on 900-series superalloys at cryogenic temperature[9] is also rarely found. In our previous work[10], we have studied the microstructures of Inco 907 under different heat treatments and their effects on tensile properties of the alloy. In this paper, we mainly discuss the effect of hydrogen on the tensile properties of Inco 907 from ambient to cryogenic temperature.

MATERIAL AND EXPERIMENT

The material used in this study was obtained by vacuum−induction melting with the composition of superalloy Incoloy 907 as listed in Table 1.

All the specimens of Incoloy 907 were solution treated at 980℃ for one hour, then air cooled, and finally about 50 μm austenite grain sizes were obtained. Aging treatments on the specimens after solution treatment were carried out at 620℃ for 1, 4, 12 and 24 hours, respectively, to obtain the different microstructures, i.e., the different sizes of the γ' phase in the alloy. The tensile specimens with gage section of 5 mm and 25 mm in length were hydrogen charged under the condition of 10 MPa gaseous hydrogen with 99.99% purity at 300℃ for 14 days and hydrogen contents in specimens were analysed to be 16.5 ppm wt.% on average. Tensile tests from room temperature to liquid nitrogen(LN) temperature(−196℃) with a strain rate of $6.7 \times 10^{-4} s^{-1}$ were performed for specimens with and without hydrogen charging, and the reduction of area (RA) loss was used to determine the susceptibility to hydrogen embrittlement. The fracture surfaces after tensile testing were studied by SEM to further understand the effect of microstructure as well as temperature on hydrogen embrittlement in Incoloy 907.

RESULTS AND DISCUSSION

Microstructures of Inco 907

In our previous work, microstructures of Inco 907 under different aging treatments were investigated systematically[10]. For 620℃ aging treatment, only a large amount of the γ' phases with a dispersed distribution, which are considered to be the main strengthening phases in Inco 907, was found in the alloy, and there was no evidence of γ'' phase formation in the alloy, though Inco 907 has a similar composition of strengthening elements as that of Inco 718. As the aging time increases, the size of γ' phases increases and its appearance changes gradually from global to cubic to lower the strain energy of γ' phase−matrix boundaries. The measurement of the γ' phases in the alloy under such aging treatments showed that the sizes are in the order of one to ten nm. The X−ray diffraction results indicated that the lattice parameter of the matrix in the alloy aging treated does not change with the increase of aging time, which implies that the growth of the γ' phase in the alloy shoud be the dominant process once the precipitation of the γ' phase occurs.

From the observation it was also found that for the aging up to 100 hours at 620℃, there is still no sign of the ε phase precipitation, which usually has a needle−like appearance and is considered as the product of overaging[3,9]. Therefore, the variation of microstructures of the alloy in this study should correspond to the variation of the sizes of the γ' phases in the alloy.

Tensile Properties and IHE Tendency of Inco 907 from Ambient to Cryogenic Temperature

Figure 1 shows the variation of tensile properties of Inco 907 alloy under different aging treatments mentioned above in the temperature range of room temperature and LN temperature. From Figure 1 it is seen that as the aging time increases for the aging treatment at 620℃ from 1 hour to 24 hours, the strength of the alloy, i.e., both the yield strength(σ_{ys}) and the ultimate strength(σ_{uts}), increases, but the ductility, i.e., the

Table 1. The Composition of the Material Used in This Study

Ele.	Ni	Co	Nb	Ti	Al	Mn	Si	C	S	P	B	Fe
wt.%	37.4	12.9	4.77	1.47	0.08	0.36	0.21	0.02	.004	.002	.002	bal.

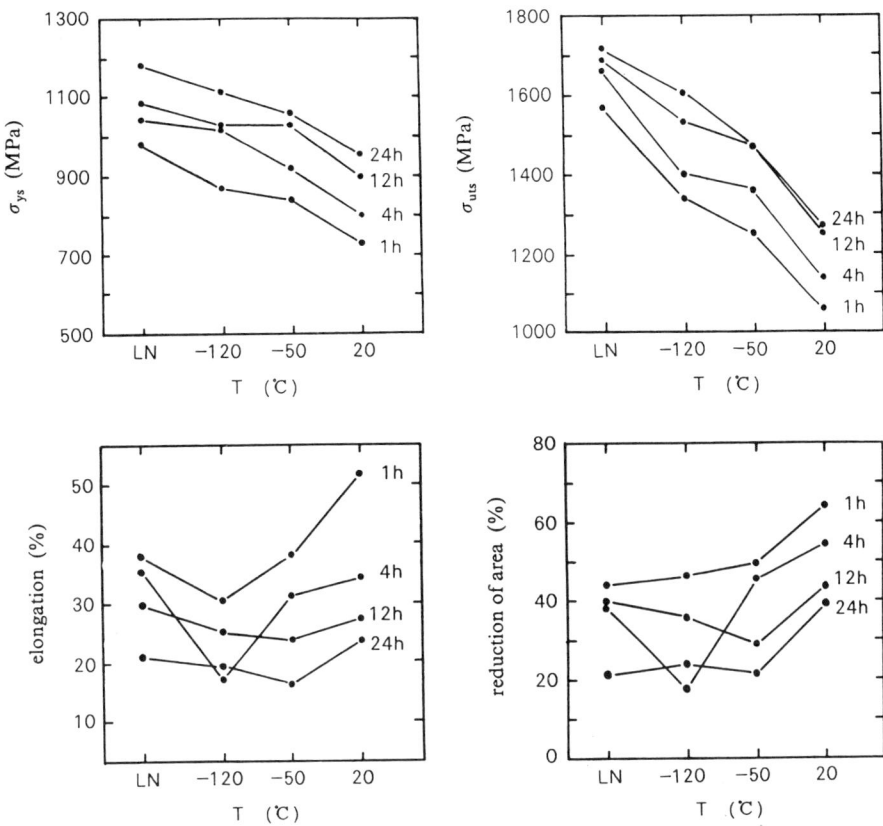

Figure 1. Tensile Properties of Inco 907 from Room to LN Temperature

elongation and reduction of aera, decreases in all the studying temperature range. This result is related to the fact that the size as well as the amount of the γ' phases in the alloy increases continuously with the increase of the aging time at this aging temperature[10], which gradually enhances the strengthening effect of the γ' phase and makes the strength increase and the ductility decrease with the aging time.

As to the effect of temperature, Figure 1 shows that as the temperature lowers, the strength of the alloy for all the aging treatments increases continuously, but the decrease of the ductility is not monotonous and even a rising tendency occurs at cryogenic temperature. This unusual phenomenon of the ductility variation with temperature may be due to the reason that the homogeneous deformation in the alloy may become more easy at cryogenic temperature.

Figure 2 shows some representative photographs of the fracture surface of Inco 907 from ambient to cryogenic temperature. Typical dimpled surfaces of microvoid coalescence are seen for the 1 hour aging treatment, which corresponds to a microstructure of the fine γ' phases distributed in the matrix, in the experimental temperature range(see Figure 2 (a)–(d)). Since the larger size of the γ' phase has stronger strengthening effect, the alloy should gradually become brittle corresponging to the increase of the strength as the aging time increases. Figure 2(e) shows that unlike the fracture surface in Figure 2(b), an obviously large part of brittle fracture, intergranular fracture and intragranular cleavage, exists on the fracture surface for the 12 hours aging. But from Figure 2(f) it can be seen that the brittle fracture tendency clearly becomes lower as the temperature decreases, which is well consistent with the result shown in Figure 1 that the ductility has a rising tendency at cryogenic temeprature. Similar variations on the fracture surfaces were also found for the other aging treatments.

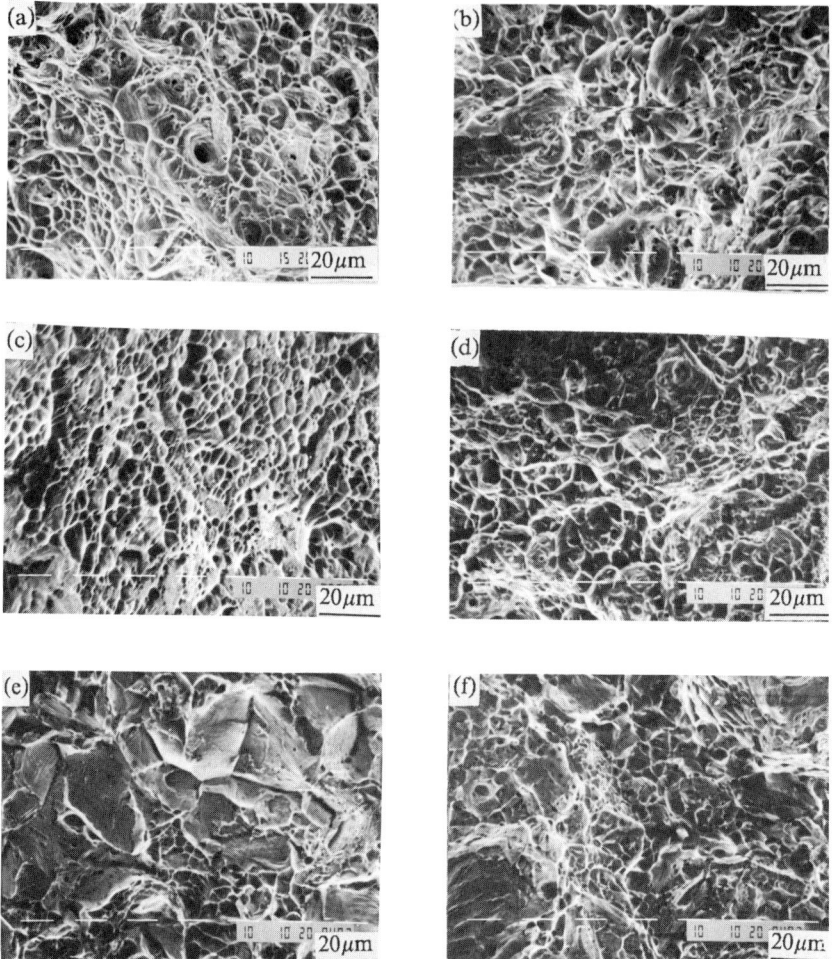

Figure 2. SEM Photographs of Tensile fracture Surface of Inco 907
(a) 1 hour aging, room temperature; (b) 1 hour aging, −50℃;
(c) 1 hour aging, −120℃; (d) 1 hour aging, LN;
(e) 12 hours aging, −50℃; (f) 12 hours aging, −120℃;

The RA loss caused by hydrogen charging, which reflects the susceptibility to IHE, of Inco 907 under different aging treatments in the experimental temperature range is shown in Figure 3. In Figure 3 it is shown that in the uper range of the experimental temperature, serious IHE occurs in Inco 907 alloy and the longer the aging time, the more serious the IHE in the alloy. A maximum IHE tendency is found at the temperature of −50℃ for all the aging treatments, and the IHE tendency decreases sharply at cryogenic temperature. The above results indicate that the susceptibility of Inco 907 to IHE is not only affected by the microstructure of the alloy, but also dependent on the temperature.

The representative photographs of fracture surface of Inco 907 hydrogen charged are shown in Figure 4. It clearly shows that the fracture surface of the alloy hydrogen charged obviously becomes brittle compared to those in Figure 2, which indicates that IHE does occur in Inco 907 alloy. For the aging of 1 hour, there is clear intragranular cleavage or quasi–cleavage fractures on the fracture surfaces(see Figure 4 (a)–(d)), and the fracture surface of the −50℃ aging that corresponds to the maximum of IHE tendency in Figure 3 shows the largest part of brittle fracture. An almost complete intergranular fracture, which is clearly more brittle thanthe formers, is found on the

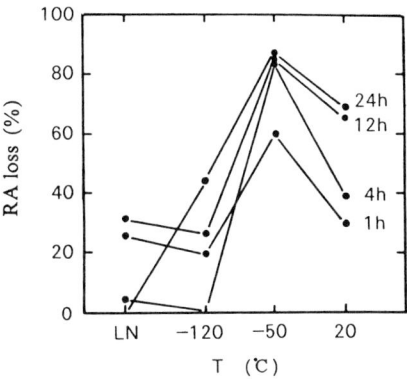

Figure 3. Reduction of Area Loss in Inco 907 for Hydrogen Charging.

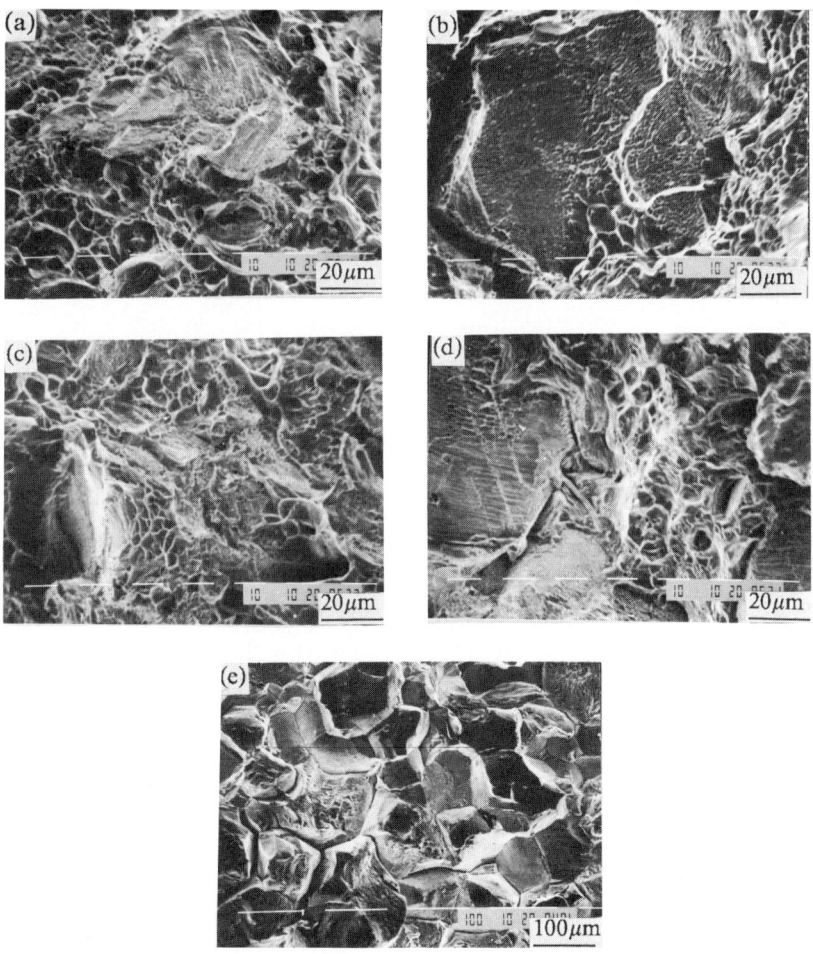

Figure 4. SEM Photographs of Tensile Fracture Surface of
Inco 907 Hydrogen Charged
(a) 1 hour aging, room temperature; (b) 1 hour aging, −50℃;
(c) 1 hour aging, −120℃;   (d) 1 hour aging, LN;
(e) 12 hours aging, −50℃;

fracture surface of the −50℃ aging for 12 hours(figure 4(e)), which indicates that the effect of microstructure on the susceptibility of the alloy to IHE.

Like the other 900−series superalloys, In907 alloy has been reported[8] to have a good resistance to embrittlement in the environment of high pressure hydrogen. But in the present work the results show that an obvious IHE tendency exists in In907 and is affected by both the microstructure and the testing temperature. This difference may be caused by the low hydrogen diffusivity in the alloy that makes it difficult for hydrogen to permeate from outside into the central area of specimens in short times or at low temperatures, e.g., room temperature. It should be pointed out that a similar situation was also found for In903 [4,5] in which obvious IHE occurs.

For the phenomenon of IHE in Inco 907 we considered that the hydrogen−dislocation interaction in the alloy should play the most important role, because the low hydrogen diffusivity should make it difficult for hydrogen to diffuse and accumulate in the alloy. It was suggested by Tien et al[11] that for the alloy hydrogen charged, dislocations in the alloy can easily trap hydrogen atoms to form Cottrell atomspheres around them. The trapped hydrogen can move with dislocations under tensile deformation at the proper strain rate, usually a slow strain rate. When movements of dislocations carrying hydrogen are obstructed by some γ' particles and piled up, accumulations of hydrogen may occur in local areas. The hydrogen accumulated can greatly assist the brittle fracture to occur in these local areas, which indicates that hydrogen embrittlement can happen in the alloy that is hydrogen charged. It can be considered that the larger the γ' particles, the larger the possibility of obstruction to dislocations and then the more the oppotunity of hydrogen accumulation, the more severe the hydrogen embrittlement in the alloy, which implies that the microstructure should have strong effect on hydrogen embrittlement in In907.

From the point of the mobility of the hydrogen atoms in the alloy, we considered that the hydrogen atoms should be more mobile at higher temperature, but decrease the mobility at lower temperature. The higher mobility of hydrogen decreases the hydrogen trapping ability by dislocation and thus lowers the number of hydrogen the dislocation can carry in movement. The lower mobility of hydrogen decreases the critical strain rate below which dislocation can carry the hydrogen to move and thus lowers the number of the dislocation carrying hydrogen at a given strain rate. From the above analysis it can be concluded that the hydrogen embrittlement tendency should decrease both at higher temperature and lower temperature and there should exist a maximum susceptibility to hydrogen embrittlement at a certain temperature for a given strain rate of deformation. This can well explain the present experimental result that the IHE in Inco 907 is temperature dependent and reaches the maximum at temperature of −50℃.

CONCLUSION

The microstructure of Inco 907 superalloy aged at 620℃ for differnt times corresponds to the fine γ' phases with dispersed distribution in the austenitic matrix.

From ambient to cryogenic temperature there exists an obvious IHE tendency in Inco 907 under the above aging treatment, which is affected by the microstructure of the alloy. The longer the aging time, the more serious the IHE in the alloy. A maximum of IHE susceptibility in the alloy was found at the temperature of −50℃ for all the aging treatments.

The obstruction of the γ' phases in the alloy to the movement of dislocations carrying hydrogen, which can results in the accumulation of hydrogen in local areas, plays the most important role in the process of IHE in Inco 907. The variation of hydrogen mobility with temperature results in a maximum of IHE in the alloy to occur in a given temperature range.

ACKNOWLEDGEMENT

The authors gratefully acknowledge the support by Chinese National Natural Sciences Foundation of No. 5901146.

REFERENCES

1. H.W. Carpenter, Alloy 903 Helps Space Shuttle Fly, Metals Progress, (8):25(1976).
2. D.F. Smith and E.F. Clatworthy, The Development of High–Strength, Low–Expasion Alloys, Metals Progress, (3):32(1981).
3. S.C. Ernst, W.A. Baeslack III and J.C. Lippold, "Weldability of High–Srength, Low–Expasion Superalloys, Welding Journal, Oct:418(1989).
4. C.G. Rhodes and A.W. Thompson, Microstructure and Hydrogen Performance of Alloy 903, Metall. Trans., 8A:949(1977).
5. N.R. Moody, R.E. Stoltz and M. W. Perra, The Effect of Hydrogen on Fracture Toughness of the Fe–Ni–Co Superalloy In903, Metall. Trans., 18A:1469(1987).
6. A.W. Thompson and J.A. Brooks, Hydrogen Performance of Precipitation–Strengthened Stainless Steels Based on A–286, Metall. Trans, 6A:1431(1975).
7. P.D. Hicks and C.J. Altstetter, Internal Hydrogen Effects on Tensile Properties of Iron– and Nickel–Base Superalloys, Metall. Trans., 21A:365(1990).
8. Incoloy Alloy 907, Alloy Digest, Feb:(1983).
9. L.T. Summer and E.N.C. Dalder, An Investigation of the Cryogenic Mechanical Properties of Low Thermal–Expansion Superalloys, Advances in Cryo. Eng. (Materials), 32:73(1985).
10. Y. Xie, K. Yang, X. Zhao, C.G. Fan and Y.Y. Li, Microstructures and Their effects on Mechanical Properties of Incoloy 907, accepted by Acta Metall. Sinica.
11. J.K. Tien. A.W. Thompson, I.M. Bernstein and R.J. Richards, Hydrogen Transport by Dislocations, Metall. Trans, 7A:821(1976).

THICK-SECTION WELDMENTS IN 21-6-9 AND 316LN

STAINLESS STEEL FOR FUSION ENERGY APPLICATIONS*

D. J. Alexander and G. M. Goodwin

Metals and Ceramics Division
Oak Ridge National Laboratory
Oak Ridge, TN 37831-6151

INTRODUCTION

The Burning Plasma Experiment (BPX), formerly known as the Compact Ignition Tokomak, will be a major advance in the design of a fusion reactor. The successful construction of fusion reactors will require extensive welding of thick-section stainless steel plates. Severe service conditions will be experienced by the structure. Operating temperatures will range from room temperature (300 K) to liquid nitrogen temperature (77 K), and perhaps even lower. The structure will be highly stressed, and subject to sudden impact loads if plasma disruptions occur. This demands a combination of high strength and high toughness from the weldments. Significant portions of the welding will be done in the field, so preweld and postweld heat treatments will be difficult. The thick sections to be welded will require a high deposition rate process, and will result in significant residual stresses in the materials. Inspection of these thick sections in complex geometries will be very difficult. All of these constraints make it essential that the welding procedures and alloys be well understood, and the mechanical properties of the welds and their heat-affected zones must be adequately characterized.

The candidate alloy for structural applications in the BPX such as the magnet cases was initially selected as 21-6-9 austenitic stainless steel, and later changed to 316LN stainless steel. This study examined several possible filler materials for thick-section (25 to 50 mm) weldments in these two materials. The tensile and Charpy V-notch properties were measured at room temperature and 77 K. The fracture toughness was measured for promising materials.

WELDMENT MATERIALS AND PREPARATION

Type 21-6-9 stainless steel [referring to its nominal composition of 21Cr-6Ni-9Mn (wt %)], also known as Nitronic 40, is one of a family of nitrogen-strengthened high-manganese austenitic alloys possessing high yield

---

*Research sponsored by the Office of Fusion Energy, U.S. Department of Energy, under contract DE-AC05-84OR21400 with Martin Marietta Energy Systems, Inc.

strengths and usually adequate base metal toughness levels. Modified composition filler metal is suggested for thick-section weldments. The Nitronic "type W" filler metals have reduced nitrogen, and a balanced chromium/nickel ratio, to assure ferrite in the weld deposit to avoid hot cracking. One might anticipate that ferrite in the weld metal would adversely affect cryogenic toughness.

Four different type 21-6-9 base metals and seven weld filler metals were included in the program. Six were added as cold filler wire using the argon shielded gas tungsten arc welding process, and ENiCrFe-3 (Inconel* 182) is a coated electrode for use with the shielded metal arc process. ERNiCrMo-3 (Inconel 625) is the Armco-recommended filler metal to use if 35W or 40W are not appropriate for some reason. ERNiCr-3 (Inconel 82) is a universal filler metal, widely used to join a variety of nickel-based alloys and numerous dissimilar metal combinations, including austenitics to ferritics. Type 21-6-9 filler is essentially a matching composition weld metal for 21-6-9 base plate. Inconel 625 PLUS is a modified Inconel 625 composition recently introduced by Carpenter Technology Corporation.

Seven welds were made for Phase I of this project. All of the base materials were prepared with a double-groove butt-weld geometry. The 25-mm-thick type 21-6-9 plate used a double-V joint design with a 45° included angle, a 1.5-mm root face, and 3-mm root opening. All other base materials used a double-U joint design with 15° included angle, 6-mm radius, 1.5-mm root face, and 3-mm root opening.

The gas tungsten arc welds were made at 10 to 14 V DCEN and 125 to 200 A with pure argon shielding gas using a stringer bead technique. The 25-mm plates required 30 to 40 passes and the 50-mm plates, 90 to 100 passes.

The Inconel 182 shielded metal arc weld in 25-mm plate was made with 3-mm electrodes at 100 A and 23 V DCEP. Eighteen passes completed the weld.

Type 316LN stainless steel does not offer strength levels as high as the 21-6-9 steel, but this steel has been widely used for cryogenic structural applications. The 316LN base plate used in this study was available from a single pedigreed heat in both 25- and 50-mm thicknesses. Four welds were produced. A submerged-arc weld made with type 316L filler was included for comparison purposes, with the presumption that its toughness properties would be low, due to the significant ferrite volume fraction. The remaining welds were produced using the flux-cored arc welding process with various filler metals. With 1.5-mm-diam wire and a current of 200 A, a double-U joint in 50-mm plate required approximately 30 passes, compared to approximately 100 passes for gas tungsten-arc.

Type 316L-T3 is a self-shielded electrode formulation designed to give ferrite contents of greater than 5 (FN). Type 316L-4K-0 is a product of Teledyne-McKay that is formulated to give intentionally low ferrite (0-2 FN) for use in cryogenic applications. Inconel 82-0 is a flux-cored arc version of the basic Inconel 82 (ERNiCr-3) composition newly introduced by Teledyne-McKay. The "O" designation indicates a self-shielded formulation.

Inspection of all welds was visual only since subsequent sectioning would reveal any possible defects.

---

*Inconel is a registered trademark of Huntington Alloys, Inc., West Virginia.

EXPERIMENTAL PROCEDURE

Test specimens were oriented transverse to and centered in the weld metal, and located near the top and bottom surfaces of the plates, to avoid the middle of the plate where the welds were thinnest and the dilution would be greatest. Two types of tensile specimens were tested. Oversize tensile specimens [gage length 38.1 mm by 5.1 mm in diameter (1.5 by 0.20 in.)] traversed the entire weld, and thus included base metal, heat-affected zone, and weld metal. Testing of these specimens provided a qualitative demonstration of the relative strengths of these different zones, and identified the weakest link in the compound structure. Miniature tensile specimens [gage length 10.2 mm by 2.5 mm in diameter (0.40 by 0.10 in.)] were machined transverse to the weld. These specimens were small enough so that the reduced diameter gage length was wholly contained in the weld metal.

Tensile testing of duplicate specimens was conducted on a screw-driven electromechanical test machine at a constant crosshead speed of 4.2×10^{-3} mm/s (0.01 in./min). This resulted in an initial strain rate of 4.2×10^{-4} s^{-1} for the small specimens. Tests at 77 K were performed with the specimens immersed in a bath of liquid nitrogen contained in a vacuum dewar. Load and crosshead displacement were recorded, and the 0.2% offset yield strength was derived from this record after allowance for the test system compliance. Uniform and total elongations were measured from the load-displacement trace.

Charpy specimens were tested at room temperature and at 77 K. The latter specimens were immersed in a bath of liquid nitrogen and then quickly transferred to the test machine with special tongs which centered the specimen. The TL specimens were notched so that the fracture would propagate in the direction of welding, while for the TS specimens the fracture would propagate through the weld thickness.

For those materials that seemed promising based on the tensile and Charpy results, 1/2T compact specimens were tested at room temperature and at 77 K to determine the fracture toughness. The specimens were oriented so that the crack growth was in the direction of welding (TL orientation). Unloading compliance was used to monitor the crack growth during the test. The specimens were fatigue precracked at room temperature, and then side-grooved 10% of the thickness on each side. Testing was conducted in general accordance with ASTM Standards E 813-89 and E 1152-87.

Sections from the welds were metallographically polished and etched to allow the different microstructures to be examined. Selected fracture surfaces were examined in a scanning electron microscope.

RESULTS

All of the welds appeared to be sound and defect-free upon visual inspection after welding. No evidence of hot cracking was observed.

The tensile data are presented in Table 1. At room temperature all of the filler metals had yield strengths which exceeded the base metal for both series of welds. However, the strength of the base metal increases rapidly as the temperature is decreased. Therefore, for the 21-6-9 series, the Inconel 625, 625 PLUS, 82, and 182 filler metals were significantly weaker than the base metal at 77 K, but the yield strength of the ferrite-containing Nitronic 35W, 40W, and the 21-6-9 filler metals exceeded that of the base metal. The Inconel 82-O and the 316L-type filler metals have

Table 1. Mechanical properties of filler and base metals

Filler metal	Welding process[a]	Temperature (K)	Strength (MPa) Yield	Strength (MPa) Tensile	Elongation (%) Uniform	Elongation (%) Total	Charpy energy (J)	Fracture toughness J_{Ic} (kJ/m²)	Fracture toughness K_J (MPa√m)
316LN Filler Metals									
316L	SA	300	421	600	25	33	8		
		77	667	1240	42	48	27		
316L-T3	FCOA	300	460	621	23	33	130		
		77	821	1248	42	49	20		
316L-4K-0	FCOA	300	571	702	8	16	117	245	209
		77	723	1187	30	31	46	133	157
Inconel 82-0	FCOA	300	413	610	27	37	155	370	277
		77	567	850	31	38	125	502	330
21-6-9 Filler Metals									
40W	GTA	300	552	724	18	26	168		
		77	1051	1358	17	17	9		
35W	GTA	300	593	762	25	39	166		
		77	979	1400	25	25	17		
21-6-9	GTA	300	579	793	26	38	202		
		77	1182	1569	27	27	28		
Inconel 625	GTA	300	514	814	27	29	36		
		77	738	1124	19	19	22		
Inconel 625 Plus	GTA	300	500	796	31	37	75		
		77	696	1110	31	32	55		
Inconel 82	GTA	300	486	714	26	33	169	714	384
		77	690	965	24	30	160	785	413
Inconel 182	SMA	300	403	631	26	34	129		
		77	527	900	31	40	111		
Base Metals									
316LN		300	280	610		57			
		77	725	1215		61			
21-6-9		300	345	690	50	65	>300		
		77	970	1510	35	40	100		

[a]SA = submerged arc, FCOA = flux-cored open arc, GTA = gas tungsten arc, SMA = shielded metal arc.

higher strengths than the 316LN base metal at room temperature. At 77 K, the 316L-type filler metals are similar in strength to the base metal, but the Inconel 82-O is much weaker.

Close examination of the oversize tensile specimens indicated that the HAZ was stronger than the base metal, at least at room temperature, as the diameter of the specimen in the HAZ area was greater than the base metal further from the weld. At 77 K the deformation was largely limited to the weld metal, and the heat-affected zone and base metal regions were unaltered. Fracture occurred in the weld metal for all of the specimens at either test temperature.

The impact properties are also shown in Table 1. The impact toughness of the base metals is very high at room temperature. Although the toughness of the base metals drops at 77 K, it is still high. All of the filler metals had good impact properties at room temperature except for Inconel 625. However, at 77 K the impact properties were very poor, except for the Inconel 82 and 182 alloys. These alloys had excellent impact properties at both temperatures, particularly Inconel 82 (both the gas tungsten arc and the flux-cored arc weldments). The 316L-4K-O had better impact properties at 77 K than the other ferrite-containing materials, but was still much worse than the Inconel 82 materials.

Only three series of fracture toughness tests were run: the Inconel 82 from the 21-6-9 weldments, and the Inconel 82-O and the 316L-4K-O from the 316LN series. The results of these tests are also in Table 1. The Inconel 82 toughnesses are very high, with the gas tungsten arc weld being much tougher than the flux-cored arc weld. The 316L-4K-O toughness is lower than the Inconel 82, but still quite high. It is intriguing to note that the toughness at 77 K exceeds the room temperature toughness for the Inconel weld materials.

The microstructures of the different weld metals were examined. The 35W, 40W, 21-6-9, 316L, and 316L-T3 filler metals had significant amounts of ferrite present, as expected. The ferrite content of the 316L-4K-O material was much lower. The Inconel-type filler metals did not contain any ferrite.

DISCUSSION

The testing conducted has shown that the filler metals with higher ferrite contents have high strengths, but suffer a severe decrease in impact properties at low temperature. The Inconel alloys are slightly weaker than the Nitronic alloys at room temperature, and much weaker at 77 K. The Inconel 82 and 182 alloys offer good impact properties, with the Inconel 82 alloy being both stronger and tougher at all temperatures. The Inconel 625 PLUS and particularly the Inconel 625 filler metal have poor impact properties regardless of temperature.

The microstructure of the ferrite-containing filler metals offers an explanation for the dramatic decrease in energy absorbed as the temperature is lowered. The welds contain about 5 to 10% ferrite phase, which is also reflected in their slight magnetism. It is believed that this ferrite phase fractures by a low energy cleavage process at low temperatures. The high volume fraction of ferrite permits the crack to move readily to nearby ferrite regions, and so the fracture process requires low energy. Any filler metal which results in a significant volume fraction of ferrite in the weld will probably show a similar low energy level for impact tests at 77 K.

The fractography of the ferrite-containing welds supports this conclusion. Fracture at room temperature occurs by a ductile microvoid coalescence process. This fracture process will be dominated by the austenitic matrix and the inclusions in the weld. However, specimens tested at 77 K display very different fracture features. The fracture surface consists of flat steps which are linked by narrow ridges of ductile tearing. It is believed that the crack preferentially follows the ferrite phase, and jumps from one island of ferrite to another. The flat regions are the result of cleavage fracture of the ferrite phase, whereas the tearing results from the crack joining these areas together by ductile tearing of the austenite matrix between the ferrite.

The Inconel alloys do not produce any ferrite phase in their welds. These alloys create a fully austenitic weldment, as indicated by their total lack of magnetism. The austenitic microstructure is not susceptible to cleavage fracture, and so the fracture process is a ductile one at either test temperature. Welding defects were noted in some of the 21-6-9 weldments, and are possibly related to the high nitrogen content of the base metal. Despite such defects, the impact energy of the Inconel 82 and 182 welds was quite high, whereas it was quite low for the Inconel 625 and 625 PLUS materials.

The absence of ferrite in the Inconel weld materials means that the fracture mode will be ductile microvoid coalescence at both room temperature and 77 K. The growth and joining of the microvoids will be very sensitive to the matrix flow properties. Greater amounts of energy will be required to deform the matrix as the temperature decreases and the flow stress rises. This explains the increase in the fracture toughness with the decrease in temperature observed for the Inconel 82 materials.

The Inconel 82 filler metal is clearly the best of the alloys examined in the 21-6-9 series of weldments. It offers excellent impact properties over the temperature range of interest, and reasonable strength. It exceeds the base metal strength at room temperature, but falls below the base metal at 77 K. It is not clear how severe a restriction this might place on the structural design.

The 316L and 316L-T3 filler metals have fairly high ferrite contents. As expected, they suffer a severe decrease in their impact properties at 77 K. The 316L-4K-0 material has a very low ferrite content. Metallographic examination showed that the ferrite was present in small apparently isolated islands within the austenitic matrix. The reduced size of the ferrite islands makes initiation of cleavage fracture more difficult. The separation of the ferrite islands makes propagation of the crack more difficult also. As a result, at 77 K the impact properties of the 316L-4K-0 material are better than any of the other ferrite-containing materials, although the energies are still much lower than the Inconel 82 material, whether gas tungsten arc or flux-cored arc. However, for applications for which higher strength in the weld material is necessary at cryogenic temperatures, the 316L-4K-0 provides an alternative to the Inconel 82 material. The impact properties will be reduced, but will be better than any of the other ferrite-containing materials.

CONCLUSIONS

Seven different filler metals have been used to produce thick-section welds in 21-6-9 stainless steel plate. Tensile and Charpy impact tests were performed at room temperature and at 77 K. These tests indicate that the Nitronic-type filler metals, which contain a significant fraction of ferrite, have strengths which exceed the base metal, but have very low impact energies at 77 K. The Inconel-type filler metals, which do not contain ferrite, are somewhat weaker than the base metal. The Inconel 82 and 182 filler metals offer good impact properties at both temperatures, and are only slightly weaker than the base metal at 77 K. The Inconel 625 and 625 PLUS materials have poor impact properties. These tests indicate that the Inconel 82 filler metal is the prime candidate material for welding 21-6-9 stainless steel in thick sections.

Four different filler metals have been used to produce welds in 316LN base metal. The 316L-type filler metals offer strength similar to the base metal, but the Inconel 82-O material is weaker. The high-ferrite weld materials 316L and 316L-T3 suffer drastic decreases in the impact properties at 77 K. The low ferrite 316L-4K-O material also shows a drop in impact properties at 77 K, but the impact energies are higher than any of the other ferrite-containing materials. The Inconel 82-O material offers excellent impact properties but low strength. For applications that demand higher weld metal strength, the 316L-4K-O material offers an alternative, with impact properties that are better than any other ferrite-containing material, but still much lower than the Inconel 82-O.

ACKNOWLEDGMENTS

We would like to acknowledge the support of the Princeton Plasma Physics Laboratory for this work. The welds were fabricated at ORNL in the Materials Joining Group by J. D. McNabb, D. W. Swaney, and D. A. Frederick. The mechanical testing was performed by J. J. Henry, Jr., and R. L. Swain. We would like to thank R. K. Nanstad for numerous useful discussions, and M. L. Santella and D. E. McCabe for helpful reviews of the manuscript, which was prepared by J. L. Bishop.

JOINING OF AUSTENITIC STAINLESS STEELS FOR CRYOGENIC APPLICATIONS

T. A. Siewert and C. N. McCowan

Materials Reliability Division
National Institute of Standards and Technology
Boulder, CO 80303

ABSTRACT

The welds that are used to fabricate a structure from wrought stainless steel subcomponents usually have poorer mechanical properties than the wrought material at cryogenic temperatures. This means that the critical fracture path in these structures could be through the welds. For many applications the welds may never be stressed to critical levels, but for very aggressive structural designs it can be a real concern. For these aggressive designs, the structural designer could place the welds in less critical regions, however, such a design philosophy might be difficult to implement. It would be better to learn how to make welds with improved properties.

We have developed quantitative data for many of the factors that influence the strength and toughness of welds, to allow more intelligent choices of welding processes and compositions for demanding applications. This paper reviews these factors and discusses the interactions between them. For example, the cryogenic strength is influenced most by the composition, with the strength being increased strongly by N addition. The toughness is decreased by residual delta ferrite (FN) and inclusions, but can be increased by addition of Ni. Recently, a gas metal arc weld with 25 wt.% Ni has produced the best combination of strength and toughness ever measured at 4 K in our laboratory. Changes in the inclusion fraction are the primary cause of differences in mechanical properties between welds produced by the various welding processes. A secondary cause of differences is a nonuniform distribution of elements in the microstructure.

STRENGTH

The strength of a weld is very similar to that of a wrought stainless steel plate of the same composition. Below room temperature, we have found no exceptions to this relationship: it seems to be true for all the austenitic compositions. As the temperature is lowered, the strength increases. Also, the increase in strength due to N becomes much more apparent as the temperature decreases. Figure 1 shows the strength of type 316L welds as a function of temperature and N.[1] At 0.20 wt.% N, the weld strength increases by a factor of 2 when the temperature is decreased from 298 to 76 K, and a factor of 2.5 when the temperature is decreased to 4 K. At 4 K, increasing the N from 0.05 wt.% (typical of a unalloyed weld) to 0.20 wt.% increases the strength by a factor of 3. Therefore, N is a very effective strengthener for 76 K service, and is even more effective at 4 K.

---

* Contribution of NIST; not subject to copyright.

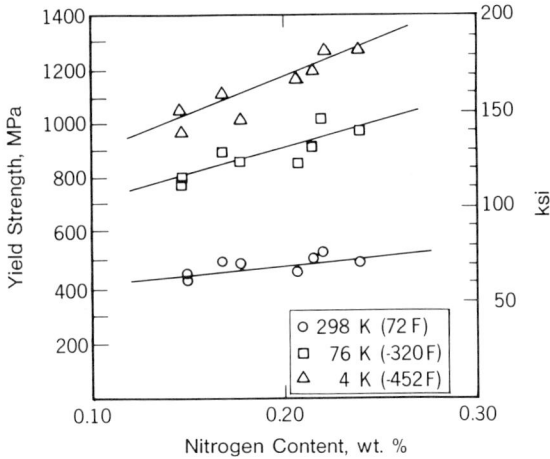

Figure 1. Yield strength versus N for type 316LN welds at 298, 76, and 4 K.

The effectiveness of N (and other elements) in changing the strengths of welds and wrought forms can be compared in the empirical equations for predicting the strength, developed from the statistical analysis of property data. Simon and Reed used regression analysis on the data from 99 tensile tests to develop the following equation for the strength of type 316 austenitic steel deposits at 4 K:[2]

$$\sigma_y \text{ (MPa)} = 316 + 2370 \text{ N} + 54 \text{ Mo} + 790 \text{ d}^{-1/2} \quad (1)$$

where N and Mo are the contents of these two elements in weight percent and d is the grain size in micrometers. The standard deviation of the data to this fit was 40 MPa, indicating that this representation is quite accurate. An equation for 18Cr-8Ni welds at 4 K has a similar form:[3]

$$\sigma_y \text{ (MPa)} = 180 + 3200 \text{ N} + 33 \text{ Mo} + 32 \text{ Mn} + 13 \text{ Ni} \quad (2)$$

and a standard deviation of 31 MPa. These two equations predict nearly equal strengths for compositions near 18Cr-8Ni indicating that the base material and weld strengths are controlled by the same metallurgical phenomena. Similar equations have been developed for other austenitic alloys, such as type 304, and are quite similar in form and coefficients.[4] The effects of other alloying elements are apparent, once the data have been normalized for the N content. Figure 2 shows the effect of Mn on the weld yield strength, at a variety of N contents.[5] Both the N and Mn trends appear linear in the Figure, supporting the linear form of Equation (2).

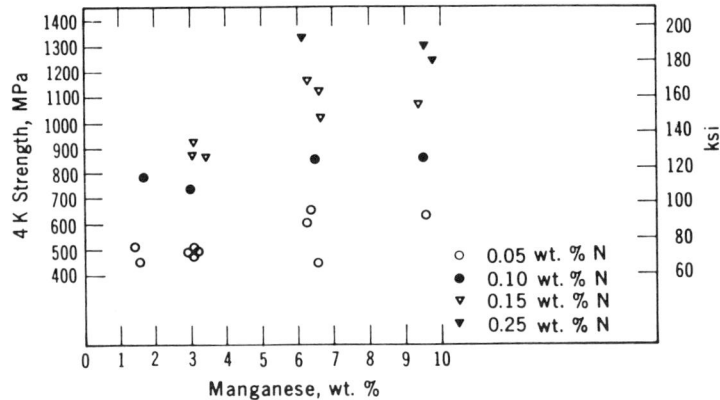

Figure 2. Yield strength versus Mn for four levels of N in type 18Cr-10Ni-2Mo welds at 4 K.

TOUGHNESS

The weld toughness is affected by many factors, including the strength. The effect of strength is shown in Figure 3 by a trend line that indicates the interrelationship between strength and toughness. This trend line, with fracture toughness on the vertical axis and tensile yield strength on the horizontal, is based on data for a 18Cr-10Ni-2Mo composition, with the strength varied by changing the N content.[6] As the strength is increased, both the wrought steel and weld toughness decrease. But, at any strength, the weld has a lower toughness than the wrought steel. This offset between the weld and the wrought steel toughness limits the design of structures that require high strength and high toughness concurrently. The offset also implies that there are fundamental differences between the two structures.

Two differences between the weld and base metals that have significant effects on toughness are the inclusion and ferrite contents. Simon and Reed were able to identify an effect of inclusion spacing in wrought steel.[4] They found a inverse correlation between the toughness and the inclusion spacing (inverse root of the inclusion area density). Since welds have a high inclusion content due to imperfect gas shielding while molten, we evaluated this effect for the weld data. Figure 4 shows the data for two gas metal arc weld compositions, at a variety of inclusion contents (varied by changing the oxidation potential of the shielding gas).[7] This weld data has the same correlation identified by Simon and Reed for the wrought steel data, indicating that the lower toughness of the welds can be explained by this one factor. Therefore, choosing welding processes that produce the lowest inclusion contents or modifying processes to reduce the inclusion content are two effective ways to improve the toughness of the welds. Welding processes such as laser, electron beam, and gas tungsten arc welding can produce welds with lower inclusion contents and produce welds with toughnesses at the upper side of the scatter band.[8]

Another measure of the toughness is the absorbed energy or lateral expansion measured during a Charpy V-notch (CVN) test. Although this test does not produce accurate data below 76 K (due to adiabatic heating and low heat capacity of the metal), it is widely used at 76 K as a screening test. For a series of austenitic steel welds, we used a stepwise statistical procedure to choose the terms that had the best correlation to the CVN absorbed energy. We allowed the regression procedure to choose independent variables, from the various elements in the composition, as well as various combined forms such as C^2 and FN. At a 95%

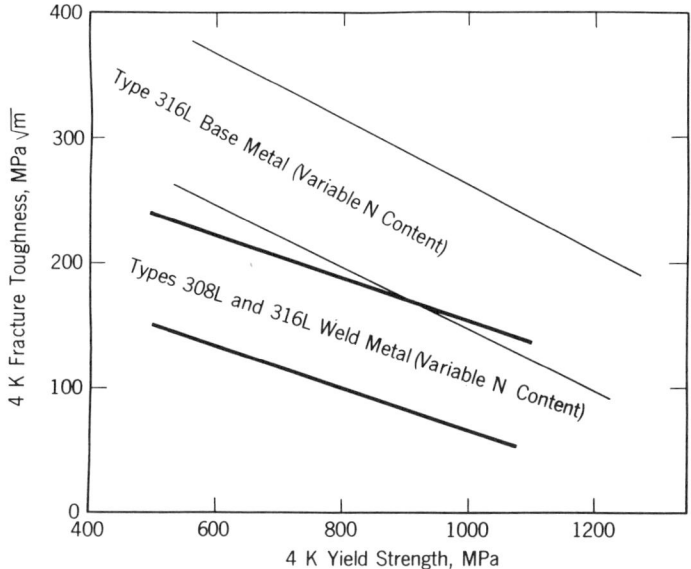

Figure 3. Fracture toughness versus yield strength at 4 K.

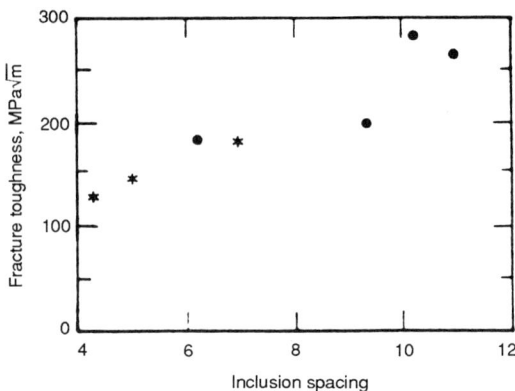

Figure 4. Fracture toughness versus inclusion spacing for type 316L welds (stars) and 18Cr-20Ni welds (circles).

confidence level, the regression analysis produced the following equation for predicting the CVN absorbed energy at 76 K:

$$\text{CVN (J)} = 19 - 1.4\ \text{FN} - 890\ C^2 + 1.4\ \text{Ni}, \tag{3}$$

where FN is the ferrite number calculated by the Schaeffler Diagram (called ferrite potential when allowed to be negative) and C and Ni are the contents of these two elements in weight percent.[9] This equation was developed only with data at low N contents, so it is not surprising that there is no N term in the equation, but the effect of strength is included as the negative coefficient for C^2. It does show that ferrite (a body centered cubic structure) and C reduce the toughness, while Ni increases it. Ni has a secondary effect because it also reduces the ferrite content. These effects of strength, FN, and Ni on the absorbed energy are supported by other studies.[10-12]

The ferrite content is particularly important. The ferrite phase occurs when the composition is adjusted so the austenite phase is metastable. A small amount of ferrite is normally desirable in stainless steel welds, because it inhibits the formation of low melting point compounds (such as FeS and FeP) that promote hot cracking in fully austenitic alloys.[13] Yet, for the best toughness in cryogenic service it should be minimized. Therefore, cryogenic grades of welding alloys have the ferrite limited to a very low value, or are ferrite-free. The ferrite-free grades are produced with strict control over the impurity elements that promote hot cracking.

Since Equation (3) and References 10 to 12 had indicated that nickel could improve the toughness, we plotted nickel content versus fracture toughness for the wide variety of compositions that we have tested at 4 K. Although there are other factors that spread the data, Figure 5 shows a clear increase in toughness with increasing nickel contents. The curve is nonlinear, with an increase in nickel content from 10 wt.% (the level in the most common austenitic alloys) to 20 wt.% giving most of the improvement. Thus, a composition with 20 wt.% Ni should exhibit a toughness very near the maximum attainable in an austenitic stainless steel weld.

Type 316LN wrought stainless steel (low C, high N version of type 316) has been proposed as the candidate material for various demanding cryogenic applications (requiring high strength and high toughness concurrently). Lead by these strength and toughness relationships that we identified for welds, we have recently been searching for a weld composition that could match the strength and toughness of type 316LN steel. Knowing that laser and electron beam welding equipment is relatively scarce and is difficult to use for large structures, we chose gas metal arc welding, a common process that has acceptable deposition rates. We chose shielding gases that were more inert than normal, to reduce the oxygen content, and so increase the

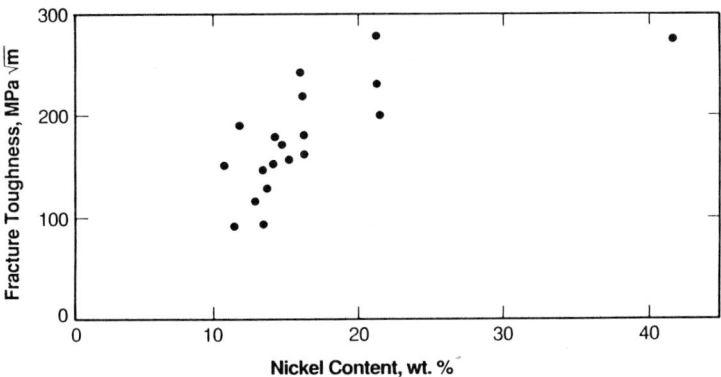

Figure 5. Fracture toughness versus nickel content for a variety of weld compositions at 4 K.

toughness of the welds. We obtained two commercially available compositions, 18Cr-20Ni-5Mn-0.16N and 20Cr-25Ni-4.5Mo.[14,15] Both met the requirements of 20 Ni for good toughness and low impurity content for good resistance to hot cracking. The first alloy had sufficient N to match the strength of the 316LN steel, while the 20Cr-25Ni-4.5Mo composition came with about 0.06 N. We were able to increase the 4 K strength of this second electrode by adding nitrogen to the weld pool through the shielding gas. Figure 6 shows the data from welds with these electrodes, superimposed on the data from Figure 3. Notice that the strength is near 1000 MPa, a value commonly used as a goal for wrought steel in high-strength structures. The toughness values are clearly higher than welds produced with 308- and 316-based compositions and standard welding procedures, and provide an acceptable match to the wrought steel toughness.

OTHER OBSERVATIONS

These relationships need to be applied with caution when predicting weld properties, because other factors may be present. In one case, we expected to obtain high toughness from

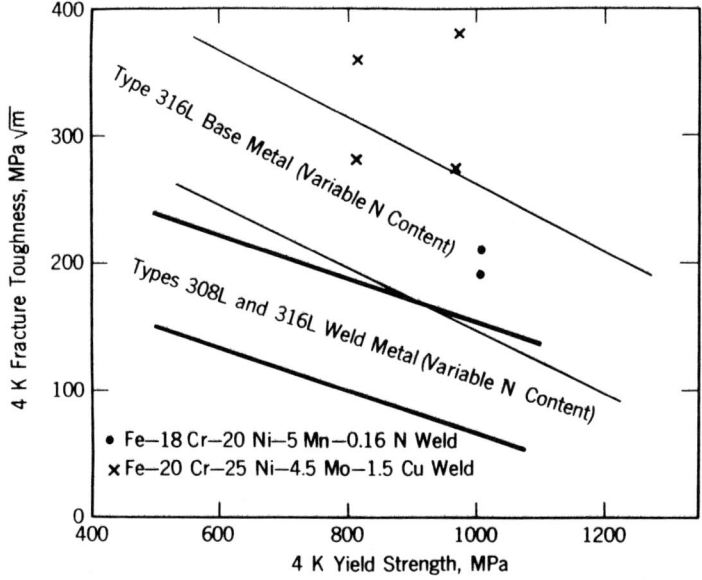

Figure 6. Fracture toughness versus yield strength trend bands for type 316L welds and base metal, showing the improvements in toughness possible with higher Ni contents.

113

a 25Cr-22Ni-4Mn-2Mo weld, but measured properties more typical of a 10 wt.% Ni weld with no changes in toughness as the inclusion content was changed.[16] Examination of the microstructure revealed an AF (primary austenite, followed by a mixed austenitic-ferritic eutectic) phase structure, with the cracks propagating through the eutectic phase. We attribute the poorer than expected properties to low toughness of this eutectic phase. The Ni was preferentially segregated to the primary austenite phase, leaving a low toughness medium through which the fracture propagated.

We have found that handbook data must not be extrapolated for structural designs. In fact, a higher yield strength steel may actually be weaker under certain service conditions. For example, one study of fatigue strength of various compositions found that a higher yield strength wrought composition (21Cr-6Ni-9Mn-0.3N) actually had a lower fatigue strength for a life of 4×10^4 cycles than a lower strength (19Cr-11Ni-2Mo-0.16N) weld.[17] The lower fatigue life is attributed to the faster crack growth rate in the higher strength steel (a manifestation of its lower toughness). Designers should be careful to consider the less obvious toughness requirements, in addition to the strength requirements, of their structural designs.

CONCLUSIONS

1. The strength of austenitic stainless steel welds is determined by the composition in the same manner as the wrought steel, so matching compositions will develop similar strength.

2. The austenitic weld toughness is lower than the wrought steel toughness, due to a higher inclusion content. Choosing welding processes that produce lower inclusion contents, or changing the welding conditions to lower the oxygen potential, increases the weld toughness.

3. Adding nickel to the weld composition also increases the toughness, so a weld composition can be designed that will match the strength and toughness of a wrought steel for very demanding applications.

ACKNOWLEDGEMENT

The authors appreciate the support of the Department of Energy and various national laboratories whose funding over a number of years made this work possible.

REFERENCES

1. T.A. Siewert and C.N. McCowan, Development of a SMA Electrode to Match Type 316LN Base Metal Cryogenic Properties, submitted to Cryogenics, 1990.

2. N.J. Simon and R.P. Reed, Design of 316LN-type Alloys, Materials Studies for Magnetic Fusion Energy Applications at Low Temperatures - XI, NBSIR 88-3082, National Bureau of Standards, Gaithersburg, Maryland, 71(1988).

3. C.N. McCowan and T.A. Siewert, Influence of Molybdenum on the Strength and Toughness of Stainless Steel Welds for Cryogenic Service, Materials Studies for Magnetic Fusion Energy Applications at Low Temperatures - X, NBSIR 87-3067, National Bureau of Standards, Gaithersburg, Maryland, 285(1987).

4. N.J. Simon and R.P. Reed, Strength and Toughness of AISI 304 and 316 at 4 K, Materials Studies for Magnetic Fusion Energy Applications at Low Temperatures - IX, NBSIR 86-3050, National Bureau of Standards, Gaithersburg, Maryland, 27(1986)

5. C.N. McCowan, T.A. Siewert, R.P. Reed, and F.B. Lake, Manganese and Nitrogen in Stainless Steel SMA Welds for Cryogenic Service, Weld. J. 66, 84-s(1987).

6. R.L. Tobler, T.A. Siewert, and H.I. McHenry, Strength-Toughness Relationship of Austenitic Stainless Steel Welds at 4 K, Cryogenics 26, 392(1986).

7. C.N. McCowan and T.A. Siewert, Fracture Toughness of 316LN Stainless Steel Welds with Varying Inclusion Contents at 4 K, Adv. Cryo. Engr. Mat. 36, 1331(1990).

8. T.A. Siewert, D. Gorni, and G. Kohn, High-energy-beam Welding of Type 316LN Stainless Steel for Cryogenic Applications, Adv. in Cryo. Engr. Mat. 34, 343(1988).

9. T.A. Siewert, Predicting the Toughness of SMA Austenitic Stainless Steel Welds at 77 K, Weld. J. 65, 23(1986).

10. T.A. Siewert, How to Predict Impact Energy from Stainless Steel Weld Composition, Weld. Des. Fab. (June), 88(1978).

11. E.R. Szumachowski and H.F. Reid, Cryogenic Toughness of SMA Austenitic Stainless Steel Weld Metals: Part 1 - Role of Ferrite, Weld. J. 57, 325-s(1978).

12. E.R. Szumachowski and H.F. Reid, Cryogenic Toughness of SMA Austenitic Stainless Steel Weld Metals: Part 2 - Role of Nitrogen, Weld. J. 58, 34-s(1979).

13. Welding Handbook, Vol. 4, Seventh Edition, American Welding Society, Miami, Florida, 103(1982).

14. T.A. Siewert and C.N. McCowan, Cryogenic Mechanical Property Data for 20Cr-25Ni-4.5Mo Gas Metal Arc Welds, Materials Studies for Magnetic Fusion Energy Applications at Low Temperatures - XIII, NISTIR 3944, National Institute of Standards and Technology, Gaithersburg, Maryland, 233(1990).

15. C.N. McCowan, T.A. Siewert, and R.L. Tobler, Tensile and Fracture Properties of an Fe-18Cr-20Ni-5Mn-0.16N Fully Austenitic Weld Metal at 4 K, J. Engr. Mat. and Tech. 108, 340(1986).

16. C.N. McCowan and T.A. Siewert, The Fracture Toughness of 25Cr-22Ni-4Mn-2Mo Stainless Steel Welds at 4 K, Materials Studies for Magnetic Fusion Energy Applications at Low Temperatures - XII, NISTIR 3931, National Institute of Standards and Technology, Gaithersburg, Maryland, 187(1989).

17. T.A. Siewert, C.N. McCowan, and D.P. Vigliotti, Cryogenic Materials Properties of Stainless Steel Tube-to-flange Welds, Cryogenics 30, 356(1990).

CREEP OF INDIUM AT LOW TEMPERATURES

R. P. Reed and R. P. Walsh

Materials Reliability Division
National Institute of Standards and Technology
Boulder, Colorado, U.S.A.

ABSTRACT

Creep tests were conducted on indium at 4 and 76 K for periods of up to one week. The initial creep strain depended on the square of the applied stress. Within this time period, conditions approaching steady-state rates were observed. Steady-state creep strain rates depended strongly on applied stress and temperature. Empirical relations were developed to predict total creep strain and strain rate as a function of temperature and stress. The activation energy obtained by assuming similar creep-deformation mechanisms at 4 and 76 K was much lower than the activation energy for self-diffusion.

INTRODUCTION

Indium and indium alloys are used extensively in aerospace applications because their excellent wetting characteristics enable reliable joining of glass, ceramics, and metals. Recent failures of indium solder joints in space applications led to this research on its low-temperature properties.

Indium is very soft. During deformation at room temperature, indium recrystallizes. The primary reason for the softness of indium at 295 K is the proximity of its melting temperature (429 K). Although the crystal structure of indium is commonly called face-centered tetragonal, its true space group is D_{4h}^{17} - I4/mmm, A = 2; thus, its structure is actually body-centered tetragonal. The nominal lattice constants are a = 3.2512 Å, c = 4.9467 Å, c/a = 1.5215. Indium is diamagnetic; it becomes superconducting at about 3.4 K.

In recent studies of the tensile behavior of indium, we identified the strain-hardening and deformation-twinning characteristics at 4, 10, and 76 K.[1] Limited compressive stress-strain characteristics of indium at low temperatures were obtained by Swenson[2] in a study of its compressibility. Mechanical twinning in indium was reported by Remaut et al.[3] and Becker et al.,[4] reviewed by Hall,[5] and studied in the In-Th alloys by Wuttig and Lin[6] and Basinski and Christian.[7] Annealing twins were previously found by Carpenter and Tamura.[8] The elastic properties of indium were recently determined by Kim and Ledbetter:[9] Young's modulus is 12.61 GPa at 300 K, 18.36 GPa at 80 K, and 19.56 GPa at 5 K. Gindin et al.[10] reported testing thin (1.5-mm-thick) sheets of high-purity indium with an average 2.5-mm grain size for 1 to 3 min. at constant load at temperatures ranging from 1.8 to 3.3 K. They found more strain per constant load increment in the superconducting state than in the normal state (obtained with a 3.5-T field at 1.8 K). They identified a logarithmic dependence of strain on holding time for both electronic states and reported a softening in the superconducting state as the temperature decreased from 3.3 to 1.8 K. This softening phenomena was attributed to the reduction of normal electrons, which reduces electron-drag effects.

The creep of indium at "high" temperatures (273, 298, 323 K) was measured under constant load by Frenkel et al.[11] In these tests, tensile loads were maintained for over 3×10^4 min., and a good correlation was found by using functional relationships between the total strain (ϵ_t) and $t \cdot e^{-Q/RT}$, where t is time, Q is activation energy for the deformation process, R is the gas constant, and T is absolute temperature. Frenkel et al. reported good experimental conformance among these variables at all temperatures and stresses. A value of 69 kJ/mol (16.5 kcal/mol, 0.71 eV) was obtained for Q. The similarity of this value with the self-diffusion activation energy of 74.8 kJ/mol (17.9 kcal/mol, 0.777 eV)[12] suggests that self-diffusion plays a major role in high-temperature creep deformation; one possible creep mechanism is dislocation climb.

We report tensile creep measurements of indium under constant-load for durations up to 1.6×10^4 min at 4 and 76 K.

EXPERIMENT

The indium was received in 3-kg ingots from a commercial supplier under the trade name Indalloy 4;* the purity was specified as 99.99 wt.%. We could not confirm this analysis: x-ray fluorescence data revealed very low spectral peaks, but the absence of an appropriate standard precluded quantitative analysis. Trace amounts of copper, manganese and iron were recorded.

Creep specimens were cast at NIST in split-die aluminum molds in a 99.95% dry-nitrogen atmosphere. The mold and indium casting temperature was 195 ±5°C. After casting, the indium was stirred, then cooled in the dry-nitrogen atmosphere. The cooling time was 2 h. The specimens had a 38-mm reduced section and were 13 mm in diameter.

The as-received ingots and as-cast specimens had average grain diameters ranging from 100 to 1000 μm; the average grain size was close to ASTM 0 (350 μm). Second-phase particles were present at what is thought to be the intercellular regions of the originally solidified cast structure. Both a cellular ghost structure and an equiaxed grain structure are observed in Fig. 1. Since the equiaxed grain boundaries are not clearly related to the arrangement of the second phase in the microstructure, secondary recrystallization probably occurred at some point during solidification. The two principal precipitate morphologies observed were rods and particles. Qualitative analysis (energy-dispersive x-ray analysis) showed the second phase to be rich in indium, with iron, copper, and sometimes nickel present as minor constituents.

The tests were performed with a commercial creep test machine that uses a lever arm and calibrated weights to apply the desired initial stress on a specimen. A cryostat fixture was mounted to the load frame to enable creep testing at low temperatures. This assembly has been described.[1]

---

*Use of this trade name is to provide accurate material identification; it does not imply NIST endorsement of this product.

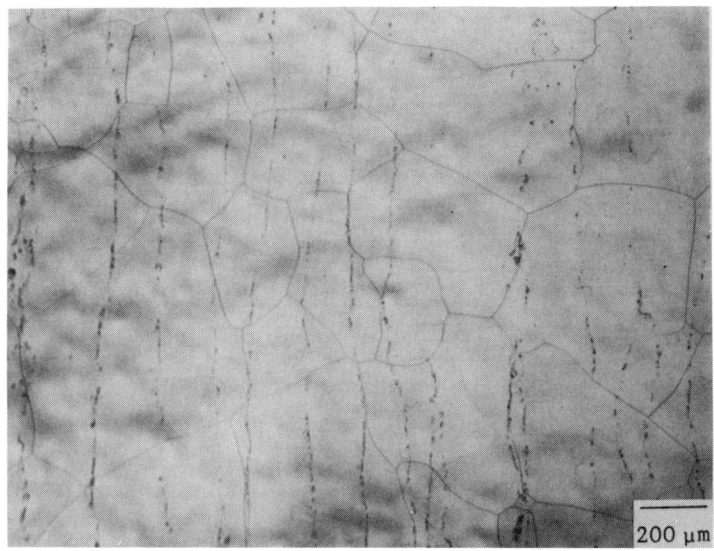

Figure 1. Optical micrograph of as-received and as-cast equiaxed grain structure with precipitates. Metallographic procedures entailed razor cutting and electrolytic polishing (with 70% methanol-30% HNO$_3$). Etching was performed electrolytically, using the same solution, at reduced current density.

In this test program, single-specimen creep tests were performed at 4 K (in liquid helium) and at 76 K (in liquid nitrogen). Specimens were 12.7-mm (0.50-in) diameter by 38-mm (1.5-in) gage-length in the as-cast condition. Specimen strain was monitored as a function of time by using a set of two clip-on strain-gage extensometers, 38 mm in gage length. These extensometers were attached between specimen grips rather than to the gage section of the specimen to ensure that the extensometer attachment did not slip during the long-term test of such a soft material.

The accuracy of the strength measurements was within ±2%, independent of temperature, but dependent on the precision of the measurement of the cross-sectional area and load-cell calibration. Typical strain resolution was 30 µm over a length of 25.4 mm; electronic drift over days is estimated as equivalent to ±60 µm.

RESULTS AND DISCUSSION

In creep tests, instantaneous specimen strain is associated with the application of the dead load. This is usually followed by three stages of creep strain: primary, secondary, and tertiary (strain leading to failure). Our creep tests at 4 and 76 K were not carried to failure. The applied stresses were in the vicinity of the yield strength and, consequently, very long times would have been required to approach the tertiary stage.

Instantaneous Strain

The square root of the instantaneous creep strain (ϵ_i) is linearly dependent on the applied stress (Fig. 2). This is expected since the tensile strain is a parabolic function of stress. The instantaneous creep strain was

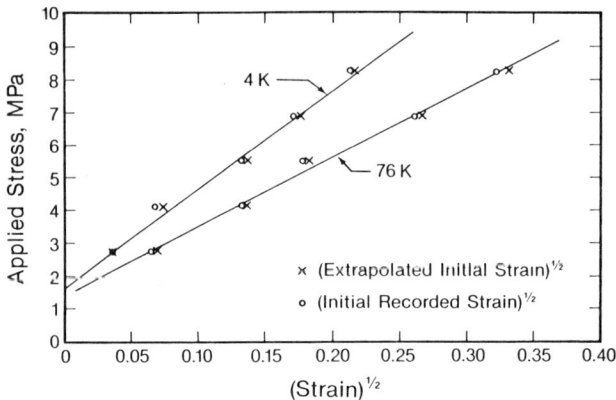

Figure 2. Effect of applied stress at 4 and 76 K on the initial creep strain (upon application of load) and on the square root of creep strain extrapolated to zero time.

obtained in two ways: from the initial recorded strain and from a linear strain-versus-time graph extrapolated to zero time. In Fig. 2, both show the same trend of ϵ_i dependence on applied stress.

The dependences of the square root of specimen strain on applied stress are different at 4 and 76 K, yet they both approach the same stress at zero strain. This suggests that the elastic limit of indium is insensitive to temperature in this range, but that strain hardening of indium is temperature sensitive. The true stress-strain curves[1] support this conclusion; there is remarkably more strain hardening at 4 K than at 76 K. Consequently, the ultimate strength is much more temperature dependent than the yield strength (even though the ductility at 4 K is less than that at 76 K).

Primary Creep

The primary creep stages at 4 and 76 K are plotted in Figs. 3 and 4, respectively, as creep strain ($\epsilon_t - \epsilon_i$) versus time, where ϵ_t is the total specimen strain. Data for the duration of each test at 4 and 76 K are plotted as ϵ_t versus log t in Figs. 5 and 6, respectively. Examination of these figures leads to several conclusions about primary creep of indium at low temperatures.

- The primary strain region is completed much earlier at 4 K (<700 min.) than at 76 K (up to 3000 min.).

- The primary strain requires more time to stabilize at higher loads at 76 K. If we accept the 4-K data at 4.1 MPa as reliable and reject the 4-K data at 8.3 MPa as representing a "cold worked" specimen, then a similar trend (but shorter in duration) occurs at 4 K.

- The total primary strains at 4 K ($<2 \times 10^{-3}$) are much less than those at 76 K ($<10 \times 10^{-3}$) at equivalent stresses, as expected.

Secondary Creep

The creep strain at 4 K is plotted versus time (t) in Fig. 7. Its dependence on t is more linear than its dependence on log t (Fig. 5). Similarly, the dependence of creep strain at 76 K is also best expressed as linearly dependent on t. The linear dependence of total specimen strain on t suggests that a steady-state deformation process controls the constant

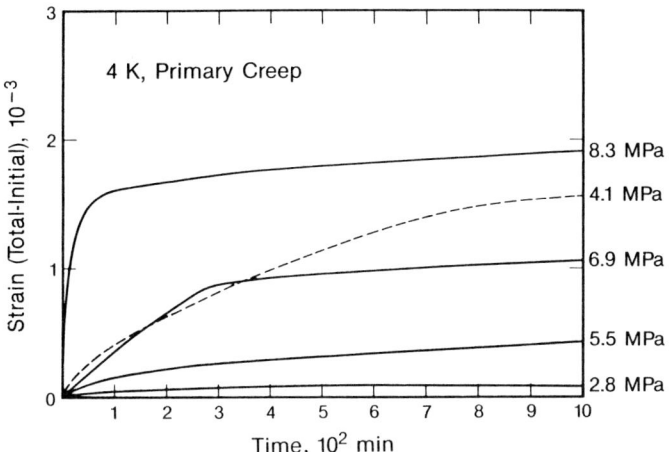

Figure 3. Primary stage of creep at 4 K for a series of applied stresses.

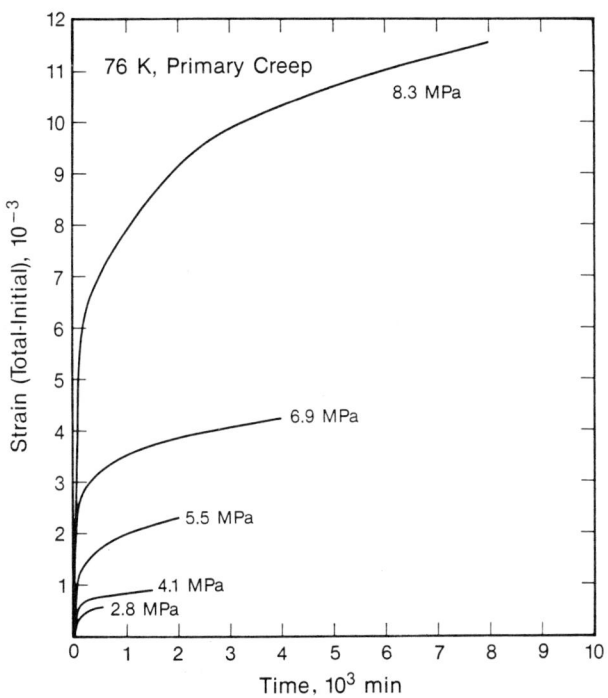

Figure 4. Primary stage of creep at 76 K for different applied stresses.

creep strain rate. The relationship between the steady-state creep rate ($\dot{\epsilon}_{ss}$) and the applied stress (σ_{app}) can be expressed as

$$\sigma_{app} = \sigma_1 + C\dot{\epsilon}_{ss}, \tag{1}$$

where σ_1 and C are constants. The steady-state creep rates at 4 and 76 K are plotted as a function of applied stress in Fig. 8. At both temperatures, there is a linear dependence, and the data fits are good.

Activation Energy

Let us assume that the same defect creep mechanism is rate controlling at each test temperature. When the creep rate is dependent on applied stress, the activation energy (Q) for a single creep mechanism can be estimated from the expression

$$\dot{\epsilon}_{ss} = A\sigma^n e^{-Q/RT}. \tag{2}$$

Here A is a constant, n is 1, T is the absolute temperature, and R equals 8.3 J/(mol·K). Usually, this expression is used to describe diffusion-controlled creep of metals at $T/T_m > 0.5$.[13] This equation has been used to describe diffusion mechanisms along dislocation cores, but the power de-

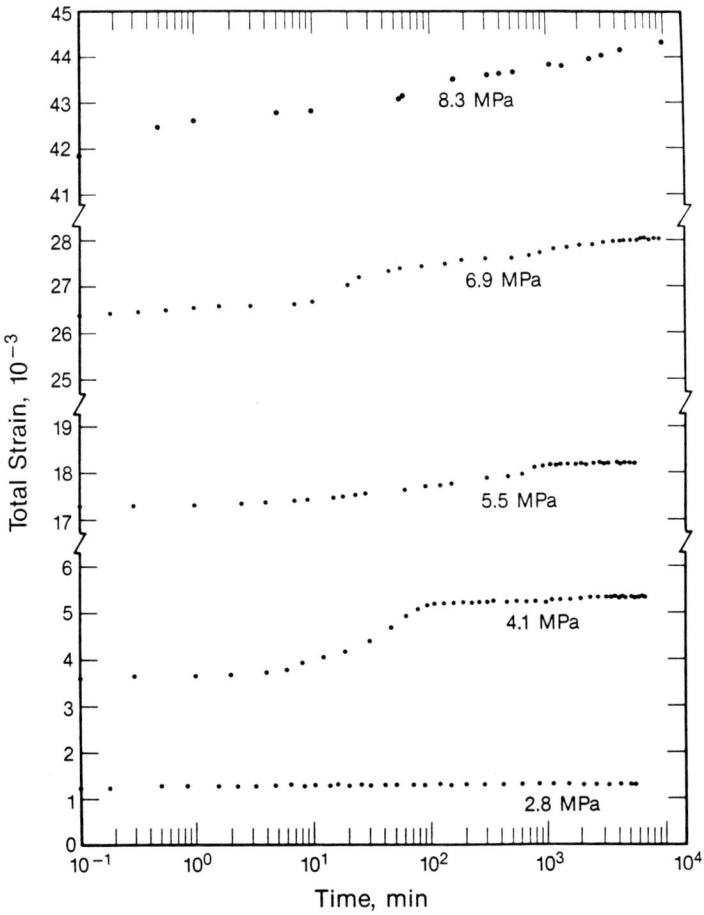

Figure 5. Total creep strain at 4 K versus time.

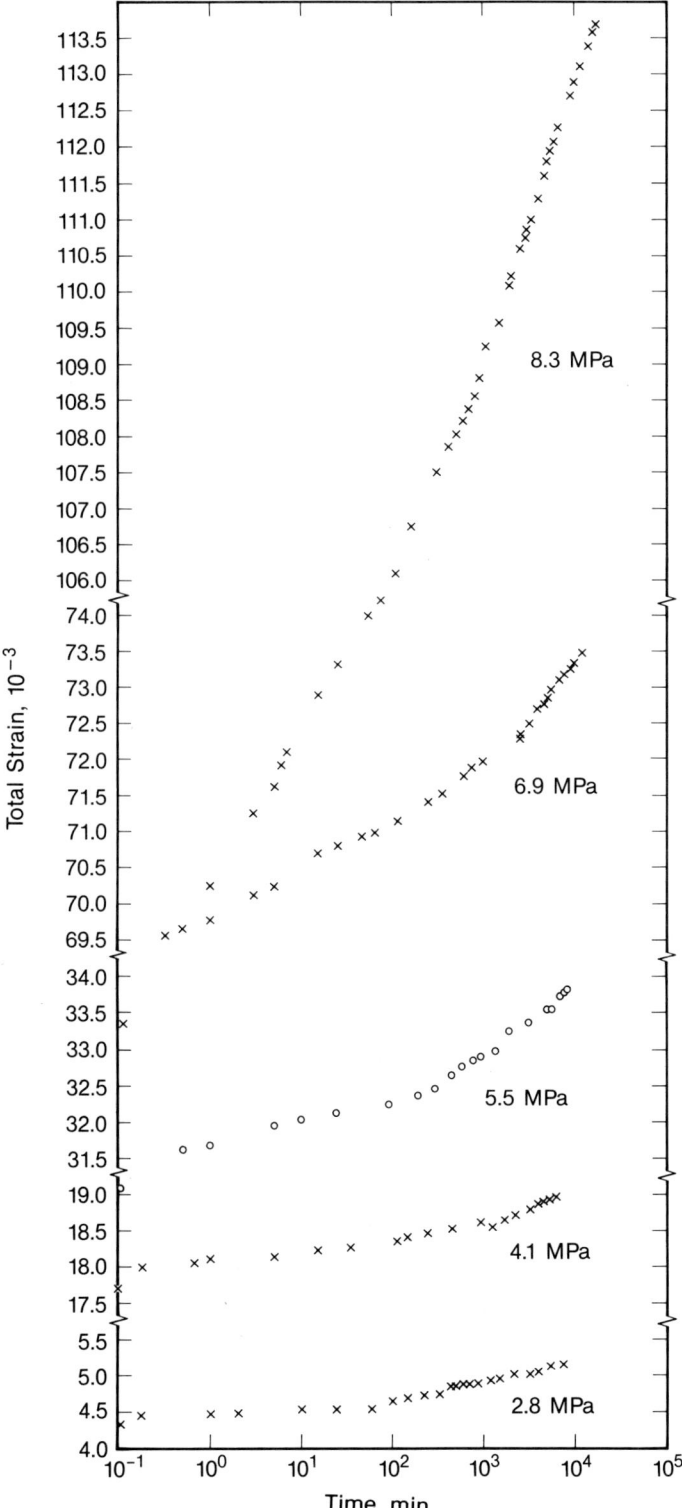

Figure 6. Total creep strain at 76 K versus time.

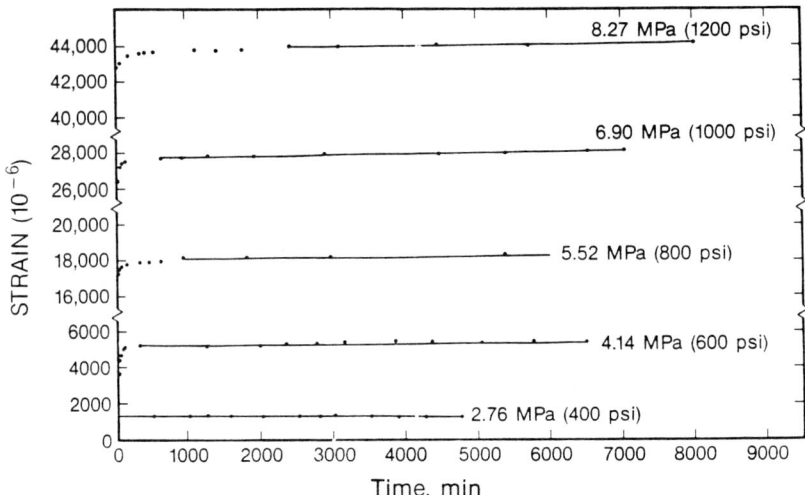

Figure 7. Total creep strain versus time (linear) for tensile creep measurements at 4 K.

pendence on stress was found to be much larger (n = 5 to 7).[14] From Eq. 2 and the slopes of applied stress versus steady-state creep rate of Fig. 8, an activation energy of 43 J/mol (10 cal/mol, 4.4×10^{-4} eV) was calculated. This very low value is three orders of magnitude less than the measured activation energy for self-diffusion of indium, which is 74.8 kJ/mol (17.9 kcal/mol, 0.77 eV).

Methods of estimating the activation energy for low-temperature creep of indium that use the Arhenius equation ($\dot{\epsilon}_{ss} = Ce^{-Q/RT}$) are not as reliable. As illustrated in Fig. 8, steady-state creep rate is a function of applied stress; calculations yield decreasing values of activation energy with increasing values of applied stress.

Thus, we have used Eq. 2 to calculate very low apparent activation energies; steady-state, constant creep rates have been assumed. Such low values, if valid, probably represent an interstitial migration mechanism. We are not aware of any lower values of activation energy for other metals that have been reported by others.

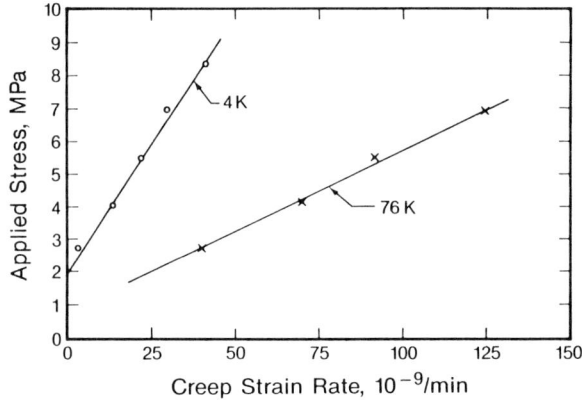

Figure 8. Influence of applied stress on steady-state creep rates at 4 and 76 K.

We urge caution: Measurements of creep at 4 and 76 K are very difficult and, therefore, the results may represent only fortuitous strain-time measurements for a limited number of specimens.

Total Creep Strain at 4 K

The constant-load creep tests were conducted in liquid helium for a period of at least one week. The best correspondence between strain and time is shown by Fig. 7, in which there is a primary creep stage for up to about 100 min. followed by secondary (steady-state) creep. The expression

$$\epsilon = \epsilon_0 + \dot{\epsilon}_{ss} t \qquad (3)$$

is appropriate if we neglect primary creep and extrapolate the steady-state range to zero time (ϵ_0). The square root of the initial strain is approximately linearly related to the applied stress (see Fig. 2). There is not much difference between the extrapolated strain and the initial strain (immediately following application of load). Thus,

$$\sigma_{app} = \sigma_0 + B(\epsilon_0)^{\frac{1}{2}}, \qquad (4)$$

where σ_0 is the applied stress extrapolated to zero strain and B is the slope of σ_{app} versus $(\epsilon_1)^{1/2}$. Figure 2 shows that σ_1 is 1.6 MPa and B is 32 MPa.

When Eqs. 1, 3, and 4 are combined, the total creep strain at 4 K is

$$\epsilon = \left(\frac{\sigma_{app} - 1.4}{32}\right)^2 + \left(\frac{\sigma_{app} - 2}{16 \times 10^7}\right) t, \qquad (5)$$

where σ_{app} is in MPa and t is in minutes.

Thus, the total creep strain at 4 K can be approximated in terms of the applied stress and time. As an example, consider the creep strain over a ten-year lifetime at a stress level equivalent to the yield strength, 3.1 MPa. This strain, calculated from Eq. 5, is 0.0380, with contributions from ϵ_0 of 0.0028 and from steady-state of 0.0352. Thus, the instantaneous strain is less than 10% of the total strain.

SUMMARY

Creep tests were conducted on polycrystalline indium at 4 and 76 K. Creep strain was found to be approximately linearly related to time; thus steady-state analysis could be used. The activation energy for creep in the temperature range 4 to 76 K (assuming a constant creep mechanism) is 43 J/mol (10 cal/mol, 4.4×10^{-4} eV), considerably less than the activation energy for self-diffusion.

Total creep strain can be estimated from analytic expressions relating applied stress to instantaneous strain and from estimates of steady-state creep contributions. At applied stresses equivalent to the yield strength at 4 K (3.1 MPa), the calculated total creep strain is 0.0352.

ACKNOWLEDGMENTS

This work was sponsored by the U.S. Air Force, Headquarters Space Division. C. N. McCowan (NIST) provided the metallographic photographs. L. A. Delgado and J. D. McColskey (NIST) assisted in specimen preparation and testing.

REFERENCES

1. R. P. Reed, C. McCowan, L. A. Delgado, R. P. Walsh, and J. D. McColskey, *Mater. Sci. Eng.* 102, 222-236 (1988).
2. C. A. Swenson, *Phys. Rev.*, 100, 1607-1614 (1955).
3. G. Remaut, A. Lagasse, and S. Amelinckx, Phys. Status Solidi, 6, 723-731 (1964).
4. J. H. Becker, B. Chalmers, and E. G. Garrow, *Acta Crystallogr. Camb.* 5, 853 (1952).
5. E. O. Hall, "Twinning and Diffusionless Transformations in Metals," Butterworth Scientific, London, 1954, pp 85-86, 139-144.
6. M. Wuttig and C. H. Lin, *Acta Metall.* 31, 1117-1122 (1983).
7. Z. S. Basinski and J. W. Christian, *Acta Metall.* 2, 101-116 (1954).
8. H. C. H. Carpenter and S. Tamura, *Proc. Roy. Soc.* A133, 161-165 (1926).
9. S. Kim and H. M. Ledbetter, National Institute of Standards and Technology, Boulder, Colorado; unpublished NIST data, 1987.
10. I. A. Gindin, B. G. Lazarev, Y. D. Starodubov, and V. P. Lebedev, *Fiz. Met. Metalloved.* 29, 862-868 (1970).
11. R. E. Frenkel, O. D. Sherby, and J. E. Dorn, *Acta Metall.* 3, 470 (1955).
12. R. E. Eckert and H. G. Drickamer, *J. Chem. Phys.* 20, 13 (1952).
13. G. E. Dieter, "Mechanical Metallurgy," McGraw-Hill, New York (1986) p. 447.
14. S. L. Robinson and O. D. Sherby, *Acta Metall.* 17, 109-125 (1969).

1100 HOUR CREEP TEST RESULTS FOR OFHC COPPER: VALIDATION OF PREVIOUSLY PUBLISHED RESULTS

L.C. McDonald and K.T. Hartwig

Texas A&M University
Department of Mechanical Engineering
College Station, TX 77843-3123

ABSTRACT

Results of 77 Kelvin creep tests on OFHC copper specimens have been obtained and are shown to support previously published results presented by researchers at the National Institute of Standards and Technology (NIST). The reproduction of the copper results obtained by NIST using the Texas A&M system provides credibility to both systems. This is particularly important when the lack of commercial cryogenic creep systems results in individualized measurement techniques and the degree of measurement difficulty is high. A suggested standard for the simplified comparison of materials which exhibit "exhaustive" creep behavior is also given.

INTRODUCTION

Researchers are becoming increasingly aware of the difficulties associated with low temperature time-dependent strain measurement. With creep rates in some materials on the order of 10^{-12} sec^{-1}, there is no easy way to establish a description of the accuracy of the equipment used for these measurements. In addition, there are no commercial cryogenic creep systems on the market, so the few systems in use are very individualistic. One solution to quantifying system accuracy is to compare independent measurement systems.

For this reason, OFHC copper creep specimens were obtained from researchers at the National Institute of Standards and Technology (NIST) and were tested at 77 Kelvin in an attempt to compare the creep measurement system in use at Texas A&M University with the one used at NIST. Similar results on the same creep samples would lend credibility to both measurement systems. This is particularly important for two reasons. First, the two measurement systems are slightly different, owing to the unavailability of a standard commercial cryogenic creep system. Secondly, the Texas A&M system was originally used for measuring creep rates in high purity aluminum, which is much more difficult due to the lower creep rates observed (an order of magnitude less than those seen in copper).

SYSTEM DESCRIPTIONS AND PROCEDURES

The systems used by NIST and Texas A&M have been described in detail in previous publications.[2,3] The primary difference between the two systems is the type of lever arm used to transfer load. The NIST system lever arm is a commercial lever designed for use in high temperature creep experiments.[2] The fulcrum is a knife edge fulcrum and the lever arm is approximately two feet long with a load ratio of approximately 5:1. The Texas A&M lever arm was constructed of 6061 aluminum and has a length and load ratio of three feet and 1:1, respectively.[3] The fulcrum in this case, however, is a low friction roller bearing pulley. The fact that the lever arms are different could be a major source of discrepancy between results since load stability is of utmost importance during low temperature creep experiments.[1-5]

Another important difference between the two systems is the strain measurement method. While both groups use strain gages mounted directly onto the specimen, the NIST researchers use 73Ni-20Cr alloy gages and a strain conditioner[2] and the Texas A&M researchers use K-alloy series gages and a Wheatstone circuit. This difference is also a possible source of discrepancy since strain stability (on the order of 5×10^{-7} sec^{-1}) and drift stability (on the order of 1×10^{-6} cm/cm per 200 hours) are necessary. A drift of one microstrain in 200 hours yields a creep rate on the order of that seen in high purity aluminum at low temperatures.

The creep specimens used for these experiments were donated by NIST and are oxygen-free high conductivity C10400 copper (Cu + Ag = 99.99 wt. %). They are the same specimens used in the creep tests reported by Reed, et al.[2] Before testing in the Texas A&M system, the samples were "reannealed" (650°C for 1 hour) at NIST to provide similar initial material characteristics.

Upon receiving the specimens, the sample surfaces were cleaned with acetone and the strain gages were mounted using a curing temperature of 150°C. The specimen dimensions were such that no alteration of geometry was necessary for measurements at Texas A&M.

RESULTS AND DISCUSSION

The results of the three samples tested were nearly identical. The total strain values at any time during the test were within 0.5% of each other and the initial strain values for the three samples were within 1.1% of each other. Representative curves of creep strain versus time and creep strain versus log time are shown in Figures 1 and 2.

To compare the results with the earlier NIST results, a curve fit such as that proposed by NIST was used.[2] The equation which describes the total strain is given by

$$\epsilon_t = \epsilon_o + a_1 \ln t + a_2 t$$

where the total strain at time, t (minutes), is equal to the initial strain plus a scaled logarithmic term plus a linear term (a_1 and a_2 are constants). The affect of the a_2 term is negligible compared to the a_1 term at times less than about 8000 minutes and is approximately negligible for times on the order

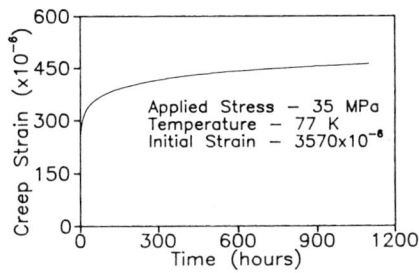

Figure 1. Representative creep curve of OFHC copper tested under 35 MPa at 77 Kelvin.

of the length of these tests.[2] Table I lists values of the a_1 term as well as values of stress level, initial strain, and test duration for the 76 Kelvin tests conducted at NIST and the 77 Kelvin tests conducted at Texas A&M.

In Figure 3, where the constant a_1 is plotted as a function of applied stress, it is easy to see that the results obtained by Texas A&M agree very well with those obtained by NIST.

In the samples tested by Texas A&M at 35 MPa, the initial strains were higher than the NIST values. However, compared with the values of initial strain in the other NIST samples, the 35 MPa NIST results appear to be low. The a_1 values obtained by Texas A&M at 35 MPa are also slightly higher than the NIST values, but the scatter in the NIST a_1 values is much greater than this difference, as can be seen in Figure 3.

ADDITIONAL COMMENTS

The fact that the creep behavior of some metals at cryogenic temperatures is exhaustive in nature, i.e., constantly decreasing, makes it difficult to compare results quantitatively. In addition, the length of the creep test can affect the steady-state approximation of the creep rate.[2] For these reasons, a standard definition for the quantification of the creep rate of a material is needed. It is therefore suggested that, for example, a "700 hour creep rate" be used to represent the approximate steady-state creep rate in the 700 hour (approximately one month) region of a creep experiment. The use of such a value could be considered conservative since extrapolations to longer times would yield a slightly higher strain than would actually occur, as shown in Figure 4. In

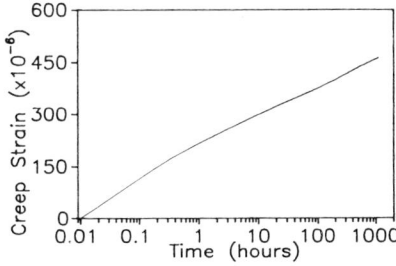

Figure 2. Dependence of creep strain on log time in the OFHC copper sample represented in Figure 1.

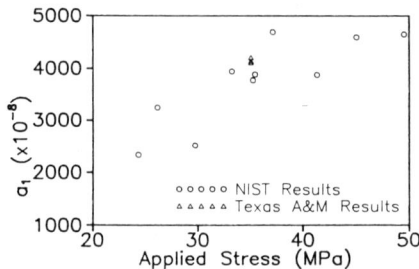

Figure 3. Logarithmic constant a_1 versus stress level for 76 and 77 Kelvin creep tests on OFHC copper.

addition, comparisons to other materials could be made using single values.

The use of a 700 hour creep rate value would serve to simplify the design criterion used by engineers. Furthermore, the error associated with such an approximation is small relative to the amount of total creep strain which occurs in a given material. Materials with lower creep rates should be expected to have "less exhaustive" (closer to steady-state) behavior, resulting in a small difference between strain values predicted using the approximation and actual strain values. In aluminum, for instance, the creep rate is so small that no exhaustive creep is detectable, resulting in an apparent steady-state behavior.[1,3] Thus the error would be zero. In materials with higher creep rates, the exhaustive behavior causes a larger discrepancy between predicted and actual values, but the difference relative to the amount of total creep strain is still relatively small.

To evaluate the 700 hour creep rates of the copper samples listed in Table I, simply divide the constant, a_1, by the time in seconds, since

$$\frac{d\epsilon}{dt} = \frac{a_1}{t}$$

and

$$\epsilon_{700hr} = \frac{a_1}{2.52 \times 10^6} \, [\sec^{-1}].$$

This calculation is listed in the last column of Table I.

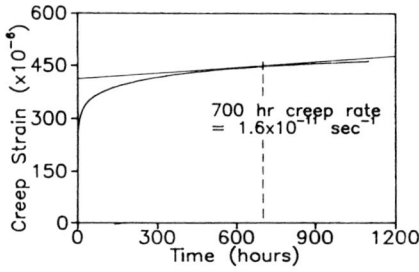

Figure 4. Schematic of a suggested standard for comparisons between materials exhibiting exhaustive creep.

Table I. Experimental Results from NIST[2] and Texas A&M.

	σ (MPa)	ϵ_o (10^{-6})	a_1 (10^{-8})	Duration (min)	$\dot{\epsilon}_{700hr}$ (10^{-11} sec^{-1})
NIST	24.3	650	2323	12591	n.a.
	26.1	1659	3233	12591	n.a.
	29.7	2609	2506	96600	0.99
	33.2	2230	3934	96600	1.56
	35.2	1188	3763	12591	n.a.
	35.4	2731	3876	96600	1.54
	37.1	4889	4684	58894	1.86
	41.3	4815	3865	36345	1.53
	45.0	5258	4588	43455	1.82
	49.5	7499	4642	12820	1.84
Texas A&M	35.0	3530	4100	66000	1.63
	35.0	3530	4100	66000	1.63
	35.0	3570	4200	66000	1.67

The selection of 700 hours for the steady-state approximation is a compromise between the need for long testing times and the economic feasibility of cryogenic creep testing. Typically, any primary creep that is observed is constrained to the first 200 hours of the test. Beyond 200 hours, materials tend to exhibit exhaustive creep (as is seen in copper) or steady-state creep (as is seen in aluminum, possibly due to strain measurement limitations). The use of time and liquid cryogen in tests longer than 700 hours may not be justified economically if steady-state approximations are adequate.

CONCLUSIONS

Based on the similarities of the results obtained by Texas A&M using the specimens provided by NIST, it is fair to conclude that the systems are both adequate. This claim can also be justified since the NIST results were initially compared to OFHC copper results obtained by Yen, et al.[2] The fact that results can be repeated on individual measurement systems when the difficulty of the experimentation is extremely high lends credibility to all of the measurement systems involved. This, in turn, provides supportive evidence of the accuracy of results previously published by both NIST and Texas A&M.

Concerning the suggested "700 hour creep rate," simplified comparisons between materials which exhibit exhaustive behavior and those which exhibit steady-state behavior could easily be made. In addition, the use of such a value would give slightly conservative estimates of long term creep strain in exhaustive-type materials.

ACKNOWLEDGEMENTS

The researchers would like to thank R.P. Reed, N.J. Simon, and R.P. Walsh of NIST for the use of their samples for these experiments.

REFERENCES

1. L.C. McDonald and K.T. Hartwig, Creep of Pure Aluminum at Cryogenic Temperatures, Adv. in Cryogenic Engr. Mat., 36:1135 (1990).

2. R.P. Reed, N.J. Simon, and R.P. Walsh, Creep of Copper: 4 to 295 K, Adv. in Cryogenic Engr. Mat., 36:1175 (1990).

3. L.C. McDonald and K.T. Hartwig, Cryogenic Creep Testing, J. Testing and Evaluation, Vol 19, No. 3, March 1991.

4. C. Yen, T. Caulfield, L.D. Roth, J.M. Wells, and J.K. Tien, Creep of Copper at Cryogenic Temperatures, J. Cryogenics, 24:381 (1984).

5. C.T. Yen, L.D. Roth, J.M. Wells, and J.K. Tien, Equipment for Long-term Creep Testing at Cryogenic Temperatures, J. Cryogenics, 24:410 (1984).

FATIGUE AND FATIGUE CRACK GROWTH PROPERTIES OF 316LN AND

INCOLOY 908 BELOW 10 K

Arman Nyilas, Jinbo Zhang*, Bernhard Obst,
and Albert Ulbricht

Kernforschungszentrum Karlsruhe
Institut für Technische Physik
W-7500 Karlsruhe 1, Germany

INTRODUCTION

The cyclic loading characteristics of Tokamak type thermonuclear machines demand for an answer towards the fatigue response of the materials used in critical components. As one of the main outstanding parts of such a device the large superconducting magnets and their superconductors will operate under cyclic mechanical stress conditions. The present paper is biased towards the current superconductor design of the NET (Next European Torus) model coil concept[1]. The superconductor of this coil will be a cable-in-conduit Nb_3Sn type with an enveloped stiff external jacket structure. The wall thickness of the jacket structure is within the range of 4-5 mm in accordance with the recent structural mechanics calculations. The manufacturing of the jacket lengths for several hundred meters require an appropriate joining process due to the prefabricated section pieces available only in short lengths of 5-7 meters. At the ongoing technical discussions the recently anticipated solution favors the flash butt welding technique, which seems to be quite reasonable considering the present industrial practice. The performance of the superconductors jacket will strongly depend on the material selection and the proper structural design according to the existing low temperature structural materials data base. The wind and react Nb_3Sn-manufacturing process must also account the materials properties after ageing. To envisage all these aforementioned factors a material test program was set up to elucidate the fatigue-life behavior and fatigue crack growth rate (FCGR) of the recently selected two candidate materials. These materials were the AISI 316LN with a specified low carbon content to avoid the embrittlement after the ageing process[2] and the material Incoloy 908. Both materials were investigated in aged condition. In addition, the 316LN material in the as received condition was also tested with respect to its fatigue-life for specimens bearing predefined flaws and cracks. For more practical engineering relevance the propagation of surface cracks at 12 K and at 295 K was characterized with non standard specimens. All these tests were performed in a newly developed cryogenic dynamic test facility under helium gas environment between

---

* Prof. Jinbo Zhang is guest scientist from General Research Institute for Non-Ferrous Materials (GRINM), Beijing.

7 K and 20 K. Using the reference growth laws obtained from all these measurements the total crack propagation starting with the initial crack length of the specimen could be predicted by numerical computation.

EXPERIMENTALS

Materials and Specimens

The 316LN material with a carbon content of 0.015 wt.% (details of the chemistry are given in 3) was supplied by Böhler company, whereas the procurement of Incoloy 908 was undertaken by Inco, Alloys, USA. The base metals and the flash butt welded plates of ~ 10 mm thickness of both materials were aged (50 hours at 700°C in air) in form of appropriate sized blocks prior to specimen machining. The fatigue life specimens were 100 mm long smooth cylindrical type with a reduced section of 20 mm length and 3.5 mm Ø. Both ends of the specimen had threads of M10 x 1,5. The specimens for the FCGR measurements were standard compact tension (CT) type of 63 x 60 x 6 mm with a starting a/W (a = crack length, W = specimen width) ratio of 0.34. The crack positions in the flash butt welded CT-specimens (narrow weld profile of ~ 0.2 mm) were in the center of the weldment in welding direction, and in transverse longitudinal plate orientation. The weldments of the fatigue-life investigations were located in the center of the 100 mm long specimens. In addition, fatigue-life tests were carried out with flaw bearing specimens of standard as received 316LN material. The size of these flawed specimens were similar to those of smooth cylindrical ones. The centerly located flaw of these specimens was machined by electro discharge method (EDM) and it was possible to machine predefined semi-circular flaws in the range of ~ 0.2 mm depth. For the surface crack growth characterization specially machined 100 mm long specimens with rectangular cross section (4 mm thick and 10 - 7 mm width) were used. Similar semi-circular flaws as in the case of the fatigue-life specimens were located in the center of the width to initiate the crack by fatigue loading.

Test Equipment

All these cyclic investigations were conducted in a helium flow cryostat with a test chamber dimension of 265 mm Ø x 210 mm depth, equipped with a servohydraulic tensile machine of ± 25 kN capacity (MTS, Modell 810). Details of this test facility are given in 3.

Test Procedure

Prior to these tests the temperature stability during the cycling was examined using specially prepared specimens with a 0.5 mm Ø horizontal bore in the high stress region with a Ni/NiCr thermocouple. For the fatigue-life measurements the specimens were sinusoidal loaded at the load ratio R = 0.1 with a frequency of 20 Hz after reaching the test temperature (e.g. 7 K). No temperature increase was detected during the cyclic loading well beyond the yield strength (~ 1.1 x σ_y) of the 316LN material. Considering this observation 20 Hz frequency was taken as a standard for all smooth specimens. The cyclic loading until failure was performed at constant load condition.

The flawed specimens for the fatigue-life tests were loaded at R = 0.1 with a higher frequency than the previous one (30 Hz), because of the lower nominal stress condition. During cycling the EDM-notch was directly observed with a built-in fiberscope (magnification 22x) to record the beginning of the crack emanation. Beside this a high resolution clamp-on extensometer mounted on the specimen was used to monitor the displacement situation on a 2 channel oscilloscope to determine the strain condition in the vicinity of the flaw region. The crack initiation and the further fatigue crack growth result in a steady displacement increase of this region at constant cyclic load. By displacement control it was possible to load the specimen at a constant stress condition with respect to the remaining non-fractured cross section. According to this technique constant stress vs. life of the material at different stress levels could be determined for the case when loading was started with an EDM-notch or with an already existing, predefined fatigue crack in the specimen.

For the FCGR investigations the CT-specimens were mirror polished and a 0.5 mm fine grid was drawn onto the specimens surface by a special micrometer device. This grid served to observe the crack propagation directly with the built-in fiberscope. To avoid any misinterpretations especially at low ΔK-regime (K = stress intensity factor, < 20 MPa$\sqrt{m}$) due to the crack monitoring only from one side of the specimen, the compliance was measured directly at 295 K with two fiberscopes inspecting both surfaces and using a high resolution extensometer attached at the front of the specimen. Altogether six specimens were measured to define the compliance curve with a high accuracy. With this technique it was possible to determine crack differences of ~ 0.1 mm. The FCGR-measurements at high ΔK-regime (> 20 MPa$\sqrt{m}$) were easily performed with the direct observation. For accuracy examination several specimens were fractured and the crack growth was related by direct stereo microscopic measurements to the fiberscopic inspections. All tests were carried out at constant sinusoidal load and at different R-ratios. Both techniques K-increasing and K-decreasing (stepwise) were used for these measurements. To establish the da/dN vs.ΔK diagram at least two identical specimens were tested (N = cycle number).

The behavior of surface cracks under fatigue loading was investigated with 4 mm thick rectangular specimens (see Fig. 1, right hand side). An

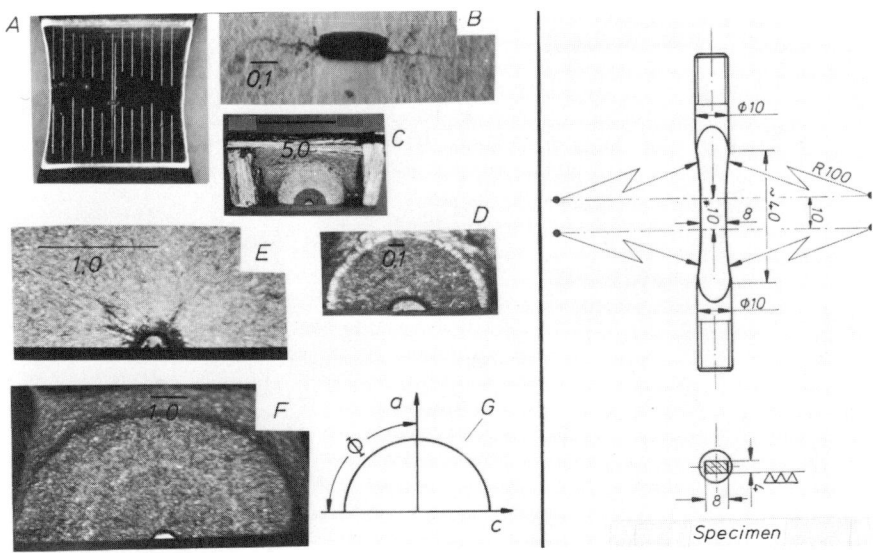

Fig. 1. Surface crack investigations. A) Specimen with the grid of 0.5 mm spacing. B) Crack penetration at the surface from the EDM-notch at 295 K. C) and D) 295 K (dark) and 12 K (light) semi-circular crack growth. E) 295 K semi-circular crack growth started from a non-circular EDM-notch. F) 295 K beach marks at different stages of crack growth. G) Surface crack growth designations according to Raju and Newman[4].

EDM-notch of ~ 0.2 mm depth served as a crack starter. The crack growth observation was performed directly with the fiberscope and the grid on the specimens surface. Several specimens were cycled at 295 K and at 12 K to see whether any difference of crack shape exists between these two temperatures. No significant difference was detected between these temperatures concerning the crack profile. The semi-circular crack shape was preserved at least until 3/4 of the specimen thickness B. This figure has been taken as a last position before fracturing the specimen. The semi-circular crack shape at different crack depths (below 3/4 thickness) could be double checked with two different techniques. First by load shedding technique (beach marks) and second by stopping the loading for the reason of marking, i.e. fatigue precracking at 295 K colors the crack surface by air oxidation dark, whereas at 12 K and at He-gas environment the crack surface remains clear (Fig. 1, see C, D, and E). The width (2b) were varied between 10 mm to 7 mm and did not show any effect on the crack shape and on the FCGR. The stress intensity factors K at c and a positions (Φ = 0° and 90°, respectively) were computed by Raju & Newman equations[4], where K is a function of Φ, a, c, b, B, and stress range $\Delta\sigma$ (see Fig. 1, G). The determination of da/dN vs. ΔK at 295 K and 12 K was performed under constant load condition (K-increasing) after precracking at 295 K. The crack length was measured incrementally during the fatigue loading by monitoring the mirror polished surface via the built-in fiberscope with an accuracy better than 0.1 mm. The crack surface of the fractured specimens was inspected under high resolution stereo microscope concerning their circularity, the total propagated crack length, and the dark colored initial crack length, which differs clearly from the crack propagated regime at 12 K (see Fig. 1, C, D, and E).

RESULTS AND DISCUSSION

The 7 K measurements with Incoloy 908 and 316LN concerning the fatigue-life properties are given in Fig. 2. At load ratio R = 0.1 Incoloy 908 base metal shows a significant (~ 12% at N = $5 \cdot 10^5$) higher resistance against fatigue loading as compared to 316LN base metal. The weldments of both metals have slightly lower fatigue-life performance as compared to their base metals. Increasing the load ratio from R = 0.1 to R = 0.7 shifts the fatigue life line towards high stresses (~15%) contrary to FCGR findings (details are given in [5]). Constant stress-life investigations with 316LN flaw bearing base metals show considerably lower fatigue life performance. The stress for these tests was held constant by displacement control (i.e. the stress at the remaining cross section was kept constant by examining the constant displacement level on the oscilloscope with steady load decrease). The fractured surface of the specimen (Fig. 2) indicates the constant stress condition due to the fine sectional separation. Starting with a EDM-flaw require a considerable amount of cycles to emanate the initial crack. In this context it is worth mentioning that SEM inspection of the EDM surfaces shows virtually small microcracks introduced by thermal shock process during the electro-discharge in the oil bath. This means that any other artificial flaw (e.g. scratches, etc.) will behave not so severe as the one with the EDM-flaws. Marking the penetrated crack either by beach marks at 12 K or heat tinting at 295 K and further constant stress fatigue loading at 12 K (with respect to the remaining cross section estimated after the test on the broken halves) show no differences between a crack opened at 295 K or at 12 K in this stress regime. A specimen failure with a started predefined crack (most severe case) shows approximately ~ 20% lower performance as compared to a severe artifical flaw (EDM-notch).

Figures 3 and 4 give the base and weld metal FCGR properties of 316 LN and Incoloy 908 materials. Figure 3 (case No. 1) shows the plotted

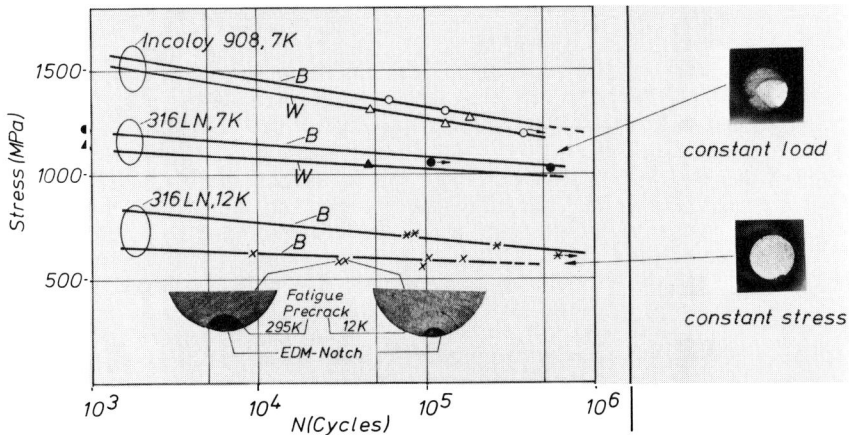

Fig. 2. 7 K fatigue-life results of smooth cylindrical specimens with 316LN and Incoloy 908 base metals (B) and their flash butt weldments (W) at R = 0.1. (*) 12 K results of 316LN base metal starting from a small EDM-notch at constant stress condition. (x) 12 K fatigue-life results of 316LN base metal starting with a predifined fatigue crack. Fatigue crack initiation performed either at 295 K or at 12 K. Surface of fractured specimens.

readings of the FCGR values, which scatter in a band of ± 40%. All other presented curves are similarly obtained as the case No. 1. This result confirms that between 20 K and 7 K no differences in FCGR exist. The weld metal FCGR for both materials show a lower FCGR as compared to the base metal. At load ratios higher than R > 0.1 the crack starts to branch into the base metal (due to narrow weld metal zone) and the FCGR increases (see Fig. 4, case No. 4). At R = 0.7 the FCGR with the welded specimen follows almost the route into the base metal (Fig. 4, case No. 5). The crack starts here to propagate from its early stage near the fusion zone along the weldment. A crack embedded fully in the weldment (Fig. 4, case No. 6) shows a significantly low FCGR compared with the flash butt weldments. This can be attributed to the existing high residual stresses along the welding direction. The weld CT-specimens after machining identify this phenomenon by closing the crack starter slot (see Fig. 4). High compressive stresses act on the crack tip in such a case and the fatigue loading does not reduce the compressive residual stresses according to these results. Thus, these stresses, which are multiaxial in their nature act on the crack tip with a load against the crack opening displacement. This means the external load must be increased for neutralizing these residual stress effects. A significant percentage of the stress intensity factor range (ΔK) is used in case of the welded CT-specimens, therefore, for opening the crack starter slot to the same level of the base metal specimen.

During the course of these investigations cryogenic FCGR-retardation below $\Delta K = 10$ MPa$\sqrt{m}$ was observed with specimens precracked at 295 K. Starting of the crack propagation below 20 K was only possible by ΔK-increase to > 15 MPa$\sqrt{m}$. Afterwards the FCGR could be determined in a standard way down to $\Delta K \sim 7$ MPa$\sqrt{m}$. This phenomenon can be explained by the stress gradient at the crack tip resulting from the cool-down of the precracked specimen. To determine the existing stresses on the specimen surface, a special specimen was prepared with 8 strain gages (1 x 1.5 mm^2) located at different positions in the vicinity of the crack tip and on reference positions. After reaching the 7 K level a compressive stress at

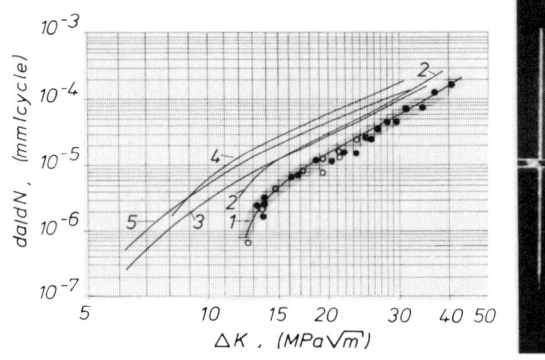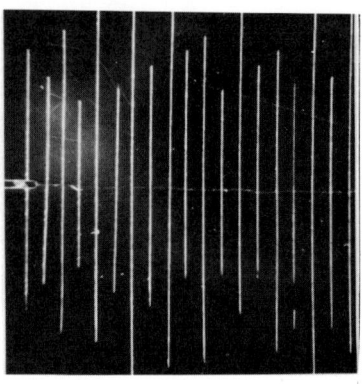

Fig. 3. Cryogenic FCGR of the base materials Incoloy 908 and 316LN in aged condition between 7 K and 20 K. 1) Incoloy 908, (o) refers to 7 K performed by K-increasing technique, (●) refers to 20 K performed by K-decreasing technique at R = 0.1,2) Incoloy 908 at R = 0.7, 3) 316LN at R = 0.1, 4) 316LN at R = 0.7, 5) 316LN at R = 0.4. On the right hand side one can see the surface view of the propagated crack and the 0.5 mm scanned grid.

the crack tip was estimated due to the output signals in the range of ~40 MPa. Figure 5 shows a model of the estimated stresses at the crack tip after cool-down and one of the investigated strain gage bonded specimen. The result given in Fig. 5 shows the significant crack retardation below $\Delta K \sim 10$ MPa$\sqrt{m}$. This effect should be taken into account in the design of cryogenic loaded structures with manufacturing related cracks and flaws below 10 MPa$\sqrt{m}$ regime.

The FCGR of surface cracks for the 316LN base metal (as received) yield between 10 to 35 MPa$\sqrt{m}$ a linear relationship at R = 0.1 in the double logarithmic da/dN vs. ΔK diagram. The 12 K results of five specimens plotted in this diagram gives a scatterband for the crack propagation law in c-direction (see Fig. 1, G) Paris law constants of n = 4.19 and C between $3.21 \cdot 10^{-11} - 5.14 \cdot 10^{-11}$. For 295 K the obtained values for n and C are n = 3.16, C between $4.0 \cdot 10^{-9} - 4.9 \cdot 10^{-9}$ (C in mm/cycle). Regarding this

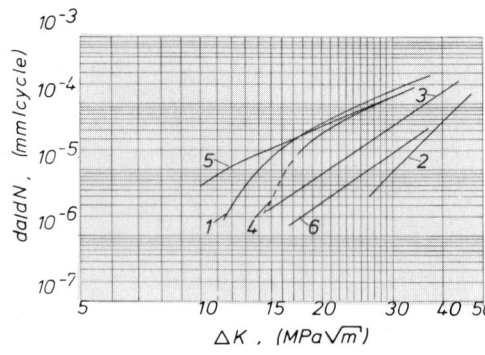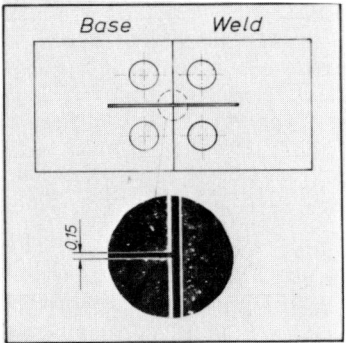

Fig. 4. Cryogenic FCGR in the flash butt welded metals of Incoloy 908 and 316LN between 7 K and 20 K in aged condition. 1) Incoloy 908 at R = 0.7, 2) Incoloy 908 at R = 0.1, 3) 316LN at R = 0.1, 4) 316LN at R = 0.4, 5) 316LN at R = 0.7 and 6) TIG-weldment of 316LN at R = 0.1. On the right hand side one may see the compressed crack starter slot of a welded CT-specimen.

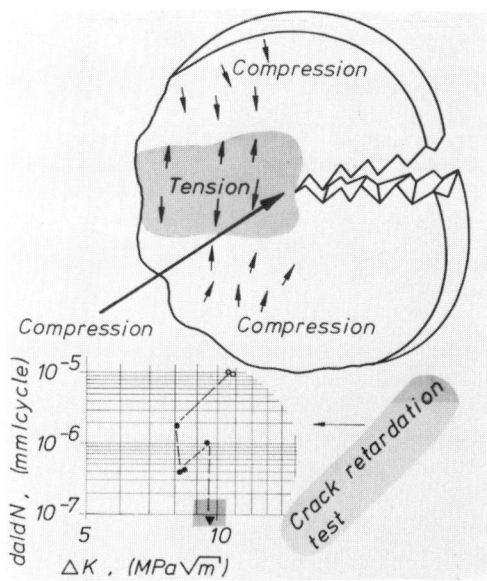

Fig. 5. Model of the stress distribution at the crack tip after cool down from 295 K to 7 K. Crack retardation test using one specimen with crack started at R =0.4 and at 7 K (o). Warm up of the specimen and crack propagation at 295 K and at R = 0.1 (●). Here note the initial anomalous very high FCGR. After propagating the crack at 295 K (●) up to $\Delta K \sim 9{,}5$ MPa$\sqrt{m}$ cool down of the specimen to 20 K (▽) and here no detectable crack propagation after $1.2 \cdot 10^6$ cycles at R = 0.1. Detection limit of ~ 0.1 mm results in a da/dN of $< 8 \cdot 10^{-8}$ mm/cycle. See also one of the strain gage monitored original specimen with the propagated crack.

fact the cryogenic environment seems to be not so harmful as compared to the 295 K case, contrary to the obtained findings with CT-specimens (embedded cracks). The 12 K FCGR-performance of the surface crack is higher than the results of the Fig. 3, case No. 3. To check the performed tests towards its consistency the cyclic life was predicted for all broken specimens concerning their initial crack a_0 and the final crack. Substituting the obtained growth law constants n and C in a system of equations as given by

$$N = \sum_{i=1}^{i=m} \Delta a / C \cdot \Delta K_i^n, \quad \Delta K_i = f(\Delta \sigma, a, c, b, B, \Phi) \qquad (Ref.\ 4)$$

$$\text{and} \qquad a = a_o + \sum_{i=1}^{i=m} \Delta a$$

a numerical computation could be carried out and it could be confirmed that for all specimens the predicted cycle numbers fit with the experimentally determined cycle numbers.

CONCLUSIONS

- Incoloy 908 base metal has a low FCGR and a high fatigue-life as compared to the 316LN base metal at cryogenic regime.

- FCGR of weldments for both metals are low as compared to their base metals.
- Residual stresses are the major responsible factor for the low FCGR in the weldments.
- Cooling down results in compressive stresses at the crack tip with significant FCGR-retardation at low ΔK-regime (< 10 MPa$\sqrt{m}$).
- In the range 20 K - 7 K FCGR of surface cracks is lower as compared to cracks propagating from a buried penny shape type crack.

ACKNOWLEDGMENT

Part of this work was performed under the NET contract NET-No. 89-201. Further we acknowledge for the valuable discussions with Dr. Neil Mitchell from the NET-Team, Garching.

REFERENCES

1. R. Annandale, L. Bottura, N. Mitchell, M. Perella, and E. Salpietro, Model coil design for verification of the coil/conductor concept for the NET coils, Proc. of Symp. on Fus. Techn. (SOFT), London, Sept. 1990, to be published.
2. W.J. Muster and J. Elster, Low temperature embrittlement after ageing stainless steels, Cryogenics, Sept. 1990, 30(9), pp. 799-802.
3. A. Nyilas, B. Obst, J. Zhang, and W.J. Muster, Cryogenic tensile, fracture and fatigue investigations of nitrogen alloyed stainless steels, Proc. of 2nd Int. Conf. on High Nitrogen Steels (HNS), Aachen, October 1990, pp. 191-196.
4. J.C. Newman and I.S. Raju, Analysis of surface crack in finite plates under tension or bending loads, 1979, NASA STP 1578.
5. S. Förster, A. Nyilas, and A. Ulbricht, Properties of Stainless Steels for Fusion Conductor Jackets and their Manufacturing Techniques, Proc. of 12th Magnet Technology (MT 12), Leningrad, June 1991, to be published.

DEFORMATION STRUCTURES IN HIGH-CYCLE FATIGUE OF 0.1N-32Mn-7Cr STEEL AT CRYOGENIC TEMPERATURES

Osamu Umezawa and Keisuke Ishikawa
National Research Institute for Metals
1-2-1 Sengen, Tsukuba, Ibaraki 305, Japan

ABSTRACT

Nitrogen-strengthened 32Mn-7Cr austenitic steel was cyclically deformed at various maximum stress amplitudes. The deformation structures induced by fatigue at cryogenic temperatures were examined by transmission electron microscopy. The dominant deformation mode was the {111}-<110> slip system, and dislocation motion was planar, especially at low stress levels. Cross slip of dislocations was strongly suppressed, and dislocation movements were restricted to their slip planes. At higher stress levels, mainly slip bands were observed. At lower stress levels, microslip or dislocation pile-ups were blocked or sharply localized at grain boundaries. Deformation twinning and ε-martensite are not presumed to play a major role in the cyclic deformation.

INTRODUCTION

Nitrogen-strengthened 32Mn-7Cr austenitic steel was designed to have high yield strength, improved toughness, and high phase stability at liquid helium temperature.[1] This steel fractured in a ductile manner with dimples,[2] but during high-cycle fatigue, subsurface crack initiation due to intergranular cracking occurred in liquid helium and liquid nitrogen.[3-4] According to the characterization of the subsurface crack initiation site and observation of the dislocation structure under the fracture surface, heterogeneous slip deformation certainly assisted an intergranular crack resulting from stress concentration at the grain boundary.[4] Moreover, at low temperatures, the 32Mn-7Cr steel containing carbon and nitrogen showed low-cycle fatigue softening.[5] That corresponded to the planar dislocation structure.[5] In these ways, the fatigue behaviors were closely related to deformation structures, but the fatigue deformation structure has not been clarified.

Therefore, in the present study, transmission electron microscopy (TEM) was used to investigate the deformation structures that developed in the solution-treated 32Mn-7Cr steel after high-cycle fatigue under various stress amplitudes and fatigue cycles at cryogenic temperatures.

Table 1. Chemical Composition of the Investigated Steel (in mass%).

C	Si	Mn	P	S	Ni	Cr	Mo	Al	N	Fe
0.14	0.60	31.58	0.022	0.006	0.23	7.04	0.04	0.012	0.133	bal.

EXPERIMENTAL PROCEDURE

Test Specimens

The chemical composition of the 32Mn-7Cr steel investigated in this study is listed in Table 1. Cast ingots were forged and rolled at 1423 K into 30-mm-thick plate, then solution-treated at 1293 K for 7.2 ks followed by a water quench. Hourglass-type unnotched specimens with a waist diameter of 6 mm (as in Fig. 1) were fatigued in liquid helium (4 K), in liquid nitrogen (77 K), and in ambient air (293 K), and examined. The longitudinal direction of the specimens was perpendicular to the rolling direction. Load-controlled test was done with a sinusoidal waveform and a minimum-to-maximum stress ratio, R, of 0.01. The investigated samples were from fatigued specimens that were used to determine the S-N curves in Ref. 3, and a specimen that had been cyclically deformed by 5 cycles at 4 K (as in Fig. 2). Plastic strain following fatigue was not measured for every specimen in this study.

Transmission Electron Microscopy

After fatigue testing, samples about 600 µm thick were sectioned from the hourglass region for TEM, as shown in Fig. 1. They were cut perpendicular to the principal stress axis and were reduced to about 100 µm in thickness by abrasion with no. 600 emery paper. The diameters of the samples were measured, and the stress level (maximum stress amplitude: $\Delta\sigma_{max}$) of each sample was determined; then TEM disks with 3-mm diameter were punched out. The stress levels and fatigue cycles of examined samples are given in Fig. 2. TEM foils were prepared by electrical jet polishing of the disk on both sides in a stirred solution of 10% acetic acid and 90% methanol at 280 K. Finally, they were rinsed in water and in methanol. A JEOL 2010 TEM equipped with a double-tilt goniometer stage was employed at 200 kV.

RESULTS AND DISCUSSION

No formation of ferromagnetic α'-phase was detected during plastic deformation in 32Mn-7Cr steel even at 4 K.[2] Hence there were three distinct deformation mechanisms in this steel at low temperatures: slip, deformation twinning, and transformation to ϵ-martensite.

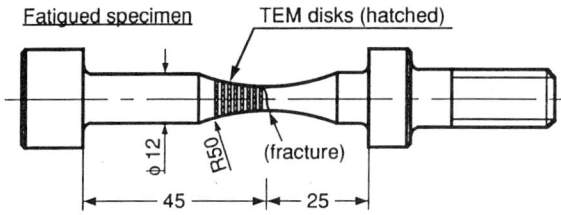

Fig. 1. Location of sections cut from fatigued specimens for TEM samples.

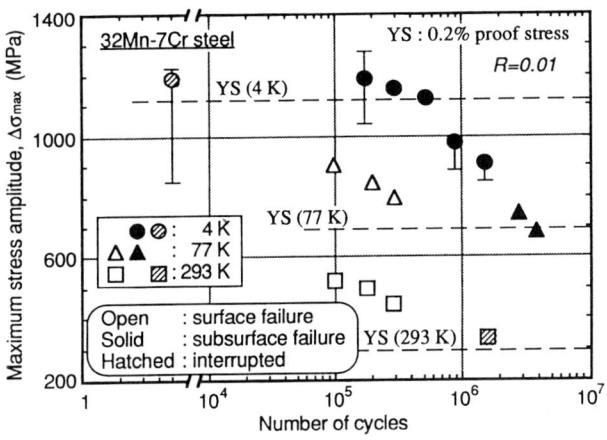

Fig. 2. Stress levels and fatigue cycles of the examined samples.

Deformation Structure at 4 K

In the planar dislocation structures at greater than 10^5 cycles, mainly slip bands were observed. The dominant deformation mode was the {111}-<110> primary slip system. Stacking faults were not observed, and ε-martensite was hardly detected. The deformation structure was changed by $\Delta\sigma_{max}$ and fatigue cycles, as mentioned below.

$\Delta\sigma_{max}$ above 0.2% proof stress (YS). Figure 3a showed straight parallel bands. They were {111} slip bands that grouped closely spaced dislocations. Figure 3b shows the slip bands and [1$\bar{1}\bar{1}$] deformation twinning. The substructure of those band structures was very dense dislocations.

$\Delta\sigma_{max}$ below YS. The cross slip of dislocations was strongly suppressed, and dislocation movements were restricted to their slip planes, as shown in Fig. 4. The appearance of secondary (conjugate) slip was observed in an undeformed region. Slip bands or high-density grouped dislocations were the main cause of structural deformation. They were widely spaced rather than at the higher stress level.

Dislocation structures fatigued by 5 cycles. Figure 5 shows the dislocation arrangements in a specimen that was cyclically deformed by 5 cycles. The stress level of the samples in Fig. 5a, 5b is almost the same as that in Fig. 3. In some grains, a cellular distribution of dislocations was recognized, as shown in Fig. 5b, and in others, a planar structure was observed, as shown in Fig. 5a. The very dense dislocations were confined to slip planes and tangled (as in Fig. 5a). Partial formation of slip band were seen (as in Fig. 5b). Deformation twinning, ε-martensite, and stacking faults were not observed. The stress level of the sample in Fig. 5c is almost the same as that in Fig. 4. In the structures fatigued at the lower stress level, dislocation arrays or pile-ups were dominant, and slip bands was not observed, as shown in Fig. 5c. The dislocations in the planar structure were confined to the slip plane, as they were at the higher stress level, and the dislocation density was somewhat lower.

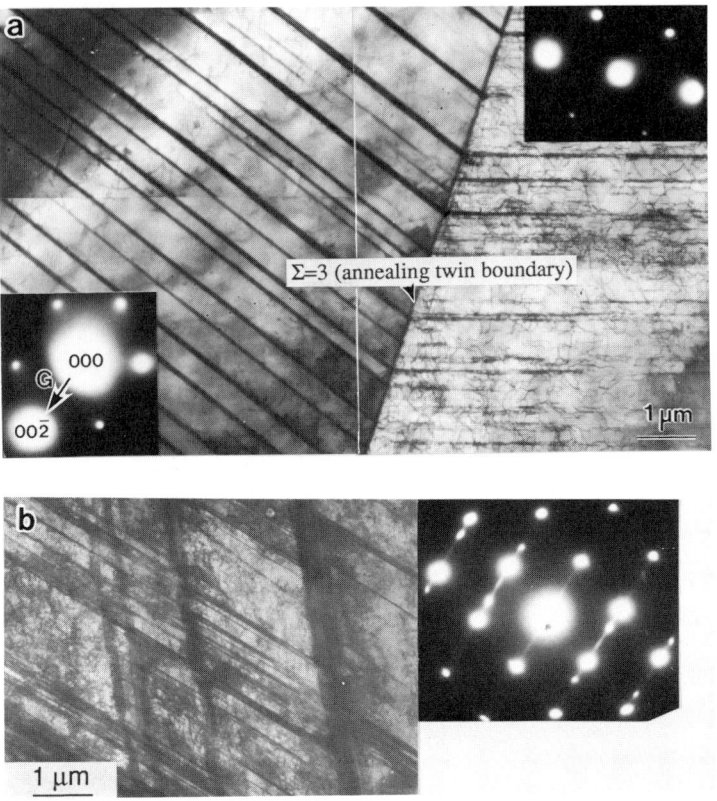

Fig. 3. Deformation structures of cyclically deformed samples after 174200 cycles at 4 K (a:$\Delta\sigma_{max}$=1224 MPa; b:$\Delta\sigma_{max}$=1280 MPa).

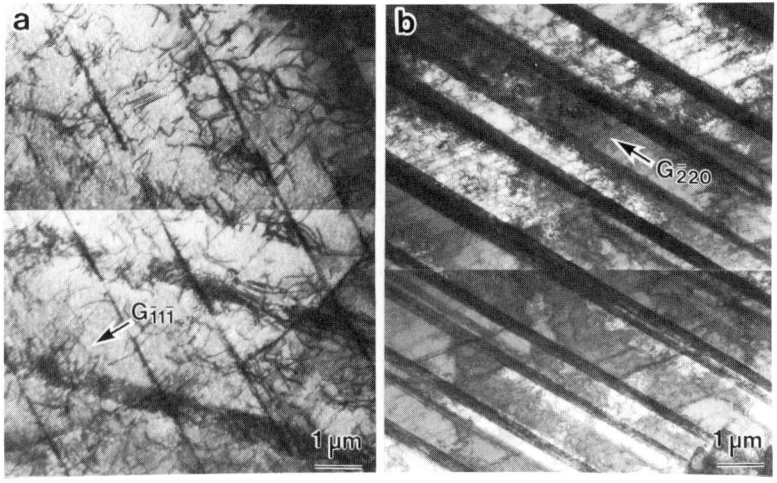

Fig. 4. Dislocation structures of cyclically deformed samples after 874010 cycles at 4 K (a:$\Delta\sigma_{max}$=992 MPa; b:$\Delta\sigma_{max}$=877 MPa).

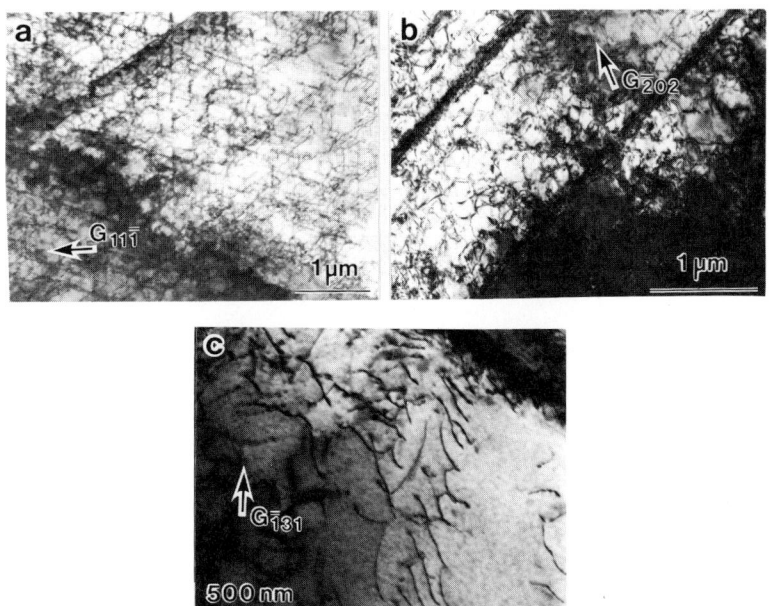

Fig. 5. Dislocation structures of cyclically deformed samples after 5 cycles at 4 K
(a,b:$\Delta\sigma_{max}$=1224 MPa; c:$\Delta\sigma_{max}$=992 MPa).

Deformation Structure at 77 K

$\Delta\sigma_{max}$ above YS. At the higher stress level, slip bands were distributed over a grain, as shown in Fig. 6. Deformation twinning was hardly observed, but ε-martensite was detected at the intersection of primary and secondary slip (as in Fig. 6).

$\Delta\sigma_{max}$ below YS. At the lower stress level, the dislocation density was less, and the plastic deformation was more distinctly localized. The dislocation arrangement was developed mainly by pile-ups or coplanar arrays (as in Fig. 7). The slip bands did not form remarkably. Neither deformation twinning nor ε-martensite was observed. In Fig. 7, the pile-up screw dislocations on slip system A in grain 1 and the transmitted dislocations on slip system B in grain 2 had a equal Burgers vectors of $a/2[10\bar{1}]$. The grain boundary was analyzed as a Σ=3 twin boundary. Slip systems A and B have a common line of intersection in the grain boundary.

Deformation Structure at 293 K

At the higher stress level, for example, $\Delta\sigma_{max}$=530 MPa, the deformation structure was similar to that at 77 K in Fig. 6. On the other hand, the dislocation configurations observed in the specimen cycled at the lower stress level were mainly dislocation pile-ups near grain boundaries, as shown in Fig. 8. Even at the lower stress level, dislocations had generated in more than one slip systems, although their density was rather low. In Fig. 8, a dislocation pile-up in slip system A transfers strain across the grain boundary with residual dislocations left in the boundary.

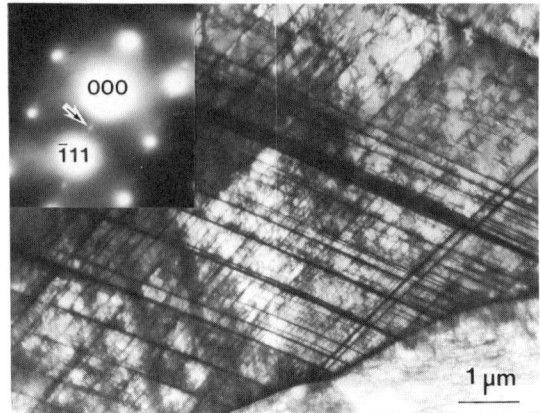

Fig. 6. Deformation structure of a cyclically deformed sample after 293680 cycles at 77 K ($\Delta\sigma_{max}$=770 MPa).

Summarized Results and Localized Deformation

The results of the deformation structure analysis by TEM are summed up in Fig. 9. Dominant deformation mode was the {111}-<110> slip system, and dislocation motion was planar, especially at the lower stress level. Deformation twinning and ε-martensite were not a major factor in cyclic deformation. At the higher stress levels, slip bands mainly were formed. At the low stress levels, microslip or dislocation pile-ups were blocked or

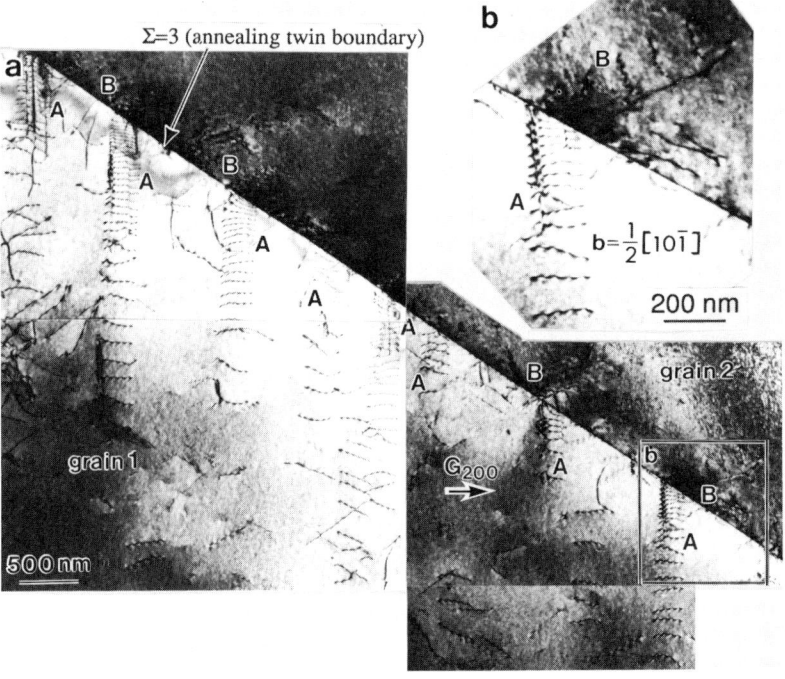

Fig. 7. Dislocation pile-ups through a grain boundary ($\Delta\sigma_{max}$=684 MPa, 3769180 cycles at 77 K). Micrograph b is enlargement of the framed part in a.

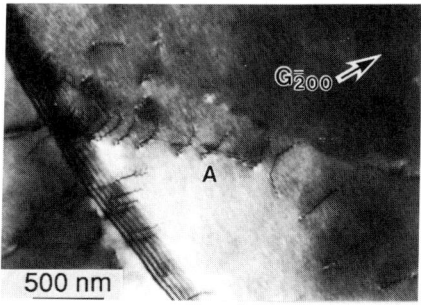

Fig. 8. Dislocation structure of a cyclically deformed sample after 1590000 cycles at 293 K ($\Delta\sigma_{max}$=322 MPa).

sharply localized at grain boundaries. Therefore, localized deformation obviously causes stress concentration at the grain boundary.

CONCLUSIONS

This study investigated the deformation structure of 0.1N-32Mn-7Cr austenitic steel that had been cyclically deformed at cryogenic temperatures. The dominant deformation mode was the {111}-<110> slip system. The dislocation arrangement was planar, and the dislocations were confined to slip planes. At a lower stress level, the deformation structure changed: slip bands, grouped dislocations on a slip plane, and dislocation pile-ups.

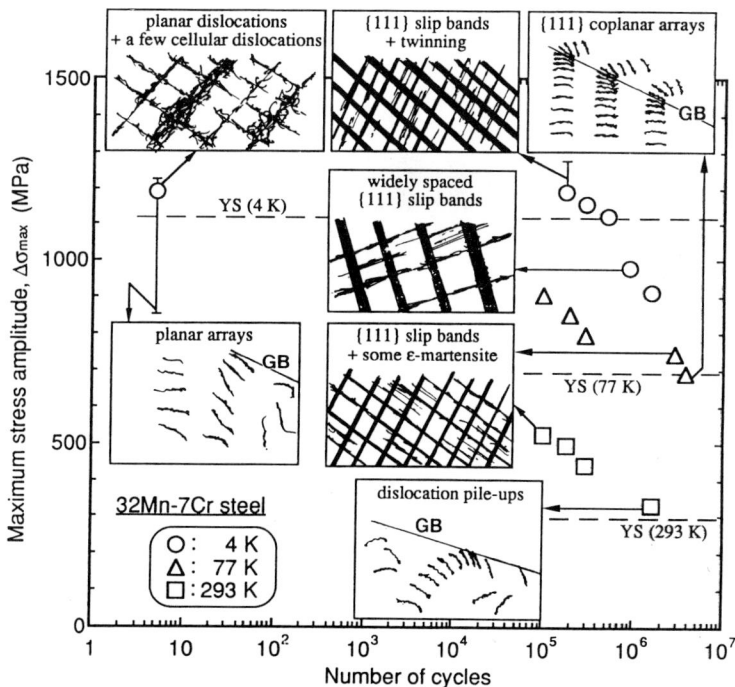

Fig. 9. Summarized results of deformation structures after high-cycle fatigue.

REFERENCES

1. R. Miura, H. Nakajima, Y. Takahashi, and K. Yoshida, 32Mn-7Cr austenite steel for cryogenic applications, in: "Advances in Cryogenic Engineering, Materials," R.P. Reed and A.F. Clark, eds., Plenum Press, N.Y., Vol.30:245-252 (1984).
2. T. Yuri, K. Nagai, and K. Ishikawa, Effect of cold rolling on mechanical properties and fracture mode of a 32Mn-7Cr steel at cryogenic temperatures, ISIJ Inter., 30:639-645 (1990).
3. T. Yuri, K. Nagai, T. Ogata, O. Umezawa, and K. Ishikawa, Cryogenic high cycle fatigue properties of high-manganese steels (in Japanese), in: "CAMP-ISIJ," Vol.2:1372 (1989).
4. O. Umezawa, K. Nagai, T. Yuri, and K. Ishikawa, Subsurface crack initiation in high cycle fatigue of 0.1N-32Mn-7Cr steel at cryogenic temperatures, in: "Proc. 6th International Conference on the Mechcanical Behavior of Materials," H. Jinno and M. Jono, eds., Pergamon Press, Oxford, (1991) in press.
5. K. Shibata and T. Fujita, Low cyclic fatigue behavior of 32% Mn nonmagnetic steel and the effects of C and N in liquid nitrogen and liquid helium (in Japanese), Testu-to-Hagane, 73:1178-1185 (1987).

NEAR-THRESHOLD FATIGUE CRACK GROWTH OF

AUSTENITIC STAINLESS STEELS AT LIQUID HELIUM TEMPERATURE

Kenichi Suzuki, Juichi Fukakura and Hideo Kashiwaya

Heavy Apparatus Engineering Laboratory
Toshiba Corporation, 2-4 Suehiro-cho, Tsurumi-ku
Yokohama, Japan

INTRODUCTION

When designing superconducting magnets for nuclear fusion equipment or superconducting generators, consideration must be made of the difficulties of in-service inspections arising from structural complications. Through paying proper consideration, it is necessary to evaluate the remaining service life by means of fatigue crack growth analyses based on preliminarily assumed initial flaws.

There exist extensive restrictions for effective utilization of data because data availability is limited on the fatigue crack growth characteristics of structural materials at cryogenic temperatures (4K, 77K) [1-7]. This lack of information is particularly true with data regarding the apparent threshold (ΔK_{th}) or the effective threshold ($\Delta K_{eff, th}$) which correct the crack closure effects. For example, most available load ratios (R) fall within the range of 0.05 - 0.1 and the data on higher ratios can hardly be found [1, 4, 7]. As for 300-series stainless steel, the data on stable stainless steel is quite scarce [5]. Also extremely limited is data concerning mechanical properties, specifically regarding yield strength effects [4]. Furthermore, we are not able to clarify influences of specimen thickness at cryogenic temperatures, although there is data on the case where ΔK_{th} at room temperature is largely varied by specimen thickness [8].

To reduce these limitations, we performed the near-threshold fatigue crack growth tests on three austenitic stainless steels at 4K and 300K.

EXPERIMENTAL PROCEDURES

Materials and Specimens. The test specimens selected were meta-stable stainless steel of 316L and 316LN and stable stainless steel of 310S. Chemical compositions and mechanical properties are shown in Table 1 and Table 2 respectively. The specimens are CT type with thickness of 12.5mm.

Loading Conditions. The fatigue crack growth tests were conducted by leaving the test specimens at atmospheric air (300K) and by keeping them immersed in liquid helium (4K) using a cryostat for 4K fatigue test [9]. Sinusoidal load waves were used and two types of R, namely 0.4 and 0.7 (0.4 only for 310S), were selected. Employed cyclic frequency was 5 - 15Hz.

On most of the test specimens, crack growth was caused from pre-fatigue crack tips along with decrease of ΔK, and on a portion of these specimens,

Table 1. Chemical compositions.

Material	C	Si	Mn	P	Ni	Cr	Mn	N
316 L	0.025	0.47	0.80	0.027	12.25	16.35	2.12	–
316 LN	0.021	0.53	0.82	0.023	11.30	17.85	2.85	0.18
310 S	0.03	1.02	1.76	0.002	20.02	24.37	–	–

(weight percent)

Table 2. Mechanical properties.

Material	Temp. (k)	Yield Strength (MPa)	Tensile Strength (MPa)
316 L	300	216	529
	77	314	1235
	4	431	1441
316 LN	300	384	735
	77	915	1505
	4	1150	1708
310 S	300	284	582
	77	602	1103
	4	818	1288

after continuing the decrease to the predetermined ΔK value, a shift was made to increase ΔK, that is, the state of constant ΔP. On the remaining test specimens, we let cracks grow from the beginning under the condition of constant ΔP. Through referring to the ASTM [10], ΔK was decreased by stepped load shedding based on $C = -0.05$ or -0.08 mm^{-1} and $\Delta a = 0.26 - 0.6$ mm on the equation : $C = (1/\Delta K)(d\Delta K/da)$.

<u>Crack Length and Closure Measurement</u>. We measured crack lengths by using the compliance method through developing a new system which employs a 16-bit computer and a microcomputer for controlling the fatigue testing equipment [1]. Figure 1 is an example of graphic output showing the relations (A) between loads and crack opening displacement (COD). Since graphic processing permits proportional expansion (B) and smoothing(C) of the nonlinear sections, the crack opening points can be easily discriminated.

TEST RESULTS

<u>Crack Growth Behavior of 316L</u>. Shown in Fig. 2 are the relations among da/dn at 4K obtained by ΔK decrease of $C = -0.08$ mm^{-1}, ΔK, and ΔK_{eff}. Also shown in this Fig. 2 for comparison purpose are the results of ΔK increase under $R = 0.4$, but no clear crack closure was observable from these results. The relations between ΔK and da/dn obtained from ΔK increase under $R = 0.4$ were equal to those resulting from ΔK decrease. During ΔK decrease, there were cases of occurrence of slight crack closure, but as a whole, the relations between da/dn and ΔK_{eff} were quite similar to those that occurred during ΔK increase. When the condition of $R = 0.7$ existed, no crack closure was found during gradual decrease of ΔK. Therefore, the relations among da/dn, ΔK, and ΔK_{eff} could be judged the same as those during ΔK increase.

Figure 3 shows the results at 300K obtained by ΔK decrease of $C = -0.08$ mm^{-1} or -0.05 mm^{-1}. Under $R=0.4$ and $C = -0.05$ mm^{-1}, the relations were akin to those during ΔK increase, but under $C = -0.08$ mm^{-1}, the lower the ΔK value became, the greater the differences of da/dn vs. ΔK_{eff} relations were observable. When ΔK decrease of $C = -0.08$ mm^{-1} was selected under $R = 0.7$ and $\Delta a = 0.26$mm on low ΔK side was used, namely, $(d\Delta K)/\Delta K$ was set at about -2%, no crack closure was observed and relations between da/dn and

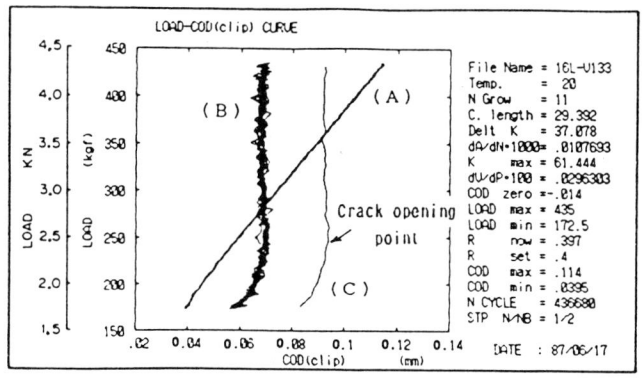

Fig. 1 CRT-image of compliance curve.

ΔK or ΔK_{eff} were equal to those during ΔK increase. Although not indicated by figures, the same results were obtained under $C = -0.05$ mm^{-1} and $\Delta a = 0.4$ mm, namely, under $(d\Delta K)/\Delta K = -2\%$. On the other hand, in case of $C = -0.08$ mm^{-1} and $\Delta a = 0.4$ mm, or $(d\Delta K)/\Delta K = -3.2\%$, crack closure was measured on the low ΔK side, and on the relations between da/dn and ΔK or ΔK_{eff}, great deviations were observed on the low ΔK side from those during ΔK increase.

Crack Growth Behavior of 316LN. Figure 4 shows the relations between da/dn and ΔK or ΔK_{eff} at 4K obtained by ΔK decrease and ΔK increase. Except the fact that crack closure was found on the low ΔK side during ΔK increase under R = 0.4, the trend identical with the results on the foregoing specimen of 316L was revealed. Shown in Fig. 5 are the test results obtained at 300K. Similarly to the case of 316L, under ΔK decrease of $C = -0.05$ mm^{-1} and $\Delta a = 0.4$ mm, the relations between da/dn and ΔK or ΔK_{eff} on R = 0.4 and R = 0.7 were identical with those during ΔK increase.

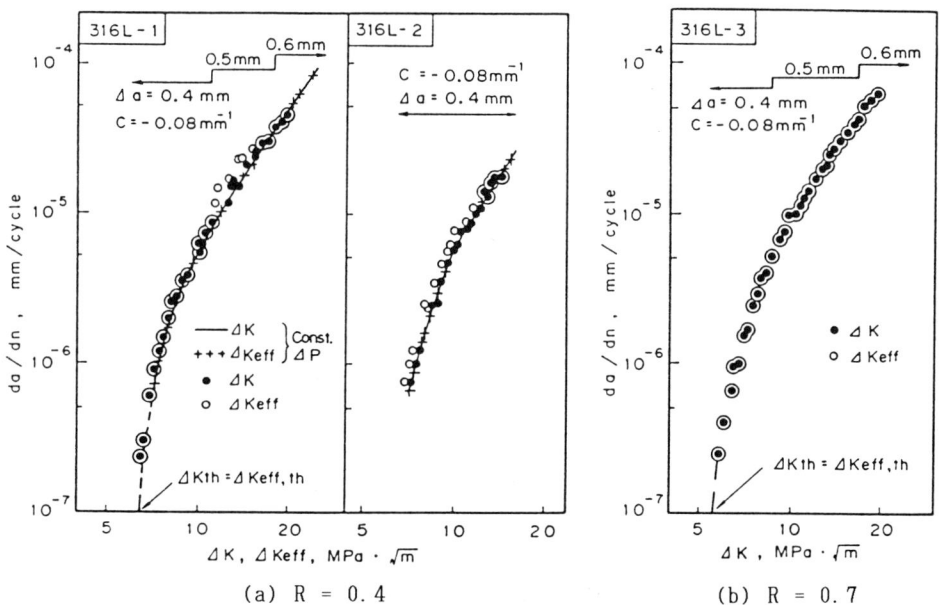

(a) R = 0.4

(b) R = 0.7

Fig. 2 Fatigue crack growth curves for 316L at 4K.

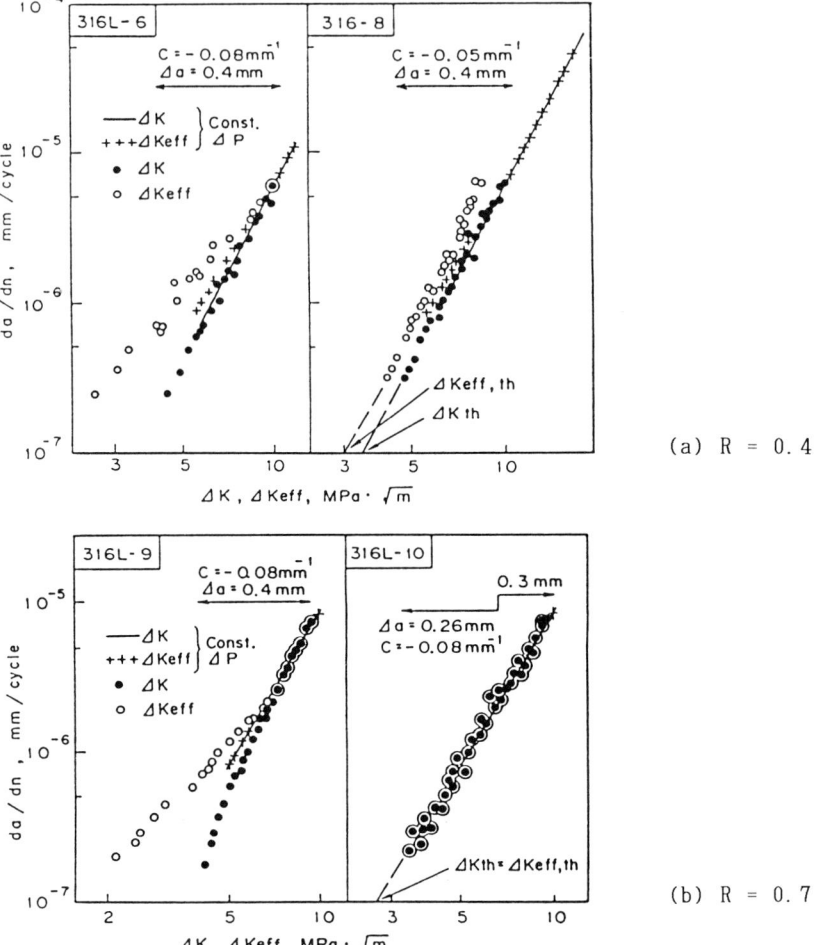

Fig. 3 Fatigue crack growth curves for 316L at 300K.

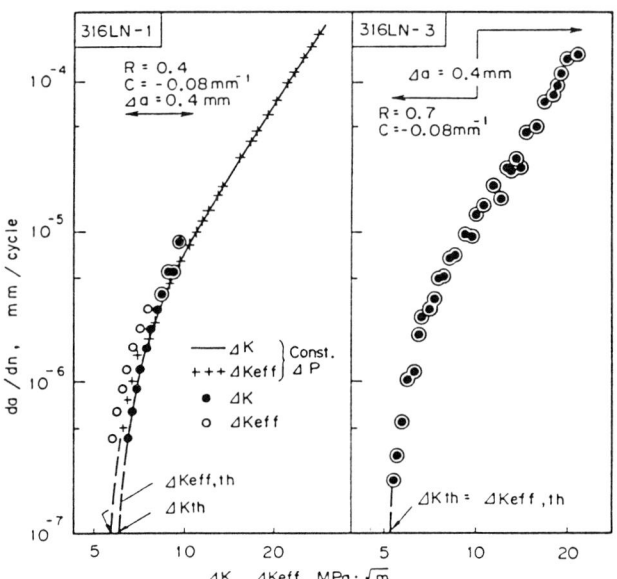

Fig. 4 Fatigue crack growth curves for 316LN at 4K.

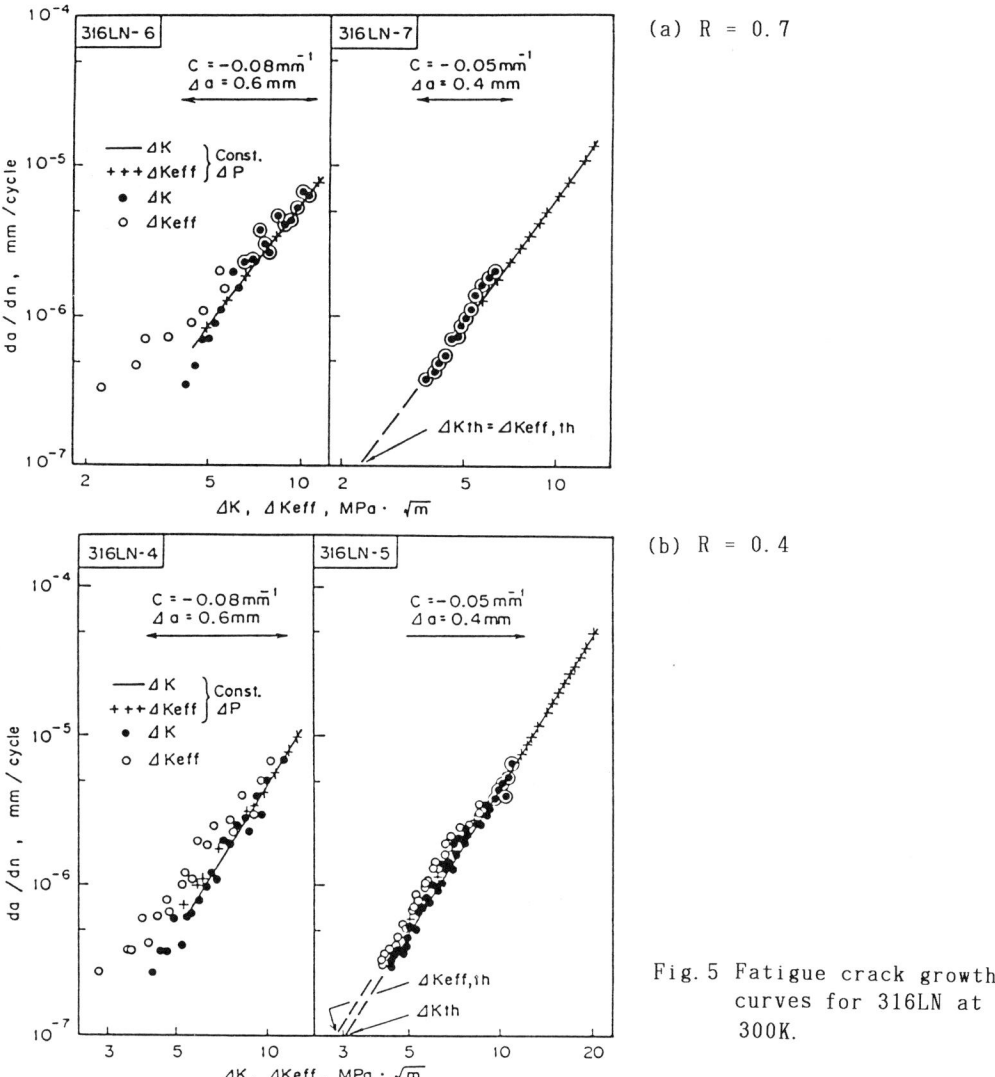

Fig. 5 Fatigue crack growth curves for 316LN at 300K.

Crack Growth Behavior of 310S. The test results at 4K and 300K are shown in Fig. 6. In view of the results on 316L and 316LN, we limited the ΔK decrease rates to $\Delta a = 0.4$ mm, $C = -0.08$ mm^{-1} (4K), and $C = -0.05$ mm^{-1} (300K). As a result, the relations between da/dn and ΔK or ΔK_{eff} created by ΔK decrease came to be equal to those by ΔK increase. At both 4K and 300K, crack closure was observed on the low ΔK side during the course of ΔK increase.

DISCUSSIONS

Near-Threshold Crack Growth Behavior

Figure 7 shows the curve lines which optimize the relations between da/dn and ΔK by summarizing the test results on 316L, 316LN, and 310S. Excluded are the data on specimens Nos. 316L-9 and 316LN-6. Since the test values were up to about da/dn = 2-3 x 10^{-7} mm/cycle, extrapolation was made up to 10^{-7} mm/cycle.

Commonly among the three types of stainless steel, da/dn vs. ΔK relations under 4K display smooth curve lines at low ΔK, while the same relations under 300K reveal a linear state. On 316L and 316LN, da/dn increases along with the increase of R at both 4K and 300K.

Influences on da/dn by temperature are dependent on steel types and ΔK levels. On 316LN, influences by temperature are shifted into the reverse direction at the turning point of da/dn = 1.6×10^{-6} mm/cycle. Influences by temperature are hardly observable on 316L in the neighborhood of da/dn = 5×10^{-6} mm/cycle. Rather, it was noted that on the lower ΔK side and the higher ΔK side, da/dn under 4K was less than da/dn at 300K. On 310S, da/dn at 4K was quite smaller than da/dn at 300K.

Apparent Threshold (ΔK_{th}) and Effective Threshold ($\Delta K_{eff,th}$)

Effect of ΔK-Decrease Rates.
As shown with the dashed lines in Figs. 3 through 6, we obtained ΔK_{th} and $\Delta K_{eff,th}$ as the values equivalent to da/dn = 10^{-7} mm/cycle by extrapolation of the relations between da/dn and ΔK or ΔK_{eff} created as a result of ΔK increase. These values are summarized in Table 3. In the case of the specimens Nos. 316L-6 and 316LN-4 under R = 0.4, the trend of causing $\Delta K_{eff,th}$ to become excessively small was observed, and this trend was believed to have resulted from apparent measurement of too large closures through being affected by influences of deformation state at positions other than front tips of cracks. On the same ΔK decrease rates as those for both specimens, ΔK_{th} also revealed the trend of marking excessively large values on the specimens Nos. 316L-9 and 316LN-6 under R = 0.7. We judge that this trend was due to the fact that the greater the value of R becomes, the greater the change rates of K_{max} (= $\Delta K/(1 - R)$) are increased. Consequently, variations of plastic zone sizes at crack front tips became greater, resulting in creating so-called delay effects.

Also, on the same ΔK decrease rates (C = -0.08 mm^{-1}, Δa = 0.4mm) as those for specimens (316L-6,9), we found the tendency of obtaining ΔK_{th} and $\Delta K_{eff,th}$ at 4K which are almost the same as the results of ΔK increase. We believe this trend was created because the yield strength was increased 2 - 3 times by temperature drop and the plastic zone sizes at crack front tips became relatively small.

Effect of Temperatures, Load Ratios(R), and Specimen Thicknesses(B).
Figure 8 shows ΔK_{th} and $\Delta K_{eff,th}$ at 4K and 300K obtained on 316L, 316LN, and 310S along with the reference data [2-8] of other 300-series stainless steel and stable austenitic alloys of precipitation hardening type. These test specimens with thin thickness (2.5 or 6.4mm) are identified with the closed symbols while thick ones (12.5 or 12.7mm) are identified with the opened symbols.

By focusing attention on 300-series stainless steel, we can note the trend that on ΔK_{th} at 300K, influences by thicknesses of specimens are greater than those by steel types, and thinner specimens indicate values approximately two times those of thicker specimens. On the other hand, for $\Delta K_{eff,th}$, influences by specimen thickness are considerably reduced. Generally, ΔK_{th} at 300K decreased along with the increase of R, but on $\Delta K_{eff,th}$, the ratios of decreases tended to become smaller. At 4K, with the exception of 310S and 304HN, influences by steel types and specimen thicknesses were relatively small against ΔK_{th} and $\Delta K_{eff,th}$. Influences of R were identical with those at 300K. As a result of temperature drop from 300K to 4K, ΔK_{th} and $\Delta K_{eff,th}$ increased on 316L, 316LN, and 310S used for our tests. However, regarding the reference data obtained with thin specimens, the results were divided into the cases of decrease and those of increase depending on steel types and R values. We can surmise the existence of such an influential element of temperature drop as intrinsic temperature effects [6, 7] related to variations of dislocation motion. Additionally, other elements that can be considered are variations of martensite transformation volumes [5, 6] and of crack front tip stress fields [8].

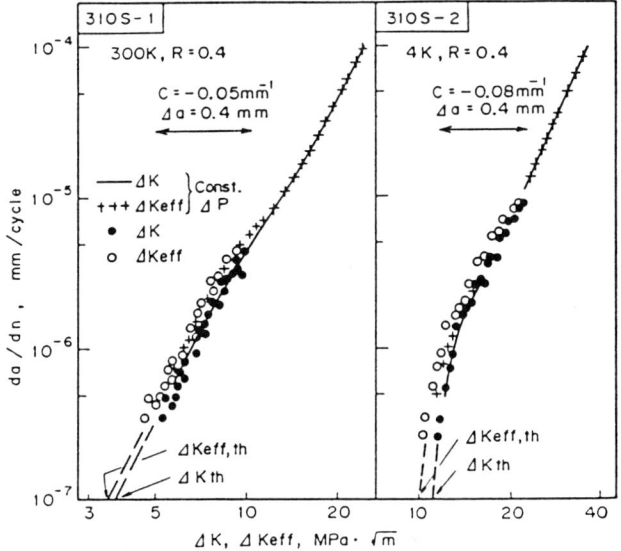

Fig. 6 Fatigue crack growth curves for 310S at 4K and 300K.

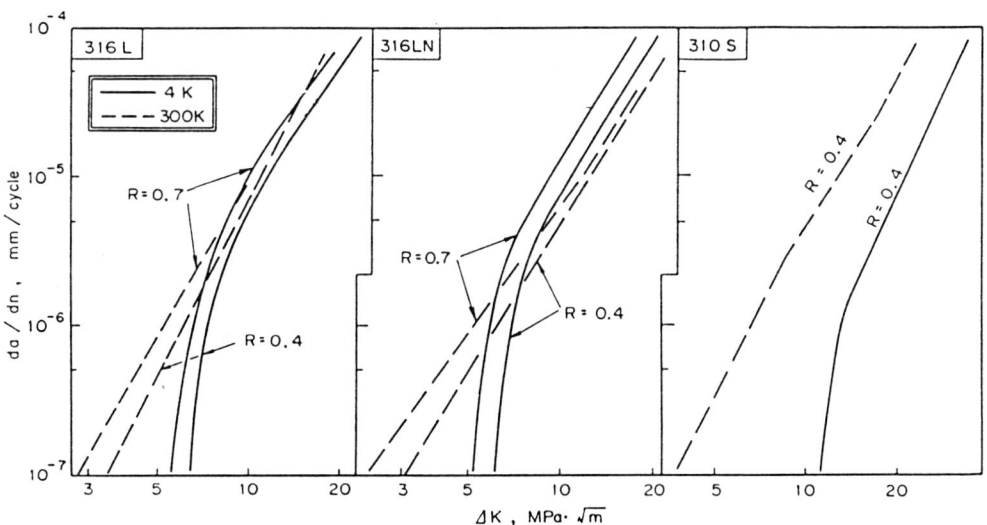

Fig. 7 Comparison of fatigue crack growth curves for 316L, 316LN and 310S at 300K and 4K.

Table 3. Apparent and effective threshold values for 316L, 316LN and 310S.

Material	Temp. (K)	ΔK_{th} (MPa·$\sqrt{m}$)		$\Delta K_{eff,th}$ (MPa·$\sqrt{m}$)	
		R=0.4	R=0.7	R=0.4	R=0.7
316 L	300	3.4	2.8	3.0	2.8
	4	6.5	5.5	6.5	5.5
316 LN	300	3.1	2.3	2.8	2.3
	4	6.0	5.2	5.7	5.2
310 S	300	3.8	–	3.5	–
	4	11.0	–	10.0	–

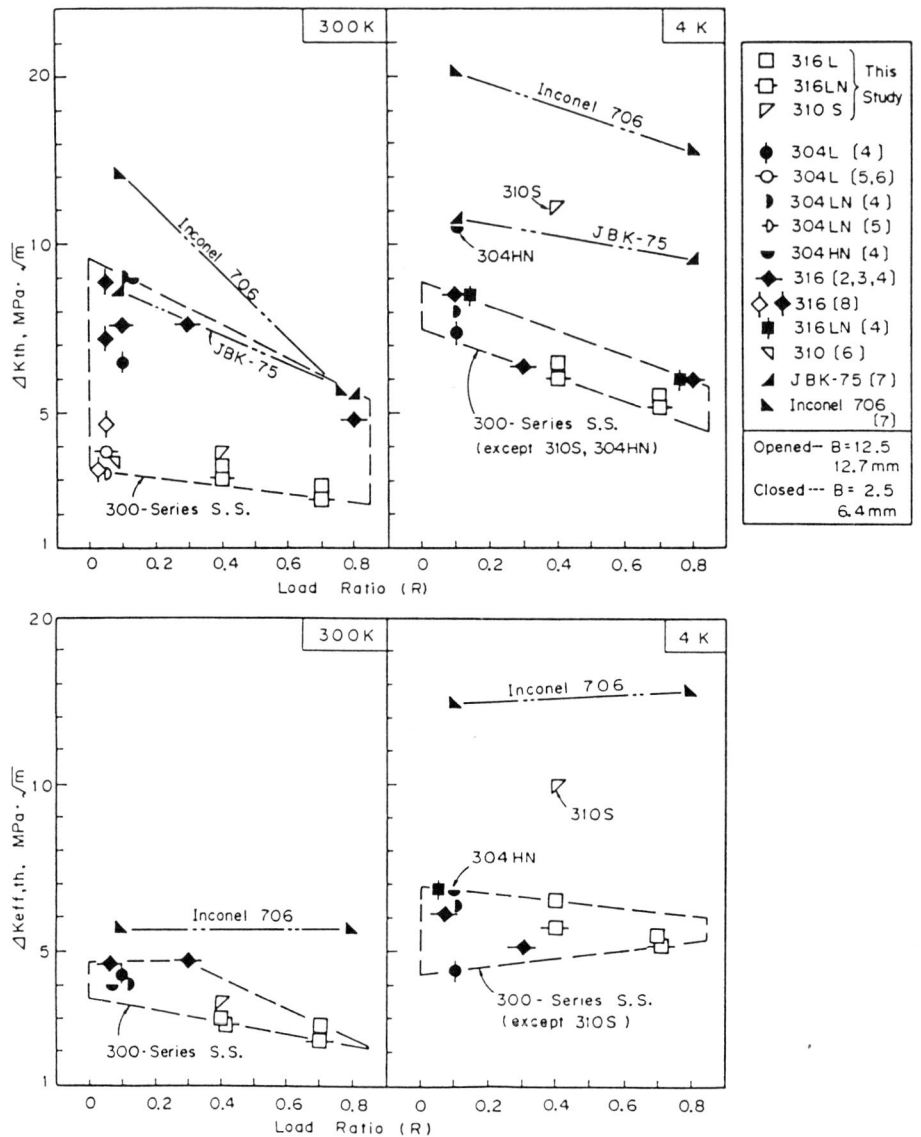

Fig. 8 Comparison of ΔK_{th} and $\Delta K_{eff,th}$ vs. R relations at 300K and 4K.

Further, we surmise that variations of crack closures would not become a direct influential element of temperature drop since temperature influences were observed also on $\Delta K_{eff,th}$ [6].

On 310S and 304HN, ΔK_{th} at 4K was higher than the values on other 300-series stainless steel, or closer to that of JBK-75, and lower than the level of Inconel 706. ΔK_{th} and $\Delta K_{eff,th}$ at room temperature were, on the other hand, at the same levels as other 300-series stainless steel. However, $\Delta K_{eff,th}$ at 4K revealed different results on 310S from those on 304HN.

Effect of Yield Strength and Microstructure. Figure 9 shows the relations between ΔK_{th} and yield strength at 4K under R = 0.1, 0.4, or 0.7 clarified based on our test results and the reference data [2-4, 7]. For information, the data under R = 0.3 and R = 0.8 are also shown as the data close to that on R = 0.4 and R = 0.7. We obtained the finding that influences by yield strength against ΔK_{th} of 300-series stainless steel excluding 310S and 304HN are rather limited. Especially, it is noteworthy that ΔK_{th} is approximately the same although the yield strength of 316LN is about 2.5 times as much as 316L.

310S, which possesses the yield strength intermediate in value between 316L and 316LN, showed ΔK_{th} at 4K about 1.7-1.8 times of these two matserals. This fact can be considered to have resulted from the points that 310S is a stable stainless steel not involving transformation to α' martensite even at 4K, that as a result or as shown in Fig. 10, the fatigue crack path of 310S at 4K is not as flat as that of 316L or 316LN, and that cracks grow with difficulty because the path zigzags considerably [6]. To support this assumption, we can cite from Table 3 the fact that $U = \Delta K_{eff,th}/\Delta K_{th}$, or the crack opening ratio, is smallest on 310S (0.91 for 310S whereas 0.95 for 316LN and 1.0 for 316L) due to roughness induced crack closure.

As for 304HN which is a metastable stainless steel, its yield strength is slightly greater than that of 316LN, but ΔK_{th} value at 4K is much higher than other types of metastable stainless steel, being closer to that of 310S and hence displaying behavior different from 316LN. The fact that $\Delta K_{eff,th}$ of 304HN is nearly the same as other metastable stainless steel makes this material a highly interesting type. Different yield strength dependence may be indicated between the 316 family and 304 family. In the future, by further accumulating data on various cryogenic structural materials including metastable stainless steel, it should become possible to accurately and systematically discuss the effects by yield strength and martensite volumes.

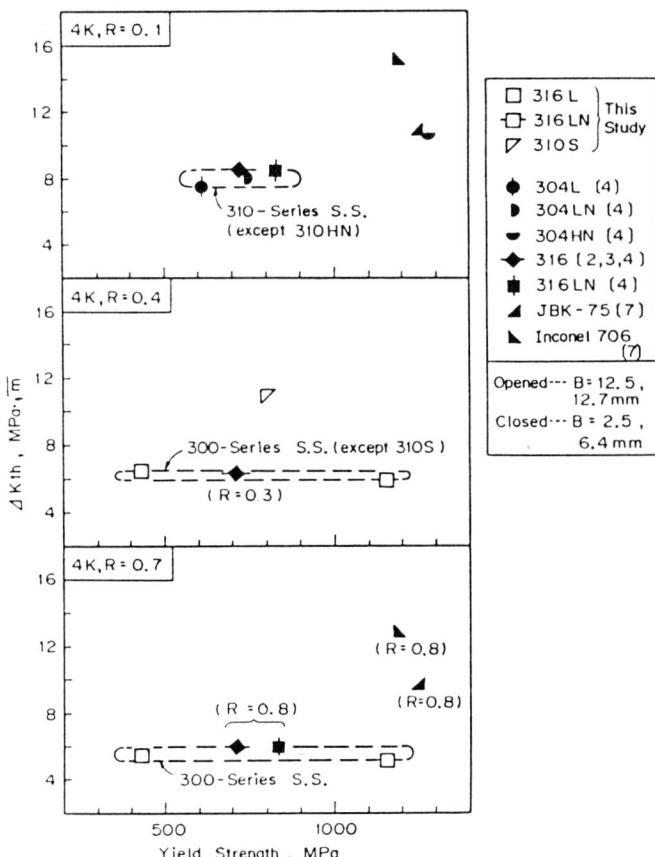

Fig. 9 Comparison of ΔK_{th} vs. yield strength relations at 4K.

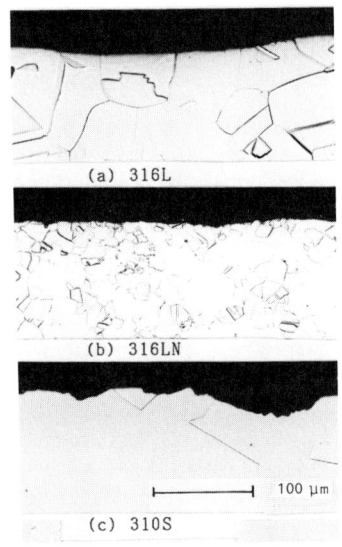

Fig. 10 Cross-sectional view of crack near threshold at 4K.

SUMMARY

On the stainless steel of 316L, 316LN, and 310S, it is possible to obtain da/dn vs. ΔK relations equivalent to those of ΔK increase tests by constant ΔP or fairly close da/dn vs. ΔK_{eff} relations through conducting ΔK decrease tests under $(d\Delta K)/\Delta K = -3.2\%$ (4K) and -2% (300K). Regarding 300-series stainless steel with the exception of 310S and 304HN, influences by specimen thicknesses are large against ΔK_{th} at 300K. On the other hand, influences are relatively small against $\Delta K_{eff,th}$ at 300K or ΔK_{th} and $\Delta K_{eff,th}$ at 4K. As for influences by load ratios (R), they are greater on ΔK_{th} rather than on $\Delta K_{eff,th}$ at both 300K and 4K. Influences by yield strength to ΔK_{th} at 4K are also relatively small. There is the trend of influences by temperature to differ depending on types of steel and thicknesses of specimens.

ΔK_{th} and $\Delta K_{eff,th}$ at 4K of 310S, a type of stable stainless steel, are at the high level of about 1.7-1.8 times that of 300-series meta-stable stainless steel excluding 304HN. This phenomenon is believed to be mainly due to nonexistence of α' martensitic transformation.

REFERENCES

1. K. Suzuki, J. Fukakura and H. Kashiwaya, Journal of the Japan Society of Materials Science, 38:1309(1989).
2. R. L. Tobler, Near-threshold Fatigue Crack Growth Behavior of AISI 316 stainless steel, in:"NBSIR 86-3050", R. P. Reed, ed., NBS(1986), p. 43.
3. R. L. Tobler and Y. W. Cheng, Int. j. Fatigue, 7:191(1985).
4. R. L. Tobler, Fatigue Crack Growth in Austenitic Stainless Steels, in:"Proceedings of the Fourth U. S. -JAPAN Workshop", JAERI(1990), p. 90.
5. Z. Mei and J. W. Morris, Jr., Metall. Tran. A, 21:3137(1991).
6. Z. Mei, J. W. Chan and J. W. Morris, Jr., Adv. Cryg. Eng., 36:1241(1990).
7. P. K. Liaw and W. A. Logsdon, Engng. Fract. Mech., 22:585(1985).
8. W. J. Mills and L. A. James, JTEVA, 15:325(1987).
9. K. Suzuki, J. Fukakura and H. Kashiwaya, JTEVA, 16:190(1988).
10. "Annual Book of ASTM Standards", E647-86a, ASTM(1983) p. 899.

LONG-CRACK FATIGUE THRESHOLDS AND SHORT CRACK SIMULATION AT LIQUID HELIUM TEMPERATURE†

R.L. Tobler, J.R. Berger,* and A. Bussiba**

Materials Reliability Division
National Institute of Standards and Technology
Boulder, Colorado 80303

ABSTRACT

A short crack simulation (SCS) test is used to characterize the near-threshold fatigue crack growth behavior of stainless steels at 4 K. The test methodology holds the maximum stress intensity factor constant while increasing the minimum stress intensity factor, thus raising the stress ratio from 0.1 at the start to about 0.8 at the end of the test. The resulting fatigue crack growth rate measurements are unaffected by crack closure, and the intrinsic threshold is directly obtained without a correction factor. Merits of the test procedure are described.

INTRODUCTION

In fracture mechanics evaluations, measurements of fatigue crack growth rate (da/dN) depend on the range of the applied stress intensity factor (ΔK), and data plotted on log-log coordinates typically approach an asymptote at very low growth rates. In this "stage I" region, the growth rates diminish to a threshold denoted ΔK_{Th} which is defined operationally as the ΔK value corresponding to a crack growth rate of 10^{-10} m/cycle.

Owing to measurement difficulties, the near-threshold database for stainless steels at cryogenic temperatures remains quite limited.[1-5] The usual problems associated with ambient testing, and more, are present in high-cycle cryogenic testing. Cost, complexity, and material variability are greater at cryogenic temperatures. Nevertheless, fatigue properties now appear to be critical to proposed fusion magnet structures subject to high-stress, high-cycle loading. Safe design requires understanding the near-threshold behavior and developing a 4-K fatigue database for austenitic alloys. To that end, we describe the usefulness of a constant K_{max} or "short crack simulation" test at cryogenic temperatures. At

---

† Contribution of NIST, not subject to copyright.
* Present affiliation: Dept. of Aerospace and Mechanical Engineering, University of Notre Dame, Notre Dame, IN.
**Visiting Scientist, on leave from Nuclear Research Center-Negev, Beer-Sheva, Israel.

Table 1. Chemical composition of test materials, mass percent.

Material	Fe	Cr	Ni	C	N	Mn	P	S	Si	Mo
AISI 316	bal	17.25	13.48	0.057	0.030	1.86	0.024	0.019	0.58	2.34
AISI 316LN	bal	16.71	13.83	0.029	0.100	1.84	0.021	0.008	0.33	2.18

Table 2. Mechanical properties at 295 and 4 K.

Material	Temperature (K)	Yield Strength (MPa)	Ultimate Strength (MPa)	Elongation (%)	Reduction of area (%)
AISI 316	295	228	576	56	73
	4	711	1301	48	57
AISI 316LN	295	267	551	53	76
	4	834	1364	45	53

ambient temperature this technique was introduced by Doker et al.,[6] and popularized by Herman et al.[7,8] In this paper, we apply the same technique at 4 K to AISI 300 series austenitic stainless steels that were formerly evaluated using the conventional test. A second paper uses the technique to efficiently characterize new superconductor sheath alloys.[9]

MATERIALS AND PROCEDURE

AISI 316 and 316LN stainless steels were tested at 295 and 4 K. The chemical compositions and mechanical properties of these steels are listed in Tables 1 and 2. AISI 316 was previously characterized using the conventional decreasing K_{max} test at 295, 77, and 4 K.[2] The compact specimens used here are identical to those used before: 6.4 mm thick and 76.25 mm wide, and notched in the T–L orientation with a clip gage mounted at the notch mouth opening. The computer-controlled 100 kN servohydraulic test machine, cryostat, and liquid helium refill system were also described before.[3] The only isolated variable considered here is the computer-programmed loading technique.

Figure 1 illustrates the two loading schemes compared in this study. In the conventional technique used previously, both K_{max} and K_{min} are progressively lowered while the R ratio (K_{min}/K_{max}) is maintained constant (Fig. 1A). The maximum stress intensity factor K_{max} determines the plastic

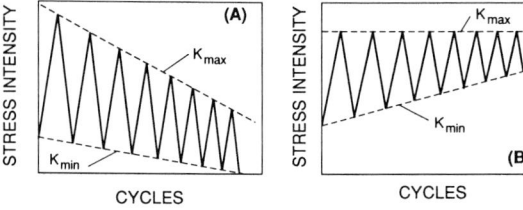

Fig. 1. Loading schemes for threshold testing: (A) conventional test (R = constant, K_{max} and K_{min} decreasing), and (B) short crack simulation with R variable, K_{max} constant and K_{min} increasing.

zone size and must be reduced gradually to avoid the specter of crack growth rate retardation. In the constant K_{max} test (Fig. 1B), ΔK is decreased by increasing the minimum stress intensity factor K_{min}. There is then no prospect of retardation, and the test duration may be shortened since ΔK can be decreased to the threshold region more rapidly. The R ratio at the end of the test is rather high, typically 0.8. Since long cracks at such high R values are like short cracks in that they are not subject to crack closure, this procedure has been described as a short crack simulation (SCS) test.[7,8]

Typically to begin an SCS test we choose K_{max} near 30 MPa·m$^{\frac{1}{2}}$ and R = 0.1 at 20 Hz. Thereafter, the computer monitors the cracking and commands loads to satisfy the expression:

$$\Delta K_i = \Delta K_0 \exp \{C(a_f - a_0)\}, \tag{1}$$

where: ΔK_i = the instantaneous stress intensity factor range,
ΔK_0 = the initial stress intensity factor range,
a_i = the initial crack length,
a_f = the final crack length, and
C = the stress intensity factor gradient $(1/K)(dK/da)$.

The parameter C determines the ΔK reduction rate and is also specified at the beginning of a test. Previous conventional tests had used C = -0.10 or less to minimize retardation. Here, C ranged from -0.10 to -0.25, and cycle frequency at lower ΔK was likewise incrementally increased. Both of these factors help to reduce the test duration (see later text).

RESULTS

Figure 2 presents fatigue crack growth rates for AISI 316 at room temperature. Conventional data at constant R (0.10) are compared to SCS data at variable R, and the fatigue crack growth rates are plotted versus the applied ΔK without corrections for crack closure. For a wide range of da/dN, the trend for the conventional technique on log-log coordinates is virtually linear with no obvious knee or asymptote at rates as low as 10^{-10} m/cycle.[2] The SCS results extend from ΔK = 26 MPa·m$^{\frac{1}{2}}$ at R = 0.1 to ΔK = 5 MPa·m$^{\frac{1}{2}}$ at R = 0.8. In the near threshold region, the SCS rates are

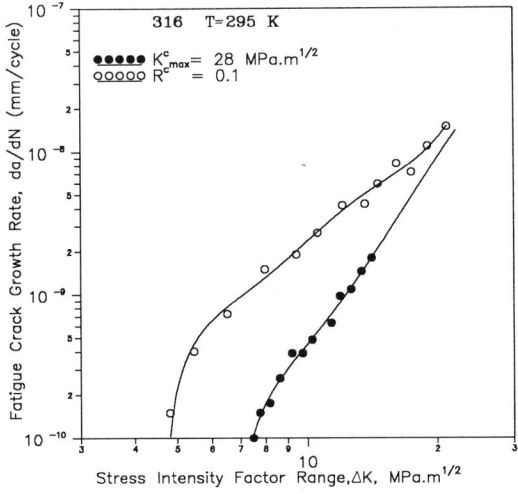

Fig. 2. Fatigue crack growth rates for AISI 316 at 295 K.

higher, there is a knee in the curve below $\Delta K = 7$ MPa·m$^{\frac{1}{2}}$, and the operationally defined threshold is 5 MPa·m$^{\frac{1}{2}}$, as compared to 7 MPa·m$^{\frac{1}{2}}$ for the conventional long crack measurements.

Figure 3 is a similar comparison of data for AISI 316LN at 4 K. As before, the conventional test maintained R = 0.1, whereas the SCS test terminated much higher at R = 0.8. Again, the SCS rates are higher and the threshold at 10^{-10} m/cycle is lower than conventional measurements. AISI 316 (data not shown) performed similarly to the 316LN at 4 K. Apparently, the differences in C + N contents (0.087% versus 1.29%) and mechanical properties (Table 2) were insufficient to affect the near-threshold performance of these steels. The thresholds for both are 6 and 8.5 MPa·m$^{\frac{1}{2}}$ for the SCS and conventional techniques, respectively.

In Figs. 2 and 3, the rates measured by both techniques agree near $\Delta K = 30$ MPa·m$^{\frac{1}{2}}$, as expected, because the initial loading conditions are the same. But with cracking in the SCS test, R increases, we move to a loading condition that obviates crack closure, and the data branch away from the conventional trend. Consequently, thresholds from the SCS test are systematically 2-3 MPa·m$^{\frac{1}{2}}$ lower, while the fatigue crack growth rates at low ΔK are higher than results at R = 0.1. This is explained by closure effects, one manifestation of which is nonlinear compliance. Oscilloscope traces of the load-displacement curves during fatigue cycling in this study confirmed that closure is significant in conventional tests at low ΔK, but not in SCS tests. Thus the results are attributed not to an inherent stress ratio effect on materials properties, but primarily to the hidden variable, crack closure.

Physical evidence of crack closure in these steels is revealed in the scanning electron micrographs of Fig. 4, which show crack surfaces formed 295 K in specimens tested by the two different techniques. We observe ductile transgranular fracture surfaces in the near-threshold region, and the basic features of morphology and failure mechanisms do not change with test techniques. However, deformities are present on the fracture surfaces after conventional tests at R = 0.1. The deformities are crushed

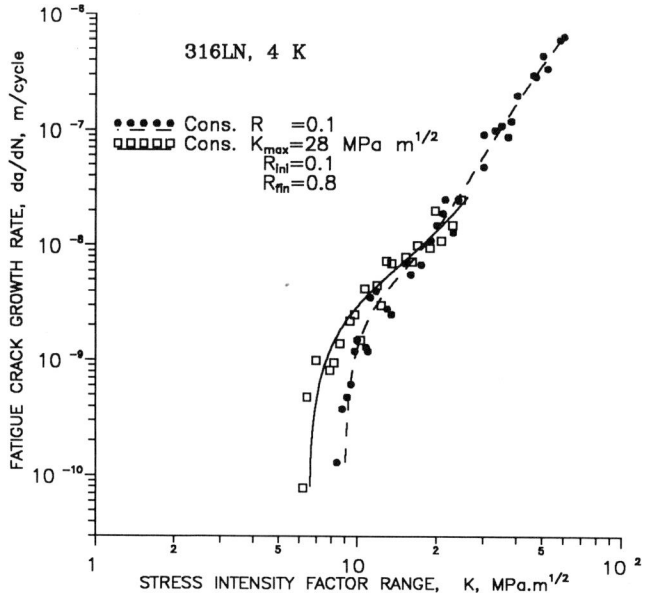

Fig. 3. Fatigue crack growth rates for 316LN at 4 K.

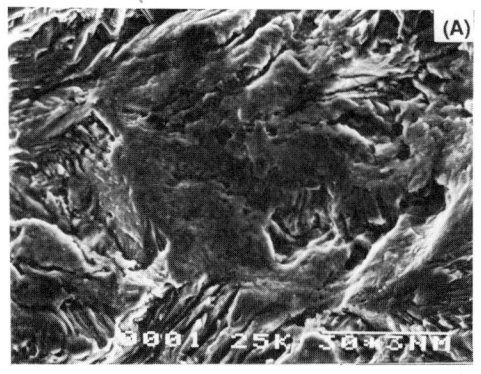 

Fig. 4. Representative 4-K fracture surfaces for steel specimens fatigued near the threshold: (A) R = 0.1 and (B) R = 0.8.

asperities that appear as flattened or smeared regions in the near threshold region. Flattening or smearing is a consequence of asperity contact during the unloading phase of fatigue cycling at low stress intensities and is a mark of roughness-induced crack closure.[10] The SCS test specimens do not show such features, nor do they show nonlinear compliance behavior during fatigue cycling.

DISCUSSION

Using on the foundation laid by Herman et al.[7,8], we interpret the 4-K results as follows. The conventional decreasing ΔK test at low R gives a traditional sigmoidal curve and represents long crack behavior where the rate measurements are subject to closure effects. The region between the SCS and conventional trendlines presumably corresponds to where the short crack growth rates would be plotted, assuming they could be measured at 4 K. The SCS trend, although derived from long crack test specimens, represents a convenient upper bound approximation for the short crack growth rates.

In long crack specimens, closure causes the crack to remain closed for a portion of the load cycle, and this portion of load does not contribute to the driving force for crack growth. In austenitic alloys, closure at low ΔK is much more pronounced at 295 K than at 4 K.[2] We believe this reflects a strong role of plasticity-induced closure, as the zone of crack-tip plastic deformation is relatively large at 295 K but restricted at cryogenic temperatures due to the increase in the yield strength. Certainly, other mechanisms contribute to closure in these materials; roughness-induced closure was cited in Fig. 4, and transformation-induced closure is a factor in unstable austenitic steels, especially grades of less-alloyed AISI 304L.[5]

Short cracks behave differently than long cracks. Short cracks are not subject to closure effects and so grow at higher rates than conventional data for long crack specimens might suggest. The difference can be significant in fracture mechanics assessments of structures. Judging from experience with other alloys, however, actual measurements for short cracks at 295 K are time consuming and highly scattered.[7,8] At 4 K, similar measurements are practically impossible; specimens in a Dewar cannot be directly observed, repeated specimen removal is prohibitive in terms of fluid costs, and the indirect compliance technique is insensitive to short crack growth. The SCS test is therefore a welcomed alternative.

Figure 5 summarizes the conventional long crack measurements for the five AISI 300 series steels tested so far.[2,3] The general performance of these steels at 4 K is similar: the apparent thresholds range from ΔK_{Th} = 7 to 11 MPa·m$^{\frac{1}{2}}$ whereas the effective thresholds range from 4 to 7 MPa·m$^{\frac{1}{2}}$. Since the apparent thresholds are affected by closure, the intrinsic metallurgical effects can be interpreted from the effective thresholds which are derived after applying the closure correction. With the closure effects thus factored out, the effective thresholds are clearly higher at cryogenic temperatures than at room temperature; this, as well as the slight dependence on yield strength at 4 K, is in keeping with expectations from dislocation dynamics theory.[1]

Regardless of the mechanisms, closure effects must be avoided or factored out to observe the intrinsic material behavior. They are avoided in the SCS technique because the test terminates at a high R ratio. In Fig. 5, the SCS threshold measurements for AISI 316 and 316LN at 4 K are plotted for comparison to the conventional results. Within experimental error, the results for long cracks at R = 0.1 after closure correction are equivalent to the SCS measurements (R = 0.8) which need no correction. There are two important implications. First, this finding implies that the near-threshold properties of the 316 and 316LN stainless steels are not inherently sensitive to mean stress. Second, it implies that the experimental procedure (which may or may not admit closure effects) is the primary factor in threshold measurement variations.

SUMMARY AND CONCLUSION

Table 3 summarizes the two test techniques. The SCS technique is a promising tool for materials evaluation at 4 K. To date, the published

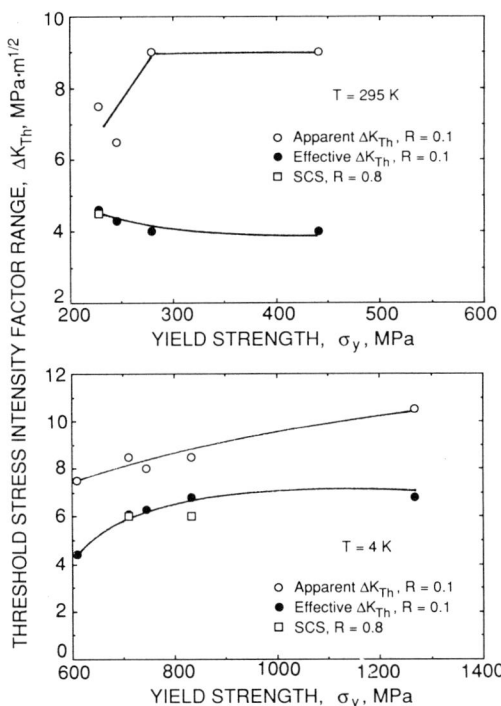

Fig. 5. Fatigue thresholds for AISI 300 series steels. (Trend lines are for constant R tests; open squares for SCS tests at constant K_{max}.

Table 3. Comparison of Fatigue Crack Growth Test Methods.

	Conventional Test	SCS Test
K_{max}	decreasing	constant
Stress Ratio	constant (0.1)	increasing (0.1–0.8)
Crack Closure	significant	insignificant
Retardation	possible	avoided
ΔK Reduction Rate	10%	no limit
C in Eq. (1)	−0.10	−0.1 to −0.25
Test Time	100 h	60–70 h

fatigue crack growth rates for the near threshold region represent measurements for long crack specimens, typically at low stress ratios. These data are not optimum for design if short crack behavior is anticipated in practice. To support material selection and specific machinery designs, both long and short crack behavior should be characterized and understood, but short crack characterization is exceptionally difficult at 4 K. The SCS technique is appealing at cryogenic temperatures because it uses long-crack specimens which are amenable to compliance test methodology. More importantly, it reduces test time and fluid costs while generating conservative data that are especially appropriate for some structures.

REFERENCES

1. P.K. Liaw and W.A. Logsdon, Fatigue Crack Growth Threshold at Cryogenic Temperatures: A Review, Eng. Fract. Mech., 22:585–594 (1985).

2. R.L. Tobler, Near-threshold Fatigue Crack Growth Behavior of AISI 316 Stainless Steel, Adv. Cryo. Eng. Maters., 32:321–327 (1986).

3. R.L. Tobler and Y.W. Cheng, Automatic Near-threshold Fatigue Crack Growth Rate Measurements at Liquid Helium Temperature, Int. J. Fat., 7:191–197 (1985).

4. Z. Mei, J.W. Chan, and J.W. Morris, Jr., The Effect of Temperature on Fatigue Crack Propagation in 310 Austenitic Stainless Steel, Adv. Cryo. Eng. Maters., 36A:1241–1247 (1990).

5. Z. Mei and J.W. Morris, Jr., Influence of Deformation Induced Martensite on the Fatigue Crack Propagation in 304-Type Steels, Metall. Trans., 21A:3137–3152 (1990).

6. H. Doker, V. Bachman, and G. Marci, A Comparison of Different Methods of Determination of the Threshold for Fatigue Crack Propagation, in: "Fatigue Thresholds", J. Backlund, A. Blom, and C.J. Beevers, eds., EMAS Ltd., United Kingdom, 45 (1982).

7. W.A. Herman, A Reevaluation of Fatigue Threshold Test Methods, in: "Fatigue 87", EMAS, Ltd., United Kingdom, 2:819 (1987).

8. W.A. Herman, R.W. Hertzberg, and R. Jaccard, A Simplified Laboratory Approach for the Prediction of Short Crack Behavior, Fat. Fract. Eng. Maters. Struct., 11:303 (1988).

9. A. Bussiba, R.L. Tobler, and J. Berger, Superconductor Conduits:

Fatigue Crack Growth Rate and Near-Threshold Behavior of Three Alloys, this volume.

10. S. Suresh and R.O. Ritchie, in: Fatigue Crack Growth Threshold: Concepts, D.L. Davidson and S. Suresh, eds., TMS-AIME Warrendale, PA, 227-261 (1984).

SUPERCONDUCTOR CONDUITS: FATIGUE CRACK GROWTH RATE AND NEAR-

THRESHOLD BEHAVIOR OF THREE ALLOYS[†]

A. Bussiba[*], R. L. Tobler, and J. R. Berger[**]

Materials Reliability Division
National Institute of Standards and Technology
Boulder, Colorado 80303

ABSTRACT

Three superconductor sheath alloys were fatigue-tested at 4 K to evaluate their suitability for International Thermonuclear Experimental Reactor magnets. The alloys tested were low carbon 316LN (0.018C), niobium-modified 316LN (0.009C, 0.05Nb) and nickel-based alloy 908. Threshold stress intensity factors ΔK_{th} at 10^{-10} m/cycle and near threshold fatigue crack growth rates were measured using the short crack simulation technique which avoids crack closure. Results indicate some advantage of 908 over the 316LN alloys. Also, microstructural effects on fatigue resistance are identified, mainly for 316LN (0.018C), where ΔK_{th} increases with decreasing grain size.

INTRODUCTION

The cable-in-conduit conductor (CICC) design used in advanced superconducting magnets for experimental fusion reactors demands cryogenic fatigue data for reliable material selection and safety. Most fatigue crack growth rate (FCGR) data da/dN available in the literature were generated using the decreasing stress intensity factor range ΔK procedure where the stress ratio R is constant at a low value, such as 0.1. This technique can lead to an overly optimistic life assessment in cases where crack closure is prominent in laboratory materials tests but not in the structure. In structures containing short cracks or residual tensile stresses developed during welding or compressive loading, crack closure may not be manifest.[1,2] In laboratory tests of long crack specimens, crack closure is especially pronounced in the low ΔK region, and it reduces the effective ΔK below the nominal applied value.

Despite many attempts, the intrinsic value of ΔK_{th} for long crack specimens is difficult to determine due to experimental inaccuracies associated with closure load measurements.[3] To overcome this problem, it

---

[†] Contribution of NIST, not subject to copyright.
[*] On leave from the Nuclear Research Center-Negev, Beer-Sheva, Israel.
[**] Present affiliation: Department of Aerospace and Mechanical Engineerng, University of Notre Dame, Notre Dame, IN.

was recently proposed that conservative data can more easily be generated using a new decreasing ΔK technique: the maximum stress intensity factor K^c_{max} is held constant while R is gradually increased from 0.1 to about 0.8.[4,5] The new technique avoids crack closure effects, gives reliable FCGR and threshold values, and is practical for cryogenic tests.[6]

In this study, we adopt the K^c_{max} technique to evaluate three austenitic alloys that are candidates for the inner poloidal field coils of superconducting magnets using the CICC design with in situ formation of Nb_3Sn filaments. The alloys were developed for high strength at 4 K, good weldability, and improved resistance to Nb_3Sn aging treatments. Aging for enough time at 650–700°C produces $M_{23}C_6$ carbide precipitates in 316 steels or $Ni_3(Al,Ti)$ precipitates in 908; these can dangerously "sensitize" the materials. Sufficient precipitation embrittles the grain boundaries and seriously degrades the mechanical properties at cryogenic temperatures. In 316LN, carbide embrittlement is diminished by reducing the C content, or by adding Nb to promote formation of NbC.[7] This study compares the three alloys after a specific aging treatment and considers the role of microstructural factors on the near-threshold properties of low carbon 316LN.

EXPERIMENTAL PROCEDURES

Table 1 lists the test materials and their chemical compositions. A 316LN (0.018C) conduit (5 mm thick wall) was tested in four conditions: (1) as received, AR; (2) as received and aged, ARA; (3) solution-treated, ST; and (4) solution-treated and aged, STA. For this steel, two solution treatment temperatures were evaluated: 1200°C for ½ h, and 1050°C for ¼ h. A Nb-modified 316LN conduit (4.7 mm thick wall) and three 908 plates (3.2, 10, and 12.5 mm thick) were tested in their ARA states only. Aging for all alloys was identical: 650°C for 180 h, in high purity argon, as currently anticipated for International Thermonuclear Experimental Reactor (ITER) magnet applications.

The FCGR tests used compact specimens of maximum available thickness stated above. Typically, at least two specimens were tested for each treatment, but for 908 only one specimen per plate was tested. The specimens in liquid He at 4 K were tested using a computer-controlled cryogenic apparatus[6] and short crack simulation test procedure described elsewhere.[8] Cracking was monitored by compliance with a clip gage mounted at the loadline or specimen edge. The nominal ΔK applied in fatigue was controlled by programming according to

$$\Delta K_i = \Delta K_0 \exp[C(a_f-a_0)], \qquad (1)$$

where ΔK_i = instantaneous stress intensity range,
ΔK_0 = initial stress intensity range,
a_i = initial crack length,
a_f = final crack length, and
C = stress intensity factor gradient $(1/K)(dK/da)$.

Table 1. Chemical compositions for the tested materials, mass percent.

Material	Fe	Ni	Cr	Mn	S	Si	P	Nb	C	N	Mo	Ti	Al
316LN(Nb)	Bal.	12.3	17.2	1.32	.004	.12	0.003	0.051	0.009	0.17	1.96	--	--
316LN (low C)	Bal.	13.1	17.2	1.51	.001	.67	0.02	--	0.018	0.14	2.63	--	--
alloy 908	40.8	48.7	4.12	0.09	.001	.17	0.002	3.04	--	--	--	1.54	1.10

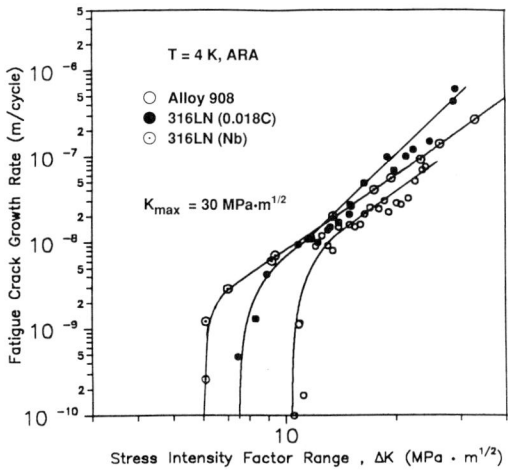

Fig. 1. FCGR data for 3 alloys.

The C value selected for these tests was initially −0.10 mm^{-1}, and K^c_{max} was either 30 or 16 MPa·m$^{\frac{1}{2}}$. At lower ΔK values, both C and the test frequency were increased according to specified programming to save time and liquid He. Fatigue fracture modes were classified by scanning electron microscopy and crack paths were examined by optical microscopy.

RESULTS

Figure 1 compares near-threshold FCGR and threshold data for the three alloys in their ARA states with identical aging. Alloy 908 (3.2 mm plate) has the highest ΔK_{th} as well as the lowest FCGR for the entire ΔK range. The 316LN (0.018C) steel has a higher threshold than Nb-modified 316LN, although the yield strengths of both materials at 4 K are nearly equal (1280 and 1237 MPa, respectively). This difference may be attributed to a grain size effect as discussed later; however, the FCGR curves intersect at a specific point ΔK_{int} and at higher ΔK the cracking rates for 316LN (0.018C) exceed those for Nb-modified 316LN.

Figure 2 shows the effect of solution treatment temperature on the 316LN (0.018C) properties. As indicated, ΔK_{th} increases while the FCGR decreases over the entire ΔK range if the solution treatment temperature is lowered from 1200 to 1050°C. For this comparison, a lower K^c_{max} (16 MPa·m$^{\frac{1}{2}}$) was used after solution treatment at 1200°C because 30 MPa·m$^{\frac{1}{2}}$ caused premature failure of one specimen shortly after a crack initiated from the machined notch. This is a consequence of accelerated cracking associated with intergranular failure at higher K^c_{max}. The curve for the ARA state indicates a lower ΔK_{th} after solution treatment at 1200°C.

Figure 3 shows microstructural effects on ΔK_{th} and FCGR for 316LN (0.018C) at K^c_{max} = 30 MPa·m$^{\frac{1}{2}}$. The ST (1200°C) condition has the lowest ΔK_{th}; ARA, the highest. The ΔK_{th} is lower for the AR condition than it is for the ARA condition. This could be an effect of residual tensile stresses or else microstructural inhomogenieties (reveafled by metallographic observations) created in the AR state by the cold forming process. In any case, the FCGR resistance is highest for the ST condition, slightly lower for the ARA, and lowest for the AR.

Figure 4 illustrates available data for aged 908 plates in three thicknesses. For the thinnest section (3.2 mm), ΔK_{th} increases and is

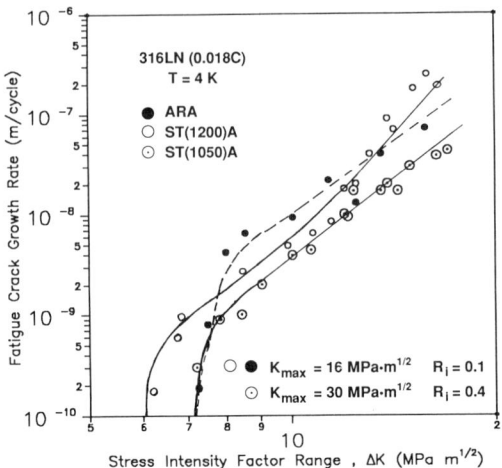

Fig. 2. FCGR data for 316LN (0.018C).

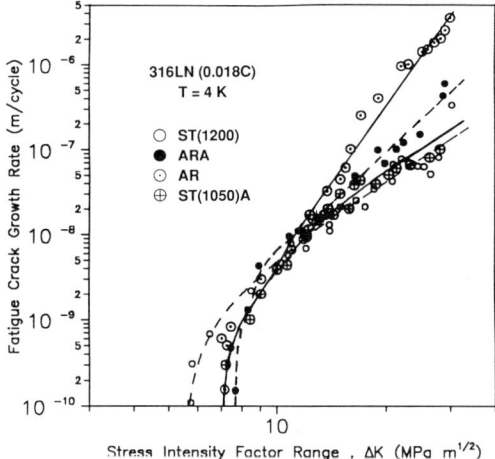

Fig. 3. Data for 316LN (0.018C).

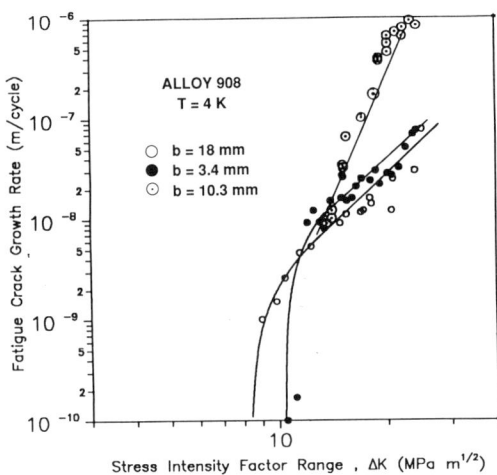

Fig. 4. Data for 908 plates.

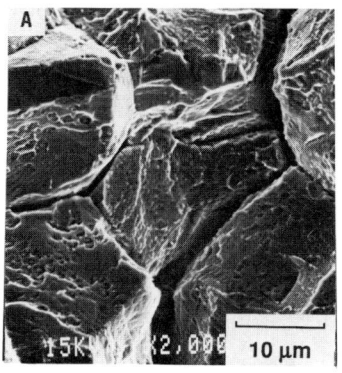 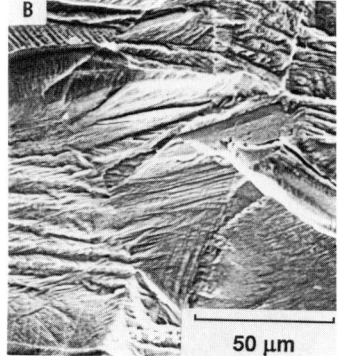

Fig. 5. IGF for 316LN (0.018C), ARA: (A) tearing, and (B) cycling mode.

superior at lower ΔK. However, at a specific point denoted ΔK_{int} the curves intersect and the ranking is reversed. FCGR data for the medium thickness (10 mm) plate are anomalously high and are associated with a brittle cracking mechanism described later.

Fractographic findings for aged materials are presented in Figs. 5-9. For 316LN (0.018C), intergranular fracture (IGF) strongly manifests itself in the static tearing mode, as well as the cycling mode (Fig. 5). IGF dominates in this alloy at higher ΔK, regardless of the solution treatment temperature; however the proportion of IGF seems to decrease slightly if solution treatment is at 1050 rather than 1200°C (Fig. 6). In contrast, a ductile mode prevails in the Nb-modified 316LN, which shows minor traces of IGF in the cycling mode only (Fig. 7). For 908, a ductile mechanism generally controls the fatigue fracture (Fig. 8A); however, a brittle mechanism operates in the 10 mm plate, as reflected by facets (Fig. 8B).

DISCUSSION

For the specific aging conditions studied here, which are relatively severe, our preliminary results indicate some potential advantage (a higher ΔK_{th}, as well as lower FCGR at lower ΔK) of alloy 908 over the two modified 316LN steels. As indicated by metallographic and fractographic

 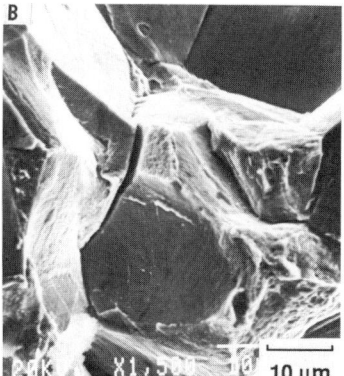

Fig. 6. Intergranular fracture for 316LN (0.018C): (A) solution-treated at 1200°C and aged, and (B) solution-treated at 1050°C and aged.

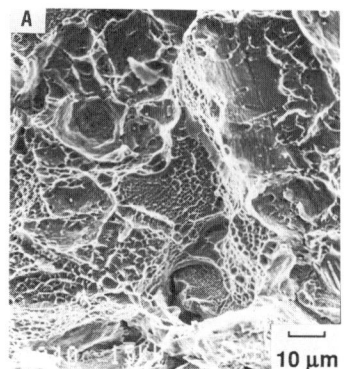 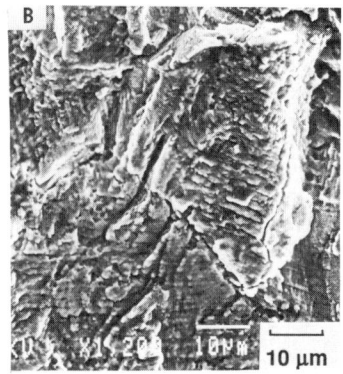

Fig. 7. Fracture mode for Nb—modified 316LN: (A) ductile fracture, and (B) traces of intergranular fracture in the ductile mode.

studies, the advantage relates to a relatively uniform distribution of γ' precipitates in 3.2 mm thick 908, in contrast to the high concentration of carbides at grain boundaries for 316LN (0.018C). The Nb—modified 316LN (0.009C) also is preferable to 316LN (0.018C) by virtue of higher resistance to intergranular fracture at 4 K for the 650°C, 180 h aging.

Despite the positive results for alloy 908, one medium thickness plate showed poorer properties than the two other plates. This may indicate a sensitivity of this complex alloy to processing and aging, since tests of the medium thickness plate in the unaged condition showed no sign of brittleness.

Other results indicate that the FCGR and threshold data obtained by the K^c_{max} technique successfully characterize the intrinsic material response by virtually eliminating extraneous crack closure and other crack tip shielding effects, including deflection and crack—wake asperity wedging. By comparing ΔK_{th} measurements for 316LN (0.018C) at $K^c_{max} = 16$ and 30 MPa·m$^{\frac{1}{2}}$ (Figs. 1 and 2), we confirmed that the threshold measurements are insensitive to K^c_{max}. This was also reported by Doker et al.,[4] who tested a titanium alloy at 295 K and concluded that ΔK_{th} was insensitive to K^c_{max} in the range 5 to 25 MPa·m$^{\frac{1}{2}}$. These findings are consistent with results from conventional constant R tests where ΔK_{th} is found to be insensitive to R when such tests are performed at high stress ratios.[9]

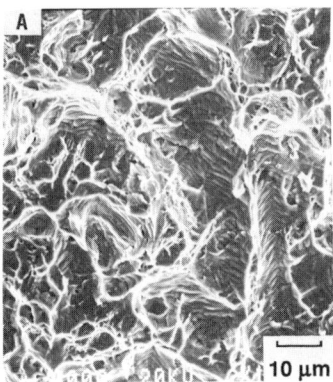 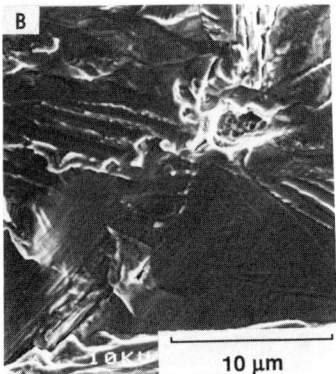

Fig. 8. Representative fractographs for 908: (A) ductile mechanism for the 3.4 and 18 mm plates, and (B) quasicleavage for 10 mm plate.

Table 2. Threshold and grain size parameters for three alloys.

Material	D µm	ΔK_{th} MPa·m$^{1/2}$	$\sqrt{\dfrac{D_2}{D_1}}$	$\dfrac{\Delta K_{th}}{\Delta K_{th}}$	$\sqrt{\dfrac{D_3}{D_2}}$	$\dfrac{\Delta K_{th}}{\Delta K_{th}}$	$\sqrt{\dfrac{D_1}{D_3}}$	$\dfrac{\Delta K_{th}^3}{\Delta K_{th}^1}$
316LN (Nb) AR	50	6	0.83	0.76				
316LN (.018C) AR	35	7.8			0.65	0.75		
908 AR	15	10.3					0.54	0.58
316LN (.018C) ARA	35	7.8	0.54	0.67				
316LN (.018C) ST	110	5.2			0.63	0.72		
316LN (.018C) STA	45	7.2					0.88	0.92

The literature apparently contains inconsistent conclusions about the influence of the grain size (D) on ΔK_{th}.[10,11] Our results support those who find

$$\Delta K_{th} \propto 1/D^{-\frac{1}{2}}. \qquad (2)$$

Table 2 summarizes relevant data for our alloys, including data for 316LN (0.018C) with two different grain sizes resulting from 1200 and 1050°C solution treatments. The 4-K yield strengths of all three alloys after aging are very close, ranging from 1237 to 1400 MPa, so this is not a strong factor. The table lists ΔK_{th}, D, and selected ratios of interest. The ratios of ΔK_{th} versus ratios of $D^{-\frac{1}{2}}$ are shown to be nearly equivalent, implying the relationship shown by Eq. (2). As explained by Hornbogen and Gahr,[12] surface observations at low ΔK support the conclusion that fatigue cracks are impeded at grain boundaries, and this provides a rationale for the favorable effect of a fine grain size.

The expected grain size effect may be reversed, however, when plasticity becomes more extensive and the process zone size increases. Indeed, in Figs. 1 and 2, the curves representing smaller grain size intersect those for the larger grain sizes, leading to accelerated FCGR at higher ΔK, as observed by Higo et al.[10] Empirically, it seems that

$$\Delta K_{int}/K^c_{max} = D_1/D_2. \qquad (3)$$

In Fig. 3 we noted a reversal of the grain size effect for 316LN (0.018C) in the sensitized state. Presumably, this relates to the inception of intergranular fracture which becomes the controlling factor when the process zone exceeds the grain size. Thus, fracture mode transitions may intercede and control the FCGR performance so that the usual grain size effect is masked.

To the first order, we can relate ΔK_{th} results from the K^c_{max} technique to results obtained using constant, low values of R. The empirical equation is

$$\frac{K^c_{th}}{K^I_{th}} = \left(\frac{K^c_{max}}{K^I_{max}}\right)^2 \cdot \frac{R^c}{R_{max}}, \qquad (4)$$

where ΔK^c_{th}, K^c_{max}, R_{max} are the parameters for variable R conditions and ΔK^r_{th}, K^r_{max}, R^c are the parameters for constant R tests. This equation was deduced using some results from other studies.[5,6] It could prove useful for design purposes since it seems reasonable to use approximate ΔK^c_{th} values which are more conservative than ΔK^r_{th} measurements.

CONCLUSION

These, our initial results, suggest that alloy 908 offers potential advantages over the modified 316LN steels for the specified aging treatment (650°C, 180 h). However, for reasons that are not clear yet, brittle cracking occurred in one 908 plate after aging, and optimum properties were not realized in that case. Intergranular fracture occurred in aged 316LN (0.018C) steel, but not in aged 316LN (0.009C) modified with 0.05Nb. Therefore 316LN (0.018C) is least desirable for the intended application, although it might be useful for other applications assuming that either the C content, the aging time, or both are reduced. Additional research is planned on these subjects.

ACKNOWLEDGMENTS

This work was funded by the U.S. Department of Energy, Office of Fusion Energy, with M. Cohen as project monitor and R.P. Reed as consultant to NIST. P. Buzzone of Asea Brown Boveri provided the 316LN (0.018C) conduit, N. Mitchell of the ITER and NET research teams provided the Nb-modified 316LN conduit, and I.S. Hwang of MIT provided the plates of alloy 908. The authors also acknowledge with thanks D. Vigliotti and R.P. Walsh of NIST for their assistance in accomplishing the tests.

REFERENCES

1. A. Ohta, E. Sasaki, M. Nikei, N. Kanao, and M. Inagaki, Int. J. Fat., 4:233-237 (1982).
2. W. Geary and J.E. King, Int. J. Fat., 9:11-16 (1987).
3. J.E. Allison and C.P. You, Fatigue 90, EMAS Ltd (1990).
4. H. Doker, V. Bachman, and G. Marci, Fatigue Thresholds, EMAS Ltd (1982).
5. R. Hertzberg, W. Herman, T. Clark, and R. Jaccard, Symposium on Small-Crack Test Methods, Nov. 14, 1990.
6. R. Tobler, J.R. Berger, and A. Bussiba, Long Crack Fatigue Thresholds and Short Crack Simulation at Liquid Helium Temperature, this volume.
7. M. Shimada and S. Tone, Adv. Cryo. Eng., 34:131-139 (1988).
8. R.L. Tobler and Y.W. Cheng, Int. J. Fat., 7:191-197 (1985).
9. J.K. Musava and J.C. Radon, Proc. 5th Int. Conf. Fract., Cannes, France, 1365 (1981).
10. Y. Higo, A.C. Pickard, and J.F. Knott, Met. Sci., 15:233-240 (1981).
11. J.E. King, Met. Sci., 16:345 (1982).
12. E. Hornbogen and K.H.Z. Gahr, Acta Metall., 24:581-592 (1976).

HIGH-CYCLE FATIGUE PROPERTIES OF TITANIUM ALLOYS AT CRYOGENIC TEMPERATURES

O. Umezawa, K. Nagai, T. Yuri, T. Ogata, and K. Ishikawa

National Research Institute for Metals
1-2-1 Sengen, Tsukuba, Ibaraki 305, Japan

ABSTRACT

High-cycle fatigue properties were investigated for Ti-5Al-2.5Sn ELI and Ti-6Al-4V alloys (in the mill-annealed condition) at cryogenic temperatures. Subsurface crack initiation occurred in the long-life range for all alloys. The rolled materials of Ti-6Al-4V had much higher fatigue strength in the long-life range than the forged materials and the Ti-5Al-2.5Sn ELI alloy. The significant difference of fatigue strength in the long-life range for Ti-6Al-4V alloys was caused by morphological changes in their microstructure.

INTRODUCTION

Titanium alloys (α and α-β type) have many advantages for cryogenic structural applications, namely, in their low specific heat, low thermal conductivity, extremely low magnetic susceptibility, high electrical resistivity, low specific gravity, and high yield strength. For cryogenic use, a good toughness as well as a high strength is indispensable. Reduced oxygen content introduced higher fracture toughness at cryogenic temperatures for the titanium alloy.[1] However, lower oxygen content produced lower yield strength, although the decreased yield strength was high enough. Fatigue properties are also a measure of the safety of the structural materials, but few data were obtained at cryogenic temperatures for these alloys. Therefore, the present authors have accumulated the cryogenic high-cycle fatigue data of Ti-5Al-2.5Sn and Ti-6Al-4V alloys with reduced oxygen contents.[2-3] Those data show the fatigue strength in long-life range of the mill-annealed alloys does not always depend on their tensile strength. Hence, the present paper reports the crack initiation behavior in high-cycle fatigue, and discusses the relation of microstructure and fatigue strength.

EXPERIMENTAL PROCEDURE

Test Materials

Test materials were a Ti-5Al-2.5Sn ELI alloy[2] and three kinds of Ti-6Al-4V alloys[3]: a normal grade (Normal), an ELI grade, and an extremely low interstitials grade (Sp.ELI). Their chemical compositions and processing are given in Tables 1 and 2. Each Ti-6Al-4V forging, called "forged material," is material forged finally in $\alpha+\beta$ region (heated to 1173 K, formed 70x70 mm bar). Then a part of the forged material was also rolled in $\alpha+\beta$ region (heated to 1173 K, formed ϕ28 mm rod); it is called "rolled material." In the rolled material, the reduction ratio in section area was 8 times that of the forged material. Finally, all materials were mill-annealed for 7.2 ks at 973 K and air cooled.

Fatigue Testing

Hourglass-type, unnotched specimens with a waist diameter of 4.5 mm (or 6 mm) were machined in the L-direction from each material. Fatigue tests were carried out in liquid helium (4 K), in liquid nitrogen (77 K), and partly in air (293 K). A cryogenic servo-hydraulic fatigue test machine with a dynamic load capacity of ±50 kN was used.[2] Load control tests were done with a sinusoidal waveform and a minimum-to-maximum stress ratio, R, of 0.01. The test frequencies of 4 Hz at 4 K, 10 Hz at 77 K, and 20 Hz at 293 K were chosen so that the specimen temperature rise was as low as possible.

RESULTS AND DISCUSSION

Microstructure

Figure 1 represents SEM micrographs of the Ti-5Al-2.5Sn ELI and Ti-6Al-4V alloys. The Ti-5Al-2.5Sn ELI alloy had an equiaxed α phase structure with a mean grain diameter of approximately 30 μm. However, this alloy had a partially duplex

Table 1. Chemical Compositions of Tested Materials.

Alloy	Composition (mass%)							
	Al	Sn	V	Fe	O	N	H	C
Ti-5Al-2.5Sn ELI	5.15	2.66	-	0.19	0.057	0.0024	0.0058	0.012
Ti-6Al-4V								
Normal	6.34	-	4.23	0.199	0.135	0.0071	0.0053	0.011
ELI	6.23	-	4.25	0.200	0.104	0.0035	0.0032	0.011
Sp.ELI	5.97	-	4.12	0.028	0.054	0.0019	0.0055	0.024

Table 2. Manufacturing Process of Tested Materials.

Material	Reduction ratio in section area ($\alpha+\beta$ region)	Mill-annealing condition
Ti-5Al-2.5Sn ELI	5	973 K - 7.2 ks, air cooling
Ti-6Al-4V forged	5	
Ti-6Al-4V rolled	37	

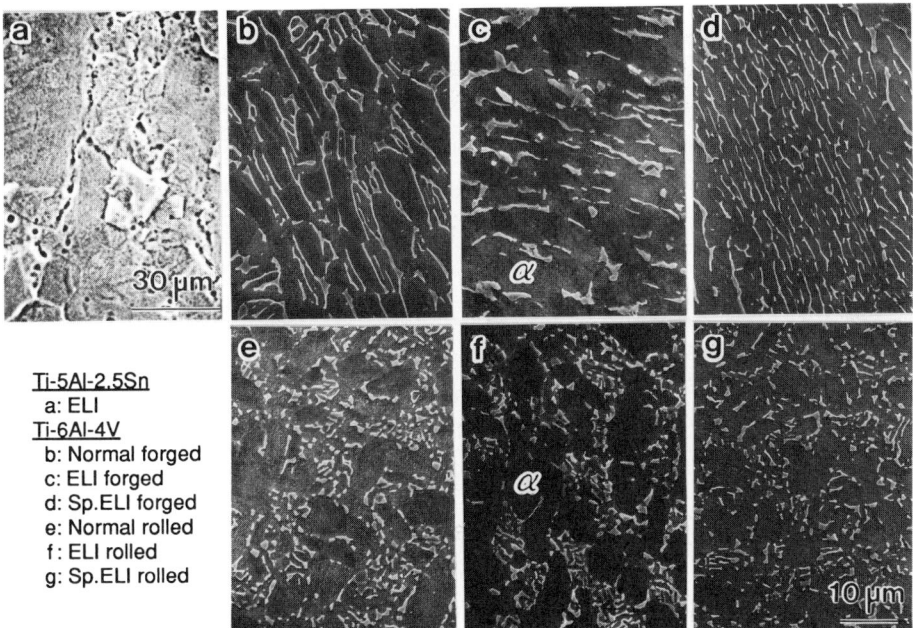

Fig. 1. SEM micrographs on the transverse section to the longitudinal direction of specimens for the Ti-5Al-2.5Sn ELI and Ti-6Al-4V alloys.

structure with α and β phases.[4] The Ti-6Al-4V forged materials were principally composed of elongated or plate-like primary α grains and β (or transformed β) platelets; they had a region in which α plates were aligned as a colony structure. The Ti-6Al-4V rolled materials were composed of globular α grains and β particles. The mean width of the α grains was almost the same for the forged and rolled materials of each alloy.[5] The Sp.ELI alloy had finer α grain structure than the other two alloys.

All the Ti-6Al-4V alloys had a similarly textured microstructure in which the prism plane was perpendicular to and the basal plane was parallel to the principal stress; that is, the c-axis was perpendicular to the principal stress.[3]

High-Cycle Fatigue Strength

The S-N curves of the forged and rolled materials for Ti-6Al-4V Normal alloy at three temperatures are described in Fig. 2, and all S-N data at 4 K are given in Fig. 3. Open symbols for each material indicate the fatigue fracture from the specimen surface, and solid symbols indicate fatigue fracture from the specimen interior. The fatigue strength of Ti-6Al-4V rolled materials was superior to that of Ti-5Al-2.5Sn ELI and Ti-6Al-4V forged materials in the long-life range at cryogenic temperatures, as shown in Figs. 2 and 3. The fatigue strength at 10^6 cycles of Ti-6Al-4V rolled materials was higher than that of austenitic steels at 4 K.[6] The fatigue strength of the Ti-5Al-2.5Sn ELI alloy and Ti-6Al-4V forged materials was as high as that of high-Mn steels at 4 K.[6]

Effect of test temperature. Temperature decrease produced an increase in strength, and generally it is said that fatigue

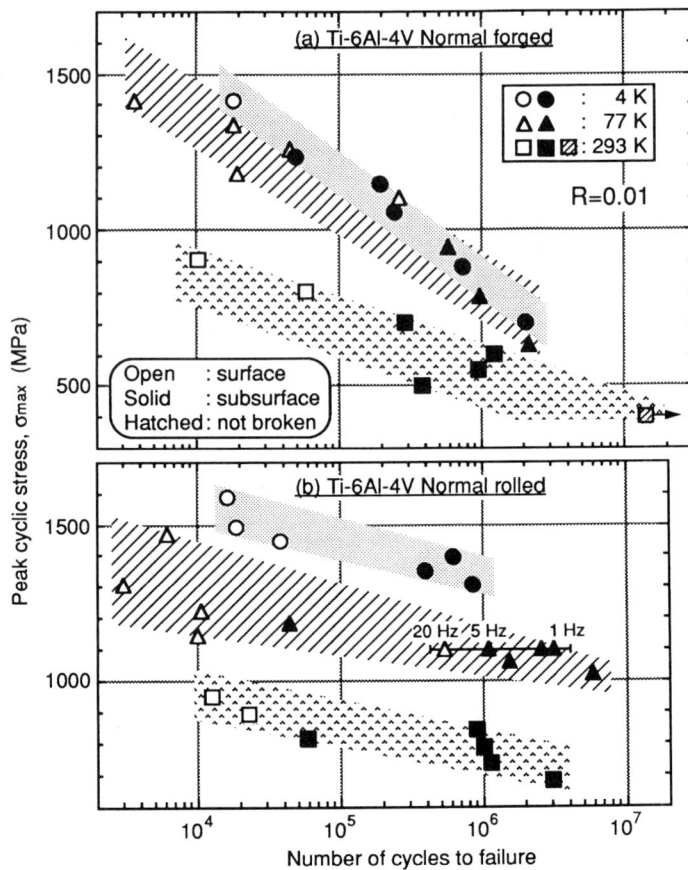

Fig. 2. S-N curves of Ti-6Al-4V Normal alloy : (a) forged and (b) rolled materials.

strength is proportional to tensile strength. Hence a simple analogy leads to the speculation that the fatigue strength increases at low temperatures. It is obviously true for the Ti-5Al-2.5Sn ELI alloy[2] and the Ti-6Al-4V rolled materials regardless of impurity level; the S-N curves shift to higher stress levels at lower temperatures, and they are almost parallel, as shown in Fig. 2(b). In the Ti-6Al-4V forged materials, on the other hand, there is hardly any gap between the S-N curves at 77 K and 4 K, as shown in Fig. 2(a). The gap between 293 K and 77 K (or 4 K) also becomes narrower as the number of cycles to failure increases. The fatigue strength of the Ti-6Al-4V forged materials was higher than that of the Ti-5Al-2.5Sn ELI alloy at 77 K, but was as high as at 4 K, as shown in Fig. 3.

Effects of strength level. Figure 4 shows the ultimate tensile strength of tested materials. The tensile strength of Ti-5Al-2.5Sn ELI alloy was lower than that of Ti-6Al-4V alloys. For each Ti-6Al-4V alloy, the tensile strength of the rolled material was higher than that of the forged one. The fatigue strength of the rolled material was also higher than that of the forged one for each alloy. However, the Sp.ELI alloys having the lowest tensile strength of the Ti-6Al-4V alloys, showed the highest fatigue strength at 10^6 cycles at every temperature, as shown in Fig. 3. Moreover, in the long-life range where

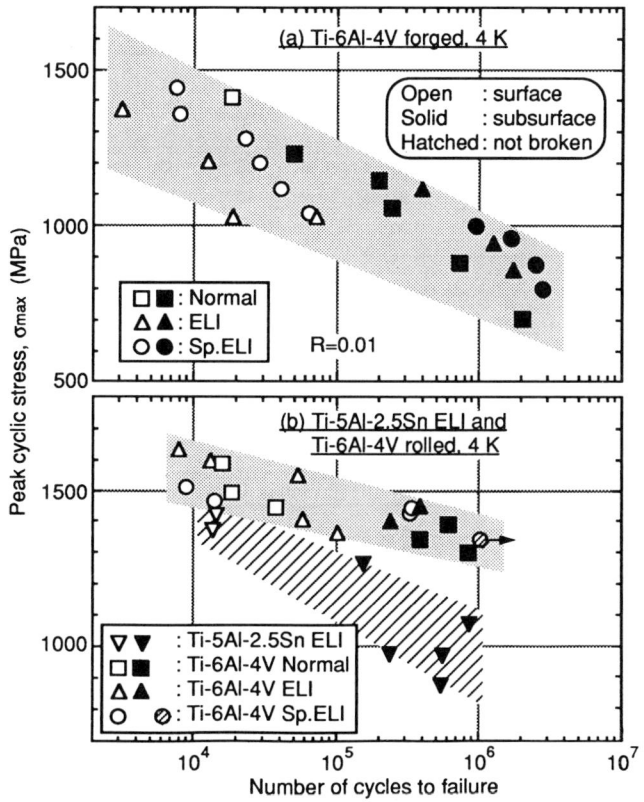

Fig. 3. S-N curves of titanium alloys at 4 K: (a) Ti-6Al-4V forged materials, and (b) Ti-5Al-2.5Sn ELI alloy and Ti-6Al-4V rolled materials.

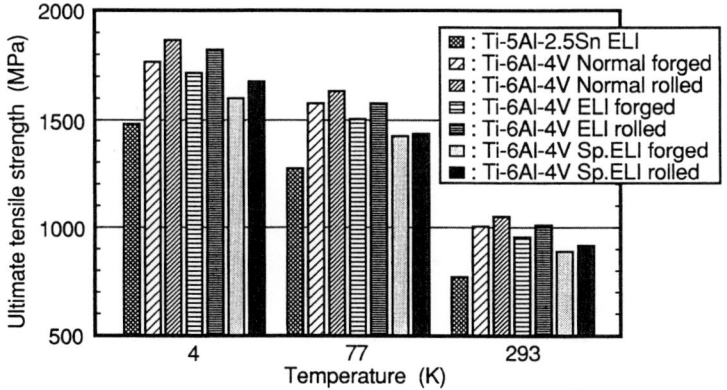

Fig. 4. Ultimate tensile strength of tested alloys.

subsurface crack initiation occurred, the fatigue strength of the forged material and Ti-5Al-2.5Sn ELI alloy was remarkably low, and their fatigue strength was almost the same [Fig. 2(a),3(a)]. Therefore, the fatigue strength in long-life range does not only depend on the tensile strength.

Subsurface Fatigue Crack Initiation

In general, fatigue crack initiation is understood to occur on the specimen surface owing to the irreversible process of extrusion and intrusion through slip deformation. In the Ti-5Al-2.5Sn and Ti-6Al-4V alloys, however, subsurface fatigue failure occurred at 4 K (shown in Fig. 3), when the number of cycles to failure was greater than about 10^5.

In Ti-6Al-4V Normal rolled materials, fatigue tests were carried out by varying the frequency at a given peak cyclic stress in the subsurface fatigue failure region at 77 K. Data points for the number of cycles to failure are scattered, as shown in Fig. 2(b), but their scatter does not reveal the time dependence. Hence, environmental effects, such as hydride precipitation and interstitial contamination, are not responsible for this subsurface crack initiation.

At low magnification, subsurface crack initiation sites appeared flat and were inclined toward the principal stress axis, as shown in Fig. 5(a). Neither defect-like inclusions nor pores were detected at subsurface crack initiation sites. The subsurface crack initiation site was composed of facets, which were identified as α phase by their morphology and composition analysis.[4-5,7] To quantify the size of the subsurface crack initiation sites of Ti-6Al-4V alloys, a crack length parameter 'f_S' was defined as the minor axis of an orthographic projection of the initiation site on the main crack propagating plane, as shown in Fig. 5(b).[5,7] Figure 6 shows the dependence of f_S on the peak cyclic stress at 4 K and 77 K.[5,7] The f_S increases as the maximum stress decreases. The subsurface crack initiation sites in these alloys were analyzed in detail elsewhere.[4-5,7]

Microstructural Factors and Fatigue Strength in Ti-6Al-4V Alloys

As mentioned above, the Ti-6Al-4V alloys had similarly textured microstructures. Hence, the difference in fatigue

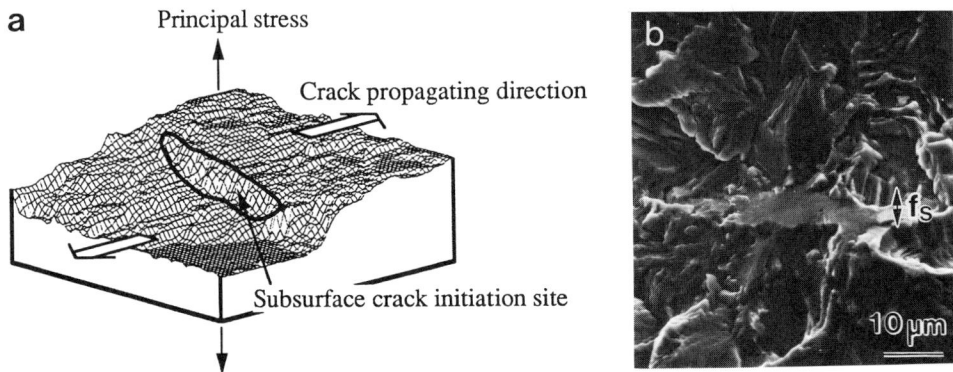

Fig. 5. Three dimensional indication (a) and SEM image (b) of subsurface crack initiation site for Ti-6Al-4V Normal forged material (4 K, σ_{max}= 1058 MPa).

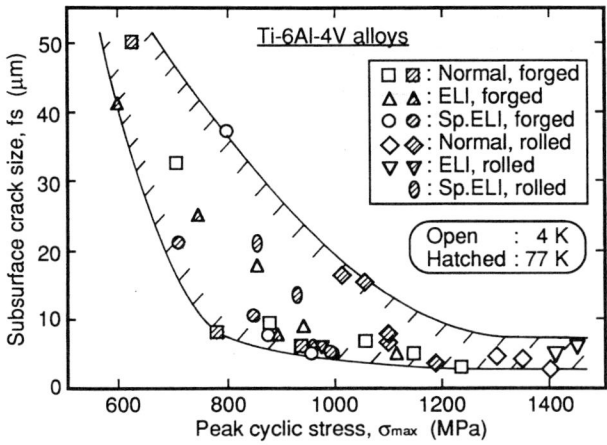

Fig. 6. Relationship between subsurface crack size and peak cyclic stress at 4 K and 77 K for Ti-6Al-4V alloys.

strength between the forged and rolled material cannot be explained in terms of texture. A significant difference between the forged and rolled materials was observed in the morphology of their microstructures, not in the primary α grain size itself, but in their shape of α grains as follows. In forged materials, elongated α platelets were aligned as a colony structure. On the other hand, in rolled materials, globular α grains were not aligned. The features of each facet in the subsurface crack initiation site corresponded to those of primary α grains in both the forged and rolled materials. In the forged material, subsurface cracking occurred at lower stress, and the size of initiation site was larger than in the rolled material of the same alloy after a similar number of cycles, as shown in Figs. 3 and 6.[5,7]

In the forged material, the primary α grains in a colony are believed to be crystallographically aligned and to act as a single path for dislocation motion. Williams and Luetjering reported that shorter slip length leads to substantial improvements in fatigue strength for the Ti-6Al-4V alloys.[8] Accordingly, the rolled material has a shorter slip length than the forged material, which may introduce lower stress localization and facilitate crack initiation at a given applied stress. Then the Sp.ELI alloy, which had finer α grain structure than other two alloys, also had a higher fatigue strength as a result of its shorter slip length.

From these considerations, we concluded that the fatigue strength in the long-life range for Ti-6Al-4V alloys depended on the morphological differences in their microstructure. The effect of reduced impurity levels has not been clear yet.

CONCLUSIONS

S-N curves for Ti-6Al-4V alloys were determined from about 10^4 to 10^6 cycles, and cryogenic fatigue properties were investigated for the Ti-5Al-2.5Sn ELI and Ti-6Al-4V alloys in the mill-annealed condition. Subsurface crack initiation occurred in the long-life range for all alloys. The rolled materials of Ti-6Al-4V had much higher fatigue strength in the long-life range than the forged materials and the Ti-5Al-2.5Sn

ELI alloy. This significant difference in the fatigue strength is ascribed to the difference in α grain structure resulting from the higher working ratio in $\alpha+\beta$ region.

ACKNOWLEDGEMENT

The authors wish to thank co-workers, Dr. T. Nishimura, Dr. T. Mizoguchi, and Dr. Y. Ito, Kobe Steel, Ltd.

REFERENCES

1. K. Nagai, T. Yuri, O. Umezawa and K. Ishikawa, The mechanical properties of Ti and its alloys, in: "Cryogenic Materials '88," R.P. Reed, Z.S. Xing, and E.W. Collings, eds., ICMC, Boulder, Vol.2:727-36 (1988).
2. K. Nagai, T. Ogata, T. Yuri, K. Ishikawa, T. Nishimura, T. Mizoguchi, and Y. Ito, Fatigue fracture of Ti-5Al-2.5Sn ELI alloy at liquid helium temperature, Trans. ISIJ, 27:376-382 (1987).
3. K. Nagai, T. Yuri, T. Ogata, O. Umezawa, T. Nishimura, T. Mizoguchi, and Y. Ito, Cryogenic mechanical properties of Ti-6Al-4V alloys, ISIJ International, (1991), in press.
4. O. Umezawa, K. Nagai, and K. Ishikawa, Subsurface crack initiation in high cycle fatigue of Ti-5Al-2.5Sn extra-low interstitial alloy at liquid helium temperature, Mater. Sci. Eng. A, 129:217-221 (1990).
5. O. Umezawa, K. Nagai, and K. Ishikawa, Subsurface crack initiation in high cycle fatigue of Ti-6Al-4V alloys at cryogenic temperatures (in Japanese), Testu-to-Hagane, 76:924-931 (1990).
6. K. Nagai, T. Yuri, O. Umezawa, T. Ogata, and K. Ishikawa, High cycle fatigue of some austenitic steels at cryogenic temperatures, in: "Stainless Steel '91," ISIJ, (1991), in press.
7. O. Umezawa, K. Nagai, and K. Ishikawa, Internal crack initiation in high cycle fatigue of Ti-6Al-4V alloys at cryogenic temperatures, in: "Fatigue 90," H. Kitagawa and T. Tanaka, eds., MCEP, Birmingham, Vol.1:267-272 (1990).
8. J.C. Williams and G. Luetjering, The effect of slip length and slip character on the properties of titanium alloys, in: "Titanium '80 Science and Technology," H. Kimura and O. Izumi, eds., AIME, N.Y., Vol.1:671-681 (1980).

COLD THERMAL FATIGUE OF AUSTENITIC STAINLESS STEEL

Arata Nishimura

National Institute for Fusion Science
Chikusa, Nagoya, Japan

Yoshihiko Mukai

Faculty of Engineering, Osaka University
Yamada-Oka, Suita, Osaka, Japan

ABSTRACT

To investigate the cold thermal-fatigue characteristics of an austenitic stainless steel (SUS304), a thermal-cycling test was conducted between ambient and cryogenic temperatures. Test specimens were tubes through which flowed liquid helium or nitrogen for cooling and nitrogen gas for warming; the specimen ends were under thermal-deformation restraint. The center of the pipe contained a thin, narrow test section, 20 mm long, to concentrate the thermal deformation. Large hysteresis loops were obtained in the test section. After 757 cycles, a fatigue crack propagated through the wall of the test section, and nitrogen gas leaked into the vacuum chamber. Cold thermal cracks were observed at a corner of the specimen test section; striations were observed on the fracture surfaces.

Cold thermal-fatigue life was estimated from strain-controlled fatigue life. Cold thermal fatigue is a significant factor in the design of cryogenic structures that will experience thermal cycles, especially those where a stress concentration exists.

INTRODUCTION

The Large Helical Device (LHD) is planned for construction at the National Institute for Fusion Science in Japan, and a superconductor will supply the magnetic field to the LHD. Since the superconductor, or coil-wound superconducting magnet, requires cryogenic temperatures to exhibit its excellent property, the equipment or support structures must also be cooled. For inspection, however, the entire facility must be warmed. Consequently, the support structures will experience thermal cycling between ambient and cryogenic temperatures.

Therefore, to maintain an accurate magnetic-field position and to ensure the soundness of this large facility, dynamic and other material properties, such as fracture toughness and fatigue, must be taken into account along with the static mechanical properties. Many studies have been made of the static mechanical properties of materials at cryogenic temperatures, but fatigue properties, especially cold thermal fatigue, have not been sufficiently characterized.

In this study, the deformation behavior of an austenitic stainless steel, under thermal-deformation restraint, was investigated during thermal cycling from 293 K to cryogenic temperatures. Cold thermal-fatigue fracture was demonstrated experimentally. The cold thermal-fatigue life was estimated on the basis of fatigue-life curves obtained from constant-strain-controlled tests.

MATERIAL AND SPECIMENS

Two kinds of SUS304 austenitic stainless steel were used in this study. Their chemical compositions and mechanical properties are shown in Tables 1 and 2.

Three specimens were made from material A, and one specimen, from material B. Each specimen had a different configuration, as shown in Fig. 1. The tube wall in the center of the specimens was machined to create a thin wall where thermal stress or strain would concentrate. The degree of stress concentration was changed by changing total specimen length and the diameter and wall thickness in the test section (T.S.). Test sections I, II, and III were machined from material A; T.S. IV, from material B.

TEST EQUIPMENT AND PROCEDURES

The cold thermal-fatigue testing equipment is illustrated in Fig. 2. The test specimen and flow tubes were connected in a vacuum chamber. The thermal contraction of the test specimen was restricted by upper and lower restraint plates and four columns made of SUS304. The specimen temperature was changed

Table 1. Chemical Compositions of Test Materials

SUS304	Chemical Composition (mass %)						
	C	Si	Mn	P	S	Ni	Cr
Material A	0.05	0.25	1.41	0.032	0.026	8.06	18.32
Material B	0.06	0.26	1.71	0.036	0.025	8.25	18.81

Table 2. Mechanical Properties of Test Materials

SUS304	E (GPa)			$\sigma_{0.2}$ (MPa)			σ_u (MPa)		
	293 K	77 K	4.2 K	293 K	77 K	4.2 K	293 K	77 K	4.2 K
Material A	188.6	197.0	187.3	277.6	388.8	491.6	715.8	1655.8	1869.7
Material B	189.3	192.1	197.3	371.8	596.6	692.1	694.0	1622.8	1823.9

SUS304	Φ^* (%)			$H'^\dagger$ (GPa)			Elongation (%)		
	293 K	77 K	4.2 K	293 K	77 K	4.2 K	293 K	77 K	4.2 K
Material A	71.6	65.0	51.0	3.54	17.07	23.58	60.9	34.1	29.1
Material B	55.4	41.9	29.3	2.83	12.52	20.30	54.7	35.0	31.6

*reduction of area; †strain-hardening coefficient

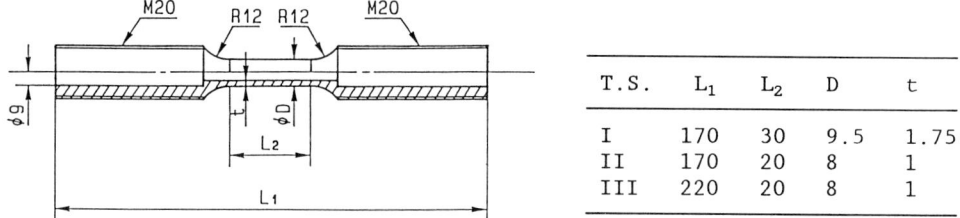

T.S.	L_1	L_2	D	t
I	170	30	9.5	1.75
II	170	20	8	1
III	220	20	8	1

(a) T.S. I, T.S. II, and T.S. III.

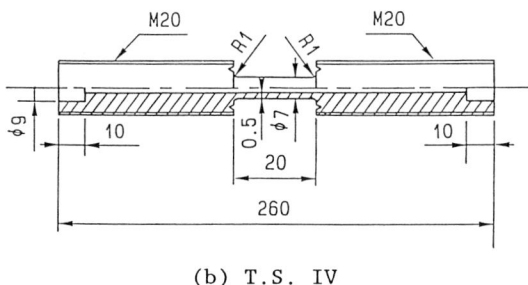

(b) T.S. IV

Fig. 1. Test specimens (dimensions in mm).

by flowing liquid helium, liquid nitrogen, or nitrogen gas through the tubes. To warm the specimen rapidly, the nitrogen gas was warmed by a code heater attached to the flow tube.

The vacuum pump operated at all times during the test to reduce heat transport by convection and to check leakage from the test specimen. Fiber-reinforced-plastic nuts were used on the SUS304 restraint plates to prevent heat transmission.

The thermal stress produced in the specimen was estimated from the strain output on the columns. The thermal strains of T.S. I, T.S. II, and T.S. III

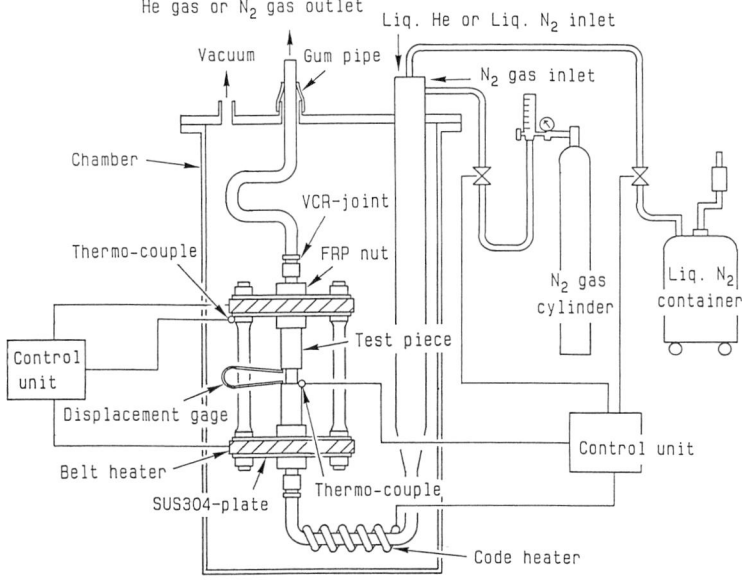

Fig. 2. Schematic drawing of cold thermal-fatigue testing equipment.

were measured by strain gages attached to the test section; for T.S. IV, a displacement gage was used because the deformation was large. During the test, the temperature of the test section was measured with an Au-0.07%Fe-Chromel thermocouple.

With this equipment in manual operation and liquid helium as the coolant, 10 cold thermal cycles (each consisting of one cooling and warming process) were applied to T.S. I, T.S. II, and T.S. III, and 36 cycles were applied to T.S. IV. After 36 cycles, T.S. IV was automatically cycled with a control system that used liquid nitrogen as the coolant. This system consisted of a thermocouple, two electromagnetic valves, a Dewar containing liquid nitrogen, and a nitrogen-gas cylinder. The thermocouple output controlled the liquid-nitrogen and nitrogen-gas valves and the code heater. The same system maintained the temperature of the restraint plates at 293 K.

THERMAL-STRESS CONCENTRATION

For this study, the specimens were designed with a small test section where the thermal stress would concentrate. The degree of thermal-stress concentration in an elastic condition was estimated from a simplified model proposed by the authors.[1]

As shown in Fig. 3, columns 1 and 2, with cross-sectional areas of A_1 and A_2 and lengths of l_1 and l_2, respectively, were connected. The deformation of this connected rod was restricted at both the top and bottom. When the rod was cooled, the cooling rate of column 1 was higher than that of column 2, and a temperature distribution formed in the longitudinal direction. If the temperature change of column 2 (ΔT_2) is k times smaller than that of column 1 (ΔT_1), the thermal-stress change of column 1 ($\Delta \sigma_1$) can be represented as

$$\Delta \sigma_1 = -E\alpha\Delta T_1 \cdot K_e^* \qquad (1)$$

$$K_e^* = A_2 (l_1 + k \cdot l_2)/(A_1 l_2 + A_2 l_1) \qquad (2)$$

where E is Young's modulus and α is the thermal-expansion coefficient. From these equations, it can be seen that the stress change of column 1 is K_e^* times larger than that of a uniform column. Since shape is a factor in determining the value of K_e^*, it is considered to be a parameter that reveals the degree of thermal-stress concentration on a certain section. The calculated values of K_e^* are 1.060, 1.361, 1.506, and 2.165 for T.S. I, T.S. II, T.S. III, and T.S. IV, respectively. For the calculation of K_e^*, k is supposed to be 0.3 on the basis of the result of a temperature-distribution experiment.

EXPERIMENTAL RESULTS AND DISCUSSION

Stress-Strain Hysteresis Curve

The variations in thermal stress and temperature versus strain during the first cycle are shown in Fig. 4. The relative size of the hysteresis curves corresponds to that of K_e^*. In the relationship between temperature and

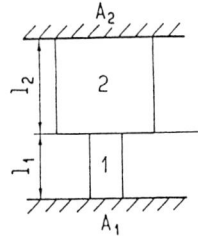

Fig. 3. A model for estimation of thermal-stress concentration.

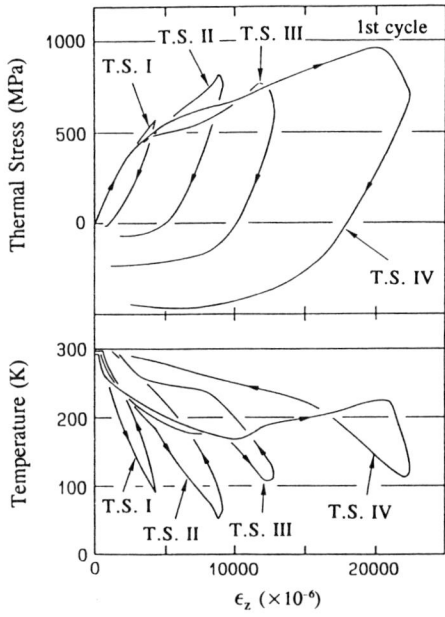

Fig. 4. Thermal stress-strain curves on the first cycle of each test specimen.

strain, the nonlinear hysteresis curves are due to the occurrence of plastic strain. This is especially noticeable in the T.S. IV curve, where a temperature rise occurred during the cooling process as a result of the large amount of heat generated by plastic strain. The following discussion is restricted to T.S. IV, since its hysteresis curve is the largest.

The hysteresis curves of cycle 100 and cycle 500 of T.S. IV are shown in Fig. 5. Clearly, the width of the hysteresis curve narrows with increasing

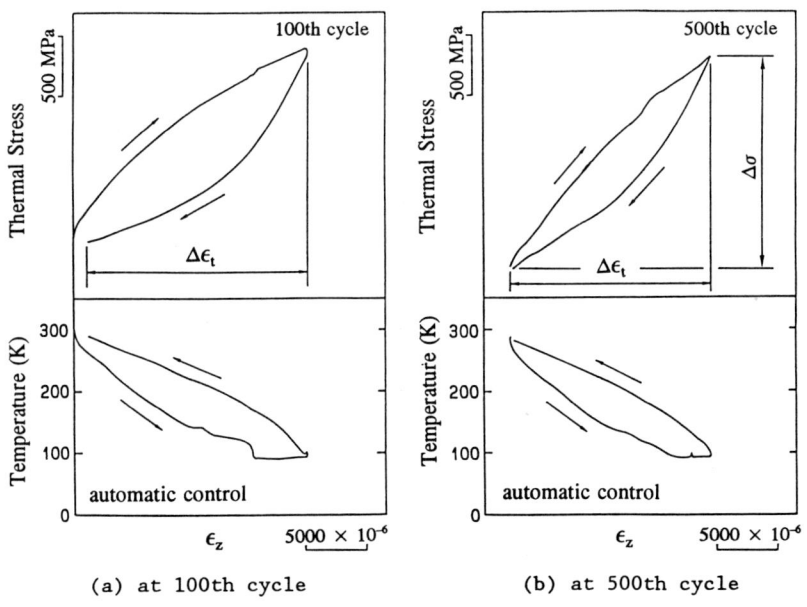

(a) at 100th cycle (b) at 500th cycle

Fig. 5. Thermal stress and temperature versus strain.

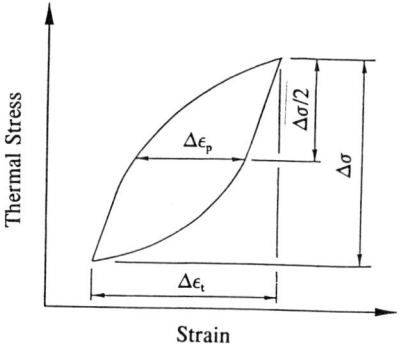

Fig. 6. Definitions of $\Delta\epsilon_t$, $\Delta\epsilon_p$, and $\Delta\sigma$.

number of thermal cycles. With the hysteresis curves defined in terms of $\Delta\epsilon_t$, $\Delta\epsilon_p$, and $\Delta\sigma$, as shown in Fig. 6, the change in these values with the number of thermal cycles is presented in Fig. 7. The hysteresis loop contracts during the initial fatigue process, and after that, $\Delta\epsilon_t$ and $\Delta\epsilon_p$ remain at about 1.6% and 0.4%, respectively. On the other hand, $\Delta\sigma$ increases and reaches about 1700 MPa. These data show that either the elastic modulus or the strain-hardening coefficient becomes large and maintains a certain condition over half or more of the total fatigue period.

Comparison with Strain-Controlled Fatigue Life

The data obtained in this study are compared with the axial strain-controlled fatigue-life curves of Suzuki et al.[2] in Fig. 8. The open circle represents the total strain range during the first cycle, and the solid circle, that of the 500th cycle, where the strain range remained constant.

The cold thermal-fatigue life of T.S. IV, which was cooled with liquid nitrogen, is shorter than that of the strain-controlled fatigue life at 77 K. For this reason, the following factors are considered:

1. The effect of mean strain: A large mean strain was exerted on this specimen, and the fatigue curve was obtained with zero mean strain.
2. The stress concentration on the corner of the test section, where a crack initiated, as discussed below.
3. The difference in test-specimen configuration: Suzuki et al. used a round, bar specimen, and this study used a tube specimen.

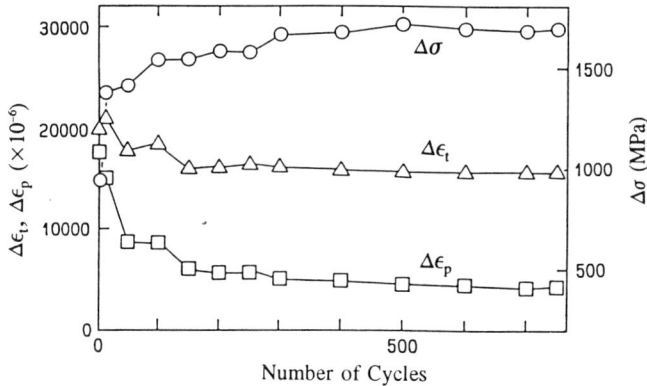

Fig. 7. $\Delta\epsilon_t$, $\Delta\epsilon_p$, and $\Delta\sigma$ versus thermal cycles of T.S. IV.

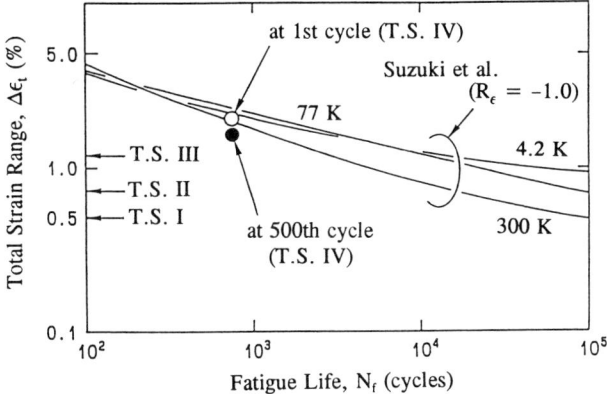

Fig. 8. Comparison of cold thermal-fatigue life with strain-controlled fatigue life.

When these factors are taken into account, the cold thermal-fatigue life should be nearly the same as the strain-controlled fatigue life at 77 K. In other words, the strain-controlled fatigue-life curve at 300 K may give a conservative fatigue life for the design curve of cold thermal fatigue. Certainly, the effect of mean strain in the strain-controlled test must be clarified, and more cold fatigue data should be obtained.

Cold Thermal-Fatigue Crack

At 757 cycles, nitrogen gas leaked through a fatigue crack in T.S. IV; the outer surface of the crack is shown in Fig. 9. The thermal-fatigue crack initiated at a corner of the test section and propagated in a circular direction. The result of SEM observation of the fracture surface is shown in Fig. 10. Many striations are evident, and the spacing between them is smaller at point A than at point B. Therefore, it is clear that the fatigue crack propagated from the outer to the inner surface.

CONCLUSIONS

In this study, the deformation behavior of an austenitic stainless steel during a cold thermal-fatigue process was investigated. The cold thermal-fatigue life was plotted on a strain-controlled fatigue-life curve, and their relationship is discussed. The main results of the study are:

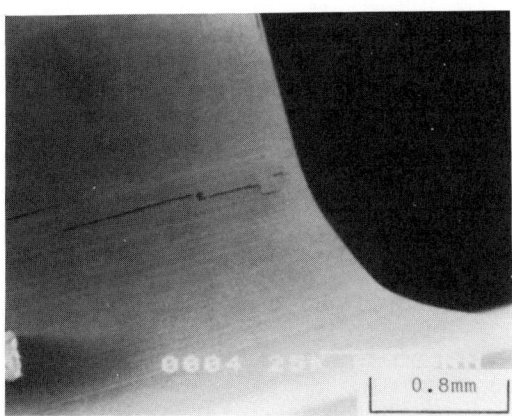

Fig. 9. Example of a cold thermal-fatigue crack on the outer surface of the specimen.

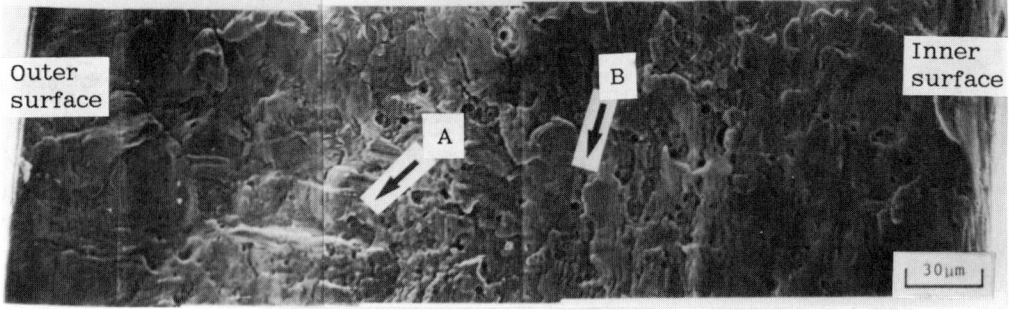

Fig. 10. SEM observation of the cold thermal-fatigue fracture surface.

1. Austenitic stainless steel under thermal-deformation restraint exhibited obvious plastic deformation during cold thermal cycles. The total cyclic strain range depended on the degree of thermal-stress concentration. The hysteresis loop contracted during the initial fatigue cycling.
2. Cold thermal-fatigue failure occurred indeed. Note that a notch can act as the initiation site of a cold thermal-fatigue crack as well as a mechanical-fatigue crack, because thermal stress and strain would also concentrate at a notch.
3. The cold thermal-fatigue life obtained in this study was shorter than that of strain-controlled fatigue life obtained under constant temperature. However, since the studies differed in mean strain, stress concentration, and specimen configuration, their correspondence has still not been clarified, but strain-controlled fatigue-life data would be useful for cold thermal-fatigue design.

ACKNOWLEDGEMENTS

This work was performed at the Low Temperature Center of Osaka University. The authors wish to express their thanks to all the staff of the center. The authors also thank Mr. M. Tanimoto, Mr. I. Maeda, and Mr. K. Aoki for their continuous cooperation in this work.

REFERENCES

1. Y. Mukai and A. Nishimura, Mechanical properties of SUS304 stainless steel under cold thermal cycles, *Proc., International Conference on Magnet Technology (MT-11)*, Elsevier, New York (1990), p. 743.
2. K. Suzuki, J. Fukakura, and T. Mori, Low-cycle fatigue properties of 304L stainless steel under axial-strain control at liquid helium temperature, *Zairyo* (in Japanese) 34-385 (1985), p. 1206.

INFLUENCE OF AGING ON THE FRACTURE TOUGHNESS OF CRYOGENIC AUSTENITIC MATERIALS, EVALUATED BY A SIMPLE TEST METHOD

Jakob Kübler, Hans-Jakob Schindler, Walter J. Muster

Swiss Federal Laboratories for Materials Testing and Research (EMPA), 8600 Dübendorf, Switzerland

ABSTRACT

The fracture toughness of heat treated austenitic stainless steel is measured. The data collected using a simple dynamic bend test at 77 K are compared and scaled with values measured on CT specimens at 4 K. It shows that the toughness depends strongly on the temperature and time of the heat treatment and that − as long as there is no need for values according to certain standards − expensive and time consuming 4 K tests on CT specimens can be substituted. The data are presented in three dimensional graphs.

INTRODUCTION

The wind-and-react technique planned for the jacket of the Nb_3Sn cable-in-conduit conductor for the inner poloidal field coils of the Next European Torus (NET) tends to embrittle the austenitic stainless steel. Previous investigations were carried out with CVN (Charpy-V-notch) tests at 77 K and CT specimens (compact tensile test) at 4 K on 316LN type materials and revealed a special sensitivity for intergranular fracture, depending on the carbon content and the time and temperature of heat treatment. The goal of this paper is the presentation of different dependences in a more general sense, focussing on three dimensional toughness graphs for several materials. To collect all the necessary data which are the basis of these graphs, many mechanical tests have to be performed. The classical methods to evaluate the fracture toughness need a lot of material and are time consuming and very expensive, especially at low testing temperatures. On the other hand, simpler tests like CVN are not supposed to deliver satisfying fracture toughness values. Thus, a fracture mechanics based testing method is proposed which enables easy derivation of approximate fracture toughness values from load-displacement curves of sharp-notched or precracked CVN-shaped bending specimens.

Table 1. Chemical composition of the investigated alloys (wt%).

	C	N	Cr	Ni	Mn	Mo	P	Nb
NET1	0.019	0.19	18.2	11.8	1.25	1.99	0.009	0.054
A410	0.022	0.17	17.2	12.8	1.5	2.69	0.024	-----
SUS	0.024	0.18	18.0	11.2	0.84	2.5	<0.03	-----

MATERIAL

Table 1 gives the compositions of the 3 investigated alloys (data of NET1 and A410 are from the materials supplier) and Table 2 gives the cryogenic mechanical properties measured at 4 K. The steel NET1 (316LN + Nb) is a modified JK1 described by Shimada and Tone [1] and tested for NET by Muster et al.[2], whereas A410 and SUS are 316LN austenitic steels that are commercially available steels used recently in a VAMAS (Versailles Projects on Advanced Materials and Standards) Round-Robin-Test [3]. All alloys were produced with modern metallurgical procedures (ESR, vacuum treating, etc.).

The specimens were machined close to their final size and afterwards heat treated. Subsequently, they were ground to their final size. The CT specimens were precracked at room temperature. The K_{Id} test specimen had a size of 10x10x55 mm and was side-grooved in accordance to ASTM E 813 to a depth of 1 mm. A first notch of 2.8 mm depth had been cut by a mill-cutting-blade of 0.4 mm width. The sharp-notch was ground using a diamond saw blade of 0.07 mm (resulting width less than 0.1 mm) to a final depth of about 3.2 mm.

TEST PROCEDURES

The evaluation of fracture toughness K_{Ic} at 4 K is very time consuming and costly. Thus, it is advantageous to strive for minimizing the number of tests. However, for the purpose of the present investigation, a large number of heat treatments had to be evaluated, leading to a sizable number of specimens. On the other hand, to compare the different heat treatments of the materials, the relative accuracy of the results is of utmost importance. For these reasons, the investigation was restricted to the determination of approximate K_{Ic}-values. The idea was to correlate the data from impact

Table 2. Cryogenic mechanical properties of the investigated alloys at 4 K (NHT: not heat treated).

	heat treatment (temp.°C/h)	YS (MPa)	UTS (MPa)	Elongation (%)
NET1	700/50	1060	1620	52
A410	700/50	980	1560	47
SUS	NHT	1080	1720	46

bending tests at 77 K with static data at 4 K. It is a well known fact that there is a similarity of the plastic deformation behavior of metals at decreasing temperature and increasing deformation rate. According to Rolfe et al.[4], the temperature shift TS due to the loading rate corresponding to the CVN test for ferritic structural steel is given by the empirical relation TS [K] = 119 - 0.12 YS [MPa]. Assuming that a similar relation also holds for the present austenitic steels, it is possible to predict the fracture behavior at approximately 4 K by increasing the loading rate at 77 K.

Furthermore, the tests were simplified by using specimens with sharp notches (root radius $p \simeq 0.05$ mm) instead of fatigue precracks. Therefore the cost of specimen preparation could be lowered significantly. Of course, a finite root radius influences the apparent J_{lc}, increasing it by a certain amount which is linear in p and the specific fracture energy [5]. However, since p is relatively small, the difference should be admissible.

Under the specific conditions - simulating 4 K by increased loading rate at 77 K and using sharp notches instead of fatigue precracks - the resulting K_{lc} are supposed to be of limited accuracy, but able to quantify correctly the differences in toughness of the various heat treatments. In order to obtain values of the correct magnitude, some standard J_{lc} tests at 4 K were performed, which could be used for scaling the J_{lc} from the impact tests to the exact values.

To obtain approximate K_{lc}-values, a method which is often practiced in materials testing is to use correlation formulas between K_{lc} and CVN. Many correlation formulas are known, most of them being of the form

$$J_{lc} = K_{lc}^2 / E = A \cdot CVN,$$

where A is a nondimensional, material-dependent constant [6]. Thus, in addition to the fracture mechanics tests, CVN tests were performed at 77 K.

The main difficulty inherent in impact bending testing is the detection of the point of crack initiation, which is necessary for deriving J_{lc} or K_{lc} values from these tests. In [5], a simple approximate method for estimating J_{lc} from a single load displacement diagram as obtained from an instrumented impact test is proposed. This method is based on a theoretical estimation of the stable crack-growth Δa_m up to the point of maximum load by

$$\Delta a_m = q\, b/(2+q)$$

where q is the apparent "hardening exponent" of the force-displacement diagram fitted in the form $F = C\, d^q$. This estimation of Δa_m differs slightly from the corresponding equation in [5], where q is approximated by the hardening exponent n. In the present case, the n values are too high for this simplification, thus q is used in its proper meaning as defined above.

TEST RESULTS

For the tensile and CT tests at 77 and 4 K a special screw driven testing machine with regulated crosshead displacement and a capacity of 200 kN was used, equipped with a He cryostat and controlled by an IBM AT O3 PC. The strain measurement for the tensile tests used two capacitance gauges. In the domain of interest the strain rate was 1.6×10^{-4} s^{-1}. The load line opening measurement of the CT specimens used a clip-on gage. The crosshead speed was lower than 0.5 mm/min for all

Table 3. Summary of the K_{Ic}, J_{Id} and CVN test results for the base and heat treated condition evaluated by 77 and 4 K (see text for explanation). (NHT: not heat treated; average of 2 tests; () linear interpolation).

	heat treatment °C/h	Column 1 4 K		Column 2 77 K	Column 3 77 K	Column 4 4 K	
		J_{Ic} J/mm	K_{Ic} MPa√m	J_{Id} J/mm	CVN J	J_{Ic} J/mm	K_{Ic} MPa√m
NET 1:	NHT	---	---	623	210[7]	214	226
	600/26	---	---	---	218	209	224
	600/50	---	---	---	(219)	210	224
	600/100	---	---	---	220	211	225
	600/200	---	---	---	(221)	212	225
	675/26	---	---	532	191[7]	183	209
	675/100	---	---	454	148	156	193
	675/200	---	---	363	111	125	173
	700/26	---	---	469	165	161	196
	700/50	125	173	371	146	127	175
	700/100	---	---	298	111[7]	102	157
	700/200	71	130[7]	211	69	72	132
	750/26	---	---	446	138	153	192
	750/100	---	---	251	86	86	144
	750/200	71	130[7]	197	57	68	127
	800/26	---	---	414	160[7]	142	184
	800/100	---	---	306	147[7]	105	159
	800/200	---	---	318	113	109	162
A410:	NHT	518	348[8]	754	274[8]	498	341
	600/26	---	---	---	(261)	434	318
	600/50	---	---	---	(249)	414	311
	600/100	---	---	---	224	372	295
	700/26	---	---	403	157	267	250
	700/50	150	187[7]	234	110[8]	155	190
	700/100	72	130[7]	110	43[7]	73	130
	750/26	---	---	307	139	203	218
	750/50	---	---	175	92	116	164
	750/100	---	---	109	37	72	130
	800/26	---	---	295	127[7]	195	213
	800/50	---	---	222	84	147	185
	800/100	---	---	144	62[7]	95	149
SUS:	NHT	331	276[3]	1004		343	281
	700/26	---	---	319		109	158
	700/50	47	104[9]	133		45	102
	700/100	---	---	48		16	62
	750/26	---	---	162		55	113
	750/50	---	---	94		32	86
	750/100	---	---	29		10	48
	800/26	---	---	183		62	120
	800/50	---	---	174		59	117
	800/100	---	---	68		23	73

tests. The K_{Ic} tests were carried out in accordance with ASTM E 813-88.

For the CVN tests a standard 300 J pendulum with a pendulum speed of 5.52 m/s was used. For the instrumented K_{Id} tests the speed was reduced to 3.19 m/s. The specimens were cooled in liquid nitrogen and tested in less than 5 s after taking them out of the cooling medium. The force vs. time curves were measured with a special semiconductor equipped tup. The signals were recorded using a fast transient recorder.

The results of the tests are summarized in Table 3. In column 1, the K_{Ic} (and the corresponding J_{Ic}) at 4 K are given, if they are known. Column 2 contains the approximate J_{Id} at 77 K determined according to [5] and the forgoing chapter. From the CVN-values (column 3) and the J_{Id} (column 2) the following correlation is found:

NET1: J_{Id} = 2.82 CVN
A410: J_{Id} = 2.52 CVN

This empirical relation is shown in Fig. 1 for NET1 material. The approximate J_{Id} (CVN) resulting from the above correlation are used to estimate the unmeasured values of column 2 (Table 3).

Obviously, there is a significant difference between the approximate J_{Id} at 77 K (column 2) and the J_{Ic} at 4 K (column 1). In order to obtain K_{Id} comparable in magnitude, the values of column 2 were scaled down to fit the values of column 1. The corresponding scaling factors $C = J_{Ic}(4 K) / J_{Id}(77 K)$ turned out to be

NET1: C = 0.34
A410: C = 0.66
SUS: C = 0.34

The resulting approximate J_{Ic} at 4 K and the corresponding K_{Ic} are shown in Table 3, column 4.

The CVN at 77 K (Table 3, column 3) as a function of temperature and duration of the heat treatment are shown graphically in Fig. 2 for the material NET1. In Fig. 3-5 the approximate K_{Ic} at 4 K is shown in function of the same parameters.

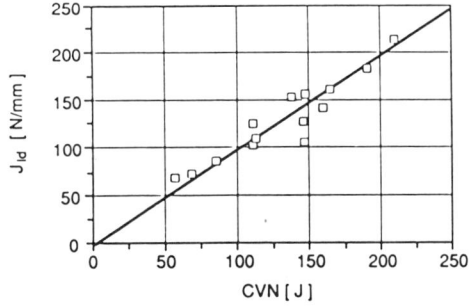

Fig. 1 CVN vs. J_{Id} values for material NET1. The good linear agreement allows the substitution of some K_{Id} by CVN tests.

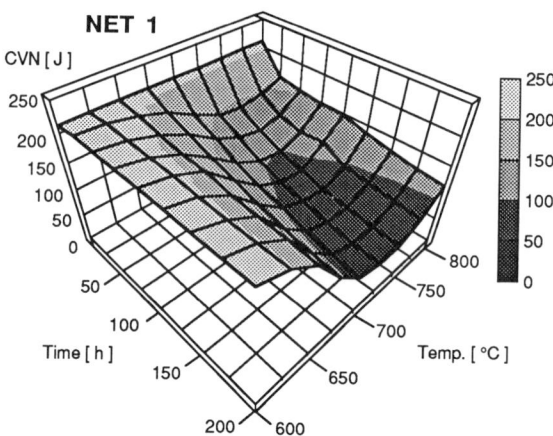

Fig. 2 CVN 3-D-graph of material NET1. Because the CVN test normally has a good sensitivity for the fracture toughness, this graph has about the same form as the corresponding K_{Ic} graphs (see Fig. 3).

DISCUSSION

The Figs. 2 and 3 show that the dependence of CVN and K_{Ic} on temperature and duration of heat treatment is generally the same. This indicates that CVN in this case is also a valuable quantitative method for characterizing the toughness and to identify the effect of heat treatment. This is supported by the good correlation between J_{Id} and CVN (Fig. 1).

However, to get more detailed information about toughness, it is necessary to consider K_{Ic}. The areas given by Figs. 3-5 show, that the effect on toughness is basically the same for all three materials. There is a certain most unfavourable temperature of heat treatment where an increasing duration leads generally to decreasing toughness.

As described previously, it was assumed that the effect of the loading rate and of temperature should - at least partly - compensate each other. Comparing column 1 and 2 indicates that this actually is not the case. The reason for the discrepancy probably is the finite root radius of the notch, which can increase the J_{Ic} for a tough material considerably [6]. However, since the amount of increase is linear in root radius and fracture energy, which is also related to fracture toughness, it seems to be reasonable to scale the J_{Id} by the material-dependent factors C down to the real J_{Ic}. At least for the materials tested here this correlation seems to hold. However, it is interesting to note that the factor C is quite different for the specific investigated materials. This may be due to different fracture mechanisms and needs further investigation.

CONCLUSION

The experiments have shown that - as long as there is no need for values according to certain standards - the expen-

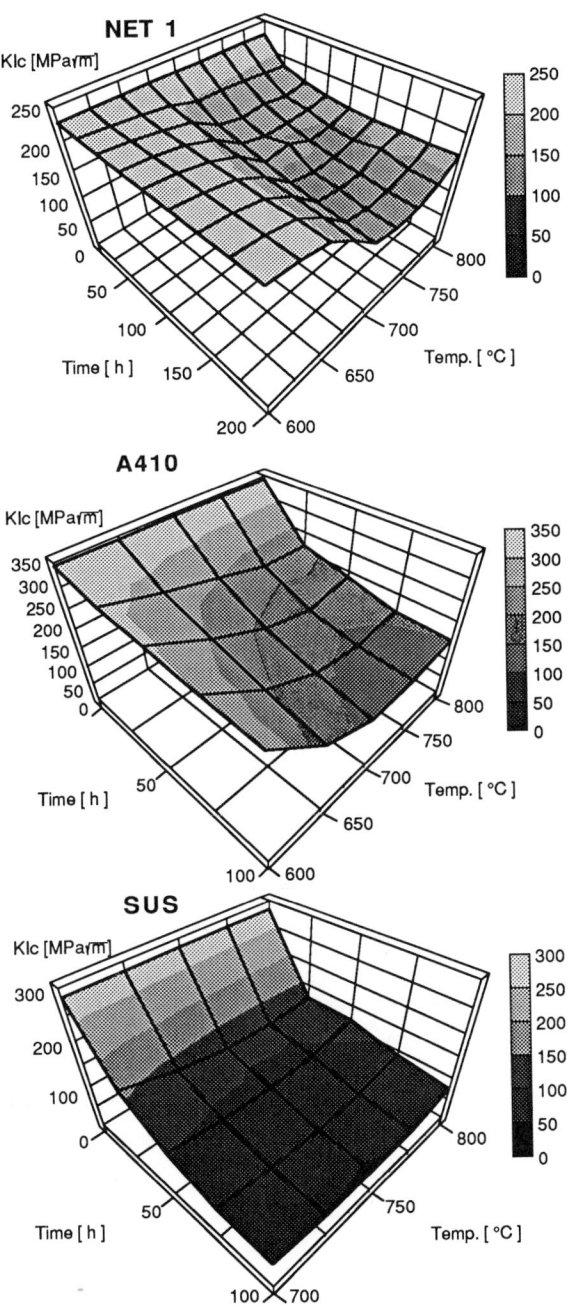

Fig. 3 to 5 K_{lc} 3-D-graphs for all three materials. The K_{lc}-values used to draw these graphs are described in the "TEST PROCEDURE" and "TEST RESULTS" and are nominally given in Table 3, column 4.

sive and time consuming fracture toughness tests on CT specimens at 4 K can be substituted by simple impact bending tests on sharp-notched CVN specimens at 77 K. Partly, the effect of the higher temperature is compensated by the increased loading rate. However, in the present case, the compensation was obviously not completed, since the dynamic fracture toughness values at 77 K were significantly higher than the static ones at 4 K. Probably this is caused by the finite root radius of the notch, so it would have been advantageous to use fatigue precracked specimens instead. Nevertheless, by scaling the values linearly down to the actual K_{Ic} at 4 K, the resulting values seem to be of reasonable accuracy.

REFERENCES

[1] Shimada M. and Tone S., Advances in Cryogenic Engineering-Materials 34 (1988), 157.
[2] Muster W.J., Kübler J. and Hochhaus Ch., Advances in Cryogenic Engineering-Materials 36A (1990), 109.
[3] VAMAS - Technical Working Area 06 - Superconducting and cryogenic structural materials, Proc. 7th Meeting Report, 1991 to be published.
[4] Rolfe S.T., Sorem W.A., Fracture control in the transition-temperature region of structural steels, J. Constr. Steel Research 12 (1989), 171.
[5] Schindler H.J., Determination of fracture mechanic material properties utilizing notched test specimens, Proc. 6th Int. Conf. Mech. Behavior, Kyoto, 1991, to be published.
[6] Schindler H.J., Morf U., An Estimation of fracture toughness from instrumented standard Charpy V-notch tests. Proc. 10th Congress on material testing, Budapest, 1991, to be published.
[7] Muster W.J, Elster J., Low temperature embrittlement after aging stainless steels, Cryogenics (May 1990).
[8] Muster W.J., Huwiler R., Kohler P., Jacket Material for NET's Wind-and React-Superconductor, Appendix to Report EMPA No. 113'362, Dübendorf (Switzerland) (1989), 5.
[9] Kübler J., Huwiler R., Pellegrini O., Fracture Toughness Test on Base and Heat Treated Cryogenic Material at 4.2 K, EMPA No. 103'378/1, Dübendorf (Switzerland) (1989), Appendix Table 1.

EFFECTS OF BORON ON INCREASING TOUGHNESS OF HIGH

STRENGTH HIGH MANGANESE NON-MAGNETIC STEELS

*Hideki Tanaka, **Kouzou Fujita and **Koji Shibata

*Materials Science, Faculty of Engineering, Graduate school
of the University of Tokyo, **The University of Tokyo
Department of Metallurgy and Materials Science, Bunkyo-ku
Tokyo 113, Japan

INTRODUCTION

High Mn non-magnetic steels, which contain above 30% Mn, are liable to embrittle at low temperatures when cooled at a slow rate from annealing temperature or reheated at about 873K due to intergranular precipitation of carbide or nitride. Therefore, these steels are generally recommended to be cooled at a fast rate from annealing temperature. Shibata et al.[1], however, observed that cryogenic intergranular fracture was enhanced by water-quenching in a 32Mn-7Cr-0.3N steel. They also observed that the steel was toughened by reheating[1] at around 773K after water-quenching or by cooling[2] at an intermediate rate from annealing temperature through suppression of the intergranular fracture. These two heat treatments which improve the cryogenic toughness will be referred in the present paper to post-annealing heat treatments. These phenomena, that is to say, the high susceptibility to cryogenic intergranular embrittlement and cryogenic toughening through the post-annealing heat treatments, were also revealed by Shibata et al.[1,2] in other high Mn non-magnetic steels, which had different content of Mn, Cr, C and N. The detailed mechanism and reason for these two phenomena are discussed in the previous papers by Shibata et al.[3,4]

As for the mechanism for cryogenic toughening through the post-annealing heat treatments in high Mn steels, Strum and Morris[5] proposed the effect of intergranular segregation of boron (B). Using steels similar to those which Shibata et al.[1,2] examined, Strum and Morris observed through Auger electron spectroscopy that B segregation to grain boundaries was remarkably enhanced during the post-annealing heat treatments. Shibata et al.[3,4] confirmed the significant effect of the intergranular segregation of B on the cryogenic toughness (Photo.1), through autoradiography; α particle fission track etching (FTE) method[6]. In the present work, in order to understand the effect of B, the variations of cryogenic toughness with the post-annealing heat treatments were investigated using 32Mn-7Cr-0.3N steels with several contents of B and phosphorus (P). Phosphorus is known as a deteriorating impurity for the strength of grain boundaries. Toughness was evaluated through Charpy absorbed energy, ductile-brittle transition curve of the absorbed energy and fracture toughness. In order to discuss the reason for the variation in toughness, the behavior of B was examined in detail.

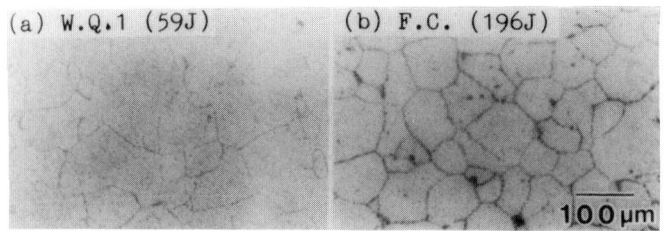

Photo.1. B-autoradiographs of 32Mn-7Cr-0.4N steel (a) water-quenched and (b) furnace-cooled from 1473K. Numerals are 193K Charpy absorbed energies of each specimens.

EXPERIMENTAL PROCEDURE

The chemical compositions of used steels are shown in Table 1. The higher content of P (0.02%) is general one of commercial steel. These steels were melted in a vacuum induction furnace using high purity alloying metals and iron, and 20kg ingots were obtained. After heating at 1473K, ingots were hot-rolled into 15mm thick plates. All specimens were annealed at 1473K for 1 hour and cooled at several rates from 1473K. Isothermal heat treatments (which will be called as reheating in the present paper) at temperatures between 573K and 973K were performed after water-quenching from 1473K. Following such heat treatments, the blanks were machined to standard sized Charpy testing specimens having T-L orientation to the rolling direction. Charpy impact tests were carried out at 77K for all specimens. In order to obtain the ductile-brittle transition curves, Charpy tests were carried out for some specimens at 4K, 120K, 193K and 273K. Elasto-plastic fracture toughness (J_{IC}) tests were performed at 4K and 77K in accordance with ASTM E813 using 1/2T compact tension specimens having T-L orientation. A fatigue precrack was introduced at room temperature. The J_{IC} value was determined by unloading compliance method. A computer-controlled electro-hydraulically actuated machine was used for tensile and J_{IC} tests. In all specimens, $K_{IC}(J)$ was calculated from JIC by the following equation.
$K_{IC}^2 = E \times J_{IC} \times (1-\nu^2)$
where, E: the Young's modulus, ν: Poisson's ratio.

The area fraction of non-metallic inclusions (the index of cleanliness) was measured by optical microscopy for water-quenched specimens without etching. The index of cleanliness was represented by the ratio of the number of the points which were on inclusions to the total point number of a 600 x 1000 grid. Magnification of optical microscopy was about 500 and the total number of observed fields was about 80 for one specimen. The observed surface was parallel to the fracture surface of Charpy impact specimens and of J_{IC} specimens. Generally-used procedure[6] was carried out for FTE method. Irradiation of thermal neutrons was performed in the reactor of Institute for Atomic Energy of Rikkyo University. Fission tracks etched under the same condition (in a 2.5N NaOH solution, 303K, 1-2 hours) were

Table 1. Chemical compositions of used steels in wt%.

Steels	C	Si	Mn	P	S	Ni	Cr	N	B	Al
AA	0.030	0.51	31.90	0.002	0.009	0.52	7.40	0.32	<0.0002	0.010
AB	0.030	0.51	31.90	0.002	0.010	0.52	7.40	0.29	0.0007	0.003
AC	0.020	0.52	32.00	0.020	0.010	0.52	7.40	0.29	<0.0002	0.008
AD	0.030	0.54	32.00	0.020	0.009	0.51	7.40	0.31	0.0009	0.004

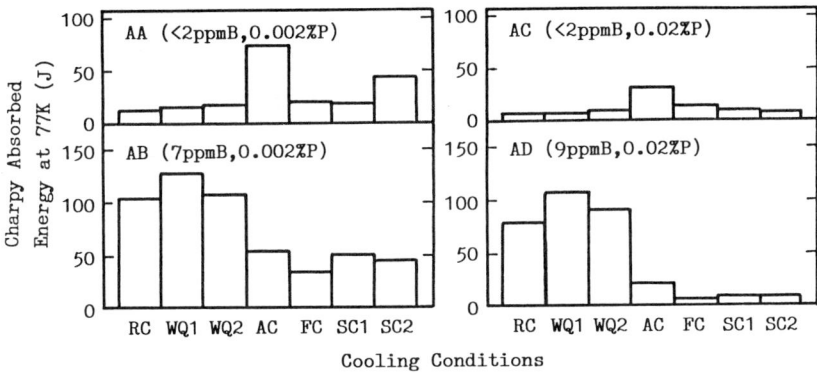

Fig.1. Variation of 77K Charpy absorbed energy with cooling conditions from 1473K. Cooling rate becomes slower toward right.

compared among specimens which had been exposed to thermal neutrons on the same occasion. In order to compare the B-autoradiograph with the optical microscopic structure of the matched area, traces by Vickers diamond were used for marks. The area fraction of large tracks in the matrix in the B-autoradiograph (this area was corresponded to non-metallic inclusions where B segregated) was investigated by the same method as that for the index of cleaniness.

RESULTS AND DISCUSSION

Charpy absorbed energy at cryogenic temperatures

Figure 1 shows the variation of 77K Charpy absorbed energy with cooling conditions from annealing temperature. In higher B content steels (AB and AD steels), the highest absorbed energy is obtained by water-quenching from 1473K. In lower B content steels, the highest one is obtained by air-cooling. All steels tend to be embrittled by cooling most rapidly and by cooling too slowly from annealing temperature and are toughened by a cooling rate control. The variation of 77K Charpy absorbed energy with reheating temperatures is shown in Fig.2. The specimens, which exhibit a low absorbed energy in the water-quenched condition, are toughened by reheating at 723-773K. On the other hand, the specimen, which shows a high absorbed energy even in the water-quenched condition, are not toughened by reheating. In

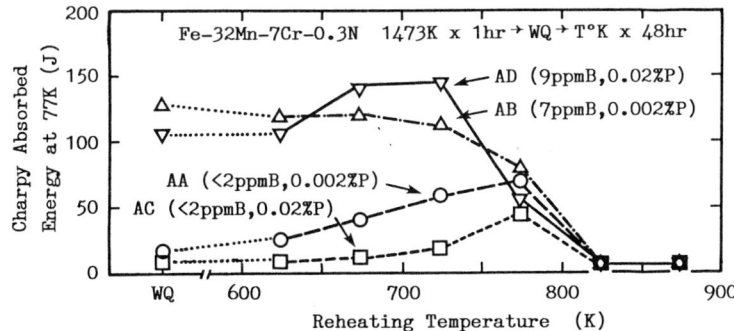

Fig.2. Variation of 77K Charpy absorbed energy with reheating temperatures (reheating time: 48hr).

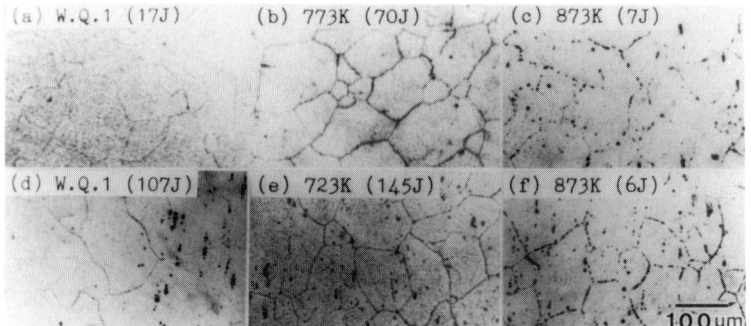

Photo.2. Boron-autoradiographs of AA (a-c) and AD (d-f) steels reheated at various temperatures for 48hr. Numerals are 77K Charpy absorbed energy.

addition, the absorbed energy of all specimens is reduced by reheating at 823K as much as by cooling at a too slow rate from annealing temperature.

Photographs 2 and 3 show B-autoradiographs and SEM fractographs of AA and AD steels following several heat treatments. In water-quenched specimens of lower B content steels, showing low absorbed energy, intergranular fracture is enhanced and the degree of the B segregation to grain boundaries is weakened. On the other hand, in specimens toughened through the post-annealing heat treatments or B addition, intergranular fracture is suppressed and the degree of the B segregation is enhanced. Furthermore, in specimens embrittled by cooling at a slow rate or reheating at higher temperatures, intergranular fracture is enhanced and B is contained in globular precipitates and the degree of the intergranular segregation of B is reduced.

Ductile-brittle transition curves

Figure 3 shows the ductile-brittle transition curves of all steels water-quenched and reheated at adequate temperatures. Two effects of B content on the ductile-brittle transition curves can be observed. One is the increase in the absorbed energy at temperatures lower than DBTT (Ductile-Brittle Transition Temperature). The other is the increase in the upper self energy. The former phenomenon is occurred not only by addition

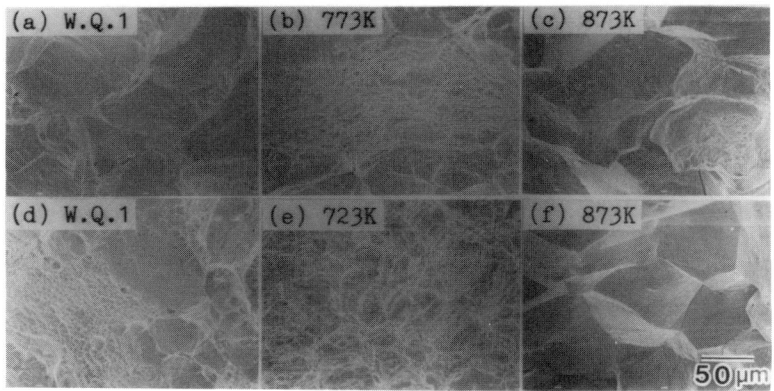

Photo.3. SEM fractographs of 77K Charpy impact specimens of (a-c) AA and (d-f) AD steels reheated at various temperatures for 48hr.

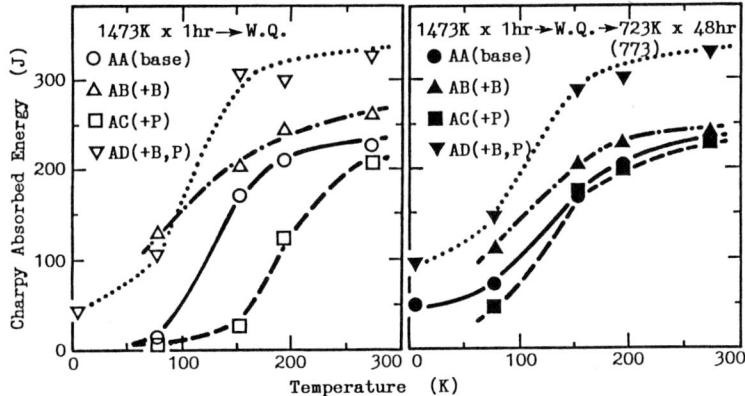

Fig.3. Ductile-brittle transition curves of steels (a) water-quenched and (b) reheated at 773K (AA, AC steels) or 723K (AB, AD steels) after water-quenching from 1473K.

of B but also by reheating at an adequate temperature. As for the latter phenomenon, in both specimens water-quenched and reheated at an adequate temperature, higher B content steels, especially higher B and higher P content steel, show higher upper shelf energy than lower B content steels. At testing temperatures higher than DBTT, that is in the upper shelf region, all specimens fractured in the ductile transgranular manner and showed the dimple appearance. Therefore, it is not considered that the increase in upper shelf energy by addition of B is due to the increase in the degree of B segregation to grain boundaries. Yield and tensile strengths do not change through the reheating or the addition of B and P.

Volume fraction of inclusions and precipitates affect the upper shelf energy. However, the increase in upper shelf energy can be observed after the heat treatment which does not occur precipitates. Therefore, the reasons concerning precipitation are negligible. On the other hand, there is no clear difference in the index of cleanliness, that is the area fraction of non-metallic inclusions, between higher B content steels and lower B content steels (Table 2). In addition, the size and the shape of inclusions were similar among them. However, in B-autoradiographs of higher B content steels, there can be observed globular tracks in the matrix (Photo.3, (a) and (d)). Boron can segregates not only at grain boundaries but also at the interfaces between non-metallic inclusions and the matrix (Photo.4). It is conceivable that B segregating at the interfaces strengthens the interfacial binding force and suppresses the nucleation of voids at the interface to increase the upper shelf energy. In fact, AD steel, which shows higher upper shelf energy than AB steel, exhibits higher area fraction of large tracks in B-autoradiograph corresponding to non-metallic inclusions

Table 2. Effects of addition of B and P on upper self energy, index of cleanliness and grain size.

Steels and compositions	Upper shelf energy (J)	Index of cleanliness (%)	Grain size (μm)
AA (<2ppmB,0.002%P)	228	0.22	94
AB (7ppmB,0.002%P)	260	0.19	54
AC (<2ppmB,0.02%P)	207	0.17	105
AD (9ppmB,0.02%P)	326	0.16	88

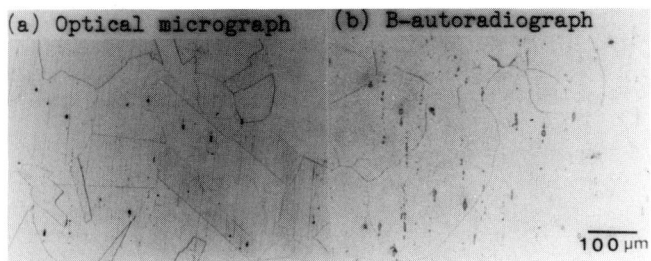

Photo.4. (a) Optical micrograph and (b) its matching B-autoradiograph of water-quenched AD steel.

in optical micrograph than AB steel (Table 3). Then, why is B easy to segregate at inclusions in higher P content AD steel. The difference of B content between these steels, 9ppm and 7ppm, may be conceived as one of the reasons, and a higher content of P may enhance the segregation or change the properties of the inclusions or the interfaces between them and the matrix. However, the reason remained unclarified and more investigation is expected.

Cryogenic fracture toughness

Figure 4 shows the effects of B and P contents and post-annealing heat treatments on load versus load-line displacement curves at 77K. In the water-quenched AA steel (a), pop-in occurs from an early stage of loading. Pop-in also appears in the reheated AA steel (b), but the higher load is needed than in the water-quenched one. On the other hand, no pop-in appears in higher B steels (c,d). $K_{IC}(J)$ of those specimens at 4K and 77K, calculated from J_{IC}, are shown in Table 4 together with Charpy absorbed energy and 0.2% yield strength. Fracture toughness, as well as Charpy absorbed energy, of these steels at the low temperatures, increases through the post-annealing heat treatments and/or B addition. The best balanced specimen between yield strength and $K_{IC}(J)$ at 4K in the present work is the AD steel reheated at 723K after water-quenching from 1473K; 64 MPa x $m^{\frac{1}{2}}$ for $K_{IC}(J)$ value and 1271 MPa for 0.2% yield strength. However, this specimen fractured partly in the intergranular manner at 4K (Photo.5). Therefore, higher toughness can be expected through grain refining by decreasing annealing temperatures.

CONCLUSION

(1) Charpy absorbed energy at temperatures lower than DBTT increased through the post-annealing heat treatments and/or B addition. It is considerable that the increase is due to intergranular segregation of B, which suppresses the intergranular fracture.

Table 3. Upper shelf energies, index of cleaniness and area fraction of large tracks in B-autoradiographs corresponding to non-metallic inclusions in optical micrographs of AB and AD steels.

Steels and compositions	Upper shelf energy (J)	Index of cleanliness (%)	Area fraction of large tracks (%)
AB (7ppmB, 0.002%P)	260	0.19	0.63
AD (9ppmB, 0.02%P)	326	0.16	1.65

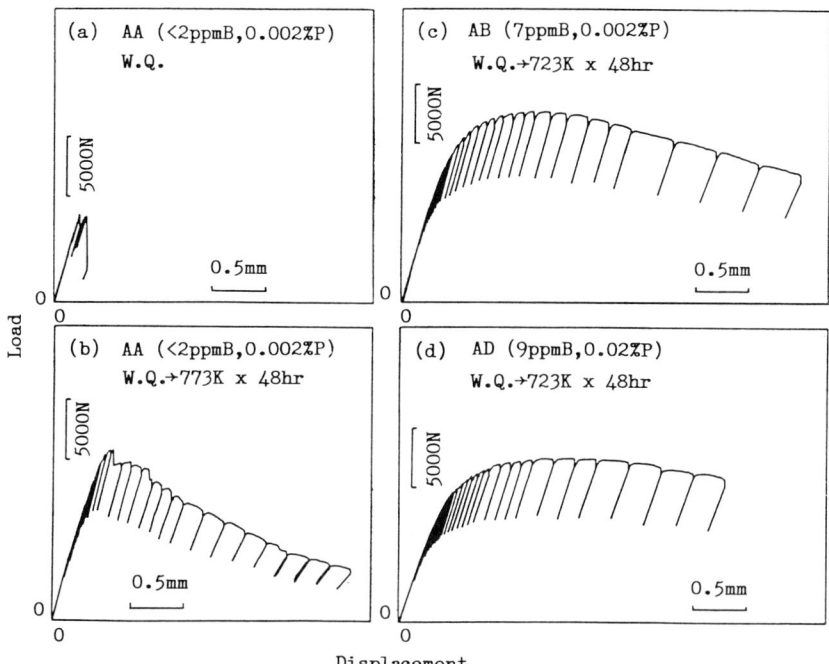

Fig.4. Effect of B and P content and post-annealing heat treatments on load versus load-line displacement curves of J_{IC} tests at 77K.

Table 4. Variation of $K_{IC}(J)$, Charpy absorbed energy (vE) and 0.2% yield strength (0.2%Y.S.) at 4K and 77K with B and P contents and post-annealing heat treatments.

Steels	Heat treatment	77K			4K		
		$K_{IC}(J)$ (MPa √m)	vE (J)	0.2%Y.S. (MPa)	$K_{IC}(J)$ (MPa √m)	vE (J)	0.2%Y.S. (MPa)
AA(base)	WQ	73	17	776	–	–	–
AA(base)	773Kx48hr	130	70	854	–	48	–
AB(+B)	723Kx48hr	249	108	843	33	–	1245
AD(+B,P)	723Kx48hr	261	141	791	64	94	1271

Photo.5. SEM fractographs of J_{IC} test specimen of AD steel reheated at 723K for 48hr.

(2) Higher B content steels, especially higher B and higher P content steels, showed higher upper shelf energy than lower B content steels. The reason for this phenomenon remained unclarified. However, it is conceivable that B segregates to the interfaces between non-metallic inclusions and the matrix and that the nucleation of voids at the interfaces is suppressed.
(3) The post-annealing heat treatments and/or B addition increased cryogenic J_{IC} value together with Charpy absorbed energy.
(4) In the present work, the steel which showed 65 MPa x $m^{1/2}$ for K_{IC} value and 1271 MPa for 0.2% yield strength at 4K, could be obtained. Through optimizing B content and grain refining, an increase in toughness can be expected.

ACKNOWLEDGMENT

The authors are grateful to Daido Steel Co., Ltd. for melting steels. They also wish to thank Institute for Atomic Energy and Dr.S.Harasawa at Rikkyo University for their assistance in autoradiography, Cryogenic Center of the University of Tokyo for assistance in low temperature experiments, and Science and Technology Agency of Japan Government for financial support by Special Coordination.

REFERENCES

1) K.Shibata, Y.Kobiki, Y.Kisimoto, and T.Fujita," Mechanical properties of high yield strength high manganese steels at cryogenic temperatures ", in:" Advances in Cryogenic Engineering Materials, vol.30 ", ed. by A.F.Clark and R.P.Reed, Plenum, New York/London(1984), pp.153-160.
2) K.Shibata, K.Fujita, and T.Fujita," Cooling conditions after solution treatment and low temperature toughness of a 32%Mn steel bearing high content of nitrogen ", Trans. ISIJ, 26(1986), B224.
3) K.Shibata, N.Kondo, K.Fujita, and H.Tanaka," Mechanical properties of high manganese steels toughened by post-annealing heat treatments ", in: " Advances in Cryogenic Engineering Materials, vol.36 ", ed. by R.P.Reed and F.R.Fickett, Plenum, New York/London(1990), pp.1257-1264.
4) H.Tanaka, N.Kondo, K.Fujita and K.Shibata," Suppression of cryogenic intergranular fracture through heat treatments and roles of boron in high manganese non-magnetic steels ", ISIJ Intern., 30(1990), pp.646-655.
5) M.J.Strum and J.W.Morris,Jr.," Influence of post-anneal cooling treatments on suppression of cryogenic intergranular fracture in experimental Ni free high Mn austenitic steels ", in:" Advances in Cryogenic Engineering Materials, vol.34 ", ed. by A.F.Clark and R.P.Reed, Plenum, New York/London(1988), pp.371-378.
6) M.Ueno and T.Inoue," Distribution of boron at austenite grain boundaries and bainitic transformation in low carbon steels ", Trans. ISIJ, 13(1973), pp.210-217.

THE CHARPY IMPACT TEST AS AN EVALUATION OF 4 K FRACTURE TOUGHNESS

H. Nakajima, K. Yoshida, and H. Tsuji

Japan Atomic Energy Research Institute
Naka-machi, Ibaraki-ken, 311-01, Japan

R. L. Tobler

National Institute of Standards and Technology
Colorado 80303, U.S.A.

I. S. Hwang, M. M. Morra, and R. G. Ballinger

Massachusetts Institute of Technology
Cambridge, Massachusetts 02139, U.S.A.

ABSTRACT

A 4 K Charpy test is defined as a Charpy test in which the initial temperature of a specimen is 4 K. Two methods, a glass Dewar method and a flow method, are compared in this study to demonstrate that both provide consistent results as long as initial specimen temperature is the same. An improved correlation between the Charpy absorbed energy and the fracture toughness at 4 K is also examined to clarify the limitation of Charpy test applications at cryogenic temperatures.

INTRODUCTION

A cable-in-conduit conductor for a superconducting magnet is cooled by supercritical helium. Therefore, a conduit which must withstand high inner pressure is defined as a pressure retaining part in the Japanese High Pressure Gas Law. This regulation specifies toughness of materials in terms of Charpy absorbed energy at the operating temperature. For superconducting magnets, Charpy tests at 4 K must be performed to ensure that the material is not brittle. Charpy tests at the operating temperature are also required to ensure the toughness of materials in other regulations such as the American Society of Mechanical Engineers (ASME) boiler and pressure vessel code and the Japanese Industrial Standards (JIS) pressure vessel code.

Strictly, it is impossible to conduct valid 4 K Charpy impact test because the specimen temperature rises due to adiabatic heating during high strain rate deformation.[1] Therefore, a Charpy test with an initial temperature of 4 K before impact is generally defined as a 4 K Charpy test. On the other hand, a poor correlation between fracture toughness

Table 1 Chemical compositions of the test materials

Material	C	Mn	Ni	Cr	Mo	N	Others
JN1	0.026	4.2	14.74	24.2	--	0.34	
JN2	0.050	22.4	3.22	13.4	0.70	0.24	Cu:0.70
A182	0.048	5.11	12.94	22.02	2.18	0.322	
							V:0.31, Nb:0.13
I908	0.01	0.04	49.7	3.83			
						Al:1.04, Nb:2.99, Ti:1.58	
A316LN	0.02	1.52	13.10	16.70	2.58	0.194	
S316LN	0.039	1.19	12.02	16.38	2.49	0.157	
304LN	0.024	3.85	7.84	19.25	0.31	0.21	
5Ni	0.08	0.60	5.03	--	0.30	0.01	Al:0.08
N33	0.050	12.66	3.42	17.30	--	0.33	
ULCS	0.007	0.021	--	--	--	0.003	Al:0.052
JK2	0.05	21.79	4.94	12.82	--	0.212	
JJ1	0.046	9.74	11.92	12.21	4.89	0.203	

JN1:JCS-CSUS-JN1; JN2:JCS-CSUS-JN2; A182:ASTM A182 FXM-19;
I908:Incoloy 908; A316LN:AISI 316LN; S316LN:SUS 316LN;
304LN:304LN(Mn); 5Ni:5Ni Steel; N33:Nitronic 33; ULCS:
Ultra-low Carbon Steel; JK2:JCS-CSUS-JK2; JJ1:JCS-CSUS-JJ1.

J_{Ic} and the absorbed energy for Charpy V-notched specimens CVN at 4 K was reported, so it is difficult to specify the 4 K fracture toughness by 4 K CVN.[2,3] Nevertheless, the Charpy test is still viewed as a useful test of cryogenic toughness from an industrial point of view because it is a simple, fast, and inexpensive. Hence it is meaningful to understand the significance and limitations of the 4 K Charpy test.

The purposes of this study are (1) to compare agreement between the glass Dewar method CVN and the flow method CVN, and (2) to investigate an improved correlation between the J_{Ic} and CVN at 4 K to clarify the value of the Charpy impact test as a toughness evaluation technique at 4 K.

TEST MATERIALS

The chemical compositions of some materials are listed in Table 1. The chemical compositions of all other materials are reported in reference 3. The twenty four Japanese Cryogenic Steels (JCS) tested in this study include materials from small trial heats in the final process of JCS development; their chemical compositions are shown in Table 1.

Eight materials were tested to calibrate a glass Dewar effect, including 6 austenitic alloys (CSUS-JN1, 316LN, 304LN, Incoloy 908,* Nitronic 33*, and ASTM A182 FXM-19) and 2 ferritic steels (5Ni steel and ultra-low carbon steel). Twelve materials were tested to compare the glass Dewar and flow methods, including 11 austenitic alloys (two heats of CSUS-JN1, two heats of 316LN, 304LN, I908, N33, ASTM A182 FXM-192, and three weld metals of CSUS-JN1, JN2) and one ferritic steel (5Ni steel).

---

*The use of trade names of specific products in this paper is essential to facilitate an understanding of the materials and procedures described. The use of trade names for this purpose in no way implies endorsement or exclusive recommendation by JAERI, NIST or MIT.

Thirty six materials were tested to study a correlation between J_{Ic} and CVN at 4 K, including 24 JCS (9 CSUS-JN1, 8 CSUS-JN2, 4 CSUS-JK2, and 3 CSUS-JJ1), 2 fully austenitic alloys (I908 and 316LN), 6 metastable austenitic stainless steels (304LN modified with 1 - 6 % Mn and two heats of N33) and 4 ferritic steels (9Ni steel, 5Ni steel, 3.5Ni steel, and ultra-low carbon steel).

4 K CHARPY IMPACT TEST

A glass Dewar method and a flow method were used to evaluate the absorbed energy at 4 K. Full size specimens (10 x 10 x 55 mm) with 2 mm V-notch in T-L orientation according to JIS Z 2202 or ASTM E 23-88 were used in this study.

Glass Dewar Method

A small vacuum-insulated glass Dewar is used to submerge a specimen in liquid helium prior to testing. Therefore, the initial temperature of 4 K is ensured but an expensive glass Dewar is needed. Liquid helium consumption is very low, about 1 liter per specimen. Since a correction for the measured CVN is needed, this method is not suitable for low toughness materials. A glass Dewar with a specimen set on a test machine is shown in Fig. 1.

Fig. 1 A glass Dewar with a specimen set on a test machine

The test procedure for the glass Dewar method is as follows:

(1) A specimen is inserted into a small vacuum-insulated glass Dewar.
(2) The glass Dewar with specimen is precooled to 77 K in a cryostat filled with liquid nitrogen.
(3) Liquid nitrogen is drained and liquid helium is filled to the cryostat. After thermal equilibrium at 4 K is reached, the glass Dewar is removed from the cryostat. The specimen remains fully submerged in a liquid helium for 2 minutes after leaving the cryostat.
(4) The glass Dewar with a specimen is placed on the anvil of a Charpy test machine such that specimen center is aligned with the center of the anvil. Then, it is struck by a hammer.
(5) The absorbed energy measurement is corrected using a calibration factor to account for the glass and the space which exists between the specimen and the anvil.

Flow Method

The expensive glass Dewar and a correction of measured CVN are not necessary in the flow method. This method, however, has disadvantages of an uncertainty of the initial temperature at 4 K and a larger liquid helium consumption (about 10 liters per specimen). Since CVN is directly measured, this method can be applied to any materials over a wide toughness range. A wrapped specimen set on a test machine is shown in Fig. 2.

The test procedure for this simple method is as follows:

(1) A polyethylene foam of 0.5 mm thickness is prepared.
(2) Two pieces of double-sided adhesive tape are attached to the polyethylene foam with a distance less than a specimen length.
(3) The specimen is fixed to the foam using the tape.
(4) The specimen is loosely wrapped and Kapton* tape is applied to the foam to prevent unwrapping. The foam is a single layer on the notch side and a double layer on the impact side.
(5) The notch position is marked on the outer surface of the foam.
(6) The specimen is set on the machine so that the mark can locate at center of the anvil.
(7) The wrapped specimen is fixed to the anvil at one end by Kapton tape. The liquid helium transfer tube is inserted at opposite end.
(8) Liquid helium is transferred from a liquid helium container and the specimen is broken after 120 s from the start of transfer.

FRACTURE TOUGHNESS J_{Ic} TEST at 4 K

In order to examine the correlation between CVN and fracture toughness, J_{Ic} was measured by single-specimen method using a computerized unloading compliance technique. The specimens were proportional CT specimens defined by ASTM E 813 with a 25 mm thickness and a 26.5 mm notch length. The notch orientation was T-L, and displacement was measured at the loadline. Specimens were precracked at room temperature using a final stress intensity factor of 33 MPa/m. The ratio of the initial crack length to the width of the specimen was about 0.6.

RESULTS AND DISCUSSION

Calibration of the Glass Dewar Effect

Since the overestimation of CVN in the glass Dewar method is caused by space between the anvil and specimen, we assume that the correction

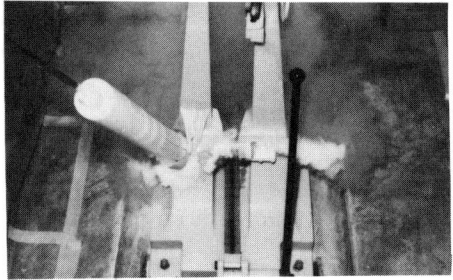

Fig. 2 A wrapped specimen cooled down by liquid helium flow

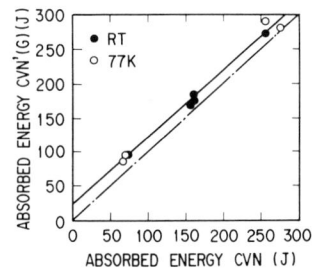

Fig. 3 Calibration curve for glass Dewar effect

coefficient does not significantly depend on temperature. Then, the calibration result obtained at room temperature or at liquid nitrogen temperature can be applied to correct the 4 K Charpy absorbed energy.

Figure 3 shows a calibration curve for the glass Dewar effect at 77 K and at room temperature. A linear relation between the uncorrected absorbed energy measured by the glass Dewar method CVN'(G) and that measured by the conventional method is observed at both temperatures for a wide range of materials. The following equation is obtained by a least square regression:

$$CVN'(G) = 0.98 \times CVN + 23.6 \ (J), \qquad (1)$$

where CVN'(G) and CVN are absorbed energy measured by the glass Dewar method (uncorrected value) and the conventional method, respectively. The corrected absorbed energy, CVN(G), measured by the glass Dewar method equals to CVN given by Equation (1).

Initial Specimen Temperature in the Flow Method

Figure 4 shows the temperature of a specimen as a function of cooldown time from the start of liquid helium transfer in the flow method. The specimen temperature was measured using a germanium thermometer. It was inserted in a hole that was centered on the side face and drilled to the depth of the V-notch of the calibration specimen. We filled the gap with copper powder and grease to enhance thermal conduction between the specimen and the thermometer. After 80 s from opening the valve, a temperature of less than 4.6 K was achieved as shown in Fig. 4. This time depends on the pressure within the liquid helium container. When the internal pressure was more than 0.02 MPa, cooling time was within 80 s. Therefore, all tests were conducted after 120 s (1.5 times 80 s) from the start of a liquid helium transfer using a transfer pressure over 0.02 MPa.

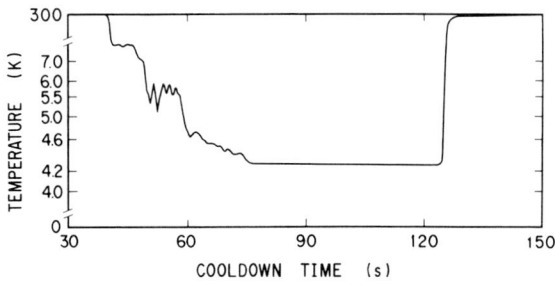

Fig. 4 Specimen temperature as a function of cooldown time

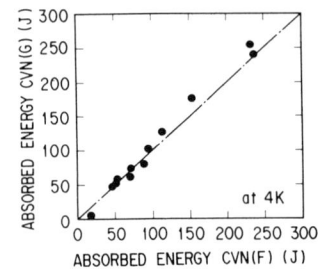

Fig. 5 Comparison between a glass Dewar method and a flow method

Comparison between the Glass Dewar Method and the Flow Method

It has not yet been demonstrated whether or not results measured by the two methods agree with each other. Therefore, it is indispensable to compare both methods in a single laboratory. Figure 5 shows that the corrected absorbed energy measured by the glass Dewar method, CVN(G), are plotted against those measured by the flow method, CVN(F), on 12 materials at 4 K. Good agreement is found over a wide range from about 20 to 250 J of absorbed energy. The two methods are consistent with each other as long as the initial specimen temperature is identical.

Correlation between CVN and J_{Ic} at 4 K

Ideally, CVN should correlate with the fracture mechanics parameter J_{Ic} or K_{Ic} if the CVN is used as a material evaluation criterion. An assumed linear relation between J_{Ic} and CVN for various materials with a wide range of toughness was checked in our previous study.[3] However, a good correlation was not found due to the specimen temperature rise.

We assume that correlation between J_{Ic} and CVN at 4 K is fit using expressions for the transition-temperature region, but not those for upper-shelf region. The following correlations between the both parameters in transition-temperature region were suggested:[4,5]

$$K_{Ic}^2/E = 2 \times CVN^{1.5}, \tag{2}$$

where K_{Ic} is in psi/inch, E is Young's modulus in psi, and CVN is in ft·lb, and

$$(K_{Ic}/100)^2 = 300 \times (CVN/\sigma_y), \tag{3}$$

where K_{Ic} is in kg/mm$^{1.5}$, CVN is in kg·m, σ_y is the uniaxial yield strength in kg/mm^2.

A linear relation between J_{Ic} and $CVN/\sigma_y^{1.5}$ is examined here because this relation has both elements expressed by Equations (2) and (3). In addition, the author already suggested that a better relation was obtained when J_{Ic} was related to $CVN/\sigma_y^{1.5}$ in austenitic stainless steels.[2]

Figure 6 shows the relation for 24 JCS alloys based on the considerations above. The 12 data of reference 3 are also plotted in this figure to check the relation for a wide range of toughness and dependence on materials. The regression line drawn in Fig. 6 is calculated using all

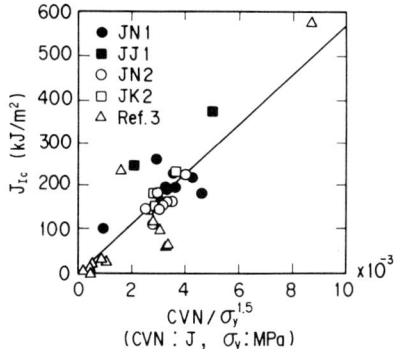

Fig. 6 Correlation between J_{Ic} and $CVN/\sigma_y^{1.5}$

data (including data of reference 3). The general formula for correlation is

$$J_{Ic} = a \times CVN/\sigma_y^{1.5} \qquad (4)$$

where J_{Ic} is in kJ/m^2, CVN is in J, and σ_y is in MPa. The slope of the regression line "a", the correlation coefficient "r^2", and the calculated maximum and minimum scatter for 6 subsets of the database vary as shown in Table 2.

The J_{Ic} - $CVN/\sigma_y^{1.5}$ correlation is better than the linear J_{Ic} - CVN correlation in all cases. The correlation coefficient r^2 depends on the toughness range of data used for regression calculation. Since the JCS have high toughness with a narrow range the correlation coefficients for cases 1, 2, and 4 become low compared with cases 5 and 6. Nevertheless, in case 3, a high correlation coefficient is obtained.

Since the scatter for cases 1, 5, and 6 is very high it is impossible to estimate the 4 K fracture toughness from the 4 K CVN. This means the 4 K Charpy test is not a reliable indicator of the 4 K fracture toughness. However, the scatter for small groups (cases 2, 3, and 4), is

Table 2 Results of least square regression

Case	a	r^2	Scatter (%)
1	5.8	0.36	+104, -32
2	5.1	0.59	+ 21, -23
3	6.0	0.89	+ 7, - 7
4	5.4	0.57	+ 19, -27
5	5.7	0.72	+155, -96
6	5.6	0.64	+111, -68

1:JCS; 2:JN2; 3:JK2;
4:JN2+JK2; 5:JCS+Ref.3(all);
6:JCS+Ref.3(316LN+304LN).

$J_{Ic} = a \times 10^4 * (CVN/\sigma_y^{1.5})$

near the J_{Ic} measurement errors reported previously: +16.5%, -12% in JK2 and +9.5%, -10.8% in JN1.[6,7] Especially, the scatter of ± 7 % in CVN for JK2 is less than that for J_{Ic} measurements. This means that it is possible to estimate the 4 K fracture toughness from CVN using Equation (4) for JK2. It is interesting that the best correlation exists for JK2 materials that were supplied from different manufacturing processes. It seems that the fracture toughness may be estimated by Charpy tests for a particular material which has the same temperature rise and "thermal-softening" effect.[3] In the case of high Mn austenitic stainless steels, case 4, it may be possible to use the 4 K CVN to estimate 4 K fracture toughness if the acceptable scatter is within ± 30 %.

CONCLUSIONS

The following conclusions regarding the use of Charpy impact test as an evaluation technique of 4 K fracture toughness are obtained:

1. The Charpy absorbed energy at 4 K can be evaluated consistently using the glass Dewar method or the flow method in accordance with procedures described here.

2. The Charpy test as an evaluation of 4 K fracture toughness is valid for limited groups of materials. However, 4 K CVN is inappropriate as an universal criterion for all materials, and it cannot be used as a substitute for fracture toughness measurements.

3. Despite its limitations, the 4 K Charpy test may be a valuable technique for estimating the 4 K toughness from an engineering point of view if it is used for specific materials where a good correlation is established between CVN and fracture toughness at 4 K.

ACKNOWLEDGMENTS

The authors would like to thank Drs. T. Iijima and S. Shimamoto for their constant encouragement of this work. JCS materials were supplied from Japan Steel Works Ltd., Kobe Steel Ltd., and Nippon Steel Corporation under collaboration contracts. The authors wish to express our gratitude to all the participants in these companies for their contributions.

REFERENCES

1. R. L. Tobler, R. P. Reed, I. S. Hwang, M. M. Morra, R. G. Ballinger, H. Nakajima, and S. Shimamoto, CHARPY IMPACT TESTS NEAR ABSOLUTE ZERO, Journal of Testing and Evaluation, Vol. 19, No.1 (1991), pp. 34-40.
2. H. Nakajima, K. Yoshida, K. Okuno, M. Oshikiri, E. Tada, and S. Shimamoto, R. Miura, M. Shimada, S. Tone, K. Suemune, T. Sakamoto, and K. Nohara, FRACTURE TOUGHNESS OF NEWLY DEVELOPED STRUCTURAL MATERIALS FOR SUPERCONDUCTING COILS OF FUSION EXPERIMENTAL REACTOR, in : "Advances in Cryogenic Engineering - Materials," Vol. 32, Plenum Press, New York (1986), pp. 347-354.
3. I. S. Hwang, M. M. Morra, R. G. Ballinger, H. Nakajima, S. Shimamoto, and R. L. Tobler, CHARPY ABSORBED ENERGY AND J_{Ic} AS MEASURES OF CRYOGENIC FRACTURE TOUGHNESS, To be published in Journal of Testing and Evaluation (1991).
4. J. M. Barsom and S. T. Rolfe, CORRELATIONS BETWEEN K_{Ic} AND CHARPY V-NOTCH TEST RESULTS IN THE TRANSITION-TEMPERATURE RANGE, in: "Impact Testing of Metals, ASTM STP 466" (1970), pp. 281-302.

5. T. Ito, K. Tanaka, and M. Sato, EFFECT OF PLATE THICKNESS ON BRITTLE FRACTURE INITIATION FROM SURFACE NOTCH IN WELD FUSION LINE AND CORRELATION BETWEEN THE RESULTS OF LARGE SCALE TEST AND THOSE OF CHARPY TEST, Journal of the Society of Naval Architecture and Ocean Engineering, No.131 (1972), pp. 335-343.
6. H. Nakajima, K. Yoshida, S. Shimamoto, R. L. Tobler, P. T. Purtscher, and R. P. Reed, ROUND ROBIN TENSILE AND FRACTURE TEST RESULTS FOR AN Fe-22Mn-13Cr-5Ni AUSTENITIC STAINLESS STEEL AT 4 K, in: "Advances in Cryogenic Engineering - Materials", Vol. 34, Plenum Press, New York (1988), pp. 241-249.
7. H. Nakajima, K. Yoshida, S. Shimamoto, R. L. Tobler, R. P. Reed, R. P. Walsh, and P. T. Purtscher, INTERLABORATORY TENSION AND FRACTURE TOUGHNESS TEST RESULTS FOR CSUS-JN1 (Fe-25Cr-15Ni-0.35N) AUSTENITIC STAINLESS STEEL AT 4 K, in: "Advances in Cryogenic Engineering - Materials", Vol. 36, Plenum Press, New York (1990), pp. 1069-1076.

CHARPY SPECIMEN TESTS AT 4 K†

R.L. Tobler and A. Bussiba*

Materials Reliability Division
National Institute of Standards and Technology
Boulder, Colorado 80303, U.S.A.

J.F. Guzzo

Welding Engineering Department
Minnesota Valley Engineering, Inc.
New Prague, MN

I.S. Hwang

Department of Materials Science and Engineering
Massachusetts Institute of Technology
Cambridge, Massachusetts 02139, U.S.A.

ABSTRACT

This paper describes nonstandard methods of testing Charpy specimens at 4 K. We show that the initial temperature can be achieved using a helium flow method with U-type as well as C-type machines. Unfortunately, heating during impact loading weakens the correlation with quasistatic fracture toughness parameters. Alternative test procedures using fatigue precracked specimens were not fully satisfactory, either: (1) Precracking reduces the impact energy by 35–54%, but the specimen heating is still significant; (2) slow bending curtails the heating most effectively, but then the desired simplicity of a screening test is lost.

INTRODUCTION

The Charpy impact test (ASTM E 23–88)[1] was originally developed as an inexpensive measure of toughness for ferritic steels exhibiting ductile-to-brittle transition behavior. In its favor, Charpy testing requires a simpler apparatus and a specimen (10 x 10 x 55 mm notched bar) of smaller size than fracture mechanics tests. Temperature control is lost, however, if the specimen is routinely transferred from a Dewar to the test machine, or fractured at impact strain rates.

---

† Contribution of NIST, not subject to copyright.
* Visiting Scientist, on leave from Nuclear Research Center–Negev, Beer-Sheva, Israel.

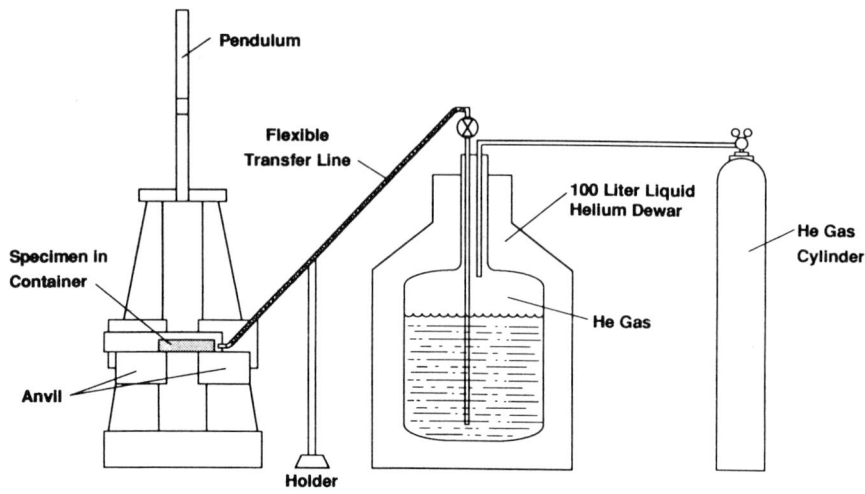

Fig. 1. Schematic diagram of a He flow method test apparatus.

At present, there is no standard procedure for impact tests at 4 K.[2] To maintain a specimen at this temperature, some kind of insulation is necessary. After reviewing a dozen alternatives in recent U.S.–Japan cooperative research,[2-4] we believe the helium flow method (HFM) will gain favor in the U.S. by virtue of its simplicity, economy, and general applicability to low and high toughness alloys. In the HFM, a specimen is wrapped in a plastic sheet, positioned on the anvil of the impact machine, and tested after cooling in a He jet. The required test apparatus is shown schematically in Fig. 1.

HFM TESTS USING U-TYPE MACHINES

Jin et al.[5] are credited with the first flow tests near 4 K, but their U-type pendulum machine and original insulation system did not accommodate a full-size specimen. Ogata et al.[6] simplified the insulation and tested full-size specimens with a C-type pendulum. In fact, C-type pendulums which are common in Japan permit more accessibility to the anvil for this type of testing. Nevertheless, the HFM also works for U-type apparatus with full-size specimens and simple insulation, as shown in the next paragraphs.

A U-type flow test apparatus was assembled recently to qualify Dewar construction materials. Because of the difference in pendulum geometries, it proves neccessary to orient the cryogenic transfer line parallel to the plane of swing, rather than transverse to it as in C-type apparatus. Therefore a U-adapter was fabricated as shown in Fig. 2. The U-adapter is secured to the anvil shroud during operation but is removable for specimen replacement. The 100-L liquid He storage Dewar is equipped with a 6 mm flexible vacuum-insulated transfer line which connects to the U-adapter. Other equipment includes a clamp to hold the insulated specimen to the boss, a U-adapter support bracket, and a compressed He gas cylinder.

Each specimen is wrapped in a clear 45 x 105 mm cellophane sheet, approximately 0.09 mm thick, and the seam is sealed with cellophane tape. Double-sided adhesive tape holds the cellophane firmly against the two sides of the specimen that must rest squarely on the anvil. Approximately 12.5 mm of cellophane extends from one open end where the boss is inserted, and standard centering tongs are used to locate the specimen on the anvil. The tongs (Fig. 2) remain attached until the U-adapter and specimen are clamped in place by tightening the caphead screws that mount the U-adapter on the shroud, and the screw that presses the boss and specimen against the anvil.

With a wrapped specimen in place, the draw tube is inserted into the He Dewar with the shut-off valve closed, and the male bayonet end is connected to the U-adapter. For the existing set-up, the gas flow must be set to stabilize the pressure in the Dewar at 54–62 kPa (8–9 psi). After 30 s, the system has cooled so that He continuously flows along two sides of the specimen. When thick white vapor emerges at the vent end of the wrap, the countdown begins. Thermometry shows that, with 54–62 kPa pressure in the Dewar, the specimen cools to 20 K or lower within 25 s and reaches 5 K within 120 s. The pendulum is then released and breaks the specimen while He is still flowing past it. These tests used 10 l of He per specimen but future tests using a more efficient transfer apparatus and less flow presssure should lower the cryogen consumption considerably.

The U-type HFM apparatus used in industry is similar to the C-type apparatus that has been used in research.[3,4,6] The model of the machine, the type of plastic insulation, and the He transfer system are the main differences. To confirm that these differences do not affect mechanical property measurements, we performed impact tests two laboratories using specimens machined from a single heat of AISI 304 stainless steel. As shown by the data listed in Table 1, comparable absorbed energy and lateral expansion measurements with sizable data scatter are realized with both apparatus.

HEATING DURING FRACTURE

Although the heating associated with specimen transfer can be eliminated using techniques such as the HFM, adiabatic heating during fracture remains a problem. Hwang et al.[4] recently correlated the absorbed impact energies and quantitative fracture toughness measurements (J_{Ic} values) for eight structural alloys at 77 and 4 K. They found that the linear correlation coefficient r^2 decreased from 0.94 at 77 K to 0.70 at 4 K. Temperature rises as high as 190 K in the "4-K" Charpy tests were cited as weakening the correlation with quasistatic fracture mechanics

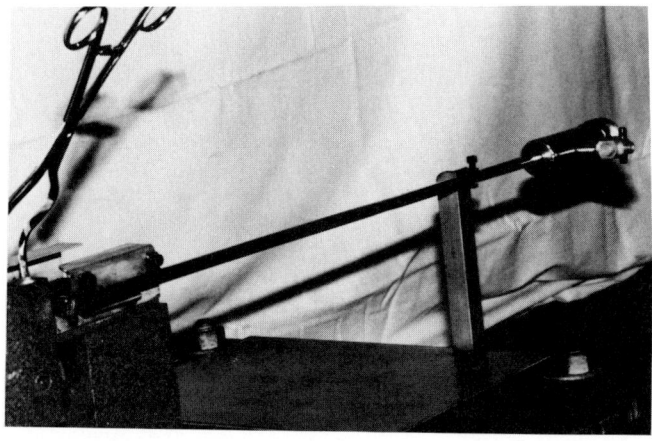

Fig. 2. Vacuum-jacketed U-shaped transfer line adapter aligned in the swing plane.

Table 1. Charpy impact data from HFM tests of AISI 304 specimens at an initial temperature near 4 K (5 ± 1 K).

Machine Type and Specimen Number	Absorbed Energy (J)	Lateral Expansion (%)
U-type Tests		
1	12.5	49
2	12.5	53
3	12.7	51
4	13.8	55
	Avg - 12.9	Avg - 52
C-type Tests		
1		
2	11.5	45
3	12.1	49
4	13.0	51
	13.0	51
	Avg - 12.4	Avg - 49

parameters at 4 K. We therefore considered two alternative test procedures to limit the heating; a brief summary of results is presented here for precracked Charpy specimens tested in impact or slow bending.

TESTS OF PRECRACKED CHARPY SPECIMENS

Procedures

Full-size specimens were efficiently fatigue-precracked using a computerized ultrasonic fatigue apparatus shown in Fig. 3. Bending loads at room temperature were applied at 150 Hz, using a stress ratio of 0.1 and a final maximum stress intensity factor of 20 MPa·m^½ to minimize warm prestress. The required fatigue cycles ranged from 0.3 to 4 million, depending on the material. The final notch-plus-precrack lengths were measured from post-test fracture surfaces at the ¼, ½, and ¾ points of thickness, and averaged.

For impact tests we precracked 4 different steels, used the HFM, and evaluated the effect of heating analytically. The absorbed energy is reduced because precracking localizes the deformation and reduces the ligament to 5 – 6 mm (versus 8 mm for notched specimens). Impact energies (C_v) are compared in terms of absorbed energy per ligament area A. The reduction of energy in percent is

$$(C_{vn}/A_n - C_{vp}/A_p)/(C_{vn}/A_n) \times 100, \qquad (1)$$

where the subscripts n and p denote notched and precracked specimens, respectively.

For slow-bending Charpy (SBC) tests, we precracked five different steels and loaded them in a three-point bend fixture at 4 K, as in the 77 K study by Witzke and Stephens.[7] Slow-bending increases the prospects of a fair correlation with J_{Ic} at 4 K, primarily because the testing rate is much lower than the impact test. We used a servohydraulic test machine in stroke control at 0.03 mm/s which is typical of the rates used for J_{Ic} measurements of these materials.[4] The load-stroke curves were recorded with an X-Y plotter, and SBC toughness was calculated in terms of approximate J-integral values. Two values were calculated: J_i, the value of J at the first pop-in or serration, and J_{max}, the value at maximum load, where

$$J = 2\alpha/Bb. \qquad (2)$$

In this expression, α is the area under the load-stroke curve, B is the specimen thickness (10 mm), and b is the uncracked ligament dimension.

Results

Figure 4 compares the normalized impact energies for notched and precracked specimens of the same steels. Only one precracked specimen of each steel was tested, but the data for the precracked and standard notched specimens indicate the same ranking of materials at 4 K. The normalized reduction of energy varies from 35 to 54%, with lower values for the lower toughness alloys.

The lower impact energies for precracked specimens imply less heat generation but the thermal response of a particular material will depend on its toughness and specific heat as a function of temperature. Using the precracked specimen data, we predicted the temperature rise for 304 stainless steel using Hwang's method.[2] Figure 5 shows the result in comparison with measured temperatures by Dobson and Johnson.[8] The calculated temperature rise for a precracked specimen during impact is 130-150 K, depending on the percent reduction of energy assumed for this alloy. This is less than the 180-190 K prediction for a standard notched specimen of the same material, but still high relative to the initial test temperature. Apparently, then, fatigue precracking does not offer much improvement over conventional notching for tests at 4 K: the temperature rise is excessive in either case.

Fig. 3. Fatigue precracking apparatus: (A) table-top ultrasonic fatigue machine and minicomputer; (B) a Charpy specimen gripped for bending at R = 0.1.

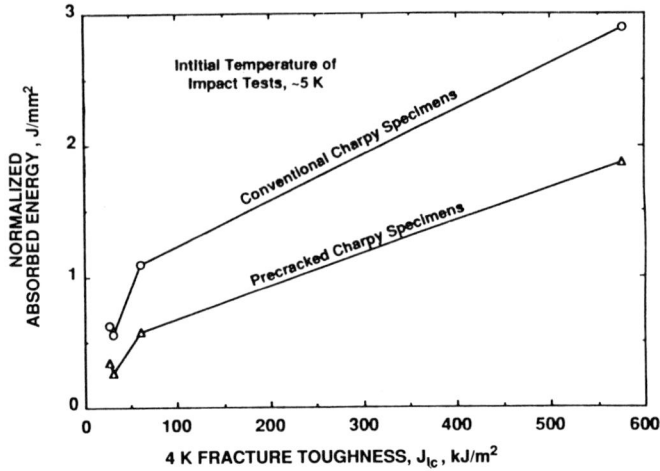

Fig. 4. Absorbed energy measurements for v-notched and fatigue precracked Charpy specimens tested near 4 K.

The slow-bending Charpy test results are shown in Fig. 6. In this graph the J_i and J_{max} parameters are plotted against the previously reported J_{Ic} values which are regarded as reliable measures of toughness at 4 K. The diagonal trendline in this figure represents what would be an ideal correlation. Over the entire low-to-high toughness range, J_{max} appears to be a better indicator of toughness than J_i. The J_i data underestimate J_{Ic}, especially for the toughest materials. By contrast, the J_{max} values tends to overestimate J_{Ic}, but the margin of disagreement is not as great.

The main uncertainty in SBC testing is the toughness measurement point. Depending on the material, fracture may begin at J_i, J_{max}, or some point between, so the J_{Ic} correlation is inherently limited by error in the SBC toughness value. Our preliminary data span a wide range of toughness, but are few in number. Therefore the correlations should be improved by more data for other materials. Regardless of the quality of potential correlations, however, the SBC test will remain at a disadvantage in terms of convenience. Considering the fatigue apparatus and time requirements, the SBC test is more demanding than a conventional impact test, and this drawback diminishes its appeal for quality control or screening purposes.

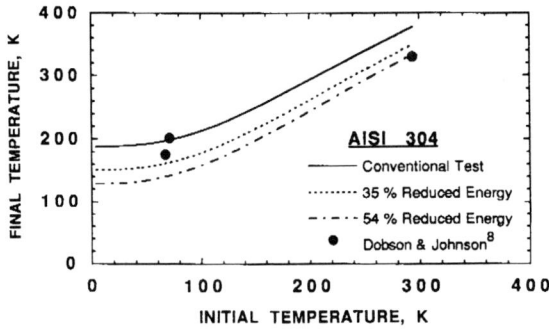

Fig. 5. Predicted temperature rise during conventional v-notched and precracked Charpy impact tests of AISI 304; precracking effects are modelled as a 35 or 54% reduction of energy.

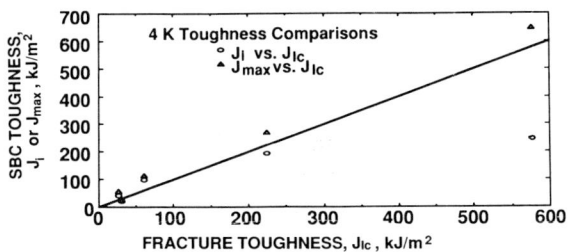

Fig. 6. Comparison of two J-based fracture parameters calculated for Charpy specimens in slow bending versus J_{Ic} measurements for 25 mm thick compact specimens.

CONCLUSION

The HFM is a one-person process requiring basic cryogenic equipment. The method yields the appropriate temperature before impact, and is adaptable to full size specimens and U-type machines commonly used in U.S. industry. Guaranteeing the initial temperature eliminates one source of experimental error, but scatter from other sources is inherent when dynamic tests are performed at cryogenic temperatures. Charpy specimens initially at 4 K actually fracture at much higher temperatures due to adiabatic heating, and the correlation between C_v and J_{Ic} suffers accordingly. Thus, at present, we have no ideal screening test at 4 K. As an improved test procedure at 4 K, fatigue cracking holds little promise since the temperature rise of tough alloys is not much curtailed. On the otherhand slow-bending certainly minimizes adiabatic heating, but the precracking and J-based measurements for that test are more involved than conventional absorbed energy measurements.

ACKNOWLEDGMENTS

Test specimens for this study were fatigue-precracked at the Nuclear Research Center-Negev. Acknowledgments with thanks are due to K. Rezac and R. Minette for help in establishing the fixturing and procedures used at MVE, and to D.P. Vigliotti for similar assistance at NIST.

REFERENCES

1. Standard Test Methods for Notched Bar Impact Testing of Metallic Materials, Designation E 23-88, in: *1990 Annual Book of ASTM Standards*, Amer. Soc. Test. Maters., Philadelphia, 03.01:197-212 (1990).

2. R.L. Tobler, R.P. Reed, I.S. Hwang, M.M. Morra, R.G. Ballinger, H. Nakajima, and S. Shimamoto, Charpy Impact Tests Near Absolute Zero, *J. Test. Eval.*, 19:34-40 (1991).

3. H. Nakajima, K. Yoshida, H. Tsuji, R.L. Tobler, I.S. Hwang, M.M. Morra, and R.G. Ballinger, The Charpy Impact Test as an Evaluation of 4-K Fracture Toughness, *Adv. Cryo. Eng.*, this volume.

4. I.S. Hwang, M.M. Morra, R.G. Ballinger, H. Nakajima, S. Shimamoto, and R.L. Tobler, Charpy Absorbed Energy and J_{Ic} as Measures of Cryogenic Fracture Toughness, *J. Test. Eval.*, accepted for publication.

5. S. Jin, W.A. Horwood, J.W. Morris, Jr., and V.F. Zackay, A Simple Method for Charpy Impact Testing Below 6 Degrees K, <u>Adv. Cryo. Eng.</u>, 19:373-378 (1974).

6. T. Ogata, K. Hiraga, K. Nagai, and K. Ishikawa, A Simplified Method for Charpy Impact Testing Near Liquid Helium Temperature, <u>Cryogenics</u>, 22:481-482 (1982).

7. W.R. Witzke and J.R. Stephens, Development of Strong and Tough Cryogenic Fe-12Ni Alloys Containing Reactive Metal Additions, <u>Cryogenics</u>, 17:681-688 (1977).

8. W.G. Dobson and D.L. Johnson, Effect of Strain Rate on Measured Mechanical Properties of Stainless Steel at 4 K, <u>Adv. Cryo. Eng.</u>, Vol. 30, 1984, pp. 185-192.

SELECTED RESIDUAL EFFECTS UPON TEMPERATURE TRANSITIONS

Yosef Katz, Moshe Kupiec and Arie Bussiba

Nuclear Research Centre - Negev
P.O. Box 9001, Beer-Sheva, 84190, ISRAEL

ABSTRACT

The present study summarizes some recent phenomenological investigations into the nature of interactive effects as originated by residuals while wandering along a temperature scale. Specifically, the low temperature regime is considered with additional complexities due to near crack-tip stress and deformation field modifications. Generally, a global physical view is attempted in a fracture mechanics framework concerning two levels;
  (i) Developments of some fundamental aspects associated with crack tip shielding potential, phase stability and damage assessments.
  (ii) Considerations as related to interactive effects and their reflection on the intrinsic mechanical properties at low temperatures.

For example, low temperature might affect the structural stability with implications on various properties including environmental susceptibility. In this context, load interactions at the higher shelf might cause significant modifications of the nominal fracture resistance values as obtained at low temperatures. As such, intrinsic properties may be masked by extrinsic influences.

Accordingly, the current investigation objectives are centered on illuminating some of these issues by following experimental findings evaluation. The experimental initiative in stationary and subcritical growing cracks include load interaction effects mainly in BCC systems and environmental interactions in metastable austenitic stainless steels.

More refined views on residual or contact stress effects, process zone formation and cyclic crack growth transients are proposed. In addition, environmental interactions are described, expressing some phase stability concerns. Finally, the role of the time factor via thermal activation approach is discussed with emphasis to low temperature material performance assessments.

INTRODUCTION

There has been a growing interest in superimposed interactions along the crack path and how it reflects on mechanical properties at the low temperature regime. Probably the most addressed phenomena regarding load interactions is the so defined Warm Prestressing (WPS) effects[1-4] in materials which are characterized by a Ductile/Brittle Transition. As confirmed experimentally, load interactions at the upper shelf modifies the instantaneous driving force for microcracking onset. These effects are manifested by substantial toughening effects even in basically cleavage fracture situations. Moreover, in a quite different and more complex case, interactive origins might affect both; i.e. the local driving force for cracking and the fracture resistance. This is associated with intrinsic microstructural stability effects, where dilational stresses are developed by phase transformations[5].

Following the aforementioned issues, some experimental steps have been initiated in WPS studies in a stable BCC steel and slow crack growth transients in 304 metastable austenitic steel, where plastic strain gradients due to overloads resulted in gradients of martensitic transformed phases. Although these systems are mainly emphasized, additional findings in other metallic systems are briefly overviewed. It seems appropriate to follow such phenomenological approach in order to assist in crystallizing a global view far beyond the exclusive WPS effects. Experimental findings gathered so far provide appropriate background to cope with the following elements;
(i) What are the dominant variables which control the near tip stress field?
(ii) What is the magnitude of the macrocrack tip shielding activated by load interactions?
(iii) Microstructural aspects and their role on the crack tip shielding including process zone/crack tip interactions.
(iv) Reflection of local interactions on the intrinsic properties at low temperatures.

As mentioned, the current elaborated description on features in steels are supplemented by other interactive examples including single crystals behavior. At this stage a general approach is presented aimed to assist in developing theoretical aspects as based on experimental findings feedback.

EXPERIMENTAL PROCEDURES

Structural Stability Effects (The 304 Stainless Steel Case)

The experimental program was focussed on mode I fatigue crack propagation rate transients activated by prior single overloads. The 304 metastable stainless steel was selected with yield stress of 416MPa at 296K. Precracked specimens were utilized in three point bending, tension-tension cyclic tests, controlled by the stress intensity range under load ratio of R ≈ 0. The crack growth was monitored by an electropotential techniques after proper calibration. This enabled crack propagation rate measurements before and after single overloads. Two different microstructural situations were investigated. First, load interaction at 296K in which the single overload resulted in crack tip plastic enclave of deformed austenite γ and subseqeuntally affecting the regular tension-tension fatigue crack propagation rates (FCPR). Second, where the plastic enclave after overloads consisted in multiphase gradients due to γ decomposition. This was achieved by overloads applied at 77K, prior to subcritical crack growth at 296K, resulting in near tip structural modifications which contained ϵ' and α' phases. (ϵ' the closed packed and α' the cubic centered martensitic phases). Here an additional step was initiated following a comparative study of delayed retardation between environmental affected material (by high fugacity hydrogen) and subcritical crack extension rate transients with no hydrogen. Thus, the experimental scheme consisted of load interactions activated by overloads with and with no environmental effects in two crack tip structural variants. The crack growth rate transient features in terms of da/dn vs. crack tip position were analyzed. Electron fractography and acoustic emission (AE) tracking were conducted for complementary findings as related to process zone nucleation and fracture modes. More about the experimental scheme has been addressed elsewhere[6].

In addition, damage assessment study in low and high fugacity hydrogen was performed in uniform uniaxial tensile specimens. The main point here followed the structural stability aspects mainly centered on ductility lose and the degree of cracking formation. High fugacity was achieved by cathodic charging, while low fugacity procedure followed gaseous charging by dry H_2 at 800 atm., at 500K for 24 h[7]. Preloading was performed at 77K far below the M_d up to 30% strain resulting in gradually increasing volume fractions of martensite phases actually controlled by the prior strain magnitudes. Here the residual ductility in low fugacity internal hydrogen was examined at 296K in terms of martensitic phases content at the crack tip interactive zone. Finally, microcracking and near surface affected zone were determined as fine scale fracture mode features to allow appropriate damage assessments.

Warm Prestressing (WPS) (The 4340 Case)

As received, 4340 steel was selected with yield stress of 350 MPa at 296K. Beyond the role of WPS on fracture resistance at the lower shelf, the state of stress effects by thickness variations between 14-50mm were investigated. Prefatigued compact tension specimens were utilized with the following load scheme. First, prestressing at 296K was applied, with stress intensity ratio of $K^{WPS}/K_c \approx 0.8$ and the fracture resistance at 77K was established in the basically cleavage mode regime. Acoustic emission tracking particularly along the WPS process was carried out in conjunction with electron fractography examinations. The process zone nature became evident as revealed by the AE feedback. Here, microcracking was confined to the local WPS interactive zone.

EXPERIMENTAL RESULTS

Structural stability effects

The post single overloads effects on the FCPR revealed the dominant role of the exact crack tip morphologies, namely, the monolithic and the multiphase variants. This was manifested by significant changes in the retardation behavior, even with the absence of environmental interactions. In fact, the thermomechanical history became significant by introducing phase stability - crack tip shielding aspects in a quite complex fashion. It

Table 1. The remote $(a-a_0)_{min}$ as a function of overload intensity ratio q with no hydrogen.

crack tip interactive zone after single overload	overload intensity ratio q	$(a-a_0)_{min}$ μm
monolithic, deformed γ	1.33	380
"	2.00	1300
"	2.80	1900
multiphase, γ, ε', α'	1.40	1300
"	2.00	850
"	2.50	180

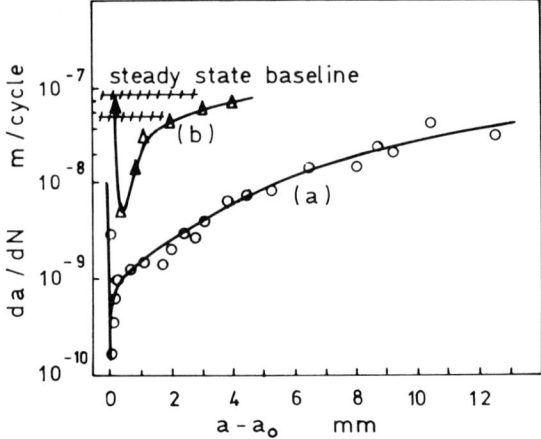

Figure 1. Retardation curves; overloads in 304 at 77K while subsequent cyclic growth at 296K: (a) no hydrogen q=2.65; (b) with high fugacity hydrogen q=2.5

was found that the whole trend of the FCPR transient behavior was altered while comparing the dilational and the dilationless crack tip morphologies. For example, consider the subcritical crack path after a single overload, a position along the crack path designated $(a-a_0)_{min}$ which corresponds to the maximum retardation value $(da/dN)_{min}$ was determined as a function of the overload intensity ratio $q = K_{OL}/K_{max}^{min}$. Table 1 indicates the consistent tendencies for both structural variations which differ substantially.

With high fugacity hydrogen the overall retardation behavior prevailed while differences were even more accentuated. Although the multiphase microstructure provided higher crack tip shielding after overloads (with and with no hydrogen effects) the macro shielding potential was reduced to about 30% by hydrogen interactions while providing an increase in the steady state FCPR baseline due to hydrogen induce cracking (Fig. 1). The main point here, to emphasize the dominant local mechanisms involved, including crack tip stress field and intrinsic fracture resistance modifications, as process zone nucleation (Fig. 2). Along these, consistent evidences were obtained by AE tracking and fractographic studies, particularly in the transformed interactive zone. Thus, competitive situation ocurred with direct effects on crack stability. As such, in some interactive situations the realization of the process zone to be an effective shielding factor becomes essential.

In this context, the uniaxial tensile results provided supplementary confirmations. In particular the low fugacity hydrogen in which prior microcracks were restricted. Here remarkable differences in terms of ductility loss were revealed while comparing monolithic and multiphase microstructures. Prior strain of 10% at 77K provided at the plastic zone 40% in volume fraction of $\alpha'+\epsilon'$ martensitic phases. Under these circumstances a reduction from 32% to 14% in uniform elongation values were experienced.

In fact, the plastic constitutive properties were altered completely caused exclusively by structural stability effects. Concerning some damage assessments considerations, cracks of about 300-600μm in scale were developed in the multiphase tensile specimens. This is an increase of more than two order of magnitude in crack length as compared to skin effects of 1μm in high fugacity hydrogen with the absence of external stress field[8].

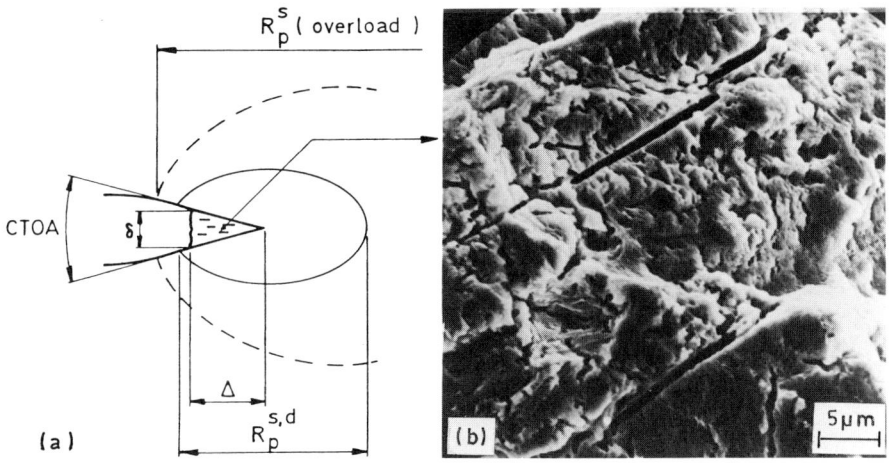

Figure 2. (a) Schematic; crack tip with process zone, plastic zone - static or dynamic and overload interactive zone; (b) Process zone in 304 with hydrogen interaction.

Warm Prestressing

Although stress state effects were identified by thickness variation studies, significant WPS effects were obtained even in thick specimens. With no WPS_1, typical critical fracture resistance values at 77K were about $30 MPam^{1/2}$ while values of $40.2 MPam^{1/2}$ were obtained for the 50mm specimens after WPS. For specimens$_2$ of 14mm in thickness some higher values were obtained up to $45.6 MPam^{1/2}$.

Under the current experimental conditions, process zone nucleation during WPS process was detected by AE continuous tracking. At the lower shelf, the micro fracture features were dominated by cleavage, still indicating a substantial crack tip shielding in the macro sense.

DISCUSSION

The current aforementioned examples demonstrate briefly the role of load interactions and phase stability effects on stationary and subcritical growing cracks. These kind of crack tip shielding effects were also expressed in the microcracking initiation time in stress corrosion cracking of high strength steel. As addressed by Nakasa et al.[9] in a blunting notch, the macroscopic compressive residual stresses produce an increase in delayed failure of more than one order of magnitude. Similar effects were established in the cyclic life for crack initiation in notched specimens of Ti-6Al-4V alloy after surface coating by plasma processes[10]. However, more on the fundamental aspects of WPS, have been investigated in Fe-3%Si single crystals. Again, after preloading at ambient temperatures higher level of stress intensity was needed at low temperature to provide a necessary driving force for crack nucleation and propagation[4]. It became apparent that variations of the crack tip stress distribution reflect on the intrinsic low temperature measurements as on other fracture processes. In order to establish some analytical foundations more is required and is clearly considered here as a long term goal. However, some progress has been achieved in identifying some of the local micromechanisms involved.

Beside extrinsic shielding effects such as crack front or branching, blunting or residual stresses by plasticity or dilational stresses, the role of process zone/crack tip interactions should be counted as shown in environmental induced microcracking. The description of the effective driving force for cracking can be refined even at this stage with some degree of success. For example, due to several beneficial conditions the overload transient effects on cyclic crack growth were addressed in Fe-3%Si single crystals[11]. There, in a combined case of mechanical and hydrogen interactions, typical {100} planar crack growth developed at low stress intensities which limited near fracture surface plasticity. Since other extrinsic shielding effects were minimized, the striation features on the cleavage plan could be tracked and compared to the overload history. Following this, particular emphasis was given to residual stress effects while exploring the high sensitivity of cyclic microcracking to the local stress field variations. Although some minor effects might be caused by other mechanisms, the excess residual compressive stress field on unloading argument appeared to be most persuasive in the Fe-3%Si crystal case. Here, a single overload with subsequent tension-tension cyclic loading in hydrogen resulted in striation spacing which initially decreased by an order of magnitude. Thus, transients which were activated by a single overload were manifested by remarkable effects on the fine scale arrest line. Following the residual stress argument only and direct X-ray measurements in single overloads[12] predicted the fracture surface striation spacing reasonably well with basically no adjustable parameters. This analysis was even more justified by the X-ray findings of Allison[13] who concluded that the residual field induced by an overload is not relaxed significantly by the subsequent fatigue crack growth.

Presently, for Fe-3%Si crystals, an alternative analytical approach is noted to be successful at least for the initial post overload cyclic crack growth. Regarding the compressive residual field due to unloading, consider the modified HRR field[14] for stationary crack with strain harden material. The effective strain at the tip exhibits a $r^{-1/1+n}$ singularity given by;

$$\epsilon_p = \alpha_1 \frac{\sigma_{ys}}{E} [(\frac{R_p}{r})^{1/n+1} - 1] \quad (1)$$

where; R_p - the plastic zone, n - the strain hardening coefficient and α_1 - a material constant.

For the current load interaction the non-linear irreversible deformed zone was activated by a single overload. Assume a material with parabolic hardening behavior;

$$\sigma_p = \sigma_{ys} [1+\alpha_2(\epsilon_p)^n] \quad (2)$$

The local stress $\sigma_p(r)$ can be formulated by combining eqs. (1) and (2);

$$\sigma_p(r) = A + B \{C [(\frac{R_{OL}}{r+r_0})^{1/1+n} - 1] -1\}^n \quad (3)$$

where; A, B and C constants, R_{OL} - plastic zone activated by single overload.

Thus, for a first order approximation, unloading residuals may be described by the following constitutive form;

$$|\sigma_R| = B'\{[C' (\frac{\dot{R}_{OL}}{r+r_0})^{1/1+n} - 1]\}^n \quad (4)$$

Table 2. Initial compressive residual field.

remote distance from the crack-tip μm	$|\sigma_R|$ calculated MPa	$|\sigma_R|$ direct X-ray measurements (ref. 10) MPa
50	256.00	260.00
150	161.28	220.00

where r_0 is finite to remove the crack tip singularity. For a plane strain situation R_{OL} can be estimated by;

$$R_{OL} = \frac{K^2}{3\pi\sigma_{ys}} \tag{5}$$

which enable to express the mechanical driving force by the applied stress intensity factor.

Following this approach, for the Fe-3%Si crystals with mechanical properties given by the constitutive flow equation[11];

$$\sigma = 1000\epsilon_p^{0.38} \text{ (MPa)} \tag{6}$$

with: $\sigma_{ys[001]}$ = 296±14 MPa, E = 1.32 10^5 MPa, n = 0.38 and $\alpha_1 \approx 1$.

and for a single overload of 26 MPam$^{1/2}$ the initial compressive residual field was calculated as given in Table 2.

Clearly the residual field varies appreciably with the depth and distance (r) from the crack tip. However, this HRR residual field estimation allows further predictions of the striation spacing as described elsewhere[11]. Accordingly, the initial post overload transient prediction from 2μm to about 0.25μm appears quite consistent with experimental observations. It should be noted that in other situations a different formulation might be well justified with emphasis to crack tip closure. This has been addressed by Fleck and Shercliff[15] while considering the modified crack tip stress distribution after load interactions. Obviously, appropriate realization of the dominat micromechanisms involved might assist in findings evaluation. In this context the partial role of closure effects have been explored while investigating influences of precracking conditions at the higher shelf on the critical fracture toughness values as measured at 4K[16]. However, regardless of the exact origin for crack tip field perturbations the basic key affecting the transient crack growth as stationary microcracking onset seems to lie in residuals even though the actual cause may be partially from closure stresses. As such, complex stress distributions might exist with closure stresses superimposed on relatively far-field residuals. This remark emphasize the possible effects in cyclic loading even at high R values.

Finally, an additional factor is briefly noted. The nature of residual stresses implies that besides their initial magnitude further understanding of their redistribution or relaxation is required. Either of these will affect the crack tip shielding potential. This issue was recently addressed developing some aspects with regard to the time factor[17]. It seems that in some situations crack tip residual fields as damage or process zone interactions are time dependent. Following some preliminary results a thermally-activated rate approach has been proposed and even included into overloads models.

The current described examples by experimental findings are centered on interactive aspects particularly as connected to the low temperature regime. While considering thermomechanical history, at least two major factors are particularly relevant in some situations to low temperatures i. e. structural stability and Ductile-Brittle transition aspects.

This implies the following conclusions;
1. Load interaction and phase stability effects have remarkable implications on intrinsic fracture resistance properties at low temperature.
2. Phenomenological studies under different circumstances develop some promise in resolving complex crack tip stress distributions affecting crack stability.
3. Crack tip field perturbations, process zone/crack tip interaction effects become significant and even accentuated by environmental induce cracking. Some of these influences have been already identified also connected to phase stability aspects.

ACKNOWLEDGEMENTS

The authors wish to thank Mr. R. Shefi and Mr. E. Woodbeker from the NRC for experimental assistance. We would also like to acknowledge Dr. R. L. Tobler from the National Institute of Standards and Technology, Boulder, Colorado, for releasing information prior to publication.

REFERENCES

1. S. Matsuoka, K. Tanaka and M. Kawahara; Engng. Fract. Mech., 8:507 (1976).
2. J. Lankford, Jr. and D. L. Davison, J. Eng. Mat. Tech., 98:17 (1976).
3. N. A. Fleck, Acta Metall., 33, 1139 (1985).
4. M. Lii, t. Foecke, X. Chen, W. Zielinski and W. W. Gerberich, Mater. Sci. and Engng, A113:327 (1989).
5. E. Hornbogen, Acta Metall., 26:147 (1978).
6. Y. Katz, A. Bussiba and H. Mathias, in: ICF7, Advences in Fracture Research Vol. 2, eds. K. Salama, K. Ravi-Chandar D. M. R. Taplin and P. Rama Rao, University of Houston (1989) p. 1947.
7. I. Gilad, and Y. Katz, Zeitschrift fur Phys. Chemie Neue Folge, 164:1561 (1989).
8. H. Mathias, Y. Katz and S. Nadiv, in: Metal-Hydrogen Systems, T. N. Veziroglu, ed., Pergamon Press, Oxford, (1982) p. 225 .
9. K. Nakasa, M. Kido and H. Takei, Trans. ISIJ, 22:106 (1982).
10. A. Bussiba, A. Raveh and Y. Katz, to be presented in the Euro-Asian Interfinish Conf., Herzlia, Israel (1991).
11. X. F. Chen, Y. Katz and W. W. Gerberich, Scripta Metall., 24:2351 (1990)
12. W. H. Sclosberg and J. B. Cohen, Metall. Trans., 13A:1987 (1982).
13. J. E. Allison, in: Fracture Mechanics, ASTM STP 677, C. W. Smith ed., American Society for testing and Materials, Philadelphia (1976) p. 550.
14. J. W. Hutchinson, J. Mech. Phys. Solids, 16:337 (1968).
15. N. A. Fleck and H. R. Shercliff, in: ICF7, Advances in Fracture Research Vol. 2, eds. K. Salama, K. Ravi-Chandar D. M. R. Taplin and P. Rama Rao, University of Houston (1989) p. 1947.
16. A. Bussiba and R. L. Tobler, unpublished results.
17. Y. Katz, A. Bussiba and W. W. Gerberich, in: Fatigue 90 Vol. 3, eds., H. Kitagawa and T. Tanaka, MCFP Pub., Birmingham, (1990) p. 1499 .

REACTOR NEUTRON AND GAMMA IRRADIATION OF VARIOUS COMPOSITE MATERIALS

N. A. Munshi

Composite Technology Development, Inc.
Boulder, Colorado, U.S.A.

H. W. Weber

Atominstitut der Österreichischen Universitäten
Vienna, Austria

ABSTRACT

Polymer matrix composites with nine different resin systems were prepared with S-2 Glass®*-fiber reinforcement. The resins evaluated were epoxy, bismaleimide, and polyimide; all are suitable for vacuum potting, resin transfer molding, filament winding, and preimpregnation. The composite specimens fabricated were 3.2-mm-diameter, unidirectionally reinforced rods. They were tested in torsion to obtain interlaminar shear properties at 295, 76, and 4 K. Specimens from the same resin systems were irradiated at three different total dose levels (4.7×10^6, 4.7×10^7, and 2.3×10^8 Gy) at 330 K and then tested in torsion at 76 K.

The results indicate that (1) properly formulated and cured liquid epoxy systems maintained a high shear strength even at the highest dose levels; (2) some polyimides lost no strength at the higher dose levels, whereas others lost considerable strength; (3) higher molecular-weight epoxy resin systems, although they had a higher strength than the bismaleimide system before radiation, lost this advantage after irradiation.

INTRODUCTION

The organic electrical insulation system of superconducting magnets for fusion reactors and accelerators is the component most sensitive to damage by high-energy radiation. For this reason, Composite Technology Development (CTD) embarked on a program to develop reliable, radiation-resistant organic insulators that will withstand high levels of radiation over several decades without substantial loss of mechanical and electrical properties. A cost-effective, integrated approach to specimen production and testing (known as CompTec®†) was developed to screen resin formulations to be used under demanding environmental conditions. A component data base that defines the cryogenic and radiation performance of composites was generated. Subsequently, these data can be used to optimize insulator fabrication, and

---

*S-2 Glass® is a registered trademark of Owens-Corning Fiberglas Corporation.

†CompTec® is a registered trademark of Composite Technology Development.

with further characterization, they will provide the engineering data base required by magnet designers.

SPECIMENS

Unidirectional, S-2 Glass®-fiber-reinforced, 3.2-mm-diameter rod specimens were produced with nine resin matrices. Vacuum impregnation was the fabrication method of choice for all but the polyimide composites, which were fabricated from preimpregnated fiber tows. The fiber volume fraction of all specimens was maintained at 55 ± 1%.

Table 1 summarizes the physical characteristics of these resins. CTD-100, CTD-101, CTD-102, and CTD-501 are low-viscosity epoxy resin-based systems suitable for vacuum impregnation, resin transfer molding, and filament winding. CTD-110 and CTD-112 are high-viscosity epoxy resin systems; CTD-200 is a bismaleimide resin system; all three are suitable for preimpregnation. Specimens were also fabricated from the commercial G-11CR epoxy resin system for baseline data comparison.

The low viscosity, long pot life, and relatively low cure temperature of CTD-101 and CTD-102 make them desirable for potting large magnet coils like those found in fusion reactors, such as the International Thermonuclear Experimental Reactor (ITER) and the Burning Plasma Experiment (BPX).

MECHANICAL TESTING

The rod specimens are ideally suited for testing in torsion. Torsion is also the most sensitive test for determining interlaminar shear properties of composite materials.[1] Torsion testing in the vertical mode was developed at the National Institute of Standards and Technology[2] and further refined at CTD[3] to test composite as well as neat resin specimens. Figure 1 illustrates the torsion test method.

The 3.2-mm-diameter rod specimens were cut into 60-mm lengths. To obtain the ultimate shear strength, ultimate shear strain, and shear modulus, each specimen was torqued to achieve 90 degrees rotation at 295, 76, and 4 K. Figure 2 is an example of a typical torsion test graph at the three temperatures.

Table 1. Physical Properties of CTD Resin Systems

CTD System	Resin Type	Pot Life (h/°C)	Viscosity (cP/°C)
100	epoxy	6/40	1100/40
101	epoxy	60/40	400/40
102	epoxy	50/40	450/40
110	epoxy	1/135	~300/135
112	epoxy	3/135	~300/135
200	bismaleimide	1.5/130	~300/135
300	polyimide	0.08/215	powder
310	polyimide	0.08/215	powder
501	epoxy	8/25	1000/40
G-11CR	epoxy	2/120	NA

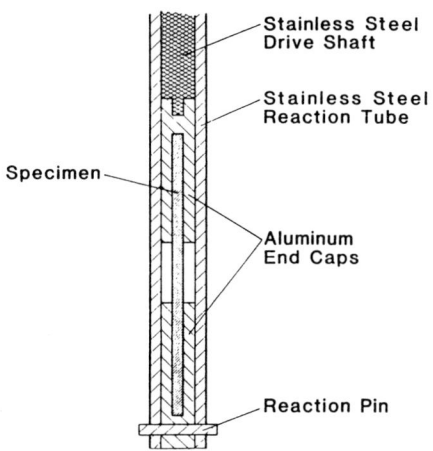

Fig. 1. Schematic of the torsion test method. End caps and drive shaft are sized to permit free rotation at cryogenic temperatures.

Table 2 summarizes the shear performance of the composite materials as a function of temperature. As seen from these data, both CTD-101 and CTD-102 compare well with the commercial high-pressure laminate material G-11CR. The bismaleimide system, CTD-200, is more brittle than the high-molecular-weight epoxy-resin systems in its class (CTD-110 and CTD-112). The polyimide system CTD-300 gains little strength between 295 and 4 K.

The fracture strength, G_{Ic}, of the composite rods was also determined. Figure 3 is an illustrative description of the test. To measure its length accurately, India ink was used to stain the crack. Figure 4 summarizes the fracture strength of the materials, which is a measure of the integrity of the fiber-matrix interface. It correlates well with the torsion shear strength results.

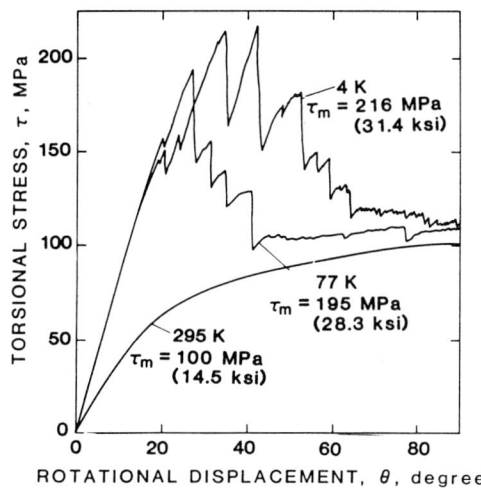

Fig. 2. Typical response of composite materials to torsion tests.

Table 2. Torsional Shear Strength of Resin Systems

Resin System	Torsional Shear Strength (MPa)		
	295 K	77 K	4 K
CTD-100	62	160	180
CTD-101	103	201	232
CTD-102	108	220	235
CTD-110	105	142	144
CTD-112	112	184	178
CTD-200	94	120	142
CTD-300	75	87	76
CTD-310	76	131	111
CTD-501	99	100	107
G-11CR	89	196	196

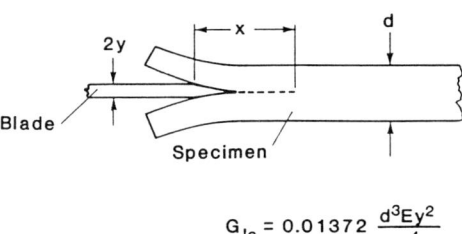

$$G_{Ic} = 0.01372 \, \frac{d^3 E y^2}{x^4}$$

Fig. 3. Schematic of the fracture-energy test method. Distance x is measured from the shoulder of the blade to the end of the crack. Dimension y is half the blade width.

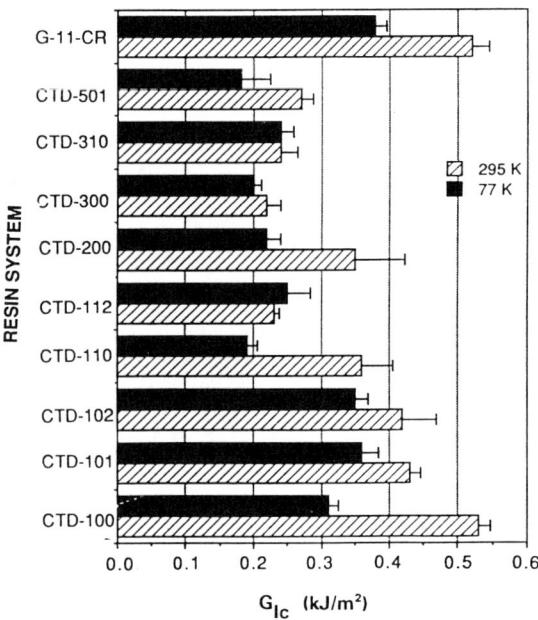

Fig. 4. Fracture strength as a function of total absorbed dose.

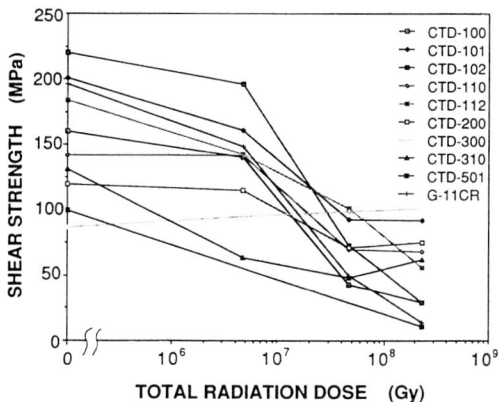

Fig. 5. Shear strength as a function of total irradiation dose at 330 K. The test temperature was 76 K.

IRRADIATION PROGRAM

The TRIGA MARK-II reactor in Vienna was used to irradiate the composite rod specimens. Weber et al. described the reactor configuration and the experimental technique used for neutron dosimetry elsewhere.[4] The fast-neutron flux density (E > 0.1 MeV) was 7.6×10^{16} m^{-2}s^{-1}. The irradiation test temperature was below 330 K.

Four specimens of each type were irradiated at three different neutron fluence levels: 1.0×10^{21} m^{-2}, 1.0×10^{22} m^{-2}, and 5.0×10^{22} m^{-2} (E > 0.1 MeV). This translates into total absorbed dose levels (neutron plus gamma) of 4.7×10^6, 4.7×10^7, and 2.3×10^8 Gy.[4,5] Note that this conversion is approximate and should be used only as a guideline.

The shear strength, shear strain, and shear modulus are plotted in Figs. 5, 6, and 7 as a function of dose level. The test temperature was 76 K. CTD-101, whose initial strength is close to G-11CR, retained nearly 80% of its strength at the lowest dose level and more than 45% of its strength at 2.3×10^8 Gy. Indeed, the strength of CTD-101 epoxy is only slightly lower than the strength of the polyimide CTD-300. All the other

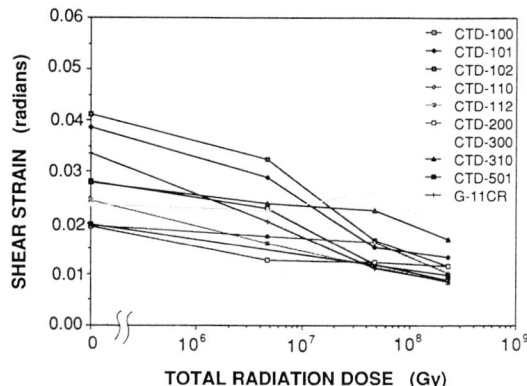

Fig. 6. Shear strain as a function of total irradiation dose at 330 K. The test temperature was 76 K.

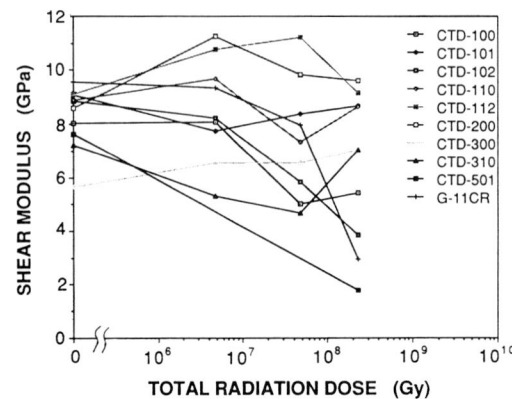

Fig. 7. Shear modulus as a function of total irradiation dose at 330 K. The test temperature was 76 K.

low-viscosity epoxy resin systems (CTD-100, CTD-102, CTD-501, and G-11CR) retained very little strength at the highest dose level; with G-11CR and CTD-501 (a room-temperature-cured resin system) becoming essentially unusable.

The bismaleimide system, CTD-200, although it has a lower strength than CTD-110 and CTD-112 before irradiation, lost only 38% of its strength at the highest dose level, whereas the two epoxies lost 48% and 70% of their strength, respectively.

One polyimide system, CTD-300, sustained the least damage due to irradiation; the other polyimide system, CTD-310, lost 53% of its initial strength after irradiation to the highest dose.

CONCLUSIONS

With proper formulation and cure, low-viscosity epoxy resins can perform as well as the best polyimide systems after irradiation to dose levels above 10^8 Gy. The ease and versatility of handling and fabrication, as well as their lower cost, make the epoxy resin systems much more desirable than polyimides or bismaleimides for magnet construction.

ACKNOWLEDGMENTS

This work was supported, in part, by the U.S. Department of Energy, SBIR contract no. DE-AC02-87ER80487, and the Austrian Bundesministerium für Wissenschaft und Forschung, contract no. 77.011/2-25/89.

REFERENCES

1. D. W. Wilson, An overview of test methods used for shear characterization of advanced composite materials, in *Advances in Cryogenic Engineering—Materials,* vol. 36, R. P. Reed and F. R. Fickett, eds. Plenum, New York (1989), p. 793.
2. M. B. Kasen, A strain-controlled torsional test method for screening the performance of composite materials at cryogenic temperatures. J. Mater. Sci. 23:813–834 (1988).

3. M. B. Kasen and N. A. Munshi, A. Cryogenic torsional shear and fracture strength of epoxy and polyimide composites, in *Cryogenic Materials '88,* International Cryogenic Materials Conference, Boulder, Colorado (1988), p. 777.
4. H. W. Weber, H. Böck, E. Unfried, and L. R. Greenwood, Neutron dosimetry and damage calculation for the TRIGA MARK-II in Vienna, *J. Nucl. Mater.* 137:236-240 (1986).
5. W. Maurer, Neutron and Gamma Irradiation Effects on Organic Insulating Materials for Fusion Magnets. KfK 3974. Kerforschungszentrum Karlsruhe, Institut für Technische Physik, Karlsruhe, Germany (1985).

RADIATION DAMAGE OF GLASS-FIBER-REINFORCED COMPOSITE MATERIALS
AT LOW TEMPERATURES

T. Okada, S. Nishijima, and T. Nishiura

ISIR Osaka University
Ibaraki, Osaka, Japan

K. Miyata and Y. Yamaoka

KURRI, Kyoto University
Kumatori, Osaka, Japan

S. Namba

College of Integrated Arts and Sciences
University of Osaka Prefecture
Sakai, Osaka, Japan

ABSTRACT

Radiation damage of glass-fiber-reinforced plastics (GFRPs) has been examined in terms of interlaminar shear strength in order to develop radiation-resistant composites. The GFRPs, containing different types and amounts of glass, were irradiated at Kyoto University Reactor below 20 K and tested at liquid-nitrogen temperature before warming to room temperature. The effect of the matrix material was also examined. After interlaminar shear tests, the specimens were examined by SEM. We found that small amounts of boron in the glass fibers significantly degraded the interlaminar shear strength in the thermal neutron environment.

INTRODUCTION

In the design of fusion reactors, radiation damage of the components must be taken into consideration. The most radiation-sensitive components in these superconducting magnets are the organic composite materials that will be used for insulation.[1,2] Consequently, the development of radiation-resistant organic composites is one of the most important aspects of research and development for fusion technology.

The degradation mechanisms of organic composites must be understood in order to develop radiation-resistant organic composites. Among their characteristics, the mechanical properties are the most important, because the electrical properties do not degrade until mechanical defects have been introduced into the materials.

We have reported that the intrinsic parameter which controls radiation-induced degradation is the interlaminar shear strength.[3,4] That was confirmed not only by theoretical calculations, but also by experiments. On the basis of this mechanism, we can estimate the degree of degradation of composite materials under three-point-bending stress conditions and the dose at which the composites start to degrade. From this information, we propose three-dimensional fiber-reinforced plastics (3D FRPs) for the insulating materials of fusion reactors; the radiation resistance of these materials has been demonstrated in previous studies.[5,6]

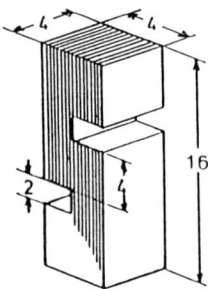

Fig. 1. Shape and size of the ILSS test specimen (dimensions are in mm).

To establish the selection standard for radiation-resistant organic composites, the effects of glass content, glass fibers, and matrix resin on radiation damage were studied only on 2D FRPs.

MECHANISM OF RADIATION-INDUCED DEGRADATION

An understanding of the mechanism of degradation is necessary for the development of radiation-resistant organic composites. In conventional FRPs, the intrinsic parameter is the interlaminar shear strength (ILSS), which is a measure of capability of the matrix resin to retain the shape of the FRP. The adhesive strength of the matrix is degraded by irradiation. Analysis of ILSS following irradiation has been reported.[3,4]

To develop radiation-resistant FRPs, we need to develop systems in which the ILSS does not degrade. As we mentioned earlier, we propose 3D FRPs. For this study, we selected materials whose ILSS was least likely to degrade during irradiation.

EXPERIMENT

Interlaminar Shear Strength

There are several methods to measure the ILSS of FRPs.[7] The shape and size of the specimens used in this study are shown in Fig. 1. The mean ILLS was obtained by compressing the specimen at a speed of 1 mm/min.

Irradiation

Low-temperature reactor irradiation was performed in the Low Temperature Loop at Kyoto University Reactor. The irradiation temperature was approximately 20 K. After irradiation, the specimens were stored in liquid nitrogen without warming to room temperature. The ILSS tests were conducted at liquid-nitrogen temperature. The absorbed dose of neutron irradiation was calculated by using a conversion factor which was determined by the neutron spectrum of the reactor.

Specimens

The specimens used in this study were basically glass-fiber-reinforced plastics (glass-cloth-reinforced laminate, 2D FRP). Five kinds of specimens were prepared that included three kinds of matrix (two kinds of epoxy and one BT-resin), two kinds of reinforcement (E- and T-glass), and two kinds of glass content (50 and 75 vol.%). The specifications of the specimens are presented in Table 1. Thus, the effects of glass fiber, resin matrix, and glass content could be examined. The glass fibers were coated with an epoxy silane finish. The epoxy resins had different curing agents.

RESULTS AND DISCUSSION

The effects of reactor irradiation on the ILSS of specimens D and E are shown in Fig. 2. The glass content of 75% in specimen E is very high compared with those of conventional glass-fiber-reinforced

Table 1. Specifications of Tested Specimens

Specimen Code	Matrix	Reinforcement	Glass Content (vol.%)
A	epoxy 1	E-glass	50
B	epoxy 1	T-glass	50
C	BT-resin	E-glass	50
D	epoxy 2	E-glass	50
E	epoxy 2	E-glass	75

plastics (GFRPs). The thermal contraction of the thickness is comparable to that of metals. Specimen D is a commercial GFRP, the same as G-10.

The ILSS degradation of specimen E was the same as that of specimen D; that means the degradation of ILSS was not affected by the glass content. Even in the highly strengthened FRP, the ILSS degradation was similar to that of a conventional FRP. In other words, the apparent amount of degradation of flexural or tensile strength in a high-strength FRP could be more than that in a conventional FRP.

The degradation of ILSS in specimens A, B, and C is demonstrated in Fig. 3. The ILSS of the composite reinforced with T-glass degraded less than that of the composite reinforced with E-glass. T-glass reinforcement increased radiation resistance more than E-glass reinforcement because the boron in the E-glass reacts with the thermal neutrons and emits alpha particles. The alpha particles reduce the adhesive strength between the epoxy and the glass fibers.

Figure 4 shows SEM photographs of the fracture surfaces of specimens A and B, which had been exposed to radiation doses of 0, 3.1, and 7.4 MGy. Before irradiation, the glass fibers of both specimens

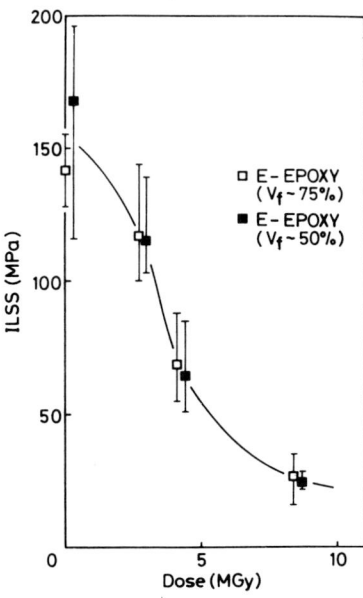

Fig. 2. Degradation of ILSS of E-glass FRPs with different glass content. □ - E-glass fiber-reinforced epoxy with 75 vol.% glass (E). ■ - E-glass fiber-reinforced epoxy with 50 vol.% glass (D).

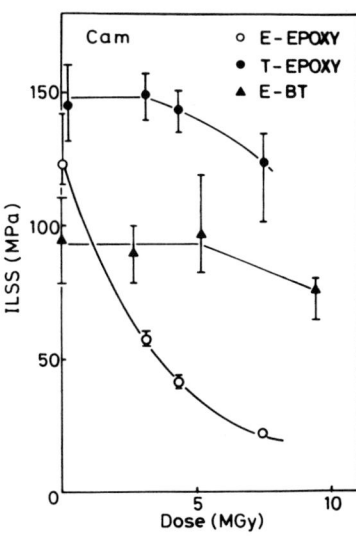

Fig. 3. Degradation of ILSS of GFRPs with different glass fiber and matrix. ○ - E-glass fiber-reinforced epoxy with 50 vol.% glass (A). ● - T-glass fiber-reinforced epoxy with 50 vol.% glass (B). ▲ - E-glass fiber-reinforced BT resin with 50 vol.% glass (C).

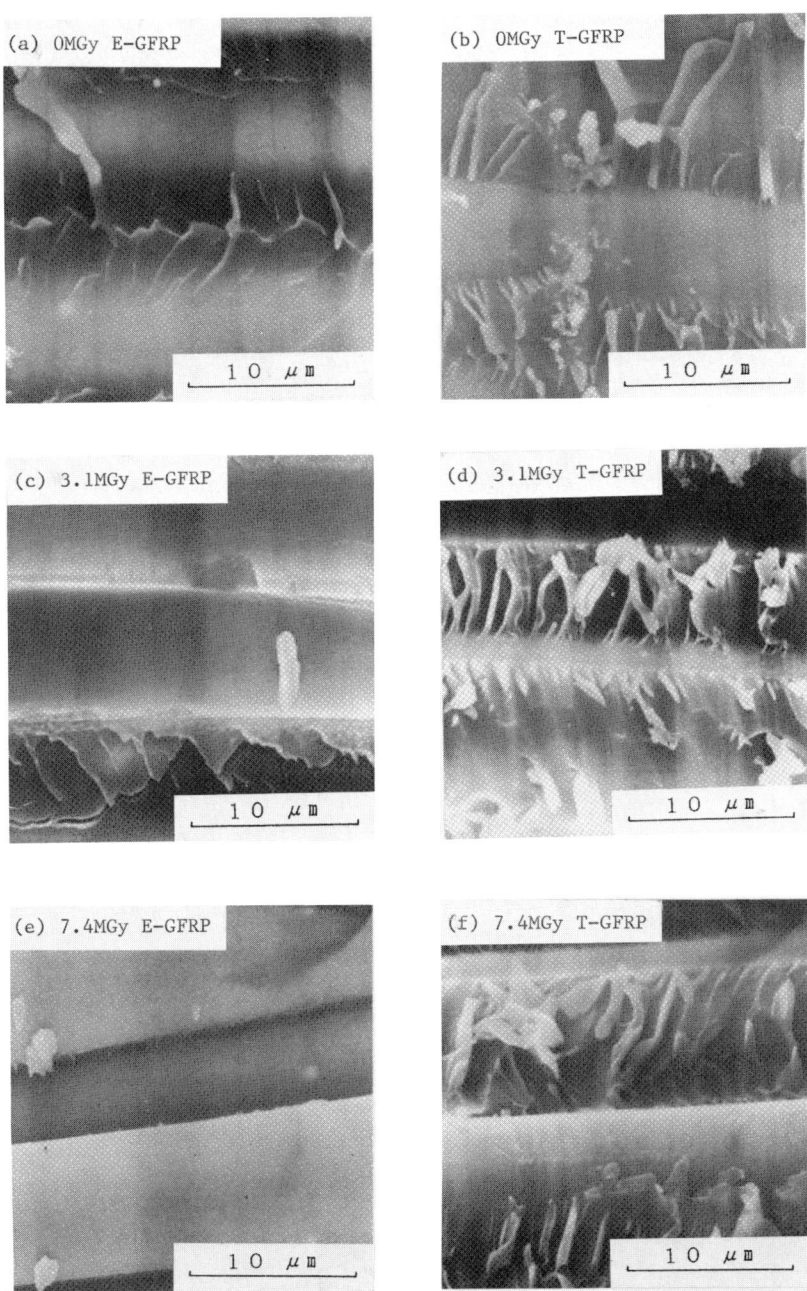

Fig. 4. SEM observation of fracture surfaces of irradiated E-glass fiber-reinforced epoxy (A, left) and T-glass fiber-reinforced epoxy (B, right).

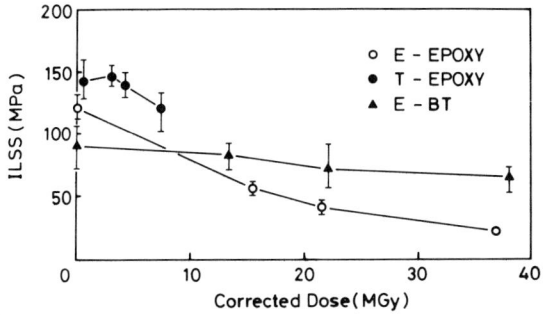

Fig. 5. ILSS versus corrected dose. ○ - E-glass fiber-reinforced epoxy with 50 vol.% glass. ● - T-glass fiber-reinforced epoxy with 50 vol.% glass. ▲ - E-glass fiber-reinforced BT resin with 50 vol.% glass.

were embedded in the epoxy matrix, rather than having the epoxy matrix adhere to the glass fibers. The cracks, therefore, run in the epoxy matrix. After irradiation to 7.4 MGy, the epoxy matrix did not adhere to the E-glass fibers at all, although it did adhere to the T-glass fibers. That is why the T-glass-reinforced plastic (B) has more radiation resistance than the E-glass reinforced plastic (A).

Although the radiation resistance was less for specimen A (reinforced with E-glass) than for specimen B (reinforced with T-glass), the radiation resistance of specimen C (reinforced with E-glass) exceeded that of specimen A (also reinforced with E-glass). The superior radiation resistance of specimen C reflects the contribution of its BT-resin matrix. Before irradiation, the ILSS of specimen C was much lower than those of epoxy-matrix composites. Therefore, BT-resin matrices should be carefully used in 2D FRPs.

The results of this study suggest that a combination of boron-free glass fibers and a BT resin produce the most radiation-resistant composite. With regard to BT-matrix composites, the 3D FRP is a much more suitable configuration because it could show a high ILSS value.

The energy transfer due to alpha particles was calculated to estimate the radiation resistance of BT T-glass composites. The corresponding nuclear reaction is

$$^{10}_{5}B + ^{1}_{0}n \rightarrow ^{7}_{3}Li + ^{4}_{2}\alpha.$$

Here, the energies of Li and α are 0.84 MeV and 1.47 MeV, respectively. The range of alpha particles in organic materials is estimated to be several micrometers. Although the calculation method has already been reported,[6] a new calculation was made that considered the effects of alpha stops in the glass fibers, the natural abundance of boron, and the neutron spectrum of the reactor. The result indicates that approximately 7 times as much energy is transferred by the alpha particles than that transferred by neutrons and gamma rays in the reactor.

Based on this calculation, the ILSS values versus corrected dose are presented in Fig. 5 for E-glass fiber-reinforced epoxy, T-glass fiber-reinforced epoxy, and E-glass fiber-reinforced BT resin. The corrected dose dependence of ILSS for the T-glass fiber-reinforced epoxy is nearly the same as that of the E-glass fiber-reinforced epoxy; these results suggest that the ILSS degradation mechanisms are identical. We can also estimate the radiation resistance of T-glass fiber-reinforced BT resin from this figure; its estimated radiation resistance is higher than that of the T-glass fiber-reinforced epoxy.

CONCLUSIONS

The radiation effects on the ILSS of several composites were examined for different glass contents, matrix resins, and glass fibers in order to establish a selection standard for the radiation-resistant composites required in fusion reactors. The following conclusions were drawn:

1. The glass content does not affect the degradation of ILSS. Although the composites with high glass content have high flexural strength, the degradation starts at a lower dose level.

2. The radiation-resistant matrix resin is suitable for the FRPs, but good adhesive strength is necessary for radiation-resistant organic composite materials.

REFERENCES

1. S. Nishijima and T. Okada, Comparative study of radiation damage and activation of superconducting magnet for fusion reactor, in: *Advances in Cryogenic Engineering Materials,* vol. 34, Plenum Press, New York (1988), p. 917.

2. W. J. Muster, J. Kubler, K. Nylund, and H. Benz, Advanced composite insulations for superconducting magnets, in: *Advances in Cryogenic Engineering Materials,* vol. 34, Plenum Press, New York (1988), p. 51.

3. T. Okada, S. Nishijima, and H. Yamaoka, Radiation damage of composite material method and evaluation, in: *Advances in Cryogenic Engineering Materials,* vol. 36, Plenum Press, New York (1990), p. 145.

4. S. Nishijima, T. Okada, K. Miyata, and H. Yamaoka, Radiation damage of composite materials at cryogenic temperatures, in: *Advances in Cryogenic Engineering Materials,* vol. 34, Plenum Press, New York (1988), p. 35.

5. J. Yasuda, T. Hirokawa, T. Uemura, S. Nishijima, T. Okada, and H. Okuyama, Cryogenic and radiation resistant properties of three dimensional fabric reinforced composite materials, *New Develop. Appl. Supercond.* 15:449 (1988).

6. S. Nishijima, T. Nishiura, T. Okada, T. Hirokawa, J. Yasuda, and Y. Iwasaki, Development of radiation resistant composite materials for fusion magnets, in: *Advances in Cryogenic Engineering Materials,* vol. 36, Plenum Press, New York (1990), p. 877.

7. H. Becker, Properties of cryogenic interlaminar shear strength testing, in: *Advances in Cryogenic Engineering Materials,* vol. 36, Plenum Press, New York (1990), p. 827.

EFFECTS OF FABRIC TYPE, SPECIMEN SIZE, AND IRRADIATION ATMOSPHERE ON THE
RADIATION RESISTANCE OF POLYMER COMPOSITES AT 77 K

S. Egusa, M. Sugimoto*, H. Nakajima*, K. Yoshida*, and
H. Tsuji*

Takasaki Radiation Chemistry Research Establishment, Japan
Atomic Energy Research Institute, Takasaki-shi, Gunma 370-12,
Japan
*Naka Fusion Research Establishment, Japan Atomic Energy Research Institute, Naka-machi, Naka-gun, Ibaraki 311-02, Japan

ABSTRACT

The radiation resistance of polymer matrix composites to the mechanical strength changes at 77 K was studied with respect to factors such as reinforcing fabric type, specimen thickness, and irradiation atmosphere. As far as the E-glass and T-glass fabric composites are concerned, the radiation resistance is almost independent of these factors, in agreement with a degradation mechanism such that the composite degradation behavior is primarily determined by a change in the matrix ultimate strain due to irradiation. The radiation-resistance evaluation was made also for the bond strength of polymer matrix composites to stainless steel at 77 K. The radiation sensitivity of the bond strength appears to depend primarily on the load transfer mode at the composite/steel interface.

INTRODUCTION

In the construction of superconducting magnets for fusion reactors, one of the important design criteria is the radiation resistance of the component materials such as superconductors, stabilizers and insulators.[1] Probably the problem of such radiation resistance is most critical for composite organic insulators, because organic materials are usually less radiation resistant than inorganic materials.

From this point of view, we have been studying the irradiation effects in polymer matrix composites since 1983, mainly with regard to the mechanical properties at low temperatures. In the course of the studies, it was found that an epoxy of tetraglycidyl diaminodiphenyl methane (TGDDM) cured with diamino diphenyl sulfone (DDS) is promising as a matrix resin of composite organic insulators to be used in fusion magnets.[2] This finding stimulated our interest in studying further the irradiation effects in the TGDDM/DDS epoxy matrix composites, with respect to factors such as reinforcing fabric type, specimen thickness, and irradiation atmosphere. This paper mainly describes the effects of these factors on the dose dependence of the composite strength tested at 77 K.

In actual fusion magnets, composite organic insulators are used to separate conductors from each other or to separate them from the magnet case.[3] As design data for fusion magnets, therefore, the radiation-resistance evaluation should be made not only for the insulator itself but also for the bond strength of the insulator to the conductor or to the magnet case. This paper describes also the dose dependence of the bond strength between polymer matrix composite and stainless steel tested at 77 K.

EXPERIMENTAL

Polymer matrix composites prepared in this work are shown in Table 1. The matrix resin was an epoxy of TGDDM/DDS or a polyimide of polyaminobismaleimide (Kerimid 601), and the reinforcing filler was E- or T-glass fabrics shown in Table 2. The E-glass fabrics of KS-1210 and KS-1600 were selected so as to differ from each other in the number of fibers in a yarn and, consequently, in the number of yarns per 25 mm in the warp and weft directions. The T-glass fabrics of WTX-116E and WTA-18W, on the other hand, were selected so as to be the counterparts of KS-1210 and KS-1600, respectively, with respect to the type of fabric weave. The composite plates thus prepared were cut into rectangular specimens of 6.4 mm width and 70 mm length, with the 70 mm axis parallel to the warp direction of the reinforcing fabrics.

Table 1. Polymer matrix composites prepared

Composite	Reinforcing fabric	Matrix resin	Volume fraction of fibers (%)	Average thickness (mm)
Glass/epoxy I	KS-1210	TGDDM/DDS	63	1.81 ± 0.07
Glass/epoxy II	KS-1600	TGDDM/DDS	66	1.98 ± 0.04
Glass/epoxy III	WTX-116E	TGDDM/DDS	63	2.12 ± 0.01
Glass/epoxy IV	WTA-18W	TGDDM/DDS	58	1.99 ± 0.02
Glass/epoxy V	KS-1210	TGDDM/DDS	67	0.97 ± 0.01
Glass/epoxy VI	KS-1210	TGDDM/DDS	65	4.98 ± 0.07
Glass/epoxy VII	KS-1210	TGDDM/DDS	55	1.93 ± 0.10
Glass/polyimide I	KS-1210	Kerimid 601	62	2.05 ± 0.06
Glass/polyimide II	WTX-116E	Kerimid 601	50	1.99 ± 0.03
Glass/polyimide III	WTA-18W	Kerimid 601	60	1.96 ± 0.01

Table 2. Reinforcing fabrics used

Reinforcing fabric[a]	Fiber type[b]	Fiber diameter (μm)	Number of fibers in a yarn	Weave style	Number of yarns per 25 mm	
					Warp	Weft
KS-1210	E-glass	7	200	Plain	53	48
KS-1600	E-glass	9	400	Plain	41	32
WTX-116E	T-glass	7	200	Plain	60	58
WTA-18W	T-glass	9	400	Plain	44	34
WPT-18D	T-glass	10	400	Satin	45	31
T-0.18S	T-glass	9	400	Satin	40	34

[a] Manufacturer: Kanebo (KS-1210, KS-1600); Nitto Boseki (WTX-116E, WTA-18W, WPT-18D); Arisawa Mfg (T-0.18S).
[b] E-glass composition (wt%): SiO_2(55.2), Al_2O_3(14.8), CaO(18.7), MgO(3.3), B_2O_3(7.3), Na_2O+K_2O(0.5), Fe_2O_3(0.3), F_2(0.3), TiO_2(0.1).
T-glass composition (wt%): SiO_2(65), Al_2O_3(23), CaO(<0.01), MgO(11), B_2O_3(<0.01), Na_2O+K_2O(<0.1), Fe_2O_3(<0.1), Zr_2O_3(<1.0).

Table 3. Prepreg plies used in preparing lap shear specimens

Manufacturer	Reinforcing fabric	Matrix resin	Volume fraction of fibers (%)	Number of prepreg plies used
Arisawa Mfg	T-0.18S	BT-A300[a]	53	3
Hitachi Chemical	WPT-18D	BT-2160[a]	63	3
Nitto Boseki	WTA-18W	TGDDM/DDS	60	5

[a] BT resins manufactured by Mitsubishi Gas Chemical for metal-to-metal adhesive (BT-A300) and tacky prepreg (BT-2160) applications.

Lap shear specimens were prepared by three different manufacturers using the prepreg plies shown in Table 3. In these prepreg plies, the reinforcing filler was T-glass fabrics shown in Table 2 and the matrix resin was bismaleimide-triazine (BT) resin or TGDDM/DDS epoxy resin. These prepreg plies were sandwiched between two semicircular rods of stainless steel and were heated under pressure, thus obtaining a lap shear specimen. The width and length of the lap (bond) area were 13 x 30 mm^2 and the thickness of the lap layer was 0.41-0.57 mm. The shape and dimensions of the stainless steel rods were exactly the same as those used by Poehlchen et al.[4]

^{60}Co γ-ray irradiations were carried out in air or in argon at room temperature. The in-air irradiation was performed with dose rates of 12-17 kGy/hr for composite specimens and 9.6 kGy/hr for lap shear specimens. The in-argon irradiation, on the other hand, was performed with a dose rate of about 13 kGy/hr at about 1.4 atm of argon in a specially constructed container of composite specimens.

Three-point bend tests were conducted at 77 K for composite specimens. The tests were made at a crosshead speed of 0.6 mm/min with a span length of 20 mm. The ultimate flexural strength was calculated from $3P_f(\ell/h)/2bh$, where P_f is the applied load at failure, ℓ is the span length, b is the specimen width, and h is the specimen depth (thickness). For lap shear specimens, on the other hand, compression tests were conducted at 77 K at a crosshead speed of 0.6 mm/min. The bond strength between polymer matrix composite and stainless steel was calculated from P_f/Lw, where L is the lap length and w is the lap width. The three-point bend tests and the compression tests were usually repeated three and four times, respectively, thus obtaining an average value and the standard deviation for each data point.

RESULTS AND DISCUSSION

Effect of Reinforcing Fabric Type

Figure 1 shows plots of the ultimate flexural strength at 77 K versus the absorbed dose due to in-air γ-irradiation for the glass/epoxy and glass/polyimide composites having different reinforcing fabrics shown in Table 2. The plots for the glass/epoxy I-IV composites in Figure 1a show that the initial strength of the KS-1210 or WTX-116E fabric composite is 28-40% higher than that of the KS-1600 or WTA-18W fabric composite, thus indicating that the initial strength is dependent on the fabric weave parameters such as the number of fibers in a yarn and the number of yarns in the warp and weft directions. The plots also show that the initial strength is less dependent on whether the reinforcing fiber is E-glass or T-glass. Following irradiation the strengths of these composites decrease monotonically with increasing absorbed dose. Roughly speaking, the dose dependence appears to follow a rather similar pattern for all of these composites, thus suggesting that the

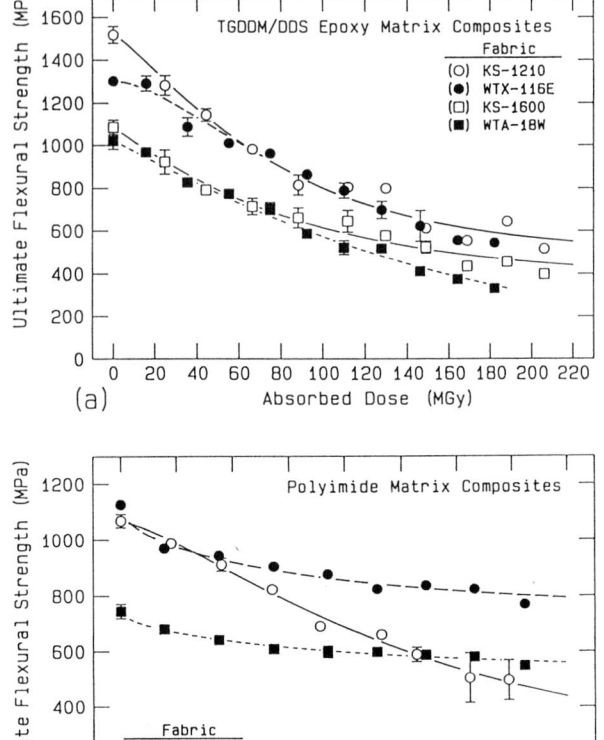

Fig. 1. Plot of the ultimate flexural strength at 77 K versus the absorbed dose due to in-air γ-irradiation for the glass/epoxy I - IV (a) and glass/polyimide I - III composites (b) having different reinforcing fabrics shown in Table 2.

degradation of the composite flexural strength depends neither on the type of fabric weave nor on the kind of glass fibers. This result is consistent with a degradation mechanism proposed by Egusa such that the dose dependence of the composite flexural strength is primarily determined by a change in the matrix ultimate strain due to irradiation.[5]

For the glass/polyimide I - III composites, comparison of the WTX-116E and WTA-18W fabric composites in Figure 1b shows that the initial strength is dependent on the type of fabric weave. Comparison of the KS-1210 and WTX-116E fabric composites, on the other hand, shows that the initial strength is almost independent of the kind of glass fibers. These results are in accord with the finding obtained for the initial strengths of the glass/epoxy composites in Figure 1a. Following irradiation, however, the strength of the KS-1210 fabric composite in Figure 1b is seen to decrease appreciably more rapidly with increasing absorbed dose compared to that of the WTX-116E or WTA-18W fabric composite. This result is in conflict with the finding obtained for the degradation behavior of the glass/epoxy composites in Figure 1a. This unexpected result for the glass/polyimide composites may be attributed, at least partly, to the fact that the KS-1210 fabric composite was prepared by a manufacturer different from that of the WTX-116E and WTA-18W fabric composites. At present, however, no clear explanation can be offered for this problem.

Whatever the reasons may be, the fact that the T-glass fiber composites are comparable or even superior to the E-glass fiber composites in the radiation resistance is of great importance from the standpoint of their applications to fusion magnets. This is because composite organic insulators in actual fusion magnets are subjected to neutrons and γ-rays simultaneously, with more than half of the total absorbed dose resulting from neutrons. It is now generally recognized that the neutron irradiation of boron-containing E-glass fiber composites produces additional radiation damage due to a $^{10}B(n,\alpha)^7Li$ reaction in E-glass fibers, thus significantly decreasing the radiation resistance of the composites towards neutrons.[6,7] For boron-free T-glass fiber composites (see Table 2), on the other hand, the extent of the radiation damage due the ^{10}B reaction will be negligible, thus leading to a higher radiation resistance towards neutrons compared to the E-glass fiber composites. These considerations lead to a conclusion that the T-glass fiber composites are recommended over the E-glass fiber composites as component materials to be used in fusion magnets.

Effects of Specimen Thickness and Volume Fraction of Fibers

Figure 2 shows plots of the ultimate flexural strength at 77 K versus the absorbed dose due to in-air γ-irradiation for the glass/epoxy V, I, and VI composites having the average thicknesses of 0.97, 1.81, and 4.98 mm, respectively. The reinforcing filler is the E-glass fabric of KS-1210 and the volume fraction of fibers is almost identical for all of these composites. The composite strength before irradiation is seen to decrease by about 26% with an increase in the specimen thickness from 0.97 to 4.98 mm. Following irradiation the strengths of these composites decrease monotonically with increasing absorbed dose. The dose dependence appears to follow an identical pattern regardless of the specimen thickness.

Figure 3 shows plots of the ultimate flexural strength at 77 K versus the absorbed dose due to in-air γ-irradiation for the glass/epoxy I and VII composites having a volume fraction of fibers, V_f, of 63 and 55%, respectively. The reinforcing filler is the E-glass fabric of KS-1210 and the specimen thickness is almost identical for the two composites. It is seen that although the composite strength before irradiation is about 12% lower for V_f = 55% than for V_f = 67%, the dose dependence follows a rather similar pattern regardless of the V_f.

This result and the composite degradation behavior independent of the specimen thickness (Fig. 2) are consistent with the above-mentioned mechanism such that the dose dependence of the composite flexural strength is primarily determined by a change in the matrix ultimate strain due to irradiation. It seems reasonable to conclude, therefore, that the specimen thickness and the volume fraction of fibers have essentially no influence on the composite degradation behavior.

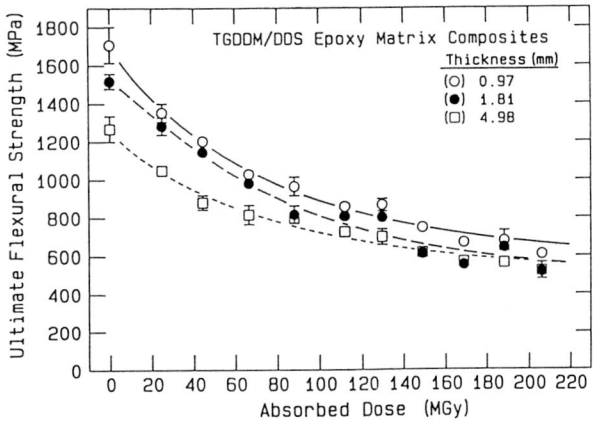

Fig. 2. Plot of the ultimate flexural strength at 77 K versus the absorbed dose due to in-air γ-irradiation for the glass/epoxy V, I, and VI composites having the average thicknesses of 0.97, 1.81, and 4.98 mm, respectively.

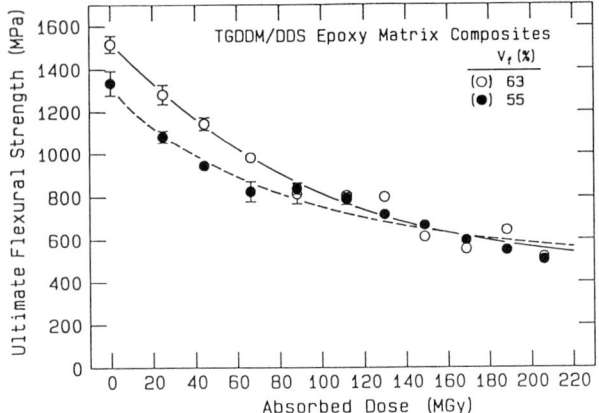

Fig. 3. Plot of the ultimate flexural strength at 77 K versus the absorbed dose due to in-air γ-irradiation for the glass/epoxy I and VII composites having a volume fraction of fibers of 63 and 55%, respectively.

Effect of Irradiation Atmosphere

Composite organic insulators used in fusion superconducting magnets are subjected to radiation in vacuo. The results obtained from irradiation in air, therefore, can not be used as design data for fusion magnets if the composite degradation behavior depends strikingly on the presence of oxygen during irradiation.

Figure 4 shows plots of the ultimate flexural strength at 77 K versus the absorbed dose due to in-argon and in-air γ-irradiations for the G-10CR, G-11CR, and TIL-G1000 composites. These are commercially available E-glass fiber composites having the matrix resin of epoxy (G-10CR, G-11CR) or polyimide (TIL-G1000). Comparison of the argon and air data points for each composite shows that the degradation behavior follows an identical pattern regardless of the irradiation atmosphere for all of these composites. It is reasonable to conclude, therefore, that the presence or absence of oxygen during irradiation has essentially no influence on the composite degradation behavior.

It should be pointed out, however, that irradiation in air versus argon must produce some differences in the radiation damage at the specimen surface by the presence and absence of oxidative degradation.[8] The fact that the composite degradation behavior is still independent of the irradiation atmosphere (Fig. 4) strongly suggests that the composite strength is fairly insensitive to small flaws which will be formed at the specimen surface by oxida-

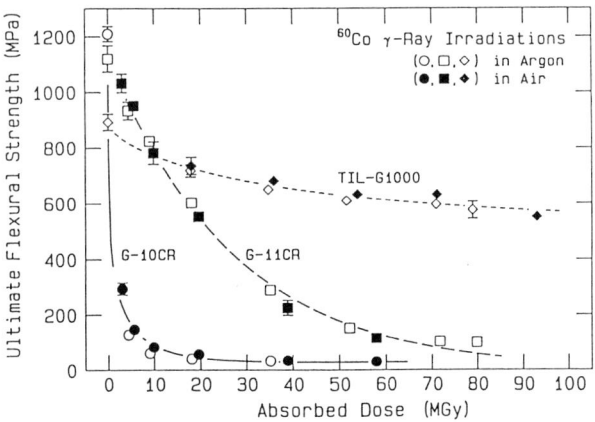

Fig. 4. Plot of the ultimate flexural strength at 77 K versus the absorbed dose due to in-argon and in-air γ-irradiations for the G-10CR, G-11CR, and TIL-G1000 composites.

tive degradation during irradiation in air. Such insensitivity is most likely ascribed to the presence of reinforcing fibers in the composite. The presence of fibers is, in fact, known to interfere with the propagation of matrix cracking in the composite.[9]

Bond Strength between Polymer Matrix Composite and Stainless Steel

Figure 5 shows plots of the bond strength between polymer matrix composite and stainless steel at 77 K versus the absorbed dose due to in-air γ-irradiation for three kinds of lap shear specimens shown in Table 3. It is seen that the initial bond strength of the Arisawa specimens is about 50% higher than that of the Hitachi or Nitto specimens. Following irradiation the bond strength of the Arisawa specimens decreases monotonically with increasing absorbed dose. For the Hitachi and Nitto specimens, on the other hand, such a decrease in the bond strength is quite small or practically nil even at 50 MGy.

Such a difference in the degradation behavior of lap shear specimens may be ascribed to differences in the mode of load transfer at the interface between polymer matrix composite and stainless steel. In general, such an interface is known to have at least two modes of load transfer, i.e., the chemical bond mode and the friction force (mechanical bond) mode. Possibly the bond strength at the composite/steel interface is due for the most part to the friction force mode, and this mode is much less sensitive to radiation compared to the chemical bond mode. This idea appears to explain why the bond strength is hardly changed by irradiation for the Hitachi and Nitto specimens. For the Arisawa specimens, on the other hand, the chemical bond mode may be added to the friction force mode, thus increasing the initial

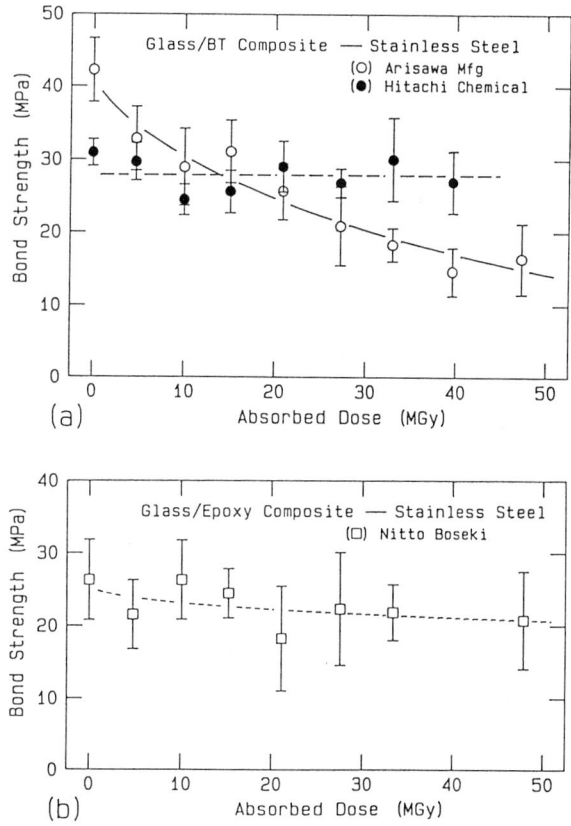

Fig. 5. Plot of the bond strength between polymer matrix composite and stainless steel at 77 K versus the absorbed dose due to in-air γ-irradiation for lap shear specimens shown in Table 3.

bond strength and, at the same time, making the bond strength more sensitive to radiation.

In this connection, it is interesting to note that the matrix resin used in the Arisawa specimens is the BT-A300 resin manufactured by Mitsubishi Gas Chemical for metal-to-metal adhesive applications (see Table 3). Then it is quite possible that the contribution of the chemical bond mode to the total bond strength at the composite/steel interface is higher for the Arisawa specimens compared to the Hitachi and Nitto specimens. At present, however, no evidence supporting this speculation is available. Further studies are required for a detailed discussion of this point.

CONCLUSIONS

This work has shown that as far as the E-glass and T-glass fabric composites are concerned, the radiation resistance to the mechanical strength changes at 77 K is almost independent of the composite specifications such as the type of fabric weave, the kind of glass fibers, the specimen thickness, and the volume fraction of fibers. This result is in agreement with a degradation mechanism such that the dose dependence of the composite flexural strength is primarily determined by a change in the matrix ultimate strain due to irradiation. It was also shown that the radiation resistance is little affected by the presence and absence of oxygen during irradiation.

As to the bond strength between polymer matrix composite and stainless steel at 77 K, this work has suggested that the radiation sensitivity of the bond strength depends primarily on the load transfer mode at the composite/steel interface. Further experiments directed toward this problem are under way at our laboratories. The study will provide a more comprehensive understanding of the radiation resistance of the composite/steel interface.

REFERENCES

1. G. L. Kulcinski, J. M. Dupouy, and S. Ishino, Key Materials Issues for Near Term Fusion Reactors, J. Nucl. Mater. 141&143:3 (1986).
2. S. Egusa, Irradiation Effects on and Degradation Mechanism of the Mechanical Properties of Polymer Matrix Composites at Low Temperatures, Adv. Cryogenic Eng. Mater. 36:861 (1990).
3. H. Nakajima, et al., Tensile Properties of New Cryogenic Steels as Conduit Materials of Forced Flow Superconductors at 4 K, Adv. Cryogenic Eng. Mater. 34:173 (1988).
4. R. Poehlchen, et al., The Mechanical Strength of Irradiated Electric Insulation of Superconducting Magnets, Adv. Cryogenic Eng. Mater. 36:893 (1990).
5. S. Egusa, Mechanism of Radiation-Induced Degradation in Mechanical Properties of Polymer Matrix Composites, J. Mater. Sci. 23:2753 (1988).
6. S. Egusa, M. A. Kirk, and R. C. Birtcher, Effects of Neutron Irradiation on Polymer Matrix Composites at 5 K and at Room Temperature. I. Absorbed-Dose Calculation, J. Nucl. Mater. 148:43 (1987).
7. S. Egusa, M. A. Kirk, and R. C. Birtcher, Effects of Neutron Irradiation on Polymer Matrix Composites at 5 K and at Room Temperature. II. Degradation of Mechanical Properties, J. Nucl. Mater. 148:53 (1987).
8. A. Chapiro, "Radiation Chemistry of Polymeric Systems," Interscience, New York (1962).
9. L. J. Broutman, Fiber-Reinforced Plastics, in: "Modern Composite Materials," L. J. Broutman and R. H. Krock, eds., Addison-Wesley, Reading, Massachusetts (1967).

A RADIATION-RESISTANT EPOXY RESIN SYSTEM FOR TOROIDAL FIELD AND OTHER SUPERCONDUCTING COIL FABRICATION

N. A. Munshi

Composite Technology Development, Inc.
Boulder, Colorado, U.S.A.

ABSTRACT

A liquid-epoxy resin system with high resistance to 4-K reactor radiation (neutron plus gamma) has been developed that meets all ITER and BPX requirements: pot life longer than 12 h at ambient temperatures; viscosity less than 1000 cP over its entire pot life; cure below 150°C; low shrinkage; no volatiles upon curing; excellent penetration; and good wetting of fibers, stainless steels, and nickel-based steels.

The strength of the epoxy resin system, CTD-101, was tested with and without S-2 Glass®* fiber reinforcement at 295, 76 and 4 K. Specimens were then irradiated at two dose levels (2.9×10^7 and 1.6×10^8 Gy) at temperatures below 6 K.

Virtually no deterioration in interlaminar shear strength at 76 K was observed after the lower dose level, which is the end-of-life dose level for ITER. After the higher dose level, the material had degraded, but its strength was still well above the BPX requirement (14 MPa in shear) at the end-of-life dose level of 1×10^8 Gy.

INTRODUCTION

The insulator requirements for superconducting magnet coils in ITER (International Thermonuclear Experimental Reactor) and BPX (Burning Plasma Experiment) are unprecedented. Radiation dose levels for insulators exceed previous design levels.

In conventional and superconducting magnets for fusion reactors, a major limitation for the use of organic materials as electrical insulation is their relatively low radiation resistance. The end-of-life dose level for ITER is 2×10^7 Gy; for BPX, it is 1×10^8 Gy. Composite insulators that are currently being used severely degrade or even disintegrate at comparable dose levels and temperatures.[1] Research is difficult because there is only one reactor in the world, the Munich Research Reactor in Garching, Germany, that has the capability for irradiation at 4 K.

---

*S-2 Glass® is a registered trademark of Owens-Corning Fiberglas Corporation.

Other factors complicating the development of new insulators are the high compression and shear stresses associated with ITER and BPX: 30 MPa shear and 450 MPa compression for ITER and 14 MPa shear with zero face compression for BPX. Low-viscosity, radiation-resistant resins with optimal properties at cryogenic temperatures are needed that have very long pot lives at room temperature, permitting efficient impregnation of the very large ITER Toroidal Field (TF) coils.

Composite Technology Development (CTD) has embarked on an aggressive program, partly funded by the U.S. Department of Energy,[2,3] to develop insulation for fusion magnets.

EXPERIMENT

The composite specimens had been made by the CompTec®† process, which was developed by CTD for rapid specimen production and testing; it is described elsewhere.[4] The specimens had been irradiated below 330 K in Vienna and tested at 76 K.[5] Materials that retained some strength after this irradiation were selected for the cryogenic (below 6 K) irradiation program at Garching.

Table 1. Properties of the CTD-101 Resin System

MIXED RESIN PROPERTIES	
Viscosity	400 cP at 40°C
Pot life	60 h at 40°C
COMPOSITE PROPERTIES (55 vol.% S-2 Glass® reinforcement)	
Torsion shear strength	201 ± 10 MPa
Torsion shear strain	0.039 ± 7%
Torsion shear modulus	9.1 ± 1.0 GPa
Compression strength	1255 MPa (L/D = 2.0)
Fracture strength, G_{Ic}	0.308 kJ·m^{-2}

The low-temperature irradiation facility at the 4-MW swimming-pool-type Munich Research Reactor offers the possibility of performing irradiation at temperatures below 6 K using liquid helium as coolant. The irradiation chamber is 16 mm in diameter, and the fast neutron flux (E > 0.1 MeV) is 3×10^{17} m^{-2}s^{-1}. The facility is described in detail by Gerstenberg and Gläser.[6]

At temperatures below 6 K, four specimens of five materials (CTD-101, CTD-102, CTD-112, CTD-200, and CTD-310) were irradiated at two fluence levels: 5.8×10^{21} and 3.3×10^{22} m^{-2}. This gives an approximate total absorbed dose (neutron plus gamma) of 2.9×10^7 and 1.6×10^8 Gy, respectively. The specimens were warmed to room temperature and then tested in torsional shear at 76 K. Of the five materials, CTD-101 and CTD-102 are candidates for vacuum impregnating large TF magnets because they have low viscosities and very long pot lives; however, CTD-101 is preferred because its strength was higher after irradiation. Its properties and physical characteristics are described in Table 1.

---

†CompTec® is a registered trademark of Composite Technology Development.

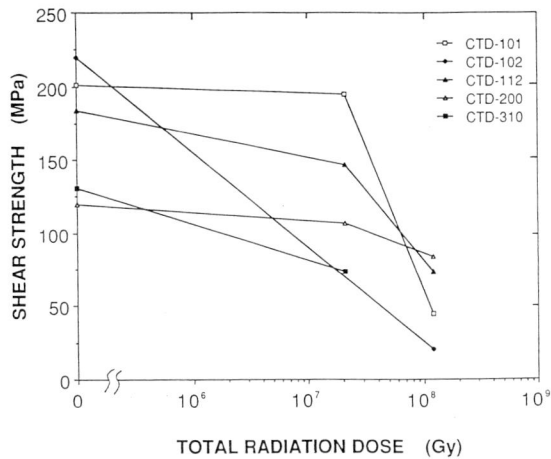

Fig. 1. Torsion shear strength as a function of dose.
Irradiation temperature: <6 K; test temperature: 76 K.

RESULTS AND DISCUSSION

The shear strength, shear strain, and shear modulus as a function of radiation dose level are plotted in Figs. 1, 2, and 3. CTD-101, which retained nearly 65% of its strength at a dose level of 2×10^7 Gy after 330-K irradiation, lost virtually no strength at the same dose level after 4-K irradiation (Fig. 4). This dose level is the expected end-of-life dose for ITER.[7] At the higher dose level of 1.6×10^8 Gy, strength dropped rapidly, and the material sustained greater damage than it did after 330-K irradiation. However, the strength retained (~50 MPa) is still sufficient for BPX.

For all the materials irradiated, this trend was observed: at the lower dose level, their strength was higher after 4-K irradiation than after 330-K irradiation; at the higher dose level, the reverse was true.

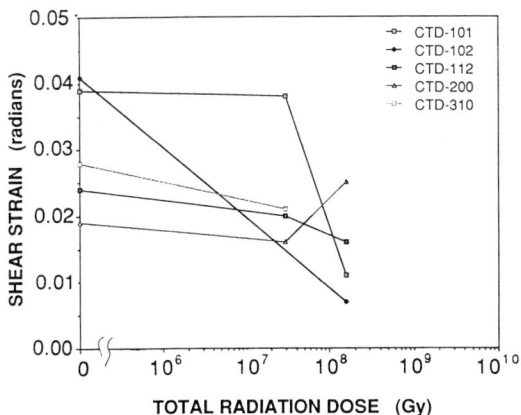

Fig. 2. Torsion shear strain as a function of dose.
Irradiation temperature: <6 K; test temperature: 76 K

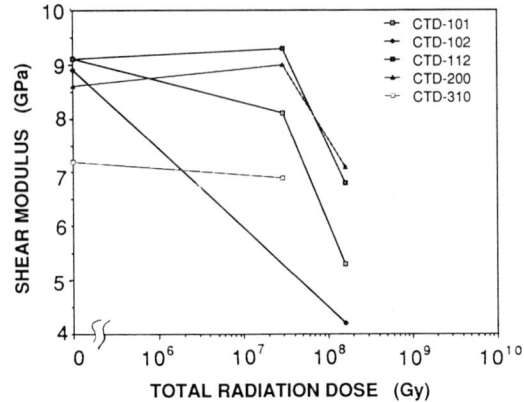

Fig. 3. Torsion shear modulus as a function of dose.
Irradiation temperature: <6 K; test temperature: 76 K

CONCLUSIONS

A low-viscosity epoxy resin system, CTD-101, has been developed by Composite Technology Development to meet all ITER and BPX requirements for a potting resin system. The material has high resistance to 330- and 4-K reactor radiation, a pot life much greater than 12 h at ambient temperatures, a viscosity less than 1000 cP, and a cure temperature of less than 150°C.

This material is being further evaluated by Oak Ridge National Laboratory (ORNL) and Princeton Plasma Physics Laboratory (PPPL) for use in the BPX experiment and by the National Institute of Standards and Technology (NIST) for ITER. ORNL and PPPL are studying the compression and shear properties of beryllium-copper chips bonded with CTD-101; NIST is studying the compression and shear properties of stainless steel chips bonded with this material.

ACKNOWLEDGMENTS

This work was supported in part by the U.S. Department of Energy, contract number DE-AC02-87ER80487. The author is indebted to Dr. H. Gerstenberg and other staff members at the Munich Research Reactor for enabling the 4-K irradiation tests.

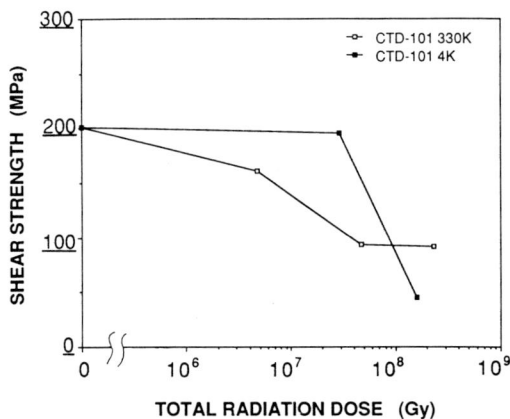

Fig. 4. Shear strength as a function of dose and irradiation temperature for the CTD-101 epoxy resin system. Specimens were tested at 76 K.

REFERENCES

1. *Insulators for Fusion Applications*, IAEA-TECDOC-417. International Atomic Energy Agency, Vienna (1987).
2. M. B. Kasen and N. A. Munshi, Cost-Effective Techniques for Development of Radiation-Resistant Organic Insulators for Superconducting Magnets, SBIR Contract No. DE-AC02-87ER80487 (1987).
3. N. A. Munshi, Effect of Polymer Additives and Residual Elements on the Cryogenic Performance and Radiation Resistance of Insulators for High-Field Magnets, SBIR Contract No. DE-FG02-90ER81073 (1990).
4. M. B. Kasen and N. A. Munshi, Cost-Effective Techniques for Development of Radiation-Resistant Organic Insulators for Superconducting Magnets, SBIR Contract No. DE-AC02-87ER80487, Final Report (1991).
5. N. A. Munshi and H. W. Weber, Reactor neutron and gamma irradiation of various composite materials, *Advances in Cryogenic Engineering - Materials, vol. 38*, R. P. Reed and F. R. Fickett, eds., Plenum, New York, submitted for publication.
6. H. Gerstenberg and W. Gläser, Neutron irradiations at temperatures below 6 K at the Munich Research Reactor (FRM), Fakultät für Physik E21, Technische Universität München, Garching, FRG (1990).
7. L. T. Summers (Lawrence Livermore National Laboratory), ITER design allowables, paper presented at Radiation Tolerant Magnet Insulation Workshop, Boulder, Colorado, June 1990.

EFFECTS OF RADIATION ON INSULATION MATERIALS

R. Pöhlchen

JET Joint Undertaking
Abingdon, Oxfordshire
England

INTRODUCTION

This presentation will concentrate on the insulation materials which are suitable for the insulation of superconducting magnets for fusion. For the next generation of fusion machines with magnetic confinement as NET and ITER general agreement exists that the insulation will consist of fibre reinforced organic matrix material, a composite.

Much effort has been put into the investigation of the radiation resistance of such materials during the last. 20-30 years, see in particular the numerous reports of accelerator laboratories on this subject. But very few of the published data are relevant for the superconducting magnets of fusion machines. Either the irradiation and testing was carried out at RT or LN_2 temperature and/or the irradiation spectrum was not representative for a fusion machine and/or the materials investigated are not applicable for the insulation of S.C. fusion magnets. Therefore test programs have been launched recently, one by the NET team.

The intention of the first chapter is to give guidance on the choice of materials which are suitable as insulation materials from a more general point of view. A good understanding of the coil manufacturing process is needed for this purpose. The second chapter explains the irradiation spectrum seen by the magnets. A third chapter does present the NET/ITER test programme. Step 1 was completed at the end of 1989, the second step will be carried out in the autumn of 1991. Finally, a general assessment of materials and testing methods will be given with recommendations for further testing.

MANUFACTURING PROCESS FOR LARGE FUSION MAGNETS

The required high mechanical strength and electrical strength of the insulation of resistive fusion magnets of present day fusion machines as JET, TFTR and ASDEX UPGRADE have been achieved via the application of a vacuum pressure impregnation (VPI) process. Dry glass tape – interleaved with Kapton when an exceptional high electric strength is required – is the basis of the turn to turn insulation and the ground insulation. These layers of tape are then impregnated with an epoxy resin system under vacuum. The winding pack and with it the layers of tape are kept under compression throughout the curing process. The result is a monolithic winding pack with a minimum content of air in the insulation and a good bond between the conductor and the insulation. Hundreds of magnets have been manufactured this way, in other words the process is well established in industry.

There is general agreement today that all future superconducting magnets of fusion machines will be cooled by a forced flow of helium through the individual conductors. Therefore, precisely the same VPI process can be employed to S.C. fusion magnets as it has been described above for resistive magnets. Already the three LCT coils which made use of a forced flow of helium instead of a helium bath cooling have been insulated by using a VPI process. Some data about these coils do illustrate the problems involved. For the European LCT coil manufactured by the Siemens company,[1] about 600 Kg of epoxy resin were required to impregnate the winding pack with a total weight of 18,000 Kg. The impregnation process took about 24 hours, it was 18 hours in the case of the Swiss LCT coil made by ABB. It is clear from these data that suitable resin systems have to fulfil first of all the following requirements:

- a vacuum process must be applicable

- low reaction shrinkage in solid phase

- viscosity has to stay low during impregnation process

- low thermal expansion coefficient

The latter is important because the maximum curing temperature may be as high as 160° C while the operating temperature is about 4K. The first condition above does exclude polyimide resins. It has to be stressed here that none of the conditions above with the exception of the first one is of any importance for the manufacture of test specimen. Many resin systems may look promising when just applied to small pieces. Therefore the only practicable way for starting a radiation resistance testing programme is to test those resin systems which have been developed by the major coil manufacturers for the impregnation of large S.C. coils. It means one has to ask these manufacturers for the manufacture of the test specimen.

IRRADIATION SPECTRUM/DEGRADATION MECHANISM

Although, the toroidal magnet is closer to the plasma than the poloidal coils the latter may be subjected as well to high irradiation levels by neutrons and γ-rays, in the vicinity of openings in the shield. In general the irradiation level depends, of course, on the shield thickness and the shield material. It is expected that the energy deposited in the coil insulation will come up to 50-80% from neutron irradiation and the balance from γ-rays. In an experimental fusion machine as JET the neutron energy spectrum during the D-T phase is expected to be in the range of 7×10^{-8} to 14.2 MeV at the vacuum vessel[2].

It is well understood in the meantime that the mechanism of energy absorption by polymer matrix composites is quite different for γ-rays, thermal neutrons and fast neutrons,[3,4]. Most of the mechanical degradation of composites by <u>fast neutron</u> irradiation is caused by the production of fast recoil protons via the collision with the hydrogen atoms of the molecules. Conversion factors for glass/epoxy and glass/polyimide composites for converting the neutron fluence into neutron absorbed dose for three different neutron energy spectra have been worked out by Egusa et al.[5]. Note that conversion factors for composites are quite different from those of pure matrix materials. The dominating effect of <u>γ-irradiation</u> is the production of free electrons (via Compton effect mainly). As many irradiation tests have been carried out on composites which were reinforced with boron containing E-glass it has to be stressed here that the deposited energy may be multiplied by Alpha and lithium particles from the nuclear reaction ^{10}B $(n,\alpha,)^{7}Li$ if the neutron energy spectrum does contain a large flux of <u>thermal neutrons</u>. There is a factor 20 for example for the H2 thimble of the Intense Pulsed Neutrons Source (IPNS)[6].

THE NET/ITER TEST PROGRAMME

Introduction

The weakest link of a monolithic winding pack is usually the shear strength of either the bond conductor jacket to the polymer matrix material or the interlaminar shear strength (ILS) of the composite. This holds as well for the windings of resistive coils where the bond of the copper turns to the interturn insulation deserves much attention when selecting the resin system and preparing the metal surfaces for impregnation. Due to the total absence of data in this field for S.C. magnets the NET team started a test program in 1989.

The selection of materials to be tested and the test specimen type used was done by following the advice of an experienced manufacturer of magnets for fusion and accelerators. "Lap Shear Specimen",[7,9] and "Symmetrical Lap Shear Specimen",[8] have been used exclusively by all European coil makers in order to establish proof of the shear strength of magnet insulation systems. Specimen size and shape had to be altered somewhat in order to cope with the space limitations imposed by the low temperature irradiation facility at the reactor of the Munich University (FRM),[10,9].

Test Specimen

The test specimen is shown in Figure 1. The insulation layer is practically a model of the turn to turn insulation

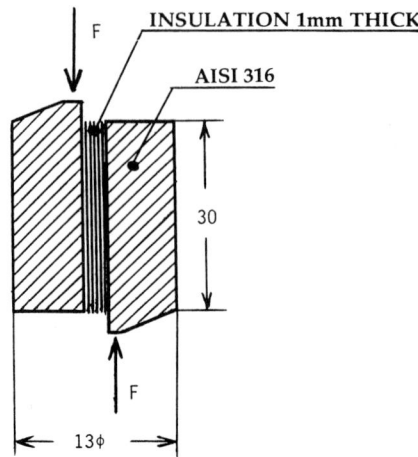

Figure 1. Lap shear test specimen

bonded to the steel surfaces of the conductor jacket. It is of crucial importance for the manufacture of the specimen that

- the impregnation takes place under vacuum
- the glass tape layers are kept under compression during curing
- there is no cutting of glass cloth after curing

It is, of course, well known that for this type of specimen the shear stress is not uniform but has peaks at the lower and upper ends. An elasto plastic analysis has shown,[11] that the deviation from average stress values is about ± 15%. Proof has been established by testing of non-irradiated samples that the results are in good agreement with earlier test results in spite of the reduction of the bonded area.

Step I of the NET/ITER Test Programme

A total of 40 specimens was irradiated with an absorbed total dose of 10^9 rad., about 50% coming from neutrons and the balance contributed by γ-rays. The irradiation was carried out at the FRM Garching at liquid helium temperature. All specimens had to be warmed up for transport to the CEN Grenoble for the mechanical testing. All R-glass tapes and ceramic tapes were subjected to a desizing heat treatment at 390° C for 5 hours, followed by a simulation of a Nb3Sn reaction heat treatment at 700° C for 50 hours.

The epoxy resin used for 35 of the specimens is an EP 311 from ABB Zürich. That is essentially ARALDITE F (CY 205, CIBA GEIGY based on Bisphenol A and cured with a Methyl Nadic Anhydride hardner, MNA or HY 906/CIBA GEIGY. One group of 5 specimens was impregnated with an OH 68 epoxy resin (ABB) which according to its structure should be more radiation resistant - with a somewhat disappointing result. For further details and the results see the following Figure 2 and the discussion in the last chapter.

Step II of the NET/ITER Test Programme

A test cryostat and a 5 ton testing machine are at present under manufacture and will be installed on the reactor bridge of the FRM Garching in summer 1991, for details see the following Figure 3 and reference 10.

Materials to be tested. All specimens will be made with EP 311 epoxy (Orlitherm type of ABB) with flexibilizer and accelerator.

Reinforcement materials to be tested:

I   R-glass thermally desized.

II   R-glass thermally desized and subject to reaction heat treatment.

III   R-glass interleaved with Kapton foil, glass fibre thermally desized.

IV   R-glass, thermally desized and subject to reaction heat treatment and both steel surfaces coated with metal oxide coating.

Group IV is expected to turn out to be the optimum solution for a radiation resistant turn to turn insulation by minimizing the amount of organic material to a very thin layer. Proof has been established earlier on, that the bond of the ceramic layers to conductor jacket and the cohesion of the ceramic is superior to the ILS of the glass/epoxy laminate,[7].

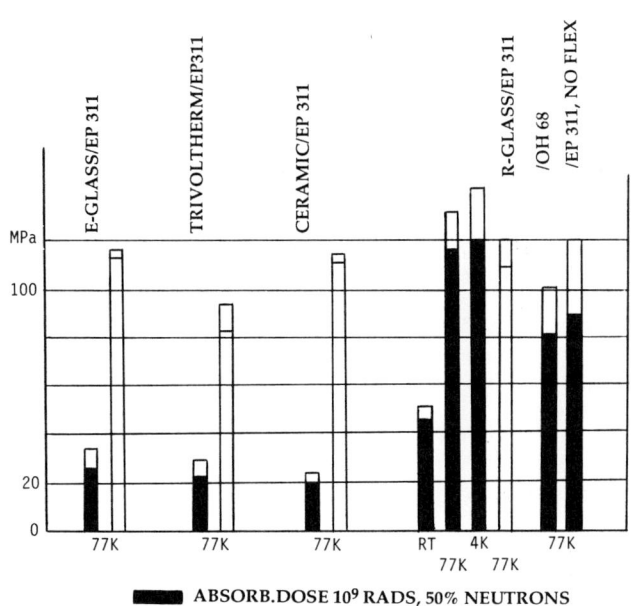

Figure 2. Shear strength of irradiated magnet insulation. Irradiation temperature 10K.

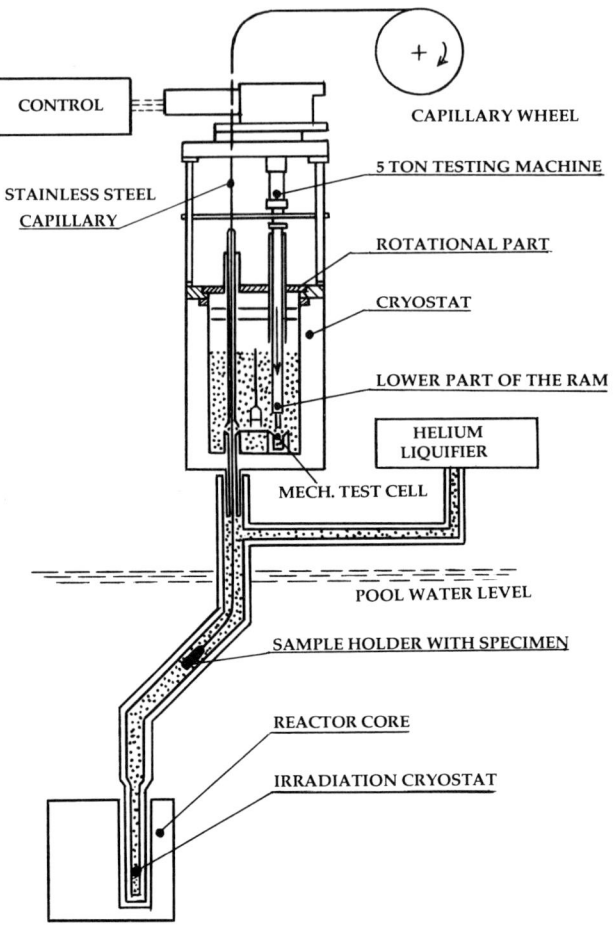

Figure 3. Schematic drawing of irradiation and shear test facility.

The Absorbed Irradiation Dose will be 5×10^9 rad, about 50% neutrons, 50% γ-rays.

<u>Irradiation and testing will be</u> carried out <u>at LHe temperature</u>, half of the specimen will see a warming up to RT in between irradiation and mechanical testing.

CONCLUSIONS

A critical assessment of materials and test methods.

Assessment of Materials

<u>Reinforcement materials</u>. Boron free glass fibres – this is R-glass in Europe, S-glass in the U.S., T-glass in Japan – appear to be the best choice for radiation resistant composites,[9,12,13]. If there is thermal neutron irradiation the boron containing E-glass has to be excluded from any further application and testing. In comparison to E-glass

fibres the R-glass fibre has also a 50% higher mechanical strength; it can stand a higher temperature (suitable to sustain a reaction treatment at 700° C without degradation) and it has the same electrical and handling properties as E-glass. The higher cost of R-glass fibres (factor 3) is only of minor importance for the overall cost of a magnet, in particular a superconducting magnet.

Composites with ceramic fibres and with Trivoltherm tape showed both a poor radiation resistance (see Figure 2), the latter, of course, because of the boron in the E-glass. In the next test series of NET, specimen with R-glass interleaved with Kapton (polyimid) foil will be tested. Although the Kapton does reduce the initial shear strength of the laminate by about 20% its beneficial effect with respect to the electric strength is still of interest for the ground insulation of high voltage coils, i.e. the poloidal coils (20KV between terminals and to ground) but also the toroidal coils in case of a fast discharge. Only one group of 5 specimens with a ceramic fibre reinforcement was irradiated in the NET test programme. It is not clear why the radiation resistance turned out to be so low. Further investigations might be useful.

Polymer Matrix Materials. There are still only the epoxy resin systems which allow a vacuum impregnation of large magnets. One should be aware of the fact that epoxy resin systems suitable for a VPI process consist in general of four components. These are the base resin, the hardener, the flexibiliser and an accelerator, for details on the chemistry see [14]. A variation of just one of these components may change the radiation resistance considerably. There are also three main groups of base resin (based on Biophenol A, or a Novolac or a Cycloaliphatic resin). Therefore general statements on radiation resistance of epoxy systems have to be taken with care, see the large variation in strength of epoxy composites with and without prior irradiation in the CERN reports,[8,14]. It has also to be stressed here that epoxy/glass sheet material is not produced by a vacuum impregnation, measurements carried out on this material are not representative for coil insulations (different resin, different glass content).

Composites made with polyimide resin as matrix material exhibit a radiation resistance which is higher than that of an epoxy resin matrix but only in a pure γ-ray irradiation environment,[15].This superiority by a factor of 5 to 10 cannot be found in the CERN results,[14] probably because of neutrons in the irradiation spectrum. But in addition, as mentioned before, polyimide resins are not suitable for a vacuum impregnation and are therefore starting with an initial mechanical strength, which is inferior to the one of epoxy composites. <u>Further testing of irradiation resistance should include all those epoxy resin systems which are in use in the field of large S.C. magnets for fusion by the major industrial coil manufacturers</u>. Research laboratories are not in a position to develop new, more radiation resistant systems by for example chemically tailoring epoxies or mixing known components to new systems and then to establish proof that the system is applicable for large magnet impregnation.

Testing Conditions

The simulation of service conditions requires irradiation at 4-6K, a warming up to room temperature, subsequent cooling down to liquid helium temperature, followed by application of the mechanical load during irradiation. Due to a lack of suitable facilities the mechanical tests are in general carried out after irradiation. There is the opinion that the irradiation temperature is not so significant for the damage caused, provided there is a warming up period before testing,[6]. The NET programme will show the difference in shear strength when there is warming up and when there is no warming up. A significant difference has to be expected. When kept cold, frozen gas stays trapped. When warmed up to RT gas (hydrogen mainly) will evolve from the matrix material and may also lead to cracks, and chemical reactions may take place in the warming up period,[3,4].

Specimen Types

Numerous papers at the 8th ICMC dealt with the advantages and disadvantages of the various types of test specimen for testing the ILS at low temperature see ref. 15, 16 for example there is no need for a repetition. For drawing up the NET/ITER test programme we made maximum use of the relevant experience in european industry. The specimens used by European coil makers for testing of shear strength of the insulation and bond to the conductor are either lap shear specimens - see Figure 1 - or symmetrical lap shear specimens,[8]. These types are simply a good model of the turn to turn insulation in a multiturn winding pack. There are shortcomings as well, the non-uniformity of the stress distribution and there is some undesirable compressive force as well. The activation of the steel pieces poses a serious handling problem. Therefore, the double notch specimen,[13,16,17] is very much preferable from the handling point of view. But cutting of the notches will cause so much damage that the measured shear strength has to expected to be below the real strength.

ACKNOWLEDGEMENTS

The author is very grateful to Mr. Koch, Mr. Rauch, and Mr. König of ASEA Brown Boveri for their continuous advice on the selection of materials to be tested and the manufacture of test specimen. The support of Professor. Dr. W. Gläser who made available the Low Temperature Irradiation Facility at the Garching reactor for the first series of tests and who is prepared to do so for the next step of the NET/ITER test programme is very much appreciated. Thanks are also due to Mr. H. Katheder at NET for the following up of the design of the new cryostat and its manufacture in industry.

REFERENCES

1. H. Hacker, C. Albrecht, L. Langenmühlen, "Epoxies for Low Temperature Application Impregnating Technology", Colorado, Springs, ICMC 1983.

2. A.F. Avery, "Dose Rates and Inventories of Active Isotopes at JET", AEEW-R 1615, Internal Rep. 1987.
3. W. Maurer, "Neutron and γ Irradiation Effects on Organic Insulating Materials for Fusion Magnets", KFK Report 3274, October 1985.
4. D. Evans, J.T. Morgan, "The Chemistry of Radiation Damage in Epoxide Resin", Advances in Cryogenic Eng. (Materials), Vol. 30/1983.
5. S. Egusa, M. A. Kirk, R. C. Birtcher, "Effects of Neutron Irradiation on Polymer Matrix Composites at 5K and at Room Temperature", Part I, Journal of Nuclear Materials 148 (1987).
6. S. Egusa, M.A. Kirk, R.C. Birtcher, "Effects of Neutron Irradiation on Polymer Matrix Composites at 5K and at Room Temperature", Part II, Journal of Nuclear Materials 148 (1987).
7. P. Bruzzone, K. Nylund, W. J. Muster, "Electrical Insulation System for Superconducting Magnets According to the Wind and React Technique", ICMC 1989.
8. G. Liptak, BBC/P. Maier, CERN/B. Haberthür, ISOLA "Radiation Tests on Selected Electrical Insulating Materials for High Power and High Voltage Application", CERN 85-02.
9. R. Pöhlchen, NET/J. Rauch, ABB/G. Claudet, CEN/I. Marangos, FRM/M. Soll, WTB, "The Mech. Strength of Irradiated Electric Insulation of Superconducting Magnets", ICMC 1989.
10. E. Krähling, H. Gerstenberg, H. Wagner, "Realisation of Shear Strength Tests on Insulation Materials after Reactor Irradiation at Temperature of 6K "Internal Report, Munich University, 1990.
11. F. Fardi, N. Mitchell, R. Pöhlchen, "Elasto Plastic Analysis of the Magnet Insulation Text Piece", MT11 1989.
12. R. E. Schmunk, L.G. Miller, H. Becker, "Tests on Irradiated Magnet Insulator Materials", Journal of Nuclear Materials 122 & 123 (1984).
13. S. Nishijima, T. Okada, T. Hirokawa, "Radiation Damage of Organic Composite Material for Fusion Magnet", Cryogenics 1991, Vol. 31.
14. H. Schönbacher, A. Stolarz-Izycka, "Compilation of Radiation Damage Test Data", CERN 79-08/ August 1979.
15. K.R. Coltmann, C.E. Klabunde, "Mech. Strength of Low Temp. Irradiated Polyimides", Journal of Nuclear Materials 103/104 (1981).
16. H. Becker, "Problems of Cryogenic Interlaminar Shear Strength Testing", ICMC 1989.
17. M.B. Kasen, "Current Status of Interlaminar Shear Testing of Composite Materials at Cryogenic Temperatures", ICMC 1989.

RADIATION EFFECTS ON HIGH CURRENT DIODES AT CRYOGENIC TEMPERATURES IN AN ACCELERATOR ENVIRONMENT

D. Hagedorn, W. Nägele

CERN, European Organization for Nuclear Research
1211 Geneva 23, Switzerland

ABSTRACT

The quench protection diodes for the proposed Large Hadron Collider at CERN will be located inside the He-II vessel of the magnet cryostat, where they could be exposed to a relatively high radiation dose. To determine the correspondence of neutron fluence results, obtained from experiments for the SSC, and the dose in an accelerator environment, several commercially available high current diodes of the DS6000 type from ABB and newly developed diodes of thin base region were irradiated at liquid nitrogen temperature in a target area of the SPS accelerator at CERN. The electrical characteristics of the diodes have been measured as a function of dose up to a maximum of about 600 Gy. After irradiation the diodes were submitted to current pulse annealing and to thermal annealing cycles up to temperatures of about 150 C. The results show that injection annealing alone can extend the service life of irradiated diodes quite substantially. The results are compared with the measurements from irradiation tests on diodes for the SSC, in a research reactor.

INTRODUCTION

Essential elements for the protection of the superconducting dipoles and quadrupoles for the proposed Large Hadron Collider at CERN are high current bypass diodes. In order to reduce the heat-load on the cryogenic system, caused by safety current leads, it is foreseen to mount these diodes inside the He-II vessel at a distance of 10 to 20 cm from the beam axis, where they could be exposed to a relatively high radiation dose. No specific simulation has been made up to now with the exact geometry, but extrapolations made from existing simulations show that a maximum dose of about 5000 Gy per year of LHC operation can be expected [1]. In recent experiments made for the SSC[2], several types of quench protection diodes were irradiated at the Texas NSC Reactor at liquid nitrogen and liquid helium temperatures. They show that the best diodes (DS6000 from ABB) are severely degraded for a neutron fluence of about $5 \cdot 10^{13}$ neutrons/cm^2. At 4.2 K the forward voltage almost doubles at 6600 A. The increase in forward voltage due to radiation will increase the heat dissipation at the junction during a quench pulse, and diode burn-out might occur before the current pulse - an exponential decay from 15 kA with a time constant of about

100s - is over[3]. In a first test run, four DS6000 diodes were exposed to a dose of about 600 Gy in a mixed secondary radiation field for about three months downstream of target T4 in the SPS area. Forward characteristic $I_f=f(U_f)$, reverse characteristic $I_r=f(U_r)$, and the zero voltage capacity of the diode were measured versus exposure. After irradiation the diodes were submitted to different type of injection annealing and temperature annealing. A second test run has started on two DS6000 diodes and six newly developed, thin base diodes[3] with a wafer of 75 mm diameter downstream target T6 in the SPS area, where a much higher dose is expected and also the different spectrum will be measured. The aim of the second test run - still at liquid nitrogen temperature - is to see, whether it is worthwhile to prepare irradiation tests on the newly developed diodes at liquid helium temperatures in an accelerator environment or at a nuclear reactor. Preliminary results from the second run are also given in this paper.

TEST SET UP

Diode assemblies

The DS6000 diodes were stacked - together with copper disks between each of the diodes acting as heat sinks and electrodes - and clamped under a force of 20 kN applied to the outside of the common holder as shown on the photo in Fig. 1.

The six newly developed diodes were clamped similarly in a separate holder under a force of 42 kN. The diode assemblies were mounted in a low loss cryostat designed to maintain all diodes at the same ambient temperature while permitting all electrical tests on individual diodes via separate leads for current pulsing and coaxial cables for voltage and capacity measurements. The leads for current pulsing were connected to the copper spacers (heat sinks) between each of the diodes and the coaxial cables directly to the anode and cathode of the housing of each diode.

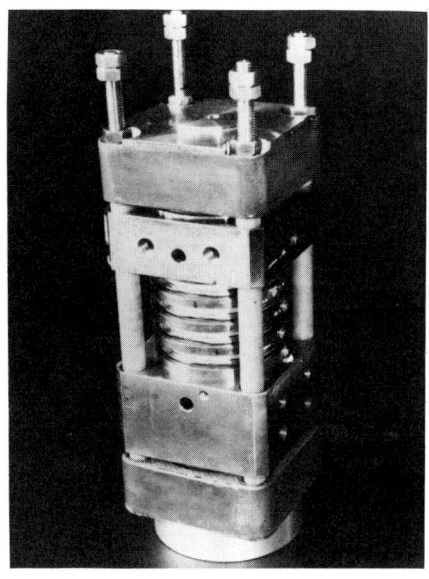

Fig. 1 : Photograph of four DS6000 assembled for irradiation tests

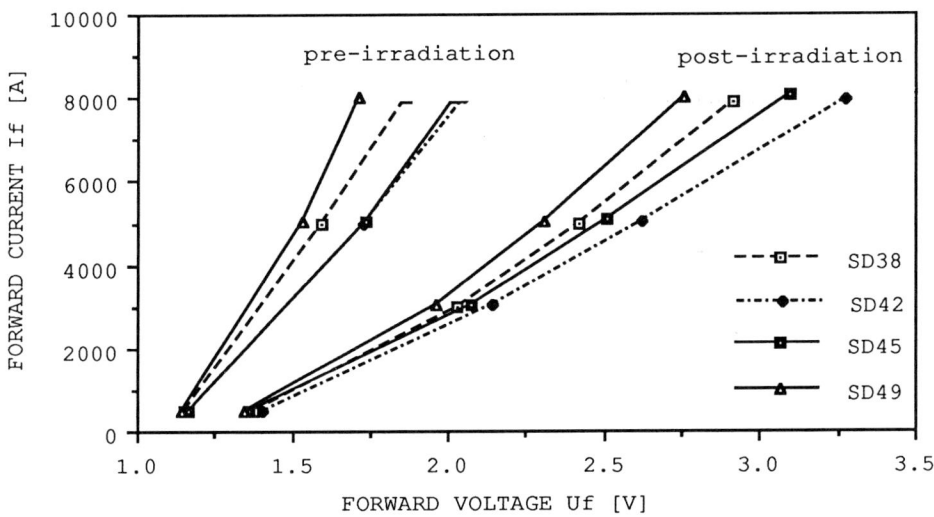

Fig. 2. Forward characteristic If = f (Uf) of four DS6000 diodes at 77 K before and after irradiation of about 600 Gy

Current pulse generator and measuring equipment

For measuring the I_f-U_f characteristic a current pulse generator of the LC-oscillator type was used providing a half-sinusoidal pulse of about 200 μs with peak current levels up to about 16 kA. Thermal model calculations have shown that the temperature rise during the current pulse does not exceed 3 K at 77K starting temperature and 1K at room temperature. A PEARSON current monitor was used to measure the current wave form with a precision of better than 1%. Both the forward voltage U_f and the forward current I_f were recorded with a vertical resolution of 12 bits and a digitizing rate of 0.5 μs per point. The junction capacitance was measured with an LCR-meter at a frequency of 10 kHz via the coax cable.

Dosimeters

For monitoring the dose during irradiation, several sets of dosimeters of the Alanine and RPL-type were used. One set consisted of 2 Alanine and 3 RPL dosimeters[4]. The dosimeter sets were mounted inside and outside the cryostat in such a way that all dosimeters had the same distance to the particle beam. One pair of dosimeter sets was analyzed after each exposure period and then replaced by new dosimeters for the following exposure. A second pair of dosimeter sets was installed right next to the first pair to integrate the dose of all the exposure periods. The irradiation non-uniformity measured with the dosimeter set outside the cryostat was less than ± 4 %.

TEST PROCEDURE

A variety of electrical tests were performed before and after each irradiation period to monitor performance degradation of the diodes as a function of dose. During the whole irradiation program of three months, the diodes were kept at liquid nitrogen temperature. After each of the six exposure periods the cryostat was removed from the radiation area, the diodes measured at liquid nitrogen temperature, and the cryostat re-installed into the radiation area. During the first test run with DS6000 diodes only, all four diodes were measured after each irradiation period and self-annealing by carrier injection (induced by the tests themselves) was observed. During the second test run with two DS6000 and six newly developed 75 mm diodes, one DS6000 and two of the other diodes were only measured before and at the end of the total exposure program to minimize the self-annealing effects. Junction capacity and reverse bias measurements were performed before applying high current pulses. The forward voltage U_f was always measured at peak current level (dI/dt=0) to avoid inductive contributions. Different current amplitudes up to about 15 kA were applied. After irradiation the four DS6000 diodes were warmed up to room temperature and re-measured at liquid nitrogen temperature before applying thermal cycles above room temperature.

TEST RESULTS ON DS6000 DIODES

Forward Voltage U_f versus Exposure

The increase in forward voltage is a sensitive indicator of radiation damage. Fig.2 shows the If-Uf-characteristic of the four DS6000 diodes at 77 K before irradiation and after the total irradiation program. The relative variation in forward voltage $\Delta U_f/U_{f_0}$, here called degradation, is given in Fig. 3 for two diodes as function of dose at three different current levels. The degradation $\Delta U_f/U_{f_0}$ due to irradiation increases with higher current levels from about 20 % at 0.5 kA to about 60 % at 8 kA indicating that small current pulse level measurements are not sufficient to predict the damage at higher current levels.

The differences among the diodes are less due to radiation non-uniformity than to manufacturing tolerances of the n-base region thickness, which has a strong influence on the forward voltage and is already visible before irradiation. The measured forward voltages are influenced by self annealing effects, so that unmeasured diodes would show a higher degradation.

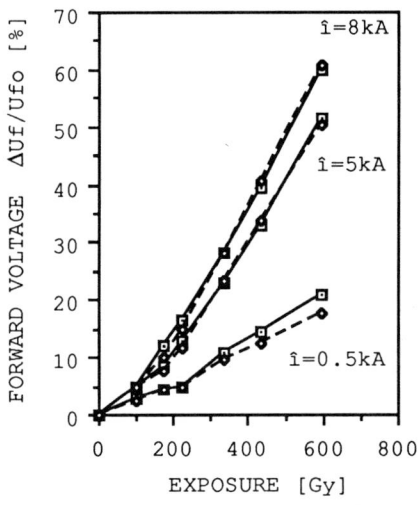

Fig. 3 Increase in forward voltage on two diodes versus exposure and current level

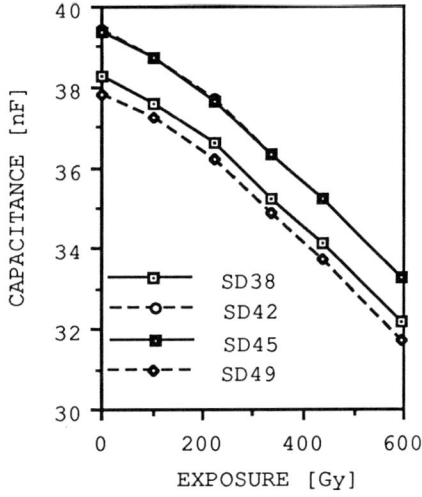

Fig. 4 Capacitance of four DS6000 versus exposure

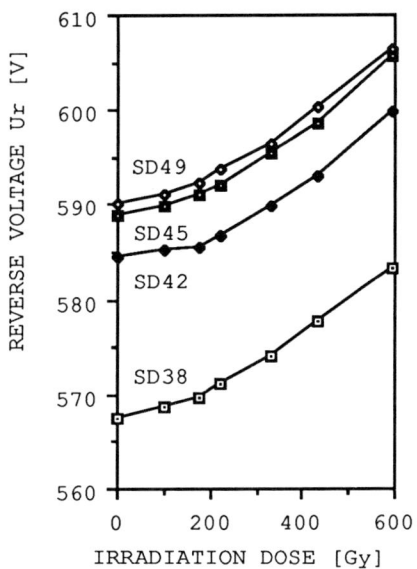

Fig. 5 Reverse voltage versus irradiation dose for the DS6000 diodes

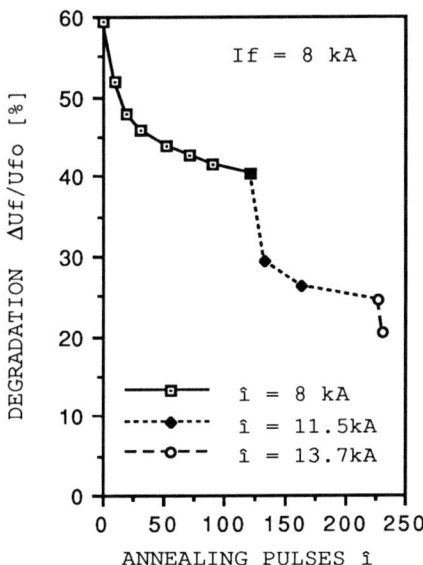

Fig. 6 Reduction of degradation $\Delta U_f/U_{fo}$ during current pulsing

Capacitance versus Exposure

The capacitance of the four diodes at zero volt is plotted in Fig. 4 as function of exposure. The relative decrease in capacitance $\Delta C/C_0$ versus dose is almost identical for all four diodes and reaches about -16 %. By measuring the capacitance, before and after the high current pulse tests, we were able to see qualitatively the self annealing. After the last irradiation period a capacitance increase of about 2 nF was observed due to current pulse measurements corresponding to about +5 % for $\Delta C/C_0$.

Reverse Voltage versus Exposure

On each diode the full reverse characteristic $I_r = f(U_r)$ was monitored as a function of exposure covering the I_r-range from about 1 µA to 3 mA. No relevant increase of the reverse leakage current I_r has been observed. The reverse voltage Ur versus exposure for the four DS6000 diodes is shown in Fig. 5. The increase amounts to approximately 3 %, almost identical for all the diodes.

Effect of Current- and Temperature Annealing

Three of the diodes were submitted to carrier injection annealing by applying high current pulses of current amplitudes between 8 kA and 13.6 kA at a rate of about one pulse per minute. The current pulse waveform was of the half sinusoidal type described already above. Fig. 6 shows the reduction of degradation in forward voltage dU_f/U_{fo} for the diode SD42 versus the number of current pulses measured at a forward current $I_f = 8$ kA. We have started with annealing current pulses of 8 kA amplitude up to about 120 pulses where the annealing became less efficient and continued with a series of about 100 pulses of 11.5 kA amplitude. After 10 pulses with 13.7 kA the annealing was stopped due to broken contact of the current leads at the heat sink. By current pulsing only, the initial degradation in forward voltage of 60% could be reduced to about 20 %, and we are sure that a further reduction could have been obtained by continuing with current pulses of higher amplitude. After warming up to room temperature for about 14 days the diodes were measured again at liquid nitrogen temperature and further reduction of degradation down to about 7% has been measured.

Fig. 7 shows the I_f-U_f-characteristic for the diode SD42 before irradiation, after irradiation at about 600 Gy, after current pulse annealing and the thermal cycle at room temperature. Further thermal cycling up to about 450 K did not show any further significant reduction of degradation. On diode SD49, which was not submitted to current pulse annealing, a reduction down to 26 % from originally 61 % has been observed at I_f = 8 kA due to thermal cycling at 300 K only. The capacitance showed a similar tendency during annealing by current pulses, but not as significant as the forward voltage. After about 220 current pulses the capacitance of SD42 increased to 35 nF, corresponding to about -11 % for $\Delta C/C_o$ compared with -16 % before annealing.

COMPARISON WITH SSC RESULTS

The comparison of the diode behavior after irradiation in a reactor and after irradiation in an accelerator environment is restricted to a few measurements where the conditions were similar, i.e. current pulses with 500 A amplitude and capacitance measurements at liquid nitrogen temperature. For the extrapolation to liquid helium temperatures care must be taken due to the different current pulse length of 300 µs, applied during SSC measurements, which certainly results in a higher temperature in the silicon wafer than for our measurements with a pulse length of 200 µs. In Fig. 8, the relative increase of the forward voltage $\Delta U_f/U_{fo}$ at 500 A and the relative decrease in capacitance $\Delta C/C_o$ are given versus exposure in Neutrons/cm^2 (from SSC measurements [2]) and in Gray respectively. After 600 Gy in an accelerator environment, the relative increase in forward voltage and the relative decrease in capacitance are almost the same,

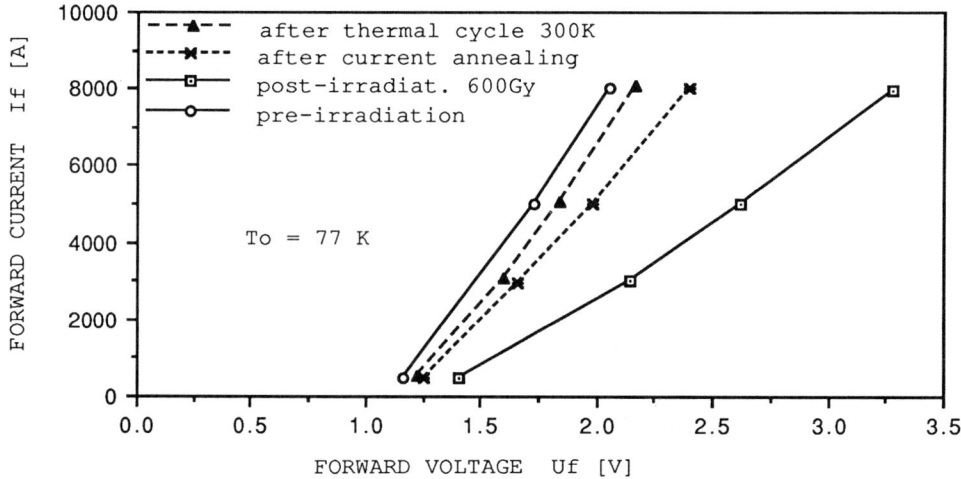

Fig. 7 If-Uf-characteristic of a DS6000 diode before and after irradiation and after current and temperature annealing

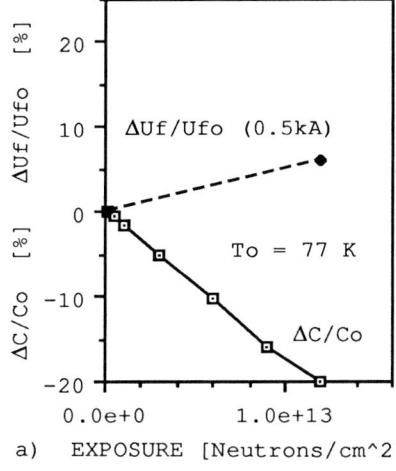 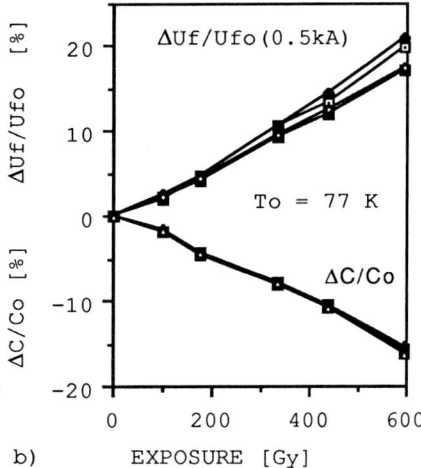

Fig. 8 Relative increase of forward voltage ΔUf/Ufo and decrease of capacitance ΔC/Co versus exposure   a) in a reactor for SSC and b) in a accelerator environment

i.e. between 16 % and 20 %, whereas after reactor irradiation the relative decrease in capacitance is four times higher than the relative increase in forward voltage. The difference is most probably due to the different radiation field and possibly the different method of capacitance measurement. Additionally, our results are also influenced by the annealing produced by the measurements themselves, since we applied current pulses up to 8 kA compared with 500 A pulses for SSC measurements.

Using the DS6000 diode as damage sensor and assuming that there is no large difference between the diodes of the same type, from Fig. 8, a correlation can be derived for the same relative forward voltage increase at 500 A and for the same relative capacitance decrease. From the capacitance measurements we obtain the equivalence of $1.5 \cdot 10^{10}$ Neutrons/cm^2 per Gray and from the forward voltage measurements about $6 \cdot 10^{10}$ Neutrons/cm^2 per Gray for the same relative degradation. For the envisaged application of the diode the increase in forward voltage - and not the decrease in capacitance - is the critical indicator in view of radiation damage.

PRELIMINARY RESULTS FROM THIN BASE REGION DIODES

Irradiation tests on six newly developed diodes with thin base region and two reference diodes of the DS6000 type are conducted at the moment and preliminary results after the first two exposure periods can be presented. After exposure to about 340 Gy, on two of the thin base diodes (E3,E10) no increase in forward voltage could be measured, whereas on one diode (M1A1) an increase of the forward voltage by about 9 % was observed as shown in Fig. 9. The DS6000 diode (A55) shows already a significant increase of the forward voltage by about 34 %. The measured capacitances show the expected tendency, i.e. no decrease for E3 and E10, 3 % decrease for M1A1 and 9 % for the DS6000 type M55. These results confirm already the necessity of a thin base region for a diode for use at low temperatures and in a high radiation environment. The present irradiation tests will continue until the end of the annual SPS physics programme with increased radiation level.

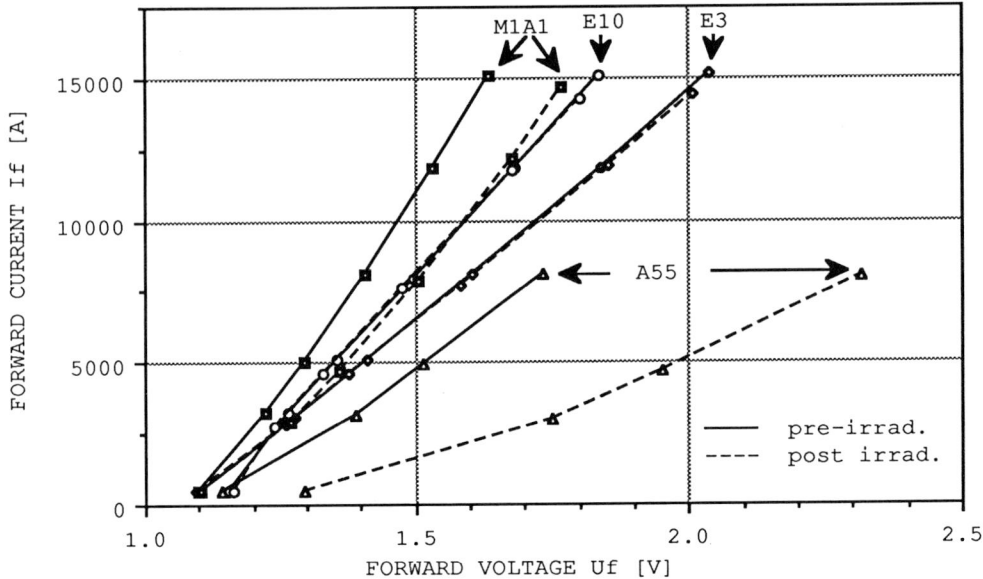

Fig. 9 If-Uf-characteristic for three thin base diodes E3, E10, M1A1, and one DS6000 diode A55 before and after irradiation of about 340 Gy

After irradiation, all six diodes will be subjected to current pulse annealing and thermal annealing. The results of these tests will influence the decision as to whether further tests on the thin base diodes, such as irradiation tests at liquid helium temperature, should be carried out.

ACKNOWLEDGEMENTS

We wish to acknowledge the support of G. Brianti and R. Perin during this work and to thank our colleagues L. Burnod, H. Schönbacher, G. R. Stevenson and M. Tavlet for many helpful discussions and assistance. The great effort of MARCONI ELECTRONIC DEVICES LTD. and EUPEC (AEG and SIEMENS) in the development of thin base diodes is gratefully acknowledged.

REFERENCES

1. L. Burnod, CERN, private communication, 1991
2. R. Carcagno, Task force on radiations effects at the SSC, SSC Central Design Group, March 1988
3. D. Hagedorn, W. Nägele, Quench protection diodes for the Large Hadron Collider at CERN, at this conference.
4. M. Coche, Comparison of high dose dosimetry systems in accelerator radiation environments, CERN TIS Commission Report, TIS-RP/205, 1988.

ALUMINA DISPERSION-STRENGTHENED COPPER ALLOY MATRIX Ti ADDED Nb_3Sn WIRE

BY THE TUBE PROCESS

S. Nakayama, S. Murase, K. Shimamura, N. Aoki*,
and N. Shiga*

Toshiba R & D Center
4-1, Ukishima-cho, Kawasaki-ku, Kawasaki City, JAPAN 210
*Showa Electric Wire & Cable Co., Ltd.
2-1-1, Odasakae, Kawasaki-ku, Kawasaki City, JAPAN 210

INTRODUCTION

Filamentary superconducting Nb_3Sn wire has become increasingly important as a high-field conductor because of its high upper critical field and critical temperature. It has applications in developing high-field and large magnets such as those used in fusion reactors, hybrid-magnet systems, and NMR. Compound Nb_3Sn, however, suffers from the disadvantage that its superconducting properties rapidly degrade under tensile and bending stress and strain due to reeling and coiling and due to the electromagnetic force generated when a coil is energized, and under thermal contraction stress upon cooling.

Some approaches to eliminate the stress and strain on Nb_3Sn have been tried, such as reinforcing materials, - stainless steel, hard copper[1], and tungsten-reinforced copper composite (W/Cu)[2], for example - in contact with the Nb_3Sn wire. The reinforcing material must have good mechanical properties, high electrical conductivity, and easy producibility. Each of these reinforcing materials has certain disadvantages; stainless steel has high electrical resistivity, hardened copper, which has somewhat low conductivity is softened by soldering, and W/Cu composite is less producible. Alumina dispersion-strengthened copper alloy[3,4] and Cu/Nb alloy, which can be co-reduced with a Nb_3Sn composite wire, are softened because they have been subjected to a reaction temperature of about 700 °C for many hours to form the Nb_3Sn layer.

In this paper, we describe a newly developed alumina dispersion-strengthened copper alloy (alumina/Cu) and Nb_3Sn wire in which the alloy forms the the matrix. It is fabricated using the tube process and has a high critical current density at high fields.

SAMPLE PREPARATION

Fabrication of Alumina Dispersion-Strengthened Copper Alloy

Internal oxidization is well known as one of the fabrication processes for alumina dispersion-strengthened copper alloys. The internal oxidization process using Cu-Al alloy powder as a base material has two disadvantages; (1) trace amounts of metallic Al in the Cu matrix decrease the electrical conductivity, and (2) in fabrication, closed packing of the powders is difficult because the oxidized Al powder in Cu is very hard, resulting in poorer mechanical properties.

Our new process for alumina dispersion-strengthened copper alloy, a selective deoxidization process[5], overcomes these two problems. The starting materials are copper oxide and alumina powders. Brittle copper oxide powder is mixed with 0.5 - 5 vol.% γ-alumina powder of 0.05 μm particle size and the mixture is then pulverized in a ball mill. The resulting powder is more homogeneous and finer than a mixture of metallic copper powder and alumina. The mixture is then heat-treated in an H_2 atmosphere and the copper oxide powder in the mixture is selectively deoxidized. The deoxidized mixture is hot-pressed at 900 °C for forming. The finished alumina dispersion-strengthened copper has a homogeneous dispersion of γ-alumina of with a size of less than 0.02 μm and at an average spacing of less than 0.1 μm.

Fabrication of $(Nb,Ti)_3Sn$ Composite Wire by the Tube Process

The reinforcement material has been applied for Ti added to the Nb_3Sn wire by the tube process[6,7,8] which is known to yield high critical current density at high fields and is used in many high-field magnets. We have selected a structure of $(Nb,Ti)_3Sn$ filaments surrounded by the alumina/Cu from many possible arrangements of the reinforcing material in the $(Nb,Ti)_3Sn$ wire. The fabrication process is shown in Fig. 1. A single core wire consists of Ti added to the Nb tube with copper sheathed tin core inside and the newly developed alumina dispersion-strengthened copper alloy pipe outside. To fabricate multifilamentary conductors, single core wires were bundled together, with 42 filaments embedded in 2 vol.% alumina in a Cu matrix group surrounding a central copper core and enclosed in an outer copper tube. The volume fraction of alumina/Cu reinforcing material is about 20 %. The composite wire was drawn down to the final size without intermediate annealing and was finally submitted to a reaction to form a $(Nb,Ti)_3Sn$ layer inside the Nb filaments. A cross-sectional view of the alumina/Cu matrix wire is shown in Fig.2. For comparison, an $(Nb,Ti)_3Sn$ wire with an all copper matrix was fabricated. Specifications of these fabricated conductors are listed in Table 1.

EXPERIMENTAL PROCEDURE

Mechanical properties, tensile strength, yield strength, and Vickers hardness were measured at room temperature for the newly developed alumina/Cu alloy under various cold-worked and heat treated conditions and for the $(Nb,Ti)_3Sn$ composite wire containing alumina/Cu heat-treated at 720 - 740 °C in vacuum to form $(Nb,Ti)_3Sn$. The distribution of elements in a cross-section of the wire was studied using CMA analysis. Critical currents (Ic) were measured in a transverse magnetic field of up to 23 T at 4.2 K using 1 μV/cm criterion. A superconducting magnet made by Toshiba Corp. was used for Ic measurements below 15 T, and a hybrid magnet at Tohoku University was used from 15 to 23 T.

RESULTS AND DISCUSSION

Workability

Workability and softening resistance was studied for the newly developed material. Figure 3 shows the effect of reducing the cross-sectional area on the Vickers hardness of the alumina/Cu. Hardness did not change as the area was reduced and it saturated at around 140 when the

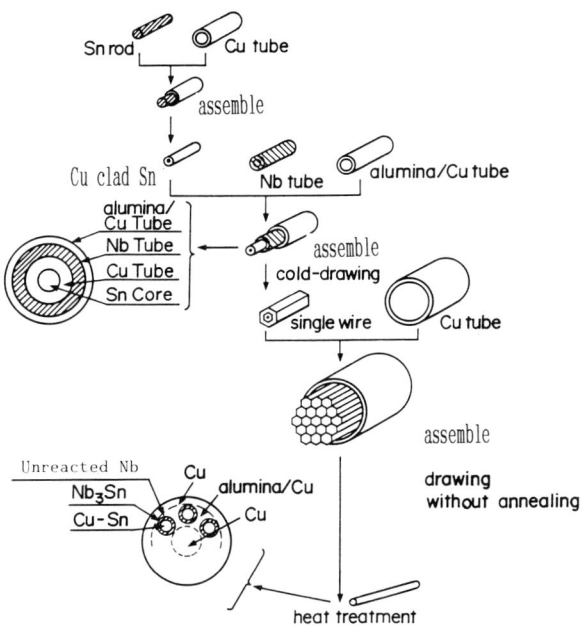

Fig. 1 Alumina dispersion-strengthened Cu matrix Ti added to the Nb_3Sn wire by the tube process

cross-sectional area was greatly reduced. The effects of annealing are shown as hardness characteristics in Fig. 3 after heat-treatment of cold-worked alumina/Cu for 74 hrs at 700 ℃. No significant softening occurred and a hardness of about 120 and good mechanical properties were maintained. These results demonstrate that the newly developed alumina dispersion-strengthened copper alloy survives great reduction in cross-sectional area and high temperature, and is thus suitable as a reinforcing material for Nb_3Sn composite wire.

Using the newly developed alumina/Cu alloy as a matrix, the composite wire was drawn down to the final size without any intermediate annealing.

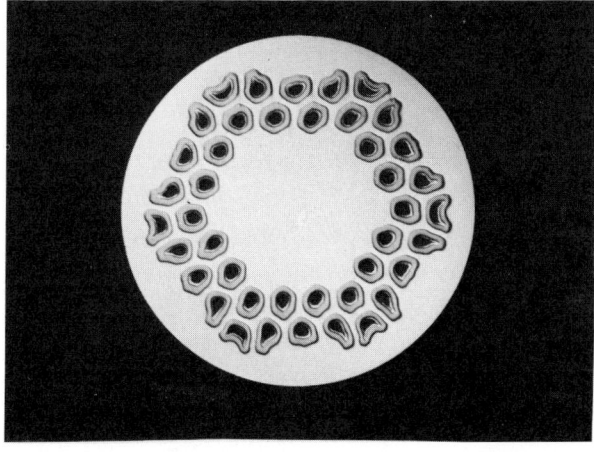

Fig. 2 A coss-sectional view of alumina/Cu matrix $(Nb,Ti)_3Sn$ wire by the tube process

Table 1 Specifications of Fabricated Conductors

	Alumina/Cu matrix	Cu matrix
Wire diameter (mm)	1.04	1.04
Filament diameter (μm)	87.5	87.5
No. of Filaments	42	42
Sn Content (wt.%)	50	50
Matrix ratio	Cu/alumina-Cu /non Cu 1.74/0.63/1	Cu/non Cu 2.37/1

The wire can be drawn down to a diameter of 0.1 mm, reducing the cross-sectional area of by 8.5×10^5 times and yielding 8.7 μm filaments without such problems as breaking and necking.

Mechanical Properties

Figure 4 shows the relationship between electrical conductivity and the yield strength of the alumina dispersion-strengthened copper alloy for various alumina contents by the new process as compared with alumina/copper alloy made by the internal oxidation process. The new alumina dispersion-strengthened copper alloy has excellent conductivity, 91 % IACS, and its yield strength is 47 kgf/mm^2 for 2 vol.% alumina in copper.

CMA analysis shows that Al did not diffuse into the central Cu core and peripheral Cu in the alumina/Cu (Nb,Ti)$_3$Sn wire after heat-treatment. However, 0.1 - 0.2 % Al was measured in the (Nb,Ti)$_3$Sn filaments. Vickers hardness was 120 -130 for the alumina/Cu section and about 60 for the Cu section. High Vickers hardness was measured for the (Nb,Ti)$_3$Sn composite wire as well as for the alumina/Cu alloy. Yield strengths (0.2 % offset) were 21.8 kgf/mm^2 and 16.4 kgf/mm^2 for the alumina/Cu matrix (Nb,Ti)$_3$Sn wire and the all-Cu matrix Nb$_3$Sn wire, respectively. This is an improvement of 30 % in yield strength for the alumina/Cu matrix

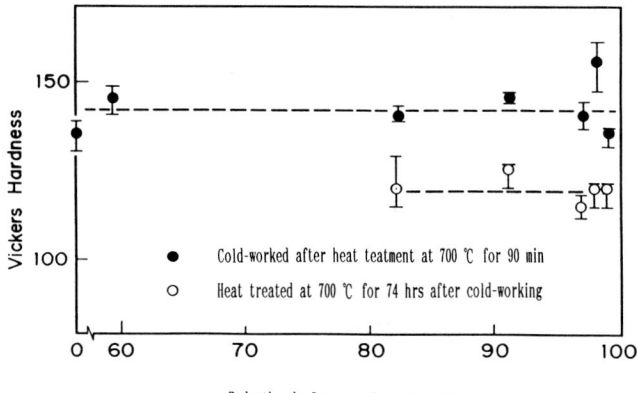

Fig. 3 Relationship between hardness and reduction in cross-sectional area of newly developed alumina dispersion-strengthened Cu alloy

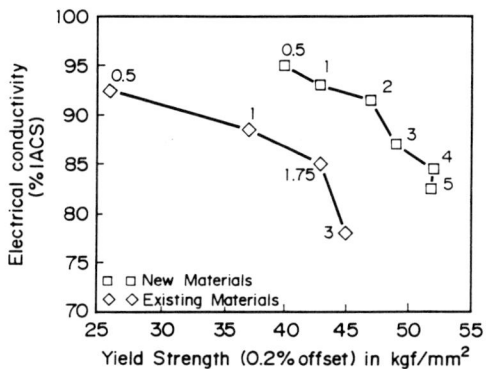

Fig. 4 Relationship between electrical conductivity and yield strength for the present materials and the conventional alumina/Cu alloy

$(Nb,Ti)_3Sn$ wire as compared with the all-Cu matrix. On the other hand, there was no difference between the two samples in tensile strength, which was about 22.5 kgf/mm^2.

Superconducting Properties

Characteristics of critical current density (Jc), which was obtained by dividing Ic by the cross-sectional area of non-copper portion, vs. magnetic field are shown in Fig. 5 for both samples. Although Al diffusion was observed in the $(Nb,Ti)_3Sn$ filaments for alumina/Cu matrix wire, Jc was as high as with the Cu matrix wire; 970 A/mm^2 at 15 T and 440 A/mm^2 at 18 T, and at 4.2 K. Ic - bending strain characteristics also showed no significant difference between these two samples. High residual resistance ratio (RRR) of about 200 was obtained for those two samples. This result shows that a center Cu core of the wire was not contaminated by Al in alumina/Cu. On the other hand, the n values of the alumina/Cu matrix wire and the Cu matrix were 37 and 19 at 15 T, respectively. The n value of the alumina/Cu matrix wire is twice as high as that for the Cu matrix wire.

The mechanical properties of the $(Nb,Ti)_3Sn$ wire with a 20 vol.% replacement of alumina dispersion-strengthened Cu alloy in Cu improved by 30 %. The material replacing alumina/Cu alloy did not affect the superconducting properties such as Jc - magnetic field and Jc - bending strain characteristics.

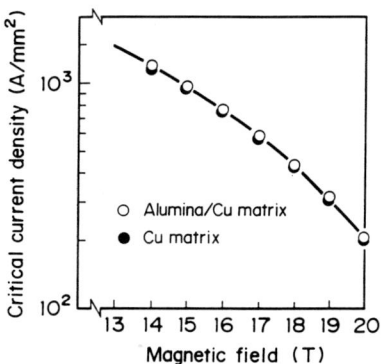

Fig. 5 Relationship between non-matrix critical current density and applied magnetic field for the alumina/Cu matrix wire and the Cu matrix wire

CONCLUSION

An alumina dispersion-strengthened copper alloy has been developed using the selective deoxidization process. The newly developed alloy has good mechanical properties which resist high temperature treatment, and it has good workability. Reinforcing alumina/Cu material containing 2 % alumina was used to replace some of the Cu stabilizer in the $(Nb,Ti)_3Sn$ wire using the tube process. $(Nb,Ti)_3Sn$ composite wire using 20 vol.% alumina/Cu alloy was successfully fabricated to 0.1 mm diameter size without any troubles. The $(Nb,Ti)_3Sn$ wire has a yield strength at room temperature of 21.8 kgf/mm^2, an increase of 30% compared with the Cu matrix wire. The alumina/Cu matrix wire demonstrated Jc as high as the Cu matrix, namely 970 A/mm^2 at 15 T and 4.2 K.

ACKNOWLEDGEMENT

The authors would like to thank Prof. N. Kobayashi, Assistant Prof. Watanabe, and Dr. Awaji for using the hybrid magnet in the High Field Laboratory for Superconducting Materials, Institute for Materials Research, Tohoku University.

REFERENCES

1. S. Murase, H. Shiraki, M. Koizumi, M. Tanaka, H. Maeda, O. Horigami, Y. Kamisada, N. Aoki, M. Ichihara, E. Suzuki, T. Ando, Y. Takahashi, M. Nishi, and S. Shimamoto, High-current MF Nb_3Sn for Up-grading of the Cluster Test Facility, Proc. of Int'l. Cryo. Mats. Conf., 215 (1982).
2. S. Murase, H. Shiraki, O. Horigami, M. Koizumi, S. Mine, H. Takeda, and H. Baba, Stress Effects on W/Cu Reinforced Nb_3Sn Composite Conductor, in:"Filamentary A15 Superconductors", M. Suenaga and A.F. Clark, ed., Plenum Press, New York (1980).
3. K. Noto, N. Konishi, A. Hoshi, K. Watanabe, M. Noguchi, and T. Fukutsuka, A New Reinforcing Stabilizer for Superconducting Wires, Proc. of 9th Int'l Conf. on Magnet Technology, 700 (1985).
4. E. Gregory, L.R. Motowidlo, G.M. Ozeryansky, L.T. Summers, High Strength Nb_3Sn Conductors for High Magnetic Field Applications, IEEE Trans. on Magn. MAG-27:2033 (1991)
5. K. Amano and K. Shimamura, High Conductivity, Alumina Dispersion-Strengthened Copper by a New Process, IEEE Tokyo Section, Denshi Tokyo, 28:94 (1989).
6. S. Murase, H. Shiraki, M. Tanaka, M. Koizumi, H. Maeda, I. Takano, N. Aoki, M. Ichihara, E. Suzuki, Properties and Performance of the Multifilamentary Nb_3Sn with Ti Addition Processed by the Nb Tube Method, IEEE Trans. on Magn., MAG-21:316 (1985).
7. H. Shiraki, S. Nakayama, M. Tanaka, S. Murase, N. Aoki, M. Ichihara, K. Watanabe, K. Noto, and Y. Muto, High-field Superconducting Properties of Ti Doped Nb_3Sn Conductor by the Nb Tube Method, MRS Int'l. Mtg. on Adv. Mats., 6:43 (1989).
8. K. Inoue, T. Takeuchi, K. Itoh, S. Murase, H. Shiraki, S. Nakayama, T. Fujioka, T. Hamajima, and Y. Sumiyoshi, High-field Superconducting Properties of a 16 T Class $(Nb,Ti)_3Sn$ Conductor by the Tube Method, Proc. of 11th Int'l Conf. on Magnet Technology, 932 (1985).

ELECTRICAL RESISTIVITY OF NANOCRYSTALLINE Ni-P ALLOYS

K. Lu, Y. Z. Wang, W. D. Wei, and Y. Y. Li

Institute of Metal Research
Academia Sinica
Shenyang, China

ABSTRACT

Nanometer-size polycrystalline Ni-P alloys that contain two phases of Ni_3P and Ni were synthesized by the crystalline process from an amorphous $Ni_{80}P_{20}$ alloy that had been isothermally annealed at certain temperatures. Nanocrystalline Ni-P samples with different mean grain sizes (10 to 100 nm) were prepared by changing the annealing temperature and time. The electrical resistivity of these nanocrystalline ribbon samples was measured with the conventional dc four-probe method from room temperature to liquid nitrogen temperature. The resistivities were linearly proportional to the absolute temperature before the onset of grain growth at about 613 K. The temperature coefficients of resistivity varied slightly with the mean grain sizes; the residual resistivities decreased significantly (from 214.1 to 65.5 $\mu\Omega \cdot cm$) as the grain size increased from 11 to 102 nm. The dependence of the residual resistivity on grain size can be reasonably correlated to the relationship between the interface volume fraction and growth size. These results indicate that the resistivity of the nanocrystalline materials is strongly dependent on the interface characteristics.

INTRODUCTION

Nanocrystalline materials are single or multiphase polycrystals in which the grain size is normally in the range 1 to 50 nm.[1-3] As a new kind of solid-state material, nanocrystalline material has attracted much theoretical and industrial interest in recent years. Because nanocrystalline materials, which are structurally characterized by ultrafine grains, contain many more interfaces than other materials, they are ideal for experiments investigating interface characteristics and the interface dependence on properties. Indeed, many properties of nanocrystalline materials are reported to be remarkably different from those of normal, coarse-grained polycrystals as well as those of glasses. One of the important physical properties, electrical resistivity, has scarcely been investigated in nanocrystalline materials. In only one paper[3] were resistivity results reported for a nanocrystalline material (pure Pd), but no details were presented.

Nanocrystalline materials are usually synthesized by in-situ compaction of ultrafine particles (UFP) prepared by using gas condensation techniques.[1] The compacted bulk sample contains many unavoidable defects, from vacancy-size defects to micrometer-size voids.[4] More or less, these microdefects would influence the mechanical and physical properties. Very recently, a new method for synthesizing nanocrystalline materials, called the *crystalline method*, has been developed;[5] it uses crystallization of amorphous alloys. The basic idea of the method is to control the crystal growth during the crystallization process by regulating the heat-treatment conditions, such as temperature and time. The advantages of this new method are its simple and convenient manufacturing process and fewer microdefects because the interfaces form naturally rather than by artificial compaction. A positron annihilation-spectrum study of these materials showed that nearly all the interfacial defects are free volumes; no microvoids or voids were detectable.[6]

In this study, the crystalline method was used to synthesize samples of nanocrystalline Ni–P alloy with various grain sizes (from 10 to 100 nm). This paper reports the first attempt to measure the electrical resistivity of Ni–P nanocrystalline samples and to study the interface dependence on electrical resistivity and the temperature coefficient of resistivity.

EXPERIMENTAL PROCEDURES

Sample Preparation

Nanocrystalline Ni–P alloys with a variation in grain size were synthesized by the crystallization method[5] from an amorphous $Ni_{80}P_{20}$ (at.%) alloy annealed at different temperatures. The Ni–P amorphous alloy ribbon was made by melt-spinning in a single-roller, rapidly quenching apparatus. The ribbon was about 0.02 mm thick and 2.1 mm wide; its amorphous nature was demonstrated by X-ray and electron diffraction. The nanocrystalline alloys were prepared by rapidly heating (at about 100 K/min) the samples to an annealing temperature (which is normally below the crystallization temperature of the glass), annealing them for a certain period, and then cooling them quickly to ambient temperature. Different annealing temperatures resulted in different grain sizes in the as-crystallized nanocrystalline samples. Preparation processes performed in a differential scanning calorimeter (DSC-II, Perkin–Elmer) showed that the amorphous alloy samples were completely crystallized during the annealing treatments. No retained amorphous regions and no oxidation were detected by transmission electron microscopy (TEM) and X-ray diffraction (XRD) of the as-crystallized nanocrystalline Ni–P samples with different grain sizes.

Grain Size Determination

Grain sizes of the nanocrystalline materials were measured quantitatively by means of both an XRD technique using the Scherrer equation and high-resolution electron microscopy (HREM). The XRD experiments were conducted in a Rigaku X-ray diffractometer with CuK_α radiation; the HREM observations, in a JEOL-200CX electron microscope at 200 kV.

Electrical Resistance Measurements

Electrical resistance of the thin-foil nanocrystalline samples was measured by the conventional dc four-probe method. For the low-temperature range, liquid nitrogen was used to cool the samples from ambient temperature to about 78 K. For the high-temperature range, a special vacuum furnace was installed to heat the specimens. During the measurement procedure, a heating rate of 10 K/min was used to raise the sample temperature.

RESULTS AND DISCUSSION

The microstructure and grain size of the as-crystallized Ni–P alloys were examined by XRD, TEM, and HREM. Both XRD and electron diffraction showed that the as-crystallized Ni–P alloy contained two phases: Ni_3P compound with a body-centered tetragonal (bct) structure and Ni austenite with a face-centered cubic (fcc) structure. The orientation relationship is <001>bct ∥ <110>fcc and <110>bct ∥ <111>fcc.[7]

For a nanocrystalline sample annealed at 325°C for 10 min, XRD measurements of the half-maximum widths led to dimensions of 7.5 nm from the (341) Ni_3P line, about 15 nm from the (202) Ni_3P line, and about 9 nm for the Ni phase. The average grain size for the two phases was about 11 nm. Observations by TEM and HREM of a selected nanocrystalline Ni–P sample revealed that the Ni_3P phase was in the form of anisotropic blocks separated by Ni austenite or Ni-rich regions, or both, which were parallel or perpendicular to the crystal-growth direction. Almost no microdefects were detected in the crystallites of the samples. Morphologies of these two crystalline phases have been described in detail in our previous work.[7] The average sizes of the Ni_3P and Ni phases determined by TEM and HREM approximately agreed with those determined by XRD.

Different grain sizes were obtained by annealing the amorphous alloy at different temperatures. Figure 1 shows a plot of grain size (determined from the 341 Ni_3P line broadening) as a function of annealing temperature, from which it can be seen that coarser grains result from higher annealing temperatures. For annealing temperatures below 330°C, the grain size of Ni_3P tends toward a constant value of 7.5 nm and an average grain size of 11 nm.

Figure 2 shows the electrical resistivity measured in a nanocrystalline Ni–P sample with an average grain size of 11 nm. It is evident that the electrical resistivity decreases linearly as the temperature decreases from 613 to 78 K. Two peaks appeared in the resistivity curve, at 613 K and at 665 K. The samples were examined by XRD at temperatures before, between, and after the two peaks. In Fig. 3, XRD patterns of three samples show that there are two crystalline phases, Ni_3P compound and Ni austenite, in each of the three samples. The only difference in these XRD patterns is the diffraction line intensity, which is closely

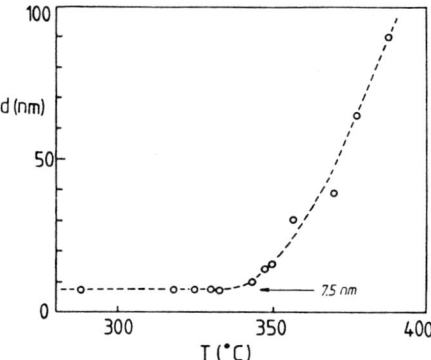

Fig. 1. Variation of grain sizes (obtained from X-ray diffraction 341 line broadening of Ni_3P) in the as-crystallized Ni–P nanocrystalline alloys as a function of annealing temperature.

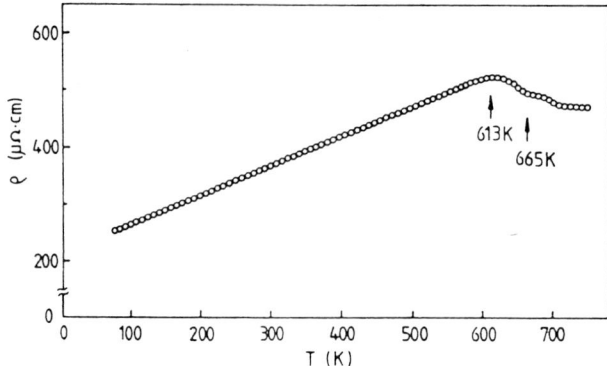

Fig. 2. Electrical resistivity curve of the nanocrystalline Ni-P alloy, with a mean grain size of 11 nm, heated at a rate of 10 K/min.

related to grain size. Measurement of the grain size revealed that after the first peak, the Ni$_3$P grain size increased significantly [from (a) to (b)], but the Ni grain size did not. During the second peak, Ni-phase crystallites grew significantly, and the Ni$_3$P grain size changed slightly. These results suggest that the first peak in resistivity is due to grain growth in the Ni$_3$P phase, and the second peak originated from Ni-austenite grain growth.

Differential scanning calorimetry (DSC) measurement of the nanocrystalline sample also exhibits two exothermal peaks corresponding to the grain growth process of Ni$_3$P and Ni crystallites.[8] Onset and maximum temperatures of the two peaks in the DSC and resistivity curves are listed in Table 1. The two sets of results are in reasonable agreement.

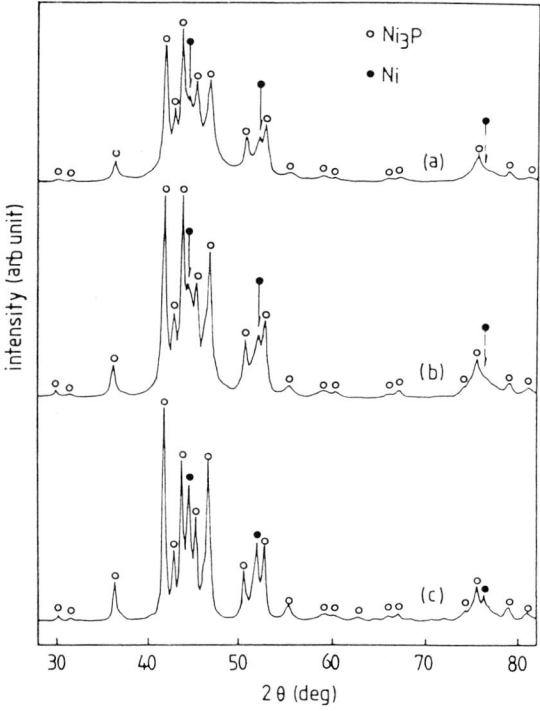

Fig. 3. X-ray diffraction spectrums of three samples heated to three temperatures (before, between, and after the two peaks in Fig. 2).

Table 1. Onset and Peak Temperatures of Grain-Growth Processes Detected by Differential Scanning Calorimetry and Resistance Measurements

	T_o	T_{p1}	T_{o2}	T_{p2}
DSC	620.4	643.7	664.0	688.4
Resistance	613	631	665	680

Figure 4 shows the plots of resistivity versus temperature (ambient to 78 K) for nanocrystalline Ni–P alloys with various grain sizes. It is evident that in this temperature range and at different grain sizes, the resistivities are linearly proportional to the absolute temperature; at a given temperature, the smaller the grain size, the higher the resistivity. By extrapolating the straight lines to absolute zero, the residual resistivities of the samples were obtained; they are plotted versus grain size in Fig. 5. The residual resistivities of nanocrystalline Ni–P alloys decreased significantly, from 220.0 to 62.0 $\mu\Omega\cdot$cm, as the average grain size increased from 11 to 102 nm.

The measured temperature coefficient of resistivity (TCR) of the nanocrystalline Ni–P alloys was found to decrease (from 2.35×10^{-3} to 1.26×10^{-3} K^{-1}) with the reduction in grain size (from 102 to 11 nm), as shown in Fig. 6. The tendency for TCR to decrease with grain size is similar to that reported by Gleiter[3] for nanocrystalline Pd materials produced from compaction of UFPs.

Nanocrystalline materials have two components: crystallites and grain boundaries. The large volume fraction of grain boundaries, which are quite different in structure from the crystallites, are responsible for the characteristic properties of nanocrystalline materials. The grain size dependence of electrical resistivity and TCR of the nanocrystalline Ni–P alloys may be understood, in terms of electron scattering inside the crystallites and scattering of electrons by the boundaries, as being similar to the electrical resistivity of thin, metallic film.[9,10]

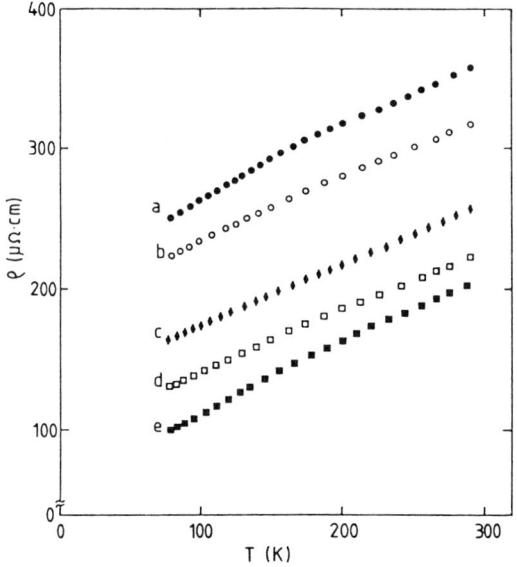

Fig. 4. Plots of resistivity versus temperature below room temperature for five nanocrystalline Ni–P samples with mean grain sizes of (a) 11 nm, (b) 14 nm, (c) 30 nm, (d) 51 nm, and (e) 102 nm.

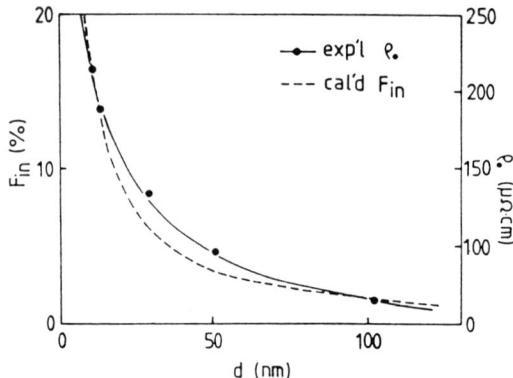

Fig. 5. Plots of residual resistivity versus the mean grain size in the nanocrystalline Ni–P alloys (solid line) and variation of interface volume fraction and the mean grain size of the samples (dashed line).

Electron scattering by a grain boundary, which can be regarded as a potential barrier with a certain width and height, decreases the conductivity of polycrystalline metallic materials with respect to the corresponding bulk single crystal. When the electron energy (E) is higher than the boundary potential barrier height (U), the electrical conductivity for a metal film with a mean grain size (d) can be approximately expressed by the following equation:[8]

$$\sigma_g = \frac{6\sigma_0 E (E - U) d}{\lambda U^2 \sin^2\{[2m(E - U)^{1/2}] a/h\}} \quad (1)$$

where σ_0 is the conductivity of the bulk metal without grain boundaries; λ, the mean free path in the metal without grain boundaries; a, the barrier width; m, the electron mass, and h, Planck's constant.

Equation 1 shows that the conductivity is proportional to the grain size. In other words, the resistivity of a polycrystalline metal is inversely proportional to its grain size. The relationship agrees with our experimental results, shown in Figs. 4 and 5.

For the nanocrystalline materials, the volume fraction (F_{in}) of the interfaces can be estimated by the ratio C/d, in which C is a constant related to the interface thickness and d is the grain size. For nanocrystalline Ni–P alloys containing two crystalline phases, Ni_3P and

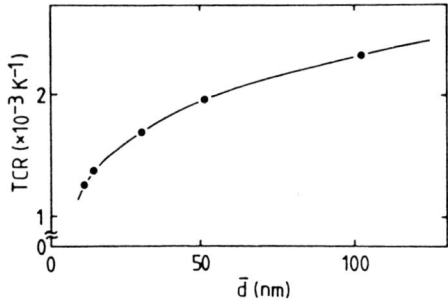

Fig. 6. Plot of the temperature coefficient of resistivity (TCR) versus the mean grain size of nanocrystalline Ni–P alloys.

Ni austenite, the volume fraction of the interfaces[11] gained was

$$F_{in} = 1.55/d \qquad (2)$$

where d is the average grain size. When F_{in} is plotted versus average grain size following Eq. 2, as shown in Fig. 5, we find that the change in interface volume fraction with grain size is quite similar to that of the residual resistivity of the nanocrystalline Ni-P alloys. The correlation of residual resistivity and the interface volume fraction actually proves the validity of the theoretical analysis of Eq. 1 for nanocrystalline materials.

A quantum mechanical calculation of the dc grain boundary resistance of polycrystalline metallic materials showed that when the grain size approaches or becomes less than the electron mean free path, the TCR decreases, even to negative values.[10] Our experimental results, that the TCR decreases with a reduction of grain size, supports the above model and indicates that scattering by grain-boundary interfaces dominates the electrical conductivity in nanocrystalline Ni-P alloys.

CONCLUSIONS

The electrical resistance of nanocrystalline Ni-P alloy samples, which were synthesized by the crystallization method, was studied. The following conclusions can be drawn:

1. Electrical resistance measurements showed that a two-step grain-growth process occurs in a nanocrystalline Ni-P alloy with an average grain size of 11 nm at temperatures to 613 K.
2. Electrical resistivities of nanocrystalline Ni-P alloys with various grain sizes are proportional to the absolute temperature from the temperature of grain-growth onset to liquid-nitrogen temperature.
3. Residual resistivity decreases with an increase in grain size in the nanocrystalline Ni-P alloys; this relationship can be correlated to the relationship between interface volume fraction and grain size.
4. The temperature coefficient of resistivity of nanocrystalline Ni-P alloys decreases with a reduction in grain size. The resistivity results for nanocrystalline Ni-P alloys agree with the theoretical analysis of electron scattering by grain boundaries in metallic systems, as reported in the literature.

REFERENCES

1. R. Birringer, U. Herr, and H. Gleiter, Nanocrystalline materials — A first report, *Trans. Jpn. Inst. Met. Suppl.* 27:43 (1986).
2. R. Birringer, Nanocrystalline materials, *Mater. Sci. Eng.* A117:33 (1989).
3. H. Gleiter, Nanocrystalline materials, *Prog. Mater. Sci.* 33:233 (1989).
4. H. E. Schaefer, R. Wurschum, R. Birringer, and H. Gleiter, Structure of nanometer-sized polycrystalline iron investigated by positron lifetime spectroscopy, *Phys. Rev.* B38:9549 (1988).
5. K. Lu, J. T. Wang, and W. D. Wei, A new method for synthesizing nanocrystalline alloys, *J. Appl. Phys.* 69(1):522 (1991).
6. M. L. Sui, L. Y. Xiong, W. Deng, K. Lu, S. Patu, and Y. Z. He, Investigation of the interfacial defects in a nanocrystalline Ni-P alloy by positron annihilation spectroscopy, *J. Appl. Phys.* 69(8) (1991), in press.
7. M. L. Sui, K. Lu, and X. Z. He, A structural investigation of the crystallization products of an amorphous Ni-P alloy, *Philos. Mag.* B63 (1991), in press.

8. K. Lu, W. D. Wei, and J. T. Wang, Grain growth kinetics and interfacial energies in nanocrystalline Ni–P alloys, *J. Appl. Phys.* 69(10) (1991), in press.
9. F. Warkusz, The scattering of electrons on grain boundaries: Electrical conductivity, *Thin Solid Films* 161:1 (1988).
10. G. Reiss, J. Vancea, and H. Hoffmann, Grain boundary resistance in polycrystalline metals, *Phys. Rev. Lett.* 56(19):2100 (1986).
11. M. L. Sui, S. Patu, and Y. Z. He, Influence of interfaces on the mechanical properties in polycrystalline Ni–P alloys with ultrafine grains, *Scripta Metall. Mater.* 25(7) (1991), in press.

MAGNETIC PROPERTY OF GADOLINIUM HYDRIDES

H. Yayama and A. Tomokiyo

Department of Physics
College of General Education
Kyushu University
Fukuoka 810, Japan

ABSTRACT

Saturated magnetic moment and magnetic susceptibility of Gd hydrides were measured in the wide range of hydrogen content. In the composition range $0 \leq H/Gd < 1.8$, the samples are ferromagnets with Curie temperature of about 291K and the magnetization decreases with increasing hydrogen content. In the range $1.8 \leq H/Gd \leq 2.3$, the hydrides exhibit antiferromagnetic behavior and the Néel point increases with increasing hydrogen content. In the range $H/Gd > 2.3$, another antiferromagnetic phase originating from $GdH_{2.9}$ coexists. The change in magnetic ordering temperature can be qualitatively explained on the basis of anionic hydrogen model and RKKY interaction.

INTRODUCTION

Rare earth metals react with hydrogen gas at elevated temperatures to form hydrides with composition up to 2 or 3 hydrogen atoms per metal atom.[1] The absorbed hydrogen changes the physical and chemical properties of the parent material. The magnetic property of heavy rare earths also strongly depends upon the amount of absorbed hydrogen.[1]

Gadolinium is a ferromagnets with Curie temperature of 291K. Addition of increasing amounts of hydrogen to Gd metal results in not only the lowering of magnetic ordering temperature but also the change of the type of magnetic ordering (ferromagnetism to antiferromagnetism).[2,3]

The magnetic susceptibility of Gd hydrides were measured by Wallace et al.[2] and Flood.[3] The samples measured so far are limited only on the stoichiometric ones GdH_2 and GdH_3. However, the Gd hydrides exhibit wide deviation of the composition from these stoichiometry.[4]

In this paper, we present the magnetic data on Gd hydrides with wide composition range. The change in magnetic ordering temperature is discussed with emphasis on the nonstoichiometric composition range in the framework of anionic hydrogen model and RKKY interaction.

SAMPLE PREPARATION AND EXPERIMENTAL

The experimental setup to prepare the samples is schematically shown in Fig. 1. The system consists of two parts. The first part is a hydrogen supplier in which Ti hydride is contained. The Ti hydride decomposes and desorbs very pure hydrogen gas at elevated temperatures above 500°C. The second part is a silica reaction tube where the temperature and the pressure can be monitored with a Pt/Pt-Rh thermocouple and a Bourdon gauge, respectively. The temperature of both parts can be controlled with electric furnaces.

The samples of Gd hydrides were prepared in the following way: A piece of Gd sample, roughly 30mg, with purity of 99.5% were placed in the silica reaction tube. A certain amount of hydrogen gas, produced by thermal decomposition of Ti hydride, was introduced into the reaction tube, after the void space of the reaction tube was evacuated by a rotary pump. This method allowed to get highly pure hydrogen gas without contamination of oxygen or water vapor. As the temperature of the sample was elevated, the reaction began slowly above 500°C and took place rapidly above 600°C. After the reaction was completed, the sample was annealed at 800°C for 15min and then cooled gradually down to room temperature.

The amount of hydrogen absorbed in Gd metal was calculated exactly from the decrease in pressure of the hydrogen gas in the known volume of the void space of the reaction tube. Since the reaction speed became very slow as the atom ratio H/Gd approached 3, the maximum hydrogen content we could obtain was H/Gd=2.76. The color of the obtained hydrides changed from metallic gray to black with increasing hydrogen content. Visually, they appeared to be stable in atmospheric environment in a few days but the samples with high hydrogen content turned into white powder in a few weeks. Therefore, we coated the samples with Apiezon N grease in order to avoid the degradation. After this treatment, no change in appearance was observed for months but magnetic measurements were performed in a few days after sample preparation to avoid the degradation.

Saturated magnetization and magnetic susceptibility were measured with SQUID susceptometer in the temperature range from 4.4 to 310K. As the preliminary measurements showed that the magnetization almost saturated at a field of 5kOe, the saturated magnetization of ferromagnetic samples were measured at this field. The magnetic susceptibility of antiferromagnetic samples were measured at 100 Oe. To minimize the effect of demagnetizing field, the measurements were made on the long thin samples (typically 0.1 x 1 x 7mm^3) laid along the direction of magnetic field.

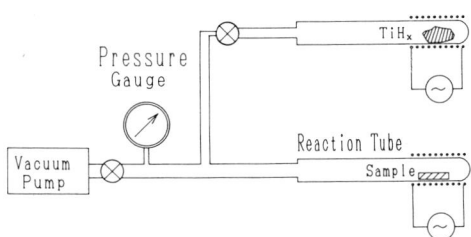

Fig. 1. Experimental setup to prepare the samples. The temperatures of reaction tube and Ti hydride bed can be controlled by electric furnace. The hydrogen gas is supplied by thermal decomposition of Ti hydride into the reaction tube. The hydrogen content can be calculated from the pressure drop in the known volume of the void space of reaction tube.

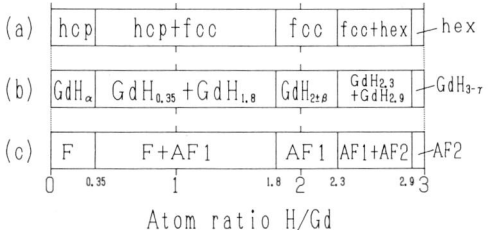

Fig. 2. (a) the crystal structure, (b) the chemical phase, (c) the magnetic phase, as a function of atom ratio H/Gd. F stands for the ferromagnet, AF1 and AF2 the antiferromagnets of di- and trihydride phases, respectively.

RESULTS AND DISCUSSION

Sturdy and Mulford presented the pressure-temperature-composition data and X-ray data for Gd-H system.[4] In their phase diagram, there exist two types of stoichiometric hydrides GdH_2 and GdH_3 the crystal structures of which are fcc and hxagonal, respectively.

From the view point of chemical phase, the full system of Gd-H can be divided into five parts according as hydrogen content. The three of them are single phase hydrides around pure Gd, stoichiometric GdH_2 and GdH_3 which have individually solubility or deficiency limits of hydrogen and the corresponding crystal structures; hcp, fcc and hexagonal, respectively. The last two are mixtures of two nonstoichiometric phases, where one is between $GdH_{0.35}$ and $GdH_{1.8}$, and the other is between $GdH_{2.3}$ and $GdH_{2.9}$. They are summarized in Fig. 2 showing the crystal structures (a) and the chemical phases (b). A remarkable feature of this system is that the limits of hydrogen solubility or deficiency in the single phase hydrides are relatively wide. As shown in Fig. 2(b), the single phase hydrides are Gd - $GdH_{0.35}$, $GdH_{1.8}$ - $GdH_{2.3}$, and $GdH_{2.9}$ - $GdH_{3.0}$. In the two phase region, for instance, the sample with H/Gd=1.0 is a mixture expressed as the mole fraction of $GdH_{0.35}$: $GdH_{1.8}$ = 55 : 45.

Our data on magnetic measurements are shown in Figs. 3 and 4, and the magnetic phase is summarized in Fig. 2(c). Figure 3 covers the composition $0 \leq H/Gd < 1.8$ where ferromagnetic behavior is predominant. The elemental Gd exhibits typical ferromagnetic behavior with Curie temperature of 291K. Except the elemental Gd, these samples are in two-phase region where $GdH_{0.35}$ and $GdH_{1.8}$ coexist (see Fig. 2(b)). With increase in hydrogen content, the

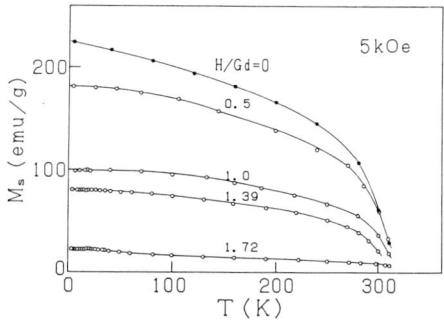

Fig. 3. Saturated magnetic moment as functions of temperature and hydrogen content. The samples show ferromagnetic behavior with Curie temperature of 291K. A small peak is seen at 19.5K on the data of sample with composition H/Gd=1.72.

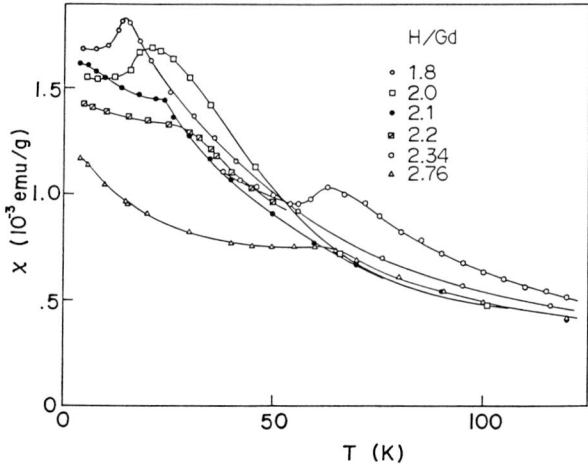

Fig. 4. Magnetic susceptibility as functions of temperature and hydrogen content. The samples exhibit antiferromagnetic behavior. The peaks correspond to Néel temperatures above which the data obey Curie-Weiss law.

magnetization decreases but the Curie temperature does not change significantly. In this region, the fraction of $GdH_{0.35}$ phase decreases as the hydrogen content increases. Since the magnetization of ferromagnetic phase is extremely larger than that of antiferromagnetic phase, even a slight trace of $GdH_{0.35}$ is sensitively detected. A small peak is seen on the data of sample with H/Gd=1.72 at 19.5K. This is due to antiferromagnetic ordering of $GdH_{1.8}$ phase.

Figure 4 covers the magnetic data of the composition $1.8 \leq H/Gd \leq 2.76$. The maxima in susceptibility are assumed to correspond to the Néel points of antiferromagnetic orderings. The data obey Curie-Weiss law above the Neel temperatures. Wallace et al. measured the magnetic susceptibility of stoichiometric GdH_2, which behaved as an antiferromagnet with a Néel point of 21K.[2] Our data on $GdH_{2.0}$ are in agreement with those of Wallace et al.[2] The Néel points obtained from Fig. 4 are plotted in Fig. 5. The samples

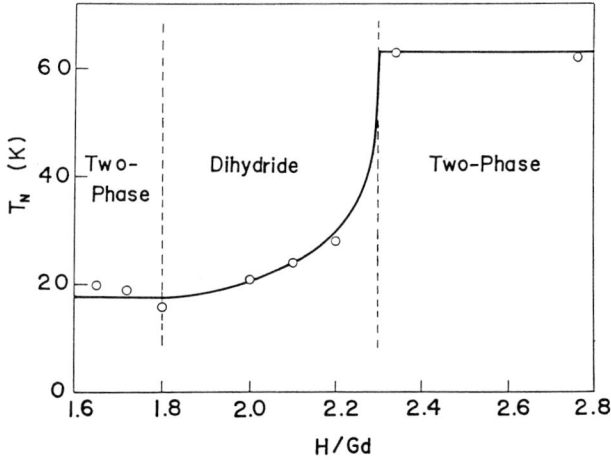

Fig. 5. Néel temperature T_N as a function of atom ratio H/Gd. T_N is constant in two-phase regions, but increases in dihydride region with increasing amounts of hydrogen.

with atom ratio from H/Gd=1.8 to 2.2 are single-phase and the Néel points of them continuously increase with hydrogen content from 16 to 28K. Both samples of H/Gd=2.34 and 2.76 are mixed phase of $GdH_{2.3}$ and $GdH_{2.9}$, and have the same Néel point at 62K originating from $GdH_{2.9}$.

Flood measured the susceptibility of GdH_3 in the temperature range 1.4 to 4.2K, and found a maximum at 3.7K.[3] He assumed it to be due to an antiferromagnetic ordering. Our data also show this tendency but unfortunately the lowest temperature we could attain was 4.4K. The increase in susceptibility as the temperature is lowered below 10K, especially apparent in H/Gd=2.76 sample, may be due to the onset of this antiferromagnetic ordering of trihydride phase with Néel point of 3.7K.

It is generally accepted that the magnetic interaction between rare earth ions is effective through the conduction electrons by RKKY mechanism.[5] The essence of RKKY interaction is as follows:[5] Exchange between the 4f and conduction electrons results in the imbalance of spin density which decreases in a damped oscillatory fashion as the distance from the magnetic ion increases. Since the amplitude and the spatial frequency of the spin density wave also strongly depend on the number of conduction electrons, the coupling of the conduction electrons with the next ion changes from positive to negative (or negative to positive) as the number of conduction electrons decreases.

Many experimental evidences show that a dissolved hydrogen in rare earth metal captures a conduction electron to form H^- ion and consequently the number of conduction electrons decreases with hydrogen content[1,2,6,7] The magnetic coupling and hence the type of magnetic ordering changes according as the hydrogen content. The change of the type of magnetic ordering from ferro in solid solution range (Gd - $GdH_{0.35}$) to antiferro in dihydride range ($GdH_{1.8}$-$GdH_{2.3}$) mainly corresponds to the decrease in the number of conduction electrons. The increase in Neel temperature with increasing amount of hydrogen in dihydride phase is also accounted for the decrease in number of conduction electrons. Our data can thus be explained qualitatively with RKKY interaction and anionic hydrogen model.

In conclusion, the chemical phases and the magnetic properties in the full system of Gd-H were made clear. Our magnetic data are consistent with the previously published ones[2,3] on the sotichiometric GdH_2 and GdH_3. In the wide range of hydrogen content of this system, our results of magnetic measurements can be explained qualitatively with RKKY interaction and anionic hydrogen model.

REFERENCES

1. W. E. Wallace, "Magnetic Property of Metal Hydrides and Hydrogenated Intermetallic Compounds", in "Hydrogen in Metals I" eds. G. Alefeld and J. Völkl, Springer-Verlag, Berlin, Heiderberg, New York (1978), chap. 7.

2. W. E. Wallace, Y. Kubota and R. L. Zanowick, Magnetic Characteristics of Gadolinium, Terbium, and Ytterbium hydrides in Relation to the Electronic Nature of the Lanthanide Hydrides, Adv. Chem. Ser. 39: 122-130 (1963).

3. D. J. Flood, Magnetization and Magnetic Susceptibility of GdH_3, Phys. Lett. 60A: 463-464 (1977).

4. G. E. Sturdy and R. N. R. Mulford, The Gadolinium-Hydrogen System, J. Am. Chem. Soc. 78: 1083-1087 (1956).

5. W. E. Wallace, "Rare Earth Intermetallics", Academic Press, Inc., New York and London (1973).

6. Z. Bieganski and B. Stalinski, Phys. Status Solidi 2: K161 (1970).

7. R. C. Heckman, Electrical Properties of the Cerium and Gadolinium Hydrogen Systems, J. Chem. Phys. 40: 2958-2963 (1964).

QUENCH PROTECTION DIODES FOR THE LARGE HADRON COLLIDER LHC AT CERN

D. Hagedorn, W. Nägele

CERN, European Organization for Nuclear Research
1211 Geneva 23, Switzerland

ABSTRACT

For the quench protection of the main ring dipole and quadrupole magnets for the proposed Large Hadron Collider at CERN two lines of approach have been pursued for the realization of a suitable high current by-pass element at liquid helium temperature. Two commercially available diodes of the HERA type connected in parallel can easily meet the requirements if a sufficient good current sharing is imposed by current balancing elements. Design criteria for these current balancing elements are derived from individual diode characteristics. Single diode elements of thin base region, newly developed in industry, have been successfully tested. The results are promising and, if the diodes can be made with reproducible characteristics, they will provide the preferred solution especially in view of radiation hardness.

INTRODUCTION

The quench protection of the main ring dipole and quadrupole magnets for the proposed Large Hadron Collider at CERN is based on self-protected magnets and high current by-pass diodes. Since there is no practical way to extract a major fraction of the stored energy from the quenching magnet, which is in a series chain of magnets, self protection is achieved by stainless steel heater strips mounted on the outer layers of each magnet. As soon as a quench is detected, capacitors are discharged on the heaters and the increasing resistance of the quenching magnet causes the magnet current to commutate over to the by-pass diodes which are mounted inside the He II vessel in order to reduce the heat load on the cryogenic system caused by safety leads. During a magnet quench the diode will conduct a peak current of about 15 kA, it will be exposed to integrated current loads $\int i^2 dt$ of about $1.2 \cdot 10^{10}$ A^2s. The copper blocks, mounted on each side of the diode and acting as heat sinks and current connections, have to absorb an energy of about 1.5 MJ, when the whole magnet ring is de-energized with a time constant of about 100 s. Two lines of approach have been pursued for the realization of a suitable current by-pass element at liquid helium temperature:

1. Parallel connection of commercially available high current diodes of the DS-6000 type from ABB (as used in HERA at DESY).
2. Development of a single diode element in collaboration with industry to avoid the problems of parallel connection.

Furthermore the diode will be subjected to a relatively high radiation dose. The degradation of diode characteristics at low temperatures in an accelerator environment is described elsewhere[1].

TESTS ON COMMERCIALLY AVAILABLE DIODES

Endurance tests

The temperature rise in the diode wafer is governed by the power generated inside the wafer and the transfer of heat to the cooler or heat sink. In the early stage of the current commutation process, a power of about 20 kW or more (depending on the forward voltage drop U_f) will be generated in the by-pass diode for the LHC magnets and must be transferred as quickly as possible to heat sinks through a sufficiently large cross-section of low contact thermal resistance. Two different types of readily available diodes have been selected for first endurance tests at 77 K starting temperature, the DS-6000 type from ABB and the JD-6000 type from PSI (USA). The DS-6000 is a heavily doped diode with a thin wafer of about 45 mm diameter only, which has already been shown to have a very low forward voltage drop U_f (on-resistance)[2]. The JD-6000 from PSI has a thicker, less heavily doped wafer of about 75 mm diameter, and thus provides larger thermal contact cross-section area for heat transfer. The DS-6000 diode was mounted in a modified diode clamp of the HERA type[2] with enlarged heat sinks. For the PSI diode a special diode clamp and heat sink were constructed. In both cases the enthalpy of the copper heat sink was large enough to absorb an energy of about 1.5 MJ without exceeding an average heat sink temperature of about 400 K. During the endurance tests, the current was ramped -by applying different current rise rates up to about 30 kA/s- to flat top values between 8 kA and 12 kA of different length. Adiabatic conditions were simulated by pulling the diode out of the liquid nitrogen bath and afterwards immediately applying the current. In the tests all DS-6000 diodes survived at constant current levels from 8 to 10 kA for up to about 170 s, whereas the PSI diode burned out (punch through) at 10 kA after 25 s. Three DS-6000 diodes burned out at 12 kA after 8.5 and 30 s however and we did not test at 15 kA. It seems that the thermal resistance at the contact area from the diode to the heat sinks is too high for a sufficient transfer of the power generated in the diode wafer. Two diodes of the DS-6000 type connected in parallel will easily meet the requirements with a sufficient reserve if the current sharing ratio is not worse than about 40% to 60%. At equal current sharing only half of the power is generated in one diode and the $\int i^2 dt$ is reduced by a factor 4.

Electrical characteristics versus temperature

In order to see whether preselected diode pairs can be simply connected in parallel or whether a current balancing network must be provided to avoid the burn out of one diode, the turn on voltage V_{to} at different voltage rise rates and the dynamic current voltage characteristic $I_f = f(U_f)$ at different current rise rates has been measured on several DS-6000 diodes at 4.2 K and at 1.8 K; results are shown in Fig. 1. Due to the small heat capacity at 1.8 K the diode wafer is rapidly heated to temperatures above 40 K as indicated by the voltage drop from $V_{to} = 2.4$ V to about $U_f = 1.2$ V. Above this temperature, the forward voltage drop U_f decreases by about 1 mV/K

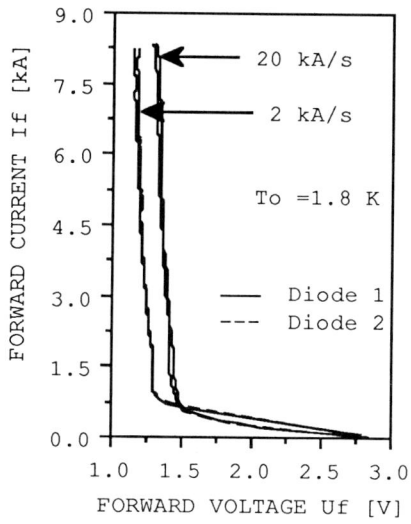
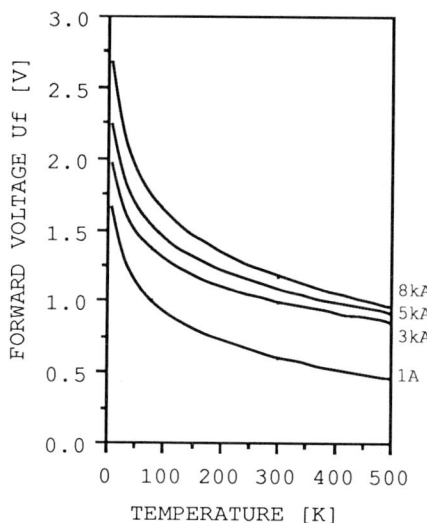

Fig. 1　Dynamic current voltage characteristic $I_f = f(U_f)$ of DS-6000 diodes

Fig. 2　Approximate forward voltage dependence on current and temperature for the DS-6000 diode

with increasing temperature. At slower current rise rate the temperature in the diode wafer is higher, which is indicated by the smaller forward voltage drop U_f. The diodes show almost identical behaviour with starting temperatures of 1.8 K and 4.2 K. Approximate forward voltage U_f versus temperature at different current levels is shown in Fig. 2.

The turn on voltage V_{to} was measured during voltage ramping and monitoring the current through the diode in forward direction. Fig. 3 shows the increase of the turn on voltage V_{to} versus voltage rise rate dU/dt measured on 6 different DS-6000 diodes at 1.8 K. The turn on voltage V_{to} varies between 2 and 3 V in the range of dU/dt typical for quenching magnets. V_{to} increases to about 4 V to 5 V at voltage rise rates in the order of 10^6 V/s. At voltage rise rates up to about 1000 V/s differences in turn on voltage up to a maximum of 600 mV were measured on diodes not originating from the same lot. For diodes from the same lot (SD4 and SD5) a maximum difference in V_{to} of about 30 mV was measured, as can be seen in Fig. 4, where the turn on voltage V_{to} versus voltage rise rate dU/dt at 1.8 K and 4.2 K is shown. These differences in turn on voltage will cause the problems for a parallel connection of diodes at liquid helium temperatures. As soon as one diode starts to conduct current -even in the 100 mA range- the small heat capacity of the silicon causes a rapid temperature increase and thus a significant decrease in forward voltage U_f. In the worst case all the current will pass through the diode which opens first and may lead to a burn out.

Parallel connection of two DS-6000

In order to observe the current sharing in a parallel connection of diodes, two diodes of the DS-6000 type with almost identical V_{to} versus dU/dt characteristic were selected and mounted in parallel without special current balancing elements. A low resistance shunt (~ 10 µΩ) made of Hastelloy C22[3] in series with each of these diodes allowed to measure the individual diode currents. Fig. 5 shows the total current and the individual diode currents versus time at 77 K

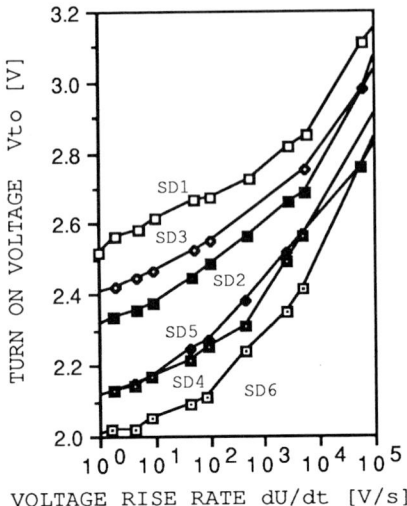
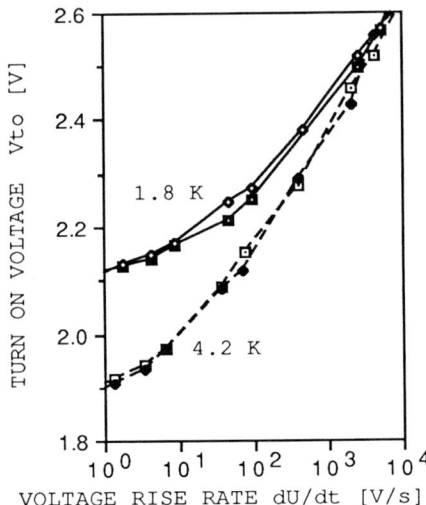

Fig. 3   Turn on voltage V_{to} versus voltage rise rate for six DS-6000 diodes at 1.8 K

Fig. 4   Turn on voltage V_{to} versus voltage rise rate dU/dt for two DS-6000 diodes

starting temperature. Both diodes open at almost the same time. The initial current sharing of about 60% to 40% improved after 70 s to about 55% to 45%. This auto-balancing effect is due to the thermal link between the two diodes and the voltage drop on the Hastelloy shunts acting as balancing resistors. The same arrangement was tested afterwards at 1.8 K starting temperature. Fig. 6 shows the individual diode currents versus time with a flat top current of about 16 kA.

Thermal model calculations for the temperature of the wafer of diode SD5 derived from temperature measurements on the heat sinks have shown that SD5 opens at about 60 K. The current sharing in fact improves with time but is still unacceptable (70% to 25%). Even optimistic extrapolations for the current versus time showed that the diode SD4 would have burned out after about 25 s. The experiment was stopped after 13.5 s for cryogenic reasons and to save the selected diode pair for an experiment with current balancing elements.

Current balancing

The following methods were investigated to improve the current balancing especially already in the very early beginning of current rising process:

a) resistors in series with each of the diodes;
b) electromagnetically coupled reactor;
c) close thermal link between the two diodes.

Since the forward voltage on the diode which opens first drops quickly to a level of about 1 V, *series resistors* must create an additional voltage up to several volts to open the second diode being still at liquid helium temperature. This resistance has to absorb energies in the range of MJ during the decay of 100 s time constant and additional heat sinks must be provided. The selection of suitable series resistors for the very beginning of the current rise process would lead to unacceptably high voltages and power consumption. In an *electromagnetically coupled reactor* the additional voltage to open the second diode is proportional to the current rise rate dI/dt in the

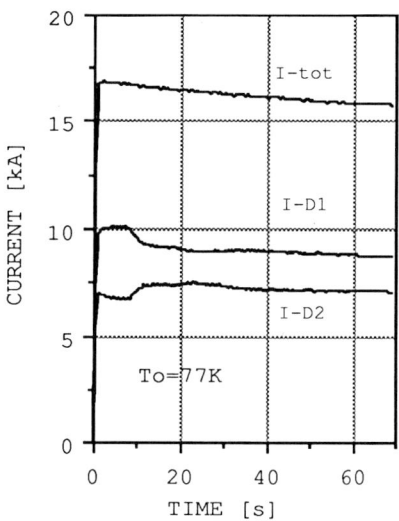
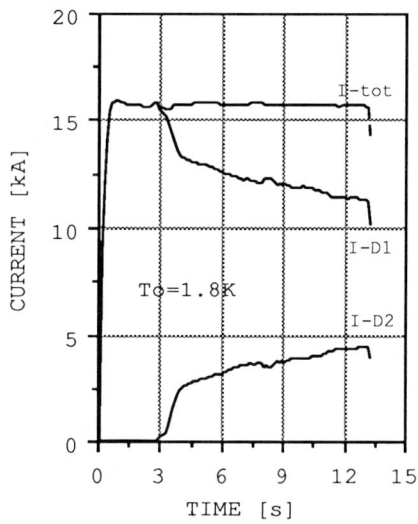

Fig. 5 Current distribution in two DS-6000 in parallel at $T_0 = 77$ K without current balancing

Fig. 6 Current distribution in two DS-6000 in parallel at $T_0 = 1.8$ K without current balancing

diode which opens first. By selecting very high permeability material for the reactor core, it is possible to create the required voltage difference of up to about 100 mV for V_{to} to open the second diode at still very low current levels (up to a few amperes) inside the first diode. From quench simulations[4] and measurements on model magnets current rise rates of about 10 A/s up to a few kA/s were estimated.

The inductance of the electromagnetically coupled reactor must be designed for the lowest current rise rate since the high permeability core will start saturating at higher current levels. For our application a coupled electromagnetic reactor of about 20 mH inductance (one turn only) saturating at a few amperes has been designed. The core material consisted of amorphous iron with a relative permeability of about 80000 at liquid helium temperatures.

To achieve a close *thermal link* between the two diodes, a minimum heat sink volume is necessary and practical limitations must be respected. The copper spacer between the two diodes acting as a heat sink has still to absorb some energy, and space must be provided for a good electrical contact to the bus bar going to the magnet. Fig. 7 shows the set-up for two diodes connected in parallel, with the electromagnetically coupled reactor -consisting of nine strip-wound cores of amorphous iron- and the series resistors for current balancing. In a special set-up, with series resistors acting as shunts to measure small currents, the efficiency of the electromagnetically coupled reactor has been verified at liquid helium temperatures. At very low current levels (< 100 mA) the current balancing was perfect, i.e. identical currents in both diodes, whereas at current levels up to 50 A the current distribution was imposed by the inequality of the resistive shunts and a current sharing ratio of about 55% to 45% was observed. For the real application at high current levels low resistive series resistors will be used in addition to the electromagnetically coupled reactor and the close thermal linking. Further tests on the parallel connection of diodes have been suspended due to the fact that meanwhile a single diode has been developed, which will eliminate the problems of parallel connection.

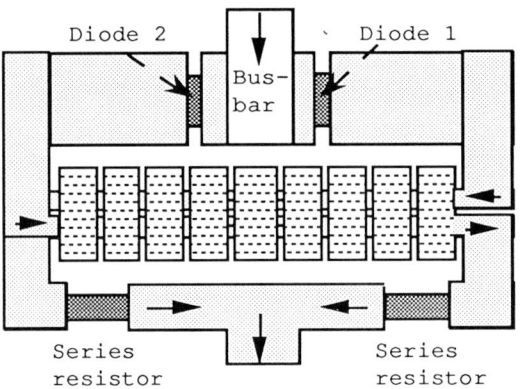

Fig. 7 Two diodes in parallel with coupled reactor and series resistors for current balancing

SINGLE HIGH CURRENT DIODE ELEMENTS

To avoid excessive temperatures inside the junction of the diode, a low forward voltage drop and a rapid heat transfer to the heat sink are essential. For a low forward voltage drop, the doping concentration in the silicon wafer should be very high and the silicon chip and the base width as thin as possible. These conditions give a diode with relatively low reverse breakdown voltage.

First estimations from the discharge circuit analysis show that the reverse voltage will normally not exceed 50 V. In close collaboration with CERN, two European firms developed prototypes of single diodes with a wafer diameter of 75 mm and a high doping concentration. They should be able to meet the above mentioned requirements with respect to current carrying capability and reverse breakdown voltage. These prototype diodes were submitted to current pulse tests at 2 K, 4.2 K, 77 K, and 300 K and to endurance tests at 77 K starting temperature.

High current pulse tests

The high current voltage characteristics $I_f = f(U_f)$ were obtained with a fast pulse power supply providing 16 kA, 200 µs half sinusoidal current pulses. The details of the measuring equipment and the test procedure are described elsewhere[1].

At room temperature and peak currents of 15 kA the temperature rise in the diode junction during the 200 µs current pulse is negligibly small, it increases by about 2 K at 77 K and by about 40 K at liquid helium starting temperature.

Fig. 8 shows the I_f-U_f characteristic at different temperatures for two different types of diode. For one of the diodes the U_f is acceptable, whereas for the other diode the increased forward voltage U_f is due to the relatively large base width. U_f increases to about 10 V at 4.2 K and 2 K.

Fig. 9 shows the reverse current I_r versus reverse blocking voltage U_r of these diodes at different temperatures. One diode shows an extremely low blocking voltage almost at all temperature ranges. On non-vented diodes reverse voltage breakdowns due to the low pressure at low temperature inside the capsule were observed above 100 V.

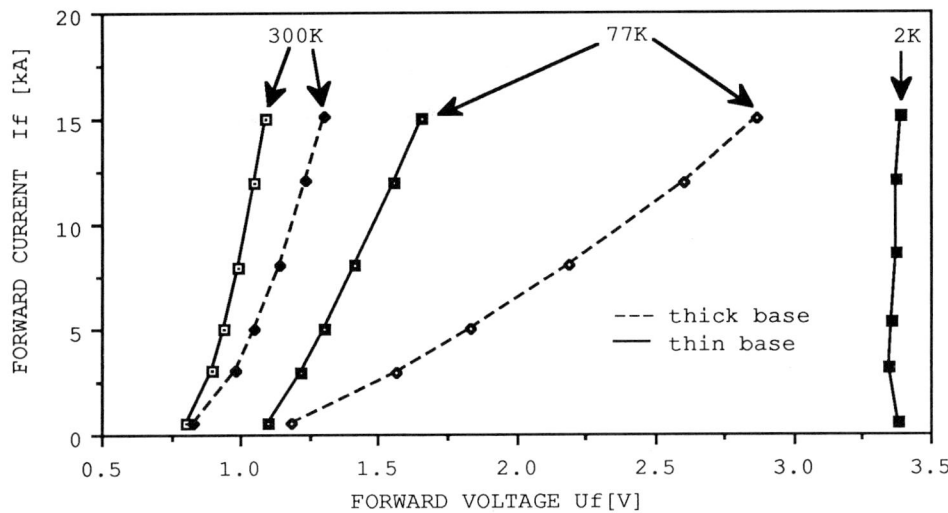

Fig. 8 $I_f = U_f$ characteristic versus temperature of diodes with thick and thin base region

Endurance tests

Endurance tests were first carried out at 77 K starting temperature and at constant current levels before submitting the diodes to endurance tests at liquid helium temperatures. In view of power generation in the diode junction, a constant current load of 15 kA during 100 s is about equivalent to an exponential decay from 15 kA with 100 s time constant. Concerning the maximum temperature rise, tests at constant current are even harder due to higher temperature gradients inside the diode assembly.

Constant currents of 15 kA for up to 170 s were applied and the temperature rise on the heat sink and diode capsule (anode and cathode) recorded versus time. The maximum wafer temperature was estimated from thermal modelling making use of the measured temperatures on the heat sinks. For the tests with 1600 cm^3 heat sink volume, a maximum junction temperature of about 490 K was estimated and with 2500 cm^3 heat sink volume T_{jmax} has not exceeded 360 K. Fig. 10 shows the measured forward voltage U_f and the diode current I_f versus time when testing in the larger heat sink.

The endurance tests at liquid nitrogen starting temperature on three different diodes mounted in different types of heat sinks have shown that the thermal contact area of about 75 mm diameter should be sufficiently large to transfer the developed power to the heat sink without overheating the wafer, if the forward voltage U_f stays reasonably small -as indicated in Fig. 10 for the thin base diode- and does not increase due to radiation, for instance. A number of diode samples will be produced and submitted to endurance tests at liquid helium temperatures in the near future.

CONCLUSIONS

Two lines of approach have been pursued successfully for the realization of a suitable current by-pass element at liquid helium temperature:

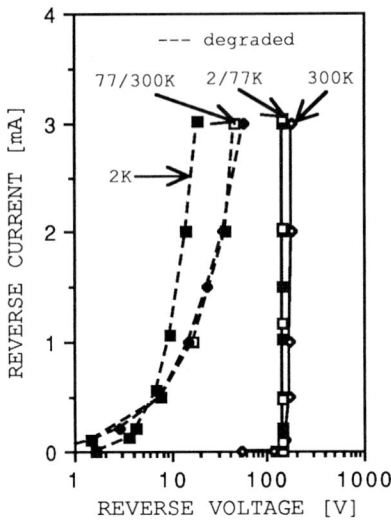
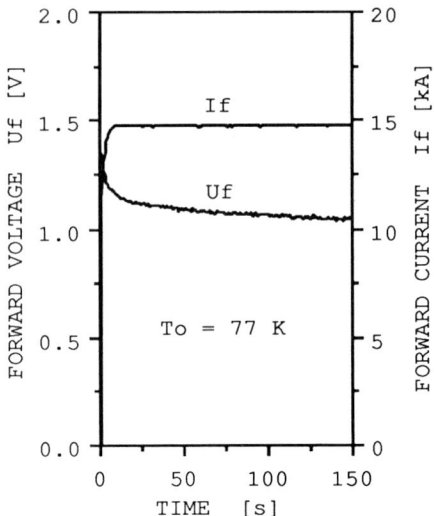

Fig. 9 Reverse bias characteristic versus temperature of two different diodes

Fig. 10 Forward voltage and current versus time during endurance test at $T_0 = 77$ K

1. Two commercially available diodes of the HERA type connected in parallel can easily meet the requirements if a sufficient good current sharing is imposed by current balancing elements. Design criteria for these current balancing elements can be derived from individual diode characteristics at low temperature.

2. Single diode elements of thin base region developed in industry have been tested. The results are promising and if the diodes can be made with reproducible characteristics they will provide the preferred solution especially in view of radiation hardness.

ACKNOWLEDGEMENTS

We wish to acknowledge the support of G. Brianti and R. Perin during this work and to thank our colleagues L. Coull, D. Leroy, and T.M. Taylor for many helpful discussions. We also thank A. Arn and J.M. Fraigne for designing and mounting the clamping systems and F. Streun for his help during testing. The great effort of MARCONI ELECTRONIC DEVICES LTD. and EUPEC (AEG and SIEMENS) for the development of thin base diodes is gratefully acknowledged.

REFERENCES

1. D. Hagedorn, W. Nägele, Radiation effects on high current diodes in an accelerator environment, at this conference.
2. K.H. Meß, Quench protection at HERA, "Proceedings of the 1987 IEEE Particle Accelerator Conference", p. 1474.
3. PARAMOUNT at Haynes International, Inc., USA.
4. D. Hagedorn, F. Rodriguez-Mateos, Modelling of the quenching process in complex superconducting magnet systems, MT-12, June 1991, Leningrad, USSR.

COEFFICIENT OF FRICTION MEASUREMENTS OF SOLID FILM LUBRICANTS AT CRYOGENIC TEMPERATURES

L. O. El-Marazki

Applied Superconductivity Center
University of Wisconsin
Madison, Wisconsin USA

ABSTRACT

Experiments were conducted to determine the coefficient of friction of the inorganic solid film lubricant TIOLUBE 1175 (T1175). The experiments were conducted first at room temperature and at atmospheric pressure. The experiments are then conducted in vacuum at room temperature, at liquid nitrogen temperature (LN_2), and at liquid helium (LHe) temperatures. The coefficient of friction is measured between two aluminum specimens coated with T1175 (self friction) at normal loads of 30 to 78 MPa, and a surface velocity of .076 to 0.3 m/min. The coefficient of friction between aluminum and aluminum coated with T1175 increases with normal load, and decreases with temperature. The heating energy generated due to friction is also determined. The experimental apparatus is discussed and the test results are presented.

INTRODUCTION

The cold to warm struts in the SMES Engineering Test Model(ETM) are subjected to combined thermal and magnetic loads due to the cooldown of the helium vessel and magnetic pressure in the radial direction. In addition, the struts are subjected to differential movement in the vertical direction. The struts are attached to the helium vessel at each end through a ball and socket mechanism. To minimize the heat generated due to friction, a thin layer of T1175 is bonded to the socket since hydrodynamic lubrication using oil or grease is ruled out for most cryogenic applications. T1175 has a low coefficient of friction which comes close to that of oil and grease especially after it is burnished.

The purpose of this test is to characterize the frictional properties of TIOLUBE 1175 which is an inorganic bonded dry-film lubricant utilizing a specially formulated molybdenum disulfide (Mos_2) pigment working in combination with other lubricant pigments.[1] An aluminum sample coated with T1175 is first tested for bond integrity by immersing it in liquid helium. Visual testing of the samples indicates that there is no sign of peeling or cracking. The coefficient of friction is first determined between two aluminum samples coated with .001 cm of T1175 (self friction). The samples are then burnished and the coefficient of friction is determined at different numbers of burnishing cycles. The coefficient of friction is then determined between aluminum and aluminum coated with T1175 at five normal loads at atmospheric pressure and in vacuum, at room, liquid nitrogen, and at liquid helium temperatures. The test specimen temperature is measured with a carbon thermometer.

EXPERIMENTAL APPARATUS AND TEST SAMPLES

Experimental Apparatus

The friction apparatus is shown in Fig. 1 and is similar to that of Stone and Young[2]. The device is self-aligning. The friction sample consists of a central pull strip with two side pads as shown in Fig. 2. The pull strip, 1.3 cm thick by 2.9 cm wide by 11.3 cm long is aluminum coated with T1175. The pull strip is connected to the load cell which in turn is connected to the actuator of the MTS machine. The two side pads, 1.3 cm thick and 2.54 x 2.54 cm square, are aluminum. These two pads are placed on opposite sides of the pull strip A and B as shown in Fig. 2, providing two frictional surfaces. The aluminum blocks into which the pads are mounted are loaded through knife edges which lie in the same plane as does the surface of the pads upon which the frictional forces act. The line connecting the two knife edges in each set bisect the horizontal centerline. The back surface of one of the pads bears against a rounded vertical line of contact with the aluminum block by which it is supported and loaded. These several articulations assure that the resultant normal forces on the two sides of the pull strip are collinear and assure that the presence of a frictional force does not change the normal force under static or quasistatic conditions.

The normal load is applied by means of mechanical linkage and dead weight outside the dewar as seen in Fig. 1. The actuator of the MTS machine which is connected to the central pull strip can be activated to move with constant velocity to apply the friction force. The friction load increases until slipping occurs. Displacement of the sample is measured by a linear variable differential transformer mounted at room temperature at the end of the actuator which is controlled by a closed loop feedback system.

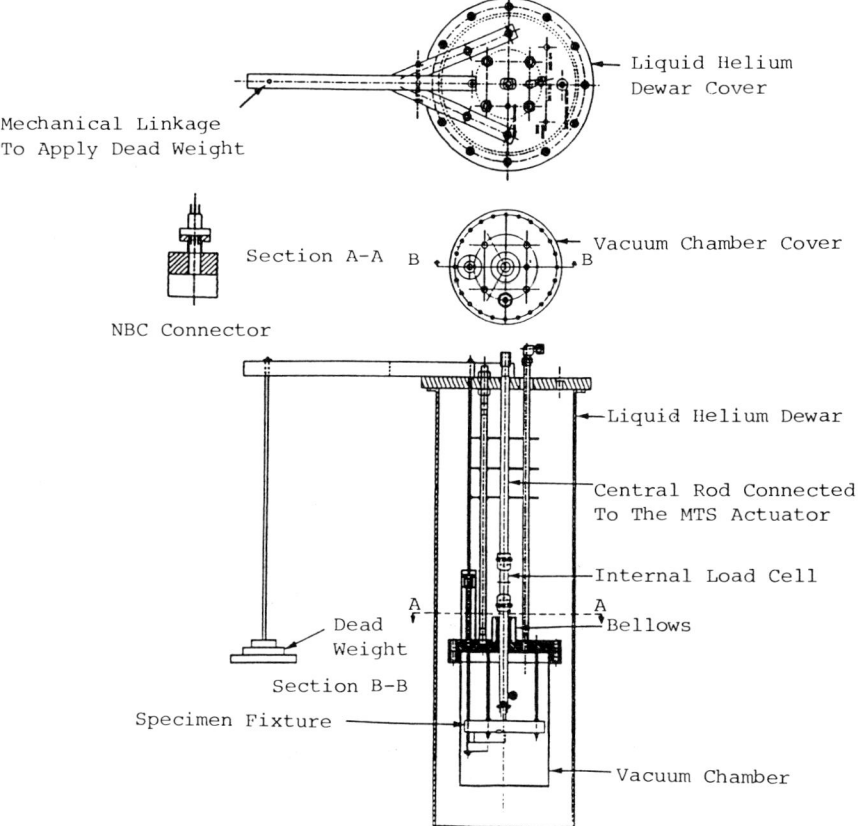

Fig. 1   Friction Test Apparatus

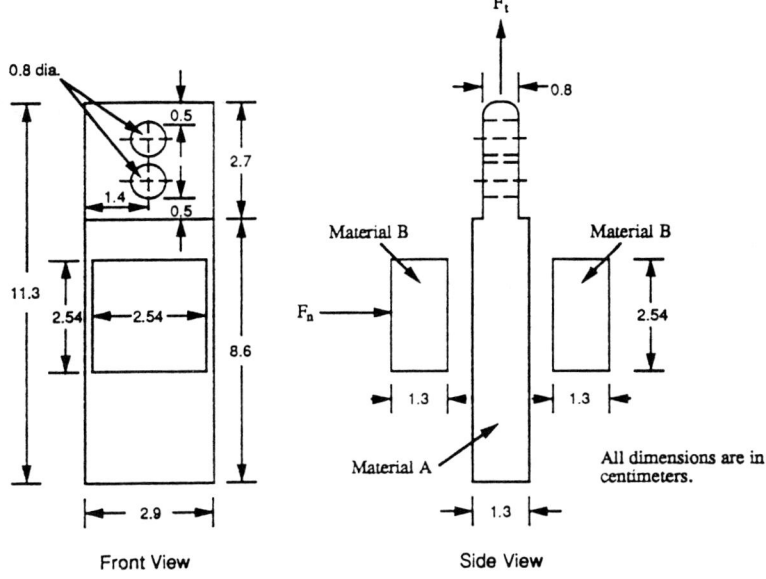

Fig. 2 Friction Sample

The friction test sample is contained in a vacuum chamber Fig. 1. To simulate the operating conditions of the struts, a minimum vacuum of 10^{-4} Torr. is maintained during the test. The friction sample holder is connected to the vacuum chamber cover with three OFHC rods which have low thermal resistance and act like cold fingers. A copper gasket between the flange and the cover of the vacuum chamber provides a metal to metal contact. Two sets of stainless steel bellows are used to maintain the appropriate vacuum, and allow for load transfer and sample displacement. The central bellows Fig. 1 are used for friction load transfer. The lever arm bellows are used for normal load application. The sample temperature is measured using a carbon thermometer and a feed through connector for electrical wire leads Fig. 1 is also used to keep the chamber vacuum tight.

Test Samples

Two sets of samples are used. The first set of samples, the sides of the pull strip, and the two side pads are coated with a thin layer of T1175. The second set of samples, the sides of the pull strip are coated with a thin layer of T1175, and 6061 aluminum with a surface finish of 3.1 and 3.3 µ in is used for the side pads. To determine the bond integrity; a piece of aluminum coated with T1175 is gradually cooled to liquid helium temperature by first exposing it to helium vapor, then immersing in liquid helium. A second specimen is quickly immersed in liquid helium. Both specimens show no cracking or peeling.

The test samples are burnished for an equivalent distance of 18.6m at normal pressure of 1.9 MPa. The coefficient of self friction is found to be .12 for samples as received and decreased to .079 after a burnishing cycle of 18.6 m.

TEST PROCEDURE

Experiments are conducted in a vacuum at room temperature, liquid nitrogen, and liquid helium temperatures. Experiments are also conducted at room temperature at atmospheric pressure. The normal load is applied after the specimen, load cell and loading mechanisms stabilized at the test temperature. The actuator, which is connected to the central pull strip through the load cell, is activated after one minute wait to apply the load to the specimen which is pulled at constant speed. The friction force versus displacement is recorded for each normal load. The vacuum chamber is burged prior to liquid helium test using room temperature helium gas to ensure no ice, solid oxygen, or solid nitrogen formation on surfaces. The specimens are cooled down to liquid nitrogen temperature before running the liquid helium test.

EXPERIMENTAL RESULTS

The pull strip and the two side pads are coated with a thin layer of T1175. The friction surfaces are burnished before the test. Five normal loads ranging from 4395 to 11298 N are used. The friction force vs. displacement curve, Fig. 3, shows an increase in friction force until slipping begins after which the force decreases. The static friction force used to calculate the static coefficient of friction is the force at the sudden transition shown in Fig. 3. The kinetic coefficient of friction is based on the friction force following the sudden change in Fig. 3 . The coefficient of friction is equal to half the friction force divided by the normal load since there are two frictional surfaces. It is found that changing both the normal load and/or the sliding velocity by 2.5 times has no effect on the self static and kinetic coefficient of friction. It is also found that the static and kinetic coefficients of friction in a vacuum are higher than at room temperature and at atmospheric pressure as shown in Table 1 for a sample speed of .076m/min.

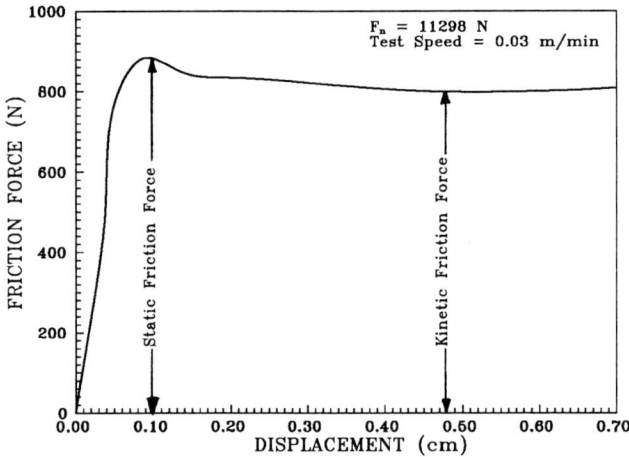

Fig. 3. Friction Force Vs. Displacement for Coated Aluminum-Coated Aluminum at Room Temperature and Atmospheric Temperature

The coefficient of friction between aluminum and coated aluminum is measured at room temperature and five normal loads at a sliding speed of .061 m/min at atmospheric pressure and in vacuum. The average coefficient of friction decreased with load as shown in Table 2. The coefficient of friction is measured at liquid nitrogen temperature at five normal loads and in vacumm of 1.7 to 2.7 $\times 10^{-6}$ Torr. The sample is contained in the vacuum chamber which is immersed in the liquid nitrogen. The sample temperature is measured with a carbon thermometer. The average coefficient of friction decreases with temperature at the same normal load, and also decreases with increase in normal load as shown in Table 2. The coefficient of friction is measured in a vacuum at five normal loads at 5.6 K. The tangential force increases with displacement after motion begins. A typical friction curve at 11, 298 Newtons is shown in Fig. 4 . The average coefficient of friction decreases also with normal load as shown in Table 2. The average value of the coefficient of friction in vacuum and at any normal load is higher at liquid helium temperature than at higher temperature as shown in Fig. 5.

Table 1. Average Value Of Static and Kinetic Coefficient of Friction at Room Temperature at Atmospheric Pressure and in Vacuum

Normal Load	Atmospheric Pressure		Vacuum (1.3 x 10^{-5} to 6 x 10^{-6}) Torr	
Newtons	μ_s (average)	μ_k (average)	μ_s (average)	μ_k
4395	.086	.076	.096	.094
6405	.089	.084	.095	.093
8398	.088	.084	.094	.094
9608	.088	.085	.094	.092
11298	.087	.084	,095	.093

Coated aluminum-coated pair
Testing speed = .076 m/min

One goal of conducting the friction test experiment is to estimate the heat energy released by friction. The heat generated due to friction is determined from the work done by friction forces on both sides of the pull strip, W = ∫ $F_t$dx, where W is the work done and F_t is the sum of the friction forces on both sides of the pull strip. Heat absorbed by the specimen holder, ΔQ is readily illustrated by the following equation

$$\Delta Q = \int_{T_i}^{T_l} m\, c_p\, dT$$

where T_i is the initial temperature of the test and T_f is the final temperature, m is the mass of the holder and the specimen and c_p is the specific heat. It is found that the heat absorbed by the sample and holder is small compared to the heat generated. This shows that the heat absorbed by the sample is small fraction of the total heat generated.

Table 2. Average Static Coefficient of Friction

Normal Load	Temperature			
	Room Temperature		Liquid Nitrogen Temperature & Vacuum (1.7-2.7) x 10^{-6} Torr	Liquid Helium Temperature & Vacuum (1.5-2.2) x 10^{-6} Torr
Newtons	Atmospheric Pressure	Vacuum (3-6) x 10^{-6} Torr		
4395	1.656E-02	2.021E-02	7.8473E-02	1.260E-01
6405	1.241E-02	1.491E-02	5.7941E-02	7.596E-02
8398	1.193E-02	1.459E-02	4.9229E-02	6.190E-02
9608	1.040E-02	1.475E-02	4.468E-02	5.33E-02
11298	9.241E-03	1.376E-02	3.790E-02	4.532E-02

Aluminum-coated aluminum pair
Testing speed = 0.061 m/min

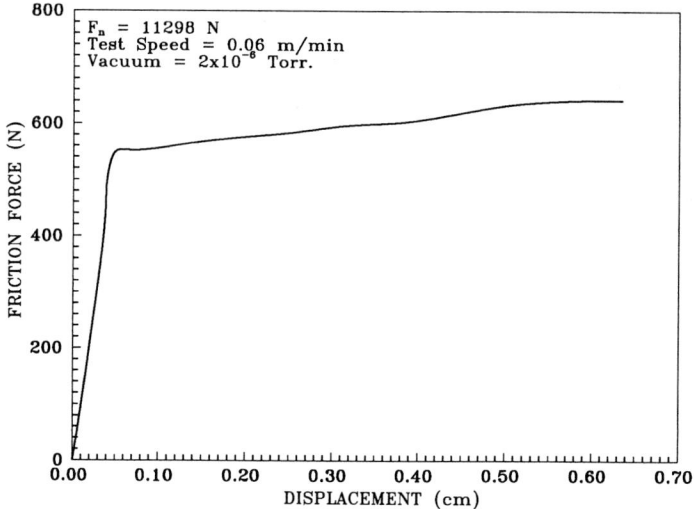

Fig. 4. Friction Force Vs. Displacement for Aluminum-Coated Aluminum
In Helium and in Vacuum

DISCUSSION of RESULTS

The friction force versus displacement for coated aluminum samples show semi-stable sliding after the friction force reaches a peak value followed by sample sliding. Sliding remains smooth and stable. The friction force versus displacement for aluminum/aluminum coated with T1175 shows stable sliding, after the sliding surfaces accelerate continuously from zero to steady-state velocity. This is desirable in cryogenic applications to avoid generating heat pulses and to allow time for heat to dissipate. The heat generated due to friction is absorbed by the sample, sample holder and the helium path. It is found that the portion absorbed by the sample and sample holder ranges from 1 to 10 percent (1-10%).

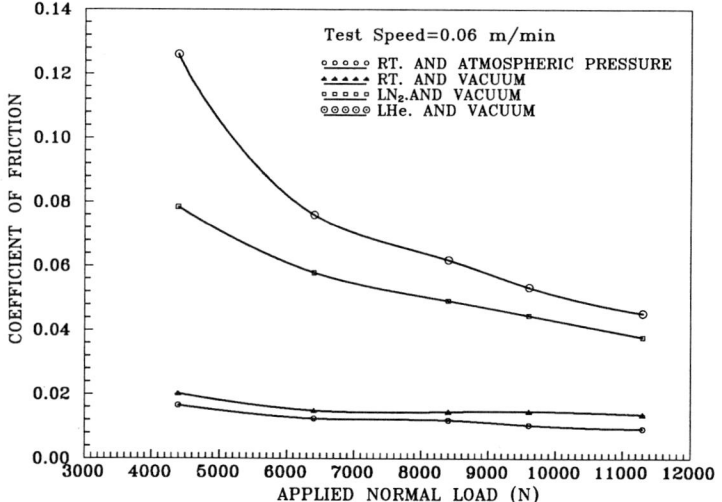

Fig. 5. Average Coefficient of Friction
For Aluminum-Coated Aluminum

312

SUMMARY AND CONCLUSION

The coefficient of friction is reproducible within the experimental error. During the liquid helium test, the fictional heat raises the specimen's temperature by 0.5 K. The self coefficient of friction does not depend on the test speed or applied normal load. However, aluminum pads versus aluminum coated with T1175 have coefficients of friction that decrease with increases in normal load and increase with decreasing temperatures. The fraction of heat absorbed by the sample is found to be small compared to the total generated friction heat.

REFERENCES

1. Technical Data Supplied by Tiodize Co., Inc.
2. E. L. Stone W. C. Young, Coefficient of Friction Measurements of Fiberglass/epoxy at Cryogenic Temperatures, in: "Advances in Cryogenic Engineering - Materials," Vol. 26, Plenum Press, New York (1979), P. 315.

MAGNETO-TRANSPORT PROPERTIES OF FILAMENTARY ALUMINUM CONDUCTORS IN MAGNETIC

FIELDS, 12 - 30 K

W. N. Lawless, C. F. Clark, and R. W. Arenz

CeramPhysics, Inc.
921 Eastwind Drive, Suite 110
Westerville, Ohio 43081

ABSTRACT

Electrical resistivity and thermal conductivity measurements are reported on a conductor containing four high-purity Al filaments in an Al-Fe-Ce alloy matrix. Measurements in both transverse and longitudinal magnetic fields up to 8 T were made in the 12 - 30 K range, and the data are analyzed according to the Wiedemann-Franz and Kohler relations. These relations are not obeyed by either the transverse or longitudinal magneto-transport components. However, phenomenological functions based on these relations are developed which describe all the experimental data very well.

INTRODUCTION

It has long been recognized that high-purity aluminum offers significant advantages for magnet applications due to a small magnetoresistance at liquid-hydrogen temperatures. For fast current penetration in pulsed-power applications, fine aluminum filaments embedded in a high strength matrix alloy are required. The breakthrough in this technology occurred in the mid-1980's with the demonstration that Al-Fe-Ce matrix alloys have a workability compatible with high-purity Al without, however, diffusion-contaminating the filaments. These alloys are rich in Al so that the lightweight advantage is not compromised.[1]

The metallurgical methods used to fabricate composite conductors containing 4 and 19 filaments of high-purity Al in Al-Fe-Ce matrix alloys have been described,[2] and measurements of the specific heat and thermal conductivity of these conductors in zero magnetic field at low temperatures indicate minimal contamination of the high purity filaments.[3] The transverse magnetoresistance of these conductors has been measured 4-30 K in fields up to 9 T,[4] as has the resistance increase with both monotonic and cyclic-strain in zero field at 4.2 and 20 K.[5] Theoretical work has dealt with computations of the ac loss in these multi-filamentary conductors[6] and the contribution of the Hall-generation loss to magnetoresistance.[7] The picture emerging from these studies is that these composite conductors appear to fulfill the desired requirements for pulsed-power magnet applications in the hydrogen-temperature range, although the magnetoresistance is larger than in pure Al.[4] There have been no measurements to date on the magnetic field dependence of the thermal conductivity of these composite conductors at low temperatures. In this paper we report measurements of both the thermal conductivity and the electrical resistivity on a 4-filament conductor in transverse and longitudinal magnetic fields, 12-30 K. The Wiedemann-Franz and Kohler relations will be examined based on these data for both the transverse and longitudinal components.

Experimental Methods and Results

The matrix alloy for the 4-filament conductor measured was 92Al-4Fe-2Ce, and samples in the form of 1-m long wires were supplied by the Alcoa Technical Center in the unannealed state. The outer diam. of the conductor was 0.0762 cm, and the filament diam. was 0.0262 cm. The measured RRR of these wire samples (after annealing) was 1430. For the transverse measurements, the wire was wrapped on a specially machined bolt to form a tight spiral; for the longitudinal measurements, a separate wire was wrapped on a nylon mandrel. In both cases, the bend diameters were 0.65 cm, and the wire samples were annealed at 410 °C for 10 min after removal from the bolt or mandrel.

Thermal conductivity and electrical resistivity measurements were made simultaneously on the same sample by a combination-measurement technique which is based on the calibration of the thermal conductance of a 38 gauge copper hookup wire as a function of temperature and field.[8]

The longitudinal portion of the wire in the transverse sample was determined from the geometry of the coil, and similarly for the transverse portion in the longitudinal sample. In both cases the orthogonal portions were small, ~ 5%.

The voltage taps were made by fusing #16 gauge copper leads to the wire using a Unitek wire bonder, and the current leads were #38 gauge copper leads soldered to the wire with a Cd-based solder (one of these current leads is the calibrated lead mentioned above). The thermometers were 1 kΩ Allen-Bradley BB resistors thermally anchored to the wire with copper wire, and a 350 Ω manganin heater (wrapped bifilarly) was attached to the free end of the wire.

Each wire sample was mounted in a single-can cryostat compatible with the bore of a Cryomagnetics 9 T superconducting magnet system. A Lake Shore Cryotronics Model DRC-91C temperature controller was used with a capacitance thermometer mounted in the reservoir post of the cryostat. The sample thermometers and the capacitance thermometer were calibrated against a Ge thermometer in zero field at 8-10 points in the 12-30 K range; subsequently, at each magnetic field level, the sample thermometers were recalibrated against the capacitance thermometer at 8-10 points. The R-T data for the sample thermometers were fitted using standard relations[9] in the data reductions.

The electrical resistivity measurements were performed with current reversal to determine thermal emf's (~5 µV), and the current levels were maintained sufficiently large to reduce these emf's relative to the total voltage signals. In the thermal conductivity measurements, the relative temperature changes $\Delta T/T$ were maintained at 2-3%, and data collections were semi-automated with computer-interfacing software.

The experimental data were separated into the purely transverse and longitudinal components, as follows: First, the experimental data at each magnetic-field level were fitted to power expansions in temperature,

$$K = \sum \alpha_n T^n, \quad \rho = \sum \beta_n T^n \qquad (1)$$

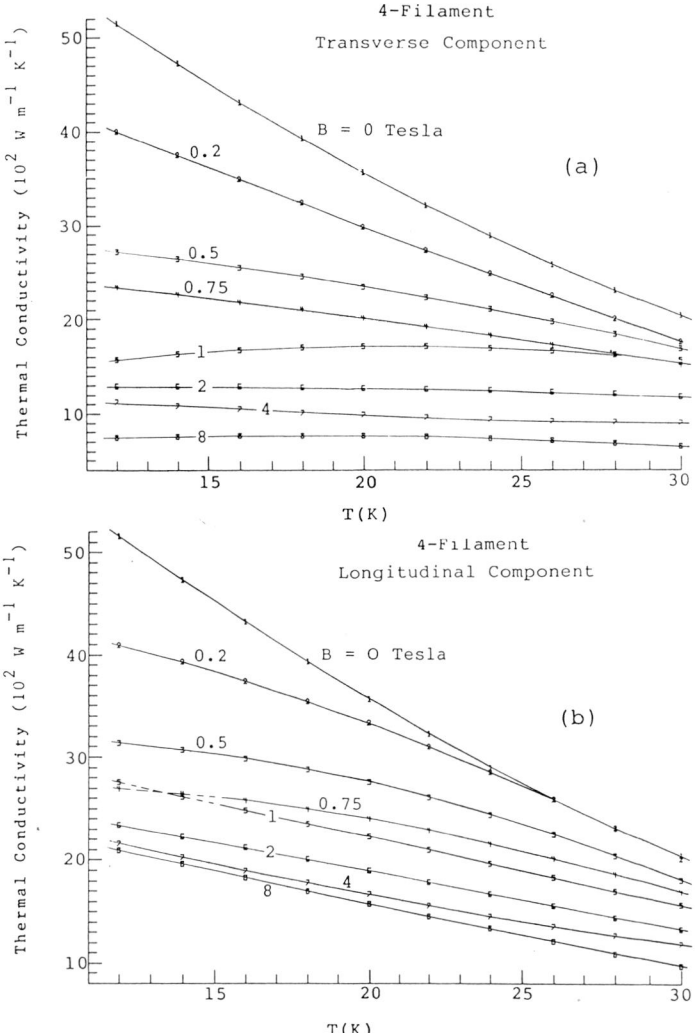

Figure 1. Thermal conductivity of the 4-filament composite conductor in transverse (a) and longitudinal (b) magnetic fields.

where K and ρ are thermal conductivity and electrical resistivity, respectively. It was found that third or fourth order fits gave excellent representations of the experimental data (residuals < 2%), and smoothed K- and ρ-data were generated at even temperature intervals (12, 14, 16,...) for each field level. Next, since both thermal and electrical resistances are additive, it is trivial to set up simultaneous equations for the purely transverse and longitudinal components, given the geometric data for the two components for the two wire samples. Solving these equations at fixed temperatures for each field level allows the separation of the purely transverse and purely longitudinal components of K and ρ at each field level at several fixed temperatures.

The separated transverse and longitudinal components of K for this composite conductor are shown in Fig. 1. The longitudinal-component data in Fig. 1(b) below 15 K are shown as dashed lines at 0.75 and 1.0 T because experimental data are missing below this temperature at these field levels.

The corresponding, separated transverse and longitudinal components of ρ for this composite conductor are shown in Fig. 2 and no experimental data are missing in these separations.

Wiedemann-Franz and Kohler Analyses

The Wiedemann-Franz law relates K and ρ to the universal function,

$$K\rho/T = L = (\pi k/e)^2/3 = 2.45 \times 10^{-8} \text{ W } \Omega \text{ K}^{-2} \qquad (2)$$

where L is the Lorentz number. The validity conditions on this law at low temperatures are that: (1) The phonon contribution to K is negligible; and (2) The conduction electrons are scattered elastically.

We can examine the first condition by writing for the phonon contribution to the thermal conductivity,

$$K_{phonon} = (310 \, K_o/GN^2\theta_D^2)T^2 \qquad (3)$$

where K_o is the thermal conductivity at high temperatures (~ 2.5 W cm^{-1} K^{-1}), G is a constant ~ 70, N is the number of free electrons per atom (1), and θ_D is the Debye temperature (~ 396 K). We find from Eq. (3) that $K_{phonon} \sim 7 \times 10^{-5} T^2$ which is negligibly small, compared even to the measured thermal conductivity at 30 K and 8 T (Fig. 1). We shall return below to the second validity condition.

Wiedemann-Franz plots of the transverse data, $K_t\rho_t/T$, and of the longitudinal data, $K_\ell\rho_\ell/T$, are plotted versus temperature in Fig. 3, and the value of the Lorentz number is also indicated. Neither the transverse nor longitudinal data follow the Wiedemann-Franz law, and there is no particular ordering of the curves in Fig. 3 by B-field level.

The spread in the data in Fig. 3 is relatively <u>small</u> -- e.g., in the worst case at 12 K, the spread is ±19% in the transverse data, ±12% in the longitudinal data. A large amount of experimental data are correlated in Fig. 3, and we interpret these spreads as due to the accumulated experimental uncertainties. We are led to this conclusion by the fact that the spreads are about the expected size of the cumulative uncertainties and are largest at the lowest temperatures where the experimental quantities have their largest values and largest uncertainties.

Given this interpretation, we look for a universal representation, and excellent fits to the <u>mean</u> values of $K\rho/T$ are obtained with

$$K\rho/T = a_0 + a_1/T + a_2/T^2. \qquad (4)$$

These fits are shown in Fig. 3, and the fitted coefficients are (for the units in Fig. 3):

<u>Transverse</u>: $a_0 = 1.268 \times 10^{-8}$, $a_1 = -3.315 \times 10^{-8}$, $a_2 = 1.966 \times 10^{-6}$

<u>Longitudinal</u>: $a_0 = 9.003 \times 10^{-9}$, $a_1 = 4.128 \times 10^{-8}$, $a_2 = 1.386 \times 10^{-6}$

Magneto-transport of electrons in metals is usually described by Kohler's law, expressed as a universal function,

$$\Delta\rho/\rho_o = f(B/\rho_o), \qquad (5)$$

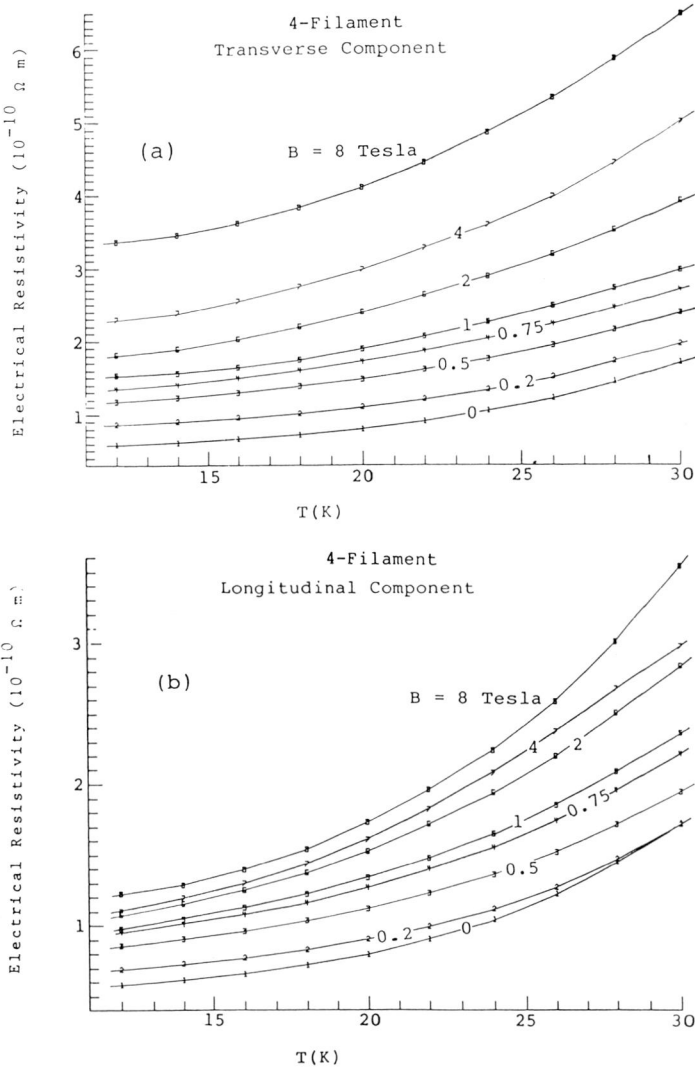

Figure 2. Electrical resistivity of the 4-filament composite conductor in transverse (a) and longitudinal (b) magnetic fields.

where $\Delta\rho$ is the change in resistivity in field B at constant temperature. Kohler's rule is strictly obeyed only when electron scattering by nonmagnetic impurities is dominant.

Kohler plots of the transverse and longitudinal data are shown in Fig. 4, at T = 12, 16, 20, and 24 K; the B/ρ_o axes in these plots reflect the area of the conductor, not the area of the filaments in the conductor. These Kohler plots are similar and approximately T-independent, particularly at low B/ρ_o values.

However, the Kohler relation is not followed in Fig. 4; rather, two regimes are indicated, above and below about 2 T which is the approximate saturation field. Two fitting regimes are suggested, and excellent fits to the data are obtained with

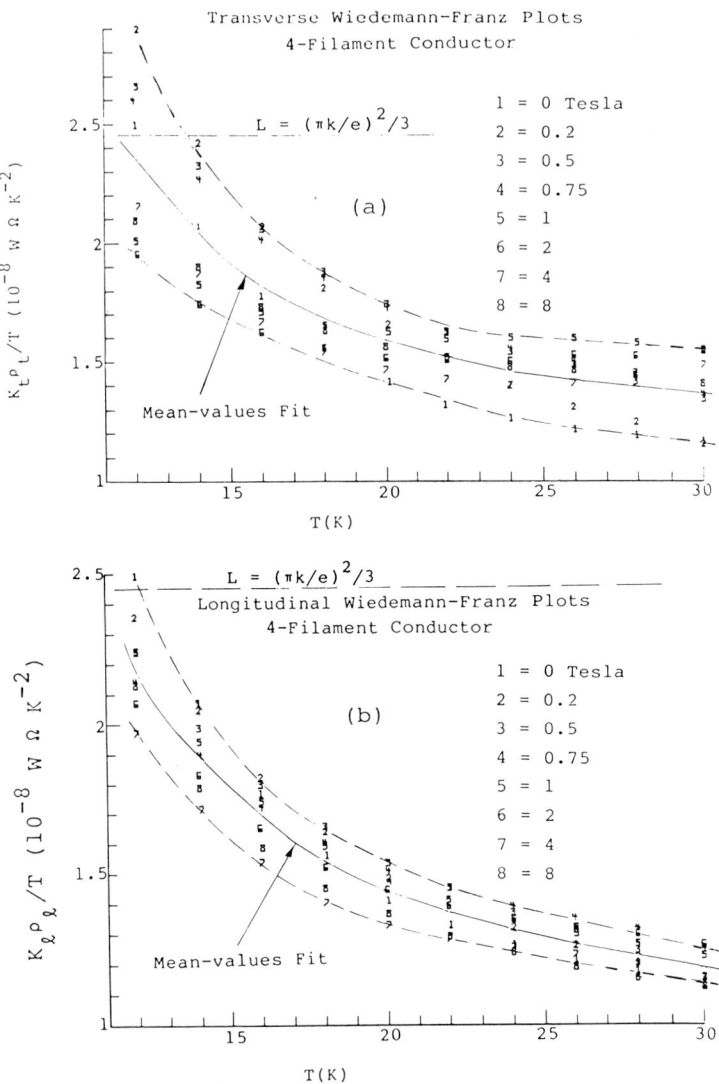

Figure 3. Wiedemann-Franz plots of the transverse-component data (a) and of the longitudinal-component data (b). The mean-value fits shown are according to Eq. (4).

$$\Delta\rho/\rho_o = b_1(B/\rho_o) + b_2(B/\rho_o)^2$$
$$(\text{small } B/\rho_o)$$

(6)

$$\Delta\rho/\rho_o = c_o + c_1(B/\rho_o)$$
$$(\text{large } B/\rho_o)$$

The two fitting regimes divide at B/ρ_o = 1.86. x 10^{10} and 2.56 x 10^{10} T Ω^{-1} m^{-1} for the transverse and longitudinal cases, respectively. The Eq.

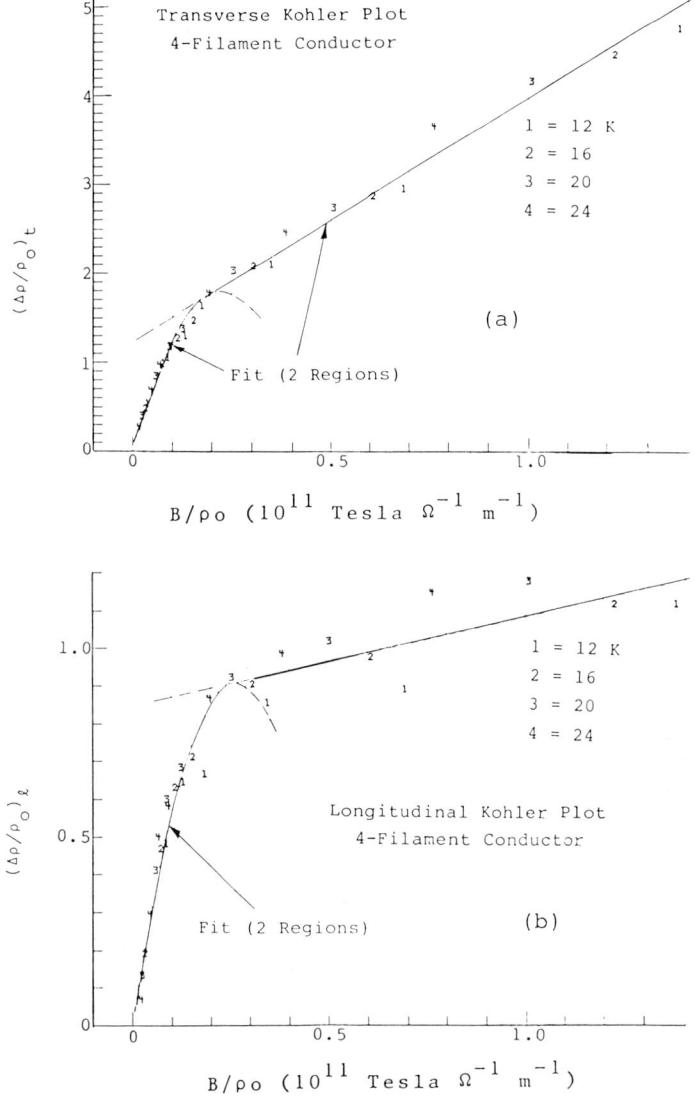

Figure 4. Kohler plots of the transverse-component data (a) and of the longitudinal-component data (b) The solid curves are the fitted data according to Eq. (6), and the break in the two fitting regimes occures at about 2 T.

(6) fits are shown in Fig. 4, and the fitted coefficients are (for the units in Fig. 4):

<u>Transverse</u>: $b_1 = 1.582 \times 10^{-10}$, $b_2 = -3.511 \times 10^{-21}$, $c_0 = 1.214$, $c_1 = 2.760 \times 10^{-11}$

<u>Longitudinal</u>: $b_1 = 6.870 \times 10^{-11}$, $b_2 = -1.304 \times 10^{-21}$, $c_0 = 0.843$, $c_1 = 2.397 \times 10^{-12}$

Discussion and Conclusions

A large amount of experimental data have been measured in both transverse and longitudinal magnetic fields on a composite conductor containing four filaments of high purity Al in an Al-Fe-Ce matrix, Figs. 1-2. The electrical and thermal transport properties of this conductor are dominated by the Al filaments.[3] Both the thermal and electrical conductivities of this wire are suppressed more by a transverse magnetic field than by a longitudinal magnetic field. Of particular importance for magnet design, the effect of a magnetic field on both K and ρ is much more pronounced at low fields, < 2 T, and, as seen in Figs. 1-2, the magneto-transport properties are essentially saturated at about 2 T in this composite conductor.

The separated transverse and longitudinal data are analyzed according to the Wiedemann-Franz law, Fig. 3, and the Kohler relation, Fig. 4. The Wiedemann-Franz law is not obeyed, and this means that electron impurity scattering does not dominate over electron-phonon scattering in this temperature range. This result also explains why the Kohler relation is not satisfied. Nonetheless, it is found that universal fitting equations describe the data within the experimental uncertainties, Eqs. (4) and (6). The practical implication here is that if ρ_o is known in the 12 - 24 K range, the transverse and longitudinal ρ-components can be found from Eq. (6) at all field levels up to 8 T, following which the transverse and longitudinal K-components can be found from Eq. (4).

This research was sponsored by the Air Force.

References

1. C. E. Oberly and J. C. Ho, "The Origin and Future of Composite Aluminum Conductors", Adv. Cryogenic Eng. 36A, 645 (Edit. R. P. Reed and F. R. Fickett, Plenum Press, New York, 1990).
2. M. K. Premkumar, F. R. Billman, D. J. Chakrebarti, R. K. Dawless, and A. R. Austen, "Composite Aluminum Conductor for High Current Density Applications at Cryogenic Temperatures", ibid. p. 733.
3. W. N. Lawless,"Thermal Properties of Composite Aluminum Conductors in the Temperature Range 13-30 K", ibid. p. 687.
4. C. A. Thompson and F. R. Fickett, "Magnetoresistance of Multi-filament Al/Al-Alloy Conductors", ibid. p. 663.
5. K. T. Hartwig and R. J. DeFrese, "Mechanical and Electrical Testing of Composite Aluminum Cryoconductors", ibid. p. 709.
6. W. J. Carr, "Ac Loss in a Composite Hyperconductor", ibid. p. 695.
7. P. W. Eckels and J. H. Parker, Jr. "Magnetoresistance Composite Conductors", ibid. p. 655.
8. W. N. Lawless, C. F. Clark, and S. K. Hampton, "Method for the Simultaneous Measurement of Electrical Resistivity and Thermal Conductivity in Magnetic Fields at Low Temperatures", Rev. Sci. Istrum. (in press).
9. See P. H. Kes, C. A. M. van der Klein, and D. de Klerk, Cryogenics 14, 168 (1974).

TRIBOLOGICAL BEHAVIOR OF 440C MARTENSITIC STAINLESS STEEL FROM -184°C TO 750°C[*]

A. J. Slifka, R. Compos, T.J. Morgan, J. D. Siegwarth

Chemical Engineering Division
National Institute of Standards and Technology
Boulder, Colorado

and

Dilip K. Chaudhuri[**]

Mechanical Engineering Department
Tennessee State University
Nashville, Tennessee

ABSTRACT

Characterization of the coefficient of friction and wear rate of 440C stainless steel is needed to understand the effects of frictional heating in the bearings of the High Pressure Oxygen Turbopump of the Space Shuttle Main Engine. The coefficient of friction and wear rate have been measured over a range of temperature varying from liquid oxygen temperature (-184°C) to 750°C. The normal load has also been varied resulting in a variation of Hertzian stress from 0.915 to 3.660 GPa while the surface velocity has been varied from 0.5 to 2.0 m/s.

INTRODUCTION

AISI 440C martensitic stainless steel is used as the High Pressure Oxygen Turbopump (HPOTP) bearing material because of its corrosion resistance. The LOX propellant , as well as the MoS_2 coating and PTFE transferred from the PTFE-filled glass bearing-cage offer little lubrication. In addition to this, the bearings are subjected to high axial loads during start-up and shut-down.[1] During normal operation of the HPOTP, microslip (≤2%) conditions exist at a typical contact stress of 2.07 GPa according to Wedeven and Miller.[2] During periods of high transient loading, more severe

---

[*] Contribution of the National Institute of Standards and Technology, not subject to copyright in the U.S.. This program funded by NASA Marshall Space Flight Center.
[**] Guest researcher at NIST.

conditions can exist, which could account for the cracking, spalling, and localized heating discovered upon examination of the HPOTP bearing balls and raceways. This essentially unlubricated, oxidative process limits the life of the bearings to about 10 percent of the 7.5 hour design life.[1] An understanding of the mechanisms involved in the tribology of 440C over a wide range of speed, load, and temperature is necessary for the improvement of the HPOTP bearings.

TEST PROCEDURE

The tribometer used to perform the tests is described in detail in a previous paper.[3] A flat specimen riding on top of three stationary balls is the geometry used for these tests. All tests are performed in an oxygen environment. The torque is measured with a strain-gauge transducer which has an uncertainty of 3%. The sliding velocity is measured with a timer and a revolution counter that have an uncertainty of 1%. The angular velocity of the disk specimen is varied to achieve sliding velocities of 0.5, 1.0, 1.5, and 2.0 m/s. The revolution counter is also used in conjunction with the average diameter of specimen contact to determine the sliding distance. The bulk temperature of the specimens is varied by means of the flow of liquid or gaseous oxygen and a heater. The temperatures reported here are bulk temperatures, ranging from liquid oxygen temperature (-184°C) to 750°C. The bulk temperatures, measured by a type-K thermocouple welded close to the ball surface on one of the ball holders, have an uncertainty of ±5°C. The coefficient of friction reported here is the average kinetic coefficient of friction for the entire data set. The variation of the coefficient of friction with sliding distance is not dealt with in this paper. The load is applied with a dead-weight loading system which has an uncertainty of ±0.5 N. Loads are used to produce Hertzian contact stress values that correspond to 0.5, 1.0, 1.5, and 2.0 times the yield stress of 440C. The uncertainty of the coefficient of friction due to instrumentation is a function of load, and is 30%, 15%, 5%, and 5%, respectively, for loads of 5.6, 44.6, 150.5, and 357.0 N.

The wear rate reported here is the average rate of wear of the three balls because it is larger and more consistent than the wear rate of the flat specimen. The balls wear faster than the flat specimens because they are in constant contact with the flat, whereas any point on the wear track of the flat specimen is only in intermittent contact with the balls. Therefore, the balls will have a slightly higher bulk temperature, causing material transfer from the balls to the flat specimen, due primarily to softening of the material with increasing temperature. The average wear rate is calculated by measuring the mass loss of the ball specimens at specified sliding distances. The mass losses are converted to volume losses by using the measured density of 440C.[4] The uncertainty of the wear rate due to instrumentation is 0.5%.

The balls used are 0.476 cm in diameter, have an average hardness of $R_c=58.5$, and a surface finish $R_a=0.07$ μm. The flat specimens are 5 cm in diameter and have an average hardness of $R_c=60.5$ and a surface finish $R_a=0.12$ um. The specimens are machined, heat-treated, ground, and metallographically polished in the shops at NIST, Boulder. The specimens are tempered at 163°C, so any tests done at higher temperatures will undergo additional tempering during the tests. We determined the chemical composition of the specimens before the tests from a representative set of specimens using electron microprobe analysis. We analyze the wear surfaces of all specimens following the tests using scanning electron microscopy (SEM), electron microprobe analysis, X-ray photoelectron spectroscopy (XPS), and Auger electron spectroscopy (AES). Metrology of the wear track will be the subject of a future paper.

RESULTS AND DISCUSSION

Friction

An increase in the bulk temperature is observed for every run due to frictional heating. The amount of frictional heating increases with both increased load and increased sliding speed. The temperature is significantly higher at the interface than in the bulk material, and this difference will be greater for high than for low sliding speeds.[5]

Figures 1a and 1b show that the coefficient of friction is insensitive to load, except for loads corresponding to contact stresses below the yield strength of the material. The frictional heating is small at low load, and therefore, the temperature at the contact is relatively low. According to Kragelsky et al, the shear resistance of the molecular bond at the contact decreases with increasing temperature.[6] Therefore, the lower the load and bulk temperature, the higher the coefficient of friction should be, if the shear resistance at the material interface is the primary mechanism for friction under the given conditions. As long as the shear-resistance mechanism dominates, the coefficient of friction should decrease as the amount of frictional heating increases. This same dependence of load on the coefficient of friction has been seen by Shih and Rigney in lead-coated 52100 bearing steels.[7]

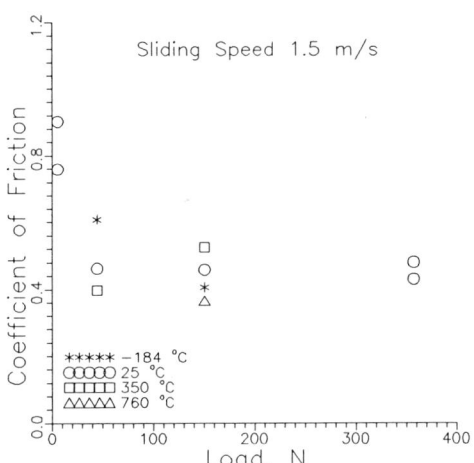
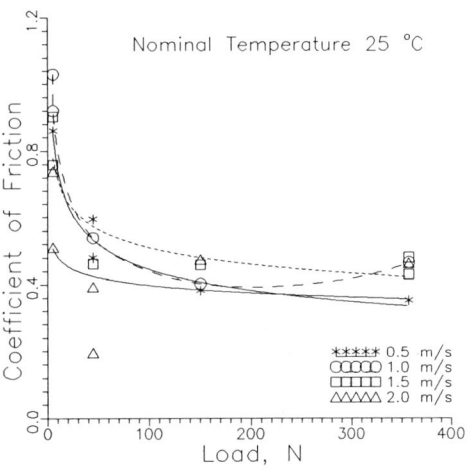

Figure 1a. Coefficient of friction versus load for various nominal temperatures at 1.5 m/s sliding speed.

Figure 1b. Coefficient of friction versus load for various sliding speeds at a nominal temperature of 25°C.

The effect of sliding speed on coefficient of friction depends on the frictional heating and chemical conditions at the sliding interface. Figures 2a and 2b show that the coefficient of friction decreases with increasing sliding speed at -184°C, and for the two lowest loads, 5.6 and 44.6 N. This is probably because frictional heating is small enough that there is not appreciable material softening at the interface. The low load may allow the decrease in shear resistance at the sliding interface to once again decrease the coefficient of friction as sliding speed increases.

The decrease in the coefficient of friction with increasing sliding speed for tests done at a nominal temperature of 760°C is shown in figure 2a. The specimens for these tests were all heavily oxidized. Possibly, the oxide strength is not sensitive to temperature, and therefore reduction in shear resistance with increasing contact temperature will control the coefficient of friction. The relatively low value of the coefficient of friction, ranging from 0.42 down to 0.29 for this set of data, supports the theory that a low-friction, wear-resistant oxide forms.

For conditions that exhibit a large amount of frictional heating, the coefficient of friction increases with increasing sliding speed. Figures 2a and 2b show a coefficient of friction increase for 25°C and 350°C tests done at 150.5 N, and for 150.5 N and 357.0 N tests done at 25°C bulk temperature. Under these conditions of moderate bulk temperature and high load, significant frictional heating probably occurs at the interface, because temperature traces from these tests show the greatest rises in temperature over the length of the tests. The frictional heating probably causes material softening at the sliding interface, which, as Kragelsky, et al, points out, increases the coefficient of friction as interfacial temperature increases, if material hardness controls friction.[6]

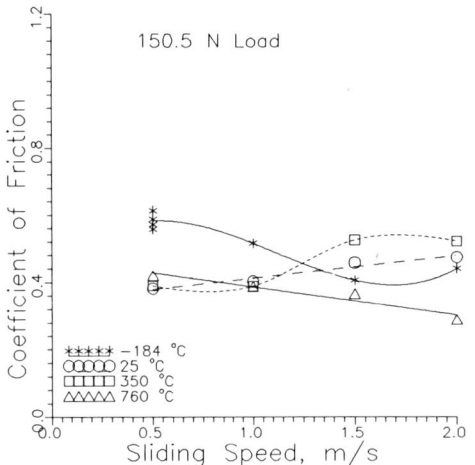 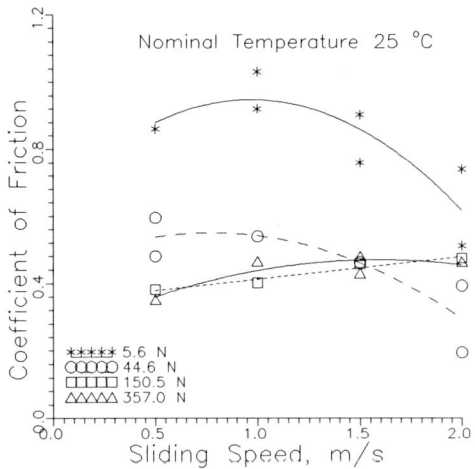

Figure 2a. Coefficient of friction versus sliding speed for various nominal temperatures at 150.5 N.

Figure 2b. Coefficient of friction versus sliding speed at various loads and 25°C.

Figure 3 shows the coefficient of friction versus bulk temperature for various sliding speeds, at 150.5 N load. Low sliding speeds show a gradual decrease in the coefficient of friction with increasing temperature. At low speeds, the shear resistance of the interface may control the coefficient of friction due to a limited frictional heating contribution. Since the tests are done in an oxygen environment, oxidation at the interface may be responsible for the decrease in shear resistance with increasing temperature. If the shear strength of the oxide interfacial bond is less than that of the metal substrate and the rate of oxidation is an exponential function of temperature, then the coefficient of friction will decrease as temperature increases. Figure 3 shows that the coefficient of friction is only a weak function of temperature, however.

For the sets at higher speeds, shown in figure 3, a different mechanism appears to dominate at low to moderate temperatures. The curves for 1.5 and 2.0 m/s are nearly identical with respect to temperature. For both sliding speeds, there is initially a rise in the coefficient of friction because a possible fatigue delamination condition overtakes the oxidation that is expected at the sliding interface. All of the tests performed at 400°C and less, at this load and 1.5 or 2.0 m/s, ran for less than 75 s, due to severe wear. The wear surfaces for these tests are bright, rather than black. Black surfaces generally coincide with the presence of oxides. Therefore, there is not enough time for significant oxidation, but the frictional heat at the interface weakens the material, causing severe fatigue delamination. The material will continue to lose strength with increasing temperature until the oxidation rate becomes high enough to generate a wear-resistant oxide layer in a short enough time. Figure 3 shows a decrease in the coefficient of friction for temperatures above 600°C. The wear tracks of the specimens run at 1.5 and 2.0 m/s and above 600°C were black with apparent oxides.

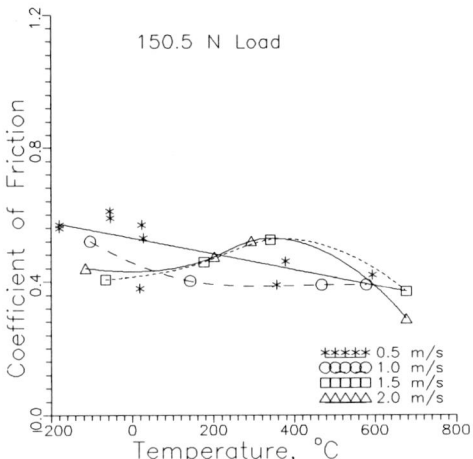

Figure 3. Coefficient of friction versus bulk temperature for various sliding speeds and 150.5 N load.

Wear

The wear rate is a strong function of load, as shown by figures 4a and 4b. At 25°C, the wear rate increases by a power law of load for all four sliding speeds, shown in figure 4a. This is contrary to the conventional theories of wear, which predict a linear dependence of load on wear.[5] A nonlinear dependence of load on wear due to fatigue and strength loss has been proposed by Kragelsky et al.[6] Fatigue damage to the wear surfaces may play a dominant role in the wear mechanism, particularly at high loads and sliding speeds, where severe wear is seen.

Figure 4b shows data only at low loads, and shows a distinct transition between low and high sliding speeds, which may be due to loss of strength at the sliding interface caused by frictional heating. The wear mechanism may be driven by oxidation for low sliding speeds, and driven by loss of strength for high sliding speeds.

Figure 5 shows the effect of sliding speed on the wear rate. For all but the highest temperature, wear rate increased with increasing sliding speed. Assuming that the oxidation rate depends exponentially on temperature, we expect the wear rate to increase with increasing sliding speed since increasing sliding speed would increase the interfacial temperature and oxidation rate, and thus wear rate. This reasoning is consistent with the Quinn model.[5] The unusually high wear rates at 1.5 and 2.0 m/s and 350°C are probably due to a condition where sufficient frictional heat at the sliding interface weakens the material, but not enough heat is available for rapid oxidation to provide a wear-resistant layer. Thus, severe wear occurs due to fatigue delamination of the weakened interfacial material.

At high temperature, the wear rate is high for 0.5 m/s, and lower but slightly increasing for higher speeds. The key to this phenomenon may lie in the coefficient of friction traces. The test conducted at 0.5 m/s exhibited an initial coefficient of friction of 0.7, which continued for about 20 s, then dropped to an average value of 0.36. Possibly, the total available energy at the sliding interface is below a threshold value for rapid oxidation at the beginning of the 0.5 m/s test, similar to the high speed, 350°C runs. The other three tests showed a very short, although high,

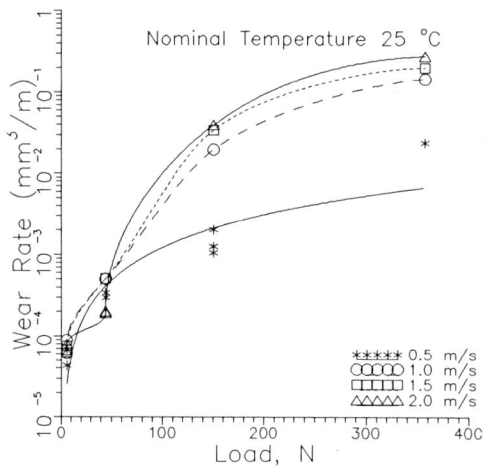

Figure 4a. Wear rate versus load for various sliding speeds at 25°C.

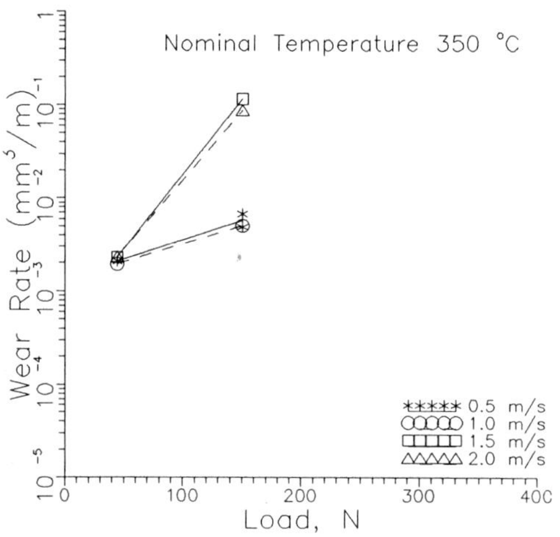

Figure 4b. Wear rate versus load for various sliding speeds at 350°C.

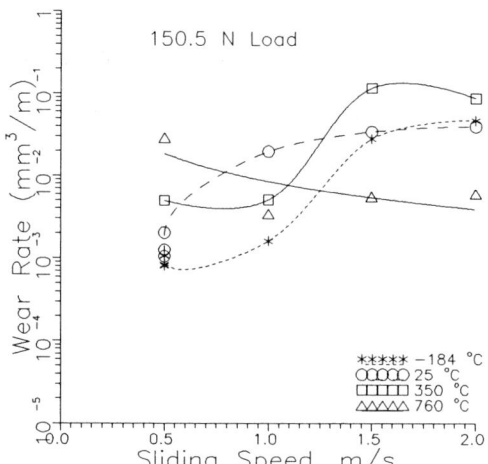

Figure 5. Wear rate versus sliding speed for various nominal temperatures at 150.5 N load.

initial coefficient of friction followed by a drop to a continuous value about 50% of the initial value. Possibly, the higher speeds, combined with the high temperatures, provide enough available energy at the interface to achieve a high oxidation rate for a critical and specific oxide that yields relatively low wear, and low coefficient of friction.

Figure 6 shows that wear rate, with respect to temperature, has a dual nature, with oxidation and loss of strength competing as temperature changes, similar to the effect of sliding speed. The 0.5 m/s data exhibit the effect of oxidation only. The increasing oxidation rate as temperature increases will cause the wear rate to steadily increase with temperature.

At 1.0 m/s and below room temperature, the material is hard and very little oxidation occurs, yielding a low wear rate. When a test is performed slightly above room temperature, loss of strength is probably the dominant wear effect, causing high wear before oxidation can get started. The wear surfaces for this test condition were bright, indicating little or no oxidation. At higher temperatures, the wear scars were black, suggesting that enough energy is available for a wear-resistant oxide to form, resulting in low wear rates.

The curves for 1.5 and 2.0 m/s are nearly identical, showing very high wear rates, increasing slightly with temperature up to 400°C. Since the wear scars for these runs are bright, the high wear rate is probably due to fatigue damage caused by loss of strength due to frictional heating, in conjunction with high contact stresses. At high temperature, above 600°C, enough energy is available for rapid oxidation, and the wear scars are dark. The appearance of the wear scars and the low wear rate suggests the rapid growth of a wear-resistant oxide layer at the sliding interface. This reasoning is the same as that given for the similar shape of the coefficient of friction curves for the same high sliding-speed data sets, shown in figure 3.

CONCLUSIONS

The coefficient of friction of 440C martensitic stainless steel is insensitive to load in an oxidizing environment, except when the contact

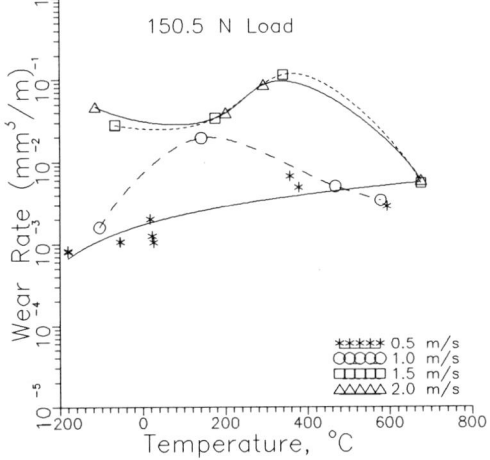

Figure 6. Wear rate versus bulk temperature for various sliding speeds and 150.5 N load.

stress is below the yield strength of 440C. The average wear rate of the ball specimens, on the other hand, is a strong function of load, although the functionality becomes less sensitive as the contact stress goes well beyond the yield strength of 440C. Both the coefficient of friction and the wear rate exhibit a dual nature with respect to sliding speed and bulk temperature by the competing effects of oxidation and loss of material strength. When loss of interfacial strength controls, both wear rate and the coefficient of friction increase with an increase in sliding speed and bulk temperature. However, when oxidation controls, an increase in wear rate with sliding speed and bulk temperature occurs, while the coefficient of friction decreases. The only time that the wear rate decreases with increasing sliding speed or bulk temperature is when a transition occurs between loss of interfacial strength and oxidation, and the oxide formed is wear-resistant with respect to the metal substrate at the given contact temperature. As more data become available, the results of the wear tests will be compared to a wear map of the kind attributable to Lim and Ashby.[8]

REFERENCES

1. B. N. Bhat and F. J. Dolan, "Past Performance Analysis of HPOTP Bearings," NASA TM-82470, March (1982).

2. L. D. Wedeven and N. C. Miller, "Material and Tribological Considerations for HPOTP Bearings," pp. 728-757, Proc. Adv. Earth-to-Orbit Propulsion Technology Conf., MSFC, AL, May (1988).

3. A. J. Slifka, J. D. Siegwarth, L. L. Sparks, and D. K. Chaudhuri, "Apparatus for Measurement of Coefficient of Friction," pp. 1119-1125, Advances in Cryogenic Engineering (Materials), Vol.36b, Plenum, NY (1990).

4. D. K. Chaudhuri and R. Verma, Engineered Materials for Advanced Friction and Wear Applications, ASM International, Metals Park, OH (1988).

5. T. F. J. Quinn, Physical Analysis for Tribology, pp. 10-31, Cambridge University Press, Cambridge (1991).

6. I. V. Kragelsky, M. N. Dobychin, and V. S. Kombalov, Friction and Wear, Pergamon Press, Oxford (1982).

7. C. Y. Shih and D. A. Rigney, "Sliding Friction and Wear of Tin, Indium, and Lead-Coated 52100 Steel," pp. 165-184, Wear, 134 (1989).

8. S. C. Lim and M. F. Ashby, "Overview #55, Wear Mechanism Maps," pp. 1-24, Acta Metall., 55:1 (1987).

SCALING TESTS ON SMOOTH AND NOTCHED SPECIMENS OF POLYIMIDE (SINTIMID)

AT CRYOGENIC TEMPERATURES*

Karl Humer[+], Harald W. Weber[+] and Elmar K. Tschegg[++]

[+]Atominstitut der Österreichischen Universitäten
A-1020 Wien, Austria
[++]Institut für Angewandte und Technische Physik
Technische Universität, A-1040 Wien, Austria

ABSTRACT

Because of applications in space and in cryogenic equipment, such as superconducting magnets, the mechanical properties and in some cases the radiation tolerance of various types of plastics have to be assessed. In the present contribution, we report on tensile strength experiments on a polyimide (SINTIMID) carried out at room temperature, 77 K and 4.2 K. Special attention was paid to "scaling" experiments, where the influence of sample size (scaled down from DIN and ASTM standards) on the elastic modulus, the ultimate tensile strength (UTS) and the failure strain was investigated. In addition, the influence of sample size on the mechanical properties in mode I have been investigated on cylindrical pre-cracked samples with a circumferential notch. The results of the tensile tests show that a variation of the sample size does not affect the UTS. Concerning the test temperature, the results show an increase of both the elastic modulus and the UTS by 40% and 60%, respectively, when decreasing the temperature to 77 K, but no further change at 4.2 K, while the failure strain decreases continuously (by about 25%) when cooling down to 4.2 K. Regarding the fracture tests no sample size dependence of the fracture properties could be detected. The fracture toughness increases continuously by about 10% with decreasing test temperature down to 4.2 K. The results will be discussed and compared with data on other materials. Additional fractographic investigations show no significant dependence of the fracture surfaces on the sample size for both the tensile and the fracture test samples.

INTRODUCTION

Because of their excellent mechanical and electrical properties as well as their low weight, plastics have found an increasing market in many technologies. New applications of various types of plastics in space and low temperature technology require adequate mechanical properties at low temperatures, and, in some cases also in gamma and/or particle radia-

---

*Work supported in part by the Federal Ministry of Science and Research, Vienna.

tion environments. Because of their high strength as well as their non-magnetic and good electrical insulation behavior, several plastics have become important as insulating and support material for the windings of superconducting magnets which will be used, e.g., in future fusion reactors.

Test results pertaining to the low temperature properties of various epoxies obtained from tensile test measurements were reported by Michael et al.[1]. Hartwig et al.[2] investigated the fracture properties of epoxy and polyethylene at low temperatures. The effect of the deformation rate on the UTS of polycarbonate, polyethylene and various epoxies at low temperatures was also studied by Hartwig et al.[3]. With regard to the influence of gamma and/or particle radiation environments on the mechanical properties of plastics, irradiation experiments from earlier studies[4,5] have shown that radiation environments introduce damage into the material, and hence, degrade the mechanical properties significantly.

Due to space limitations of existing low temperature irradiation facilities[6], all the mechanical tests prior to and after irradiation have to be done on samples which are considerably smaller than those required for standard test conditions (e.g. ASTM D638 and D3039, DIN 53455 for the tensile test). A possible influence of the sample size on the measured mechanical properties was pointed out by Kasen[7]. This has been incorporated into a test program[6] devised to study the radiation response of various plastics. First scaling results on a glass-fiber reinforced epoxy obtained from tensile measurements at room temperature, 77 K and 4.2 K have been presented recently[8]. In this study, no influence of the sample width on the UTS was noticed, but a slight reduction (15-35% at room temperature, 5-15% at low temperatures) of the UTS occured with decreasing sample thickness. It is the purpose of the present contribution to report on scaling results on a polyimide (SINTIMID) subjected to tensile and fracture tests at room temperature, 77 K and 4.2 K.

EXPERIMENTAL

Testing machine

The mechanical tests were made with a 200 kN tensile testing machine in a top loading cryostat[9] operating in the temperature range from 4.2 to 300 K. The crosshead speed was kept constant at 0.5 mm min^{-1} throughout the experiments. During the measurements, both the force and the sample elongation were recorded on an XY-recorder. After the experiments, fractographic investigations of the fracture surfaces were made in the light and/or the electron microscope.

Materials and test procedures

Tensile test samples of different sizes as shown in Fig.1 were prepared from plates (thickness 4 and 6 mm, respectively) of the polyimide SINTIMID (SINTIMID GesmbH., Reutte/Tyrol, Austria). Our test geometries include the DIN and ASTM standards. For a given sample thickness, all specimens were cut from the same plate. The tensile tests were carried out at room temperature, 77 K and 4.2 K. Based on the force/sample-elongation plot, the elastic modulus, the UTS and the failure strain were calculated. In order to determine average data, 4 measurements were made on each sample geometry and at all test temperatures.

In a recent study[10] we have shown that the fracture toughness K_{1c} of plastics can be determined from measurements on cylindrical tensile

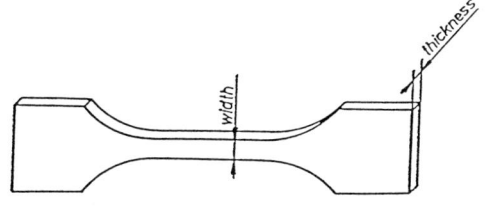

Fig. 1. Geometry and dimensions of the tensile test samples.

samples with a circumferential notch. As a further result excellent agreement of the K_{1c}-values between pre-cracked and "sharp notched" samples has been obtained. Hence, for the present contribution all fracture tests were done on sharp notched samples. We prepared tensile test samples with a circumferential notch from cylindrical rods (diameter 25, 20 and 15 mm, respectively) of SINTIMID. For a given sample diameter all specimens were cut from the same rod. The test geometry as well as the sample dimensions are shown in Fig.2. The fracture tests were carried out at room temperature, 77 K and 4.2 K. The fracture toughness was calculated from the maximum tensile load on the basis of standard equations[11]. In order to determine average data, 4 measurements were made on each sample geometry and at all test temperatures.

RESULTS AND DISCUSSION

Tensile tests

Sample dimensions, test temperatures as well as the test results of the measured stress-strain curves are summarized in Table I. As a general result we note that *for all sample geometries* (i.e. all widths and thicknesses) the elastic modulus amounts to about 2300 MPa at room temperature and to about 3200 MPa at low temperatures (4.2 and 77 K). The UTS amounts to about 115 MPa at room temperature and increases by ~60% (~175 MPa) both at 77 K and 4.2 K. On the other hand, the failure strain amounts to about 7% at room temperature, but decreases continuously with test temperature to ~6% at 77 K and to ~5% at 4.2 K.

As pointed out in the introduction, scaling results[8] on an anisotropic glass-fiber epoxy resin composite have shown no influence of the sample *width* on the UTS, but a slight reduction of the UTS with decreasing sample *thickness*. Based on tensile test results employing only standardized sample sizes (DIN and ASTM)[10] no problems concerning further scaling experiments were expected for the isotropic material SINTIMID.

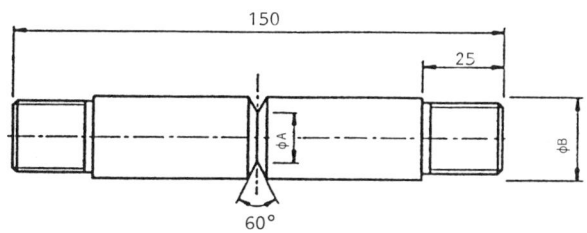

Fig. 2. Geometry and dimensions of the fracture toughness test samples.

Table I. Test temperatures, average results for the elastic modulus, the ultimate tensile strength and the failure strain for all sample geometries.

Temp. (K)	Width (mm)	Thickness (mm)	Elastic modulus (MPa)	Ultimate tensile strength (MPa)	Failure strain (%)
293	10	6	2366	111	6.4
293	10	4	2230	116	7.1
293	6	6	2260	113	6.4
293	6	4	2290	113	6.9
293	3	6	2305	118	7.0
293	3	4	2263	115	7.1
77	10	6	3220	171	6.0
77	10	4	3170	170	5.7
77	6	6	3143	167	5.9
77	6	4	3305	172	5.9
77	3	6	3140	167	6.0
77	3	4	3203	171	6.1
4.2	10	6	3136	173	4.6
4.2	10	4	3310	176	4.8
4.2	6	6	3072	179	4.7
4.2	6	4	3264	179	5.1
4.2	3	6	3303	173	5.2
4.2	3	4	3225	174	5.0

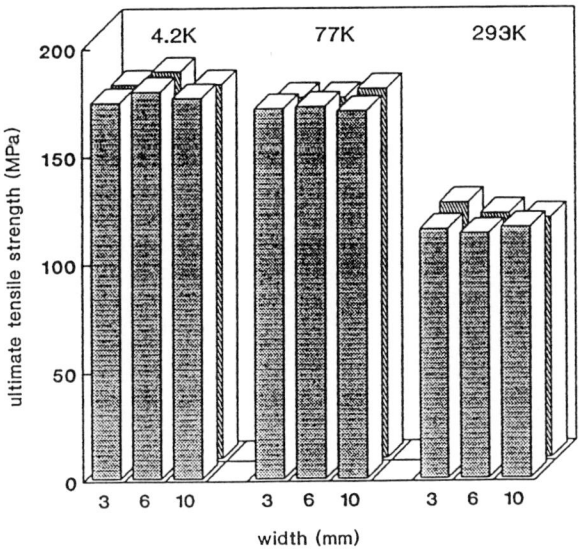

Fig. 3. Three dimensional plot of the ultimate tensile strength as a function of sample width and thickness for all test temperatures.

A graphical representation of the present scaling investigations of SINTIMID is given in Fig.3 in the form of three dimensional plots showing the measured UTS (z-axis) as a function of sample width (x-axis) and thickness (y-axis). As mentioned above, all UTS data presented both in Table 1 and in Fig.3 refer to average values. Concerning the influence of the sample width and thickness on the UTS, no significant and systematic differences were found for all three test temperatures.

Stress-strain diagrams of SINTIMID presented previously[10] have shown that *for all tensile test geometries* the stress-strain behavior is strictly linear at low temperatures (4.2, 77 K), whereas additional plastic deformation can be observed in the second part of the stress-strain curves at room temperature.

Fractographic investigations of the fracture processes and the fracture surfaces with the light microscope have led to the following results. The fracture surface of a standard tensile test sample (DIN 53455, width 10 mm, thickness 6 mm) following fracture at room temperature is shown in Fig.4. The crack initiation area with its very smooth surface and without any obvious structure is situated approximately in the middle of the fracture surface, whereas further away the fracture surface shows increasingly more structure in the form of river patterns. In addition, it should be noted that in about 50% of all measured samples the crack initiation point was situated approximately in the middle of the fracture surface, and in the rest - mostly in samples with smaller test sections - at the edge of the fracture surfaces. No influence of the test temperature on the fracture processes and the fracture surfaces could be observed.

Fracture toughness tests

Sample dimensions, test temperatures as well as the test results for the calculated average K_{1c}-values are presented in Table II. As a general result we find that *for all sample geometries* the fracture toughness increases continuously with decreasing test temperature (from 1.6 at room temperature to 1.8 MPa$\sqrt{m}$ at 4.2 K). This compares favorably with K_{1c}-values measured on another polyimide[12]. Concerning the scaling results, no systematic and significant influence of the sample geometry (diameter of the test section) on the fracture toughness was noticed at all test temperatures.

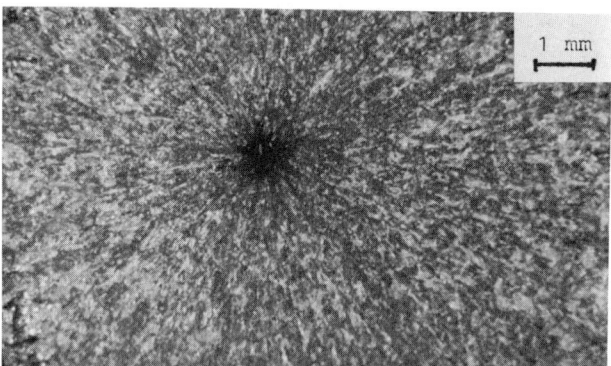

Fig. 4. Fracture surface of a 10 mm wide and 6 mm thick tensile test sample following fracture at room temperature. The crack initiation is situated approximately in the center of the fracture surface.

Table II. Test temperatures and average results for the fracture toughness of all sample geometries.

Temp. (K)	Diameter A (mm)	Fracture toughness K_{1c} (MPa$\sqrt{m}$)
293	15	1.59
293	10	1.57
293	5	1.61
77	15	1.68
77	10	1.75
77	5	1.81
4.2	15	1.77
4.2	10	1.79
4.2	5	1.85

Fracture mechanical investigations[2,3,10], especially at low temperatures, have shown that plastics - although they usually become more brittle at higher deformation rates - are more ductile at *low* temperatures. Since the specific heat of plastics is very small at low temperatures in comparison to the room temperature values, the released energy during the crack propagation can heat up the material at the crack tip and, hence, the formation of a small plastic zone around the crack tip will lead to ductile deformations in the form of river and dimple patterns. This could in fact also be seen in the present study, because high deformation rates will develop with increasing crack propagation in the fracture test samples (Fig.2) employed.

Fractographic investigations of the fracture processes and fracture surfaces have led to the following results. The crack initiation point of the final fracture is situated at the notch root. The final fracture surface shows river patterns which begin at the crack initiation point. Further along the crack propagation not only river but also dimple patterns could be observed. A fracture surface near the end of the final fracture, following fracture at 77 K, with its clearly marked river and dimple patterns can be seen in Fig.5. Enlarged dimple patterns from Fig.5 are shown in Fig.6. The formation of these patterns on the fracture

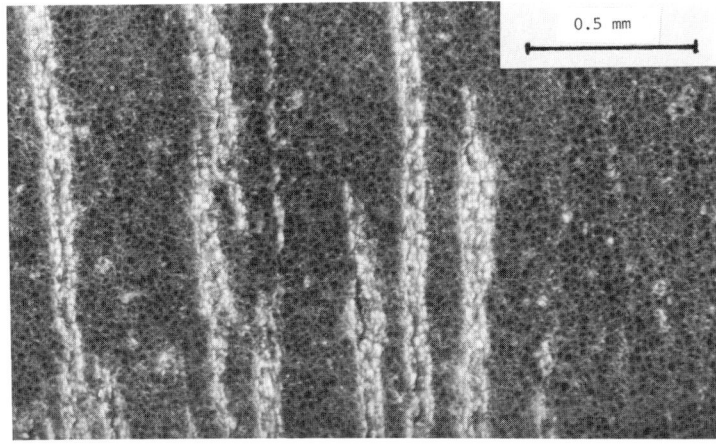

Fig. 5. Fracture surface following fracture at 77 K. River and dimple patterns near the end of the final fracture are observed.

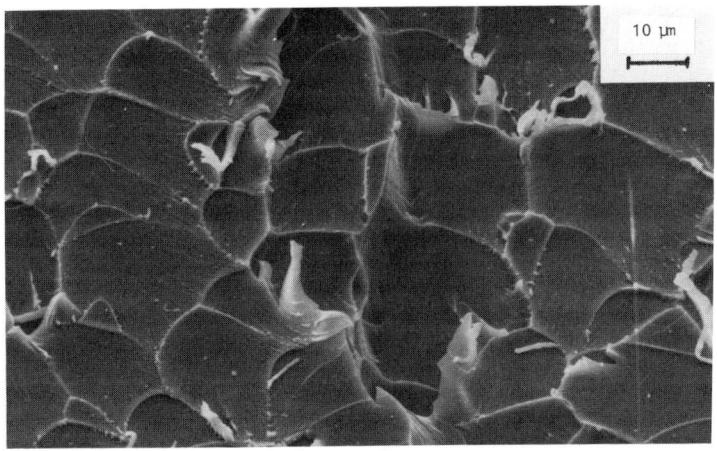

Fig. 6. Dimple pattern, enlarged from Fig. 5.

surfaces of other plastics has been observed in a similar way previously[13,14]. More details about fracture mechanical and fractographic investigations, as well as advantages and disadvantages of the test method were discussed earlier[10].

SUMMARY

As pointed out in the introduction, due to space limitations of existing low temperature irradiation facilities, only very small sample geometries can be used to study the influence of gamma and/or particle radiation environments on the mechanical properties of plastics. Hence, scaling experiments on different sample geometries were done to investigate the influence of the sample size on the measured material properties, which would permit one to deduce mechanical *material parameters* as defined by DIN and ASTM standards from measurements on much smaller test samples. The elastic modulus, the ultimate tensile strength and the failure strain were measured in the tensile test; the fracture toughness was calculated from tensile measurements on sharp circumferentially notched samples. All experiments were carried out at room temperature, 77 K and 4.2 K.

The main results of these tests, the scaling experiments and the fractographic investigations may be summarized as follows:
1. With decreasing test temperature from room temperature to 77 K, the increase of both the elastic modulus and the ultimate tensile strength amounts to about 40% and 60%, respectively, for all sample sizes. Further cooling to 4.2 K does not lead to a further increase of these parameters.
2. With decreasing test temperature from room temperature to 4.2 K, the failure strain decreases continuously by about 25% for all sample sizes.
3. With decreasing test temperature from room temperature to 4.2 K, the fracture toughness increases continuously by about 10% for all sample sizes.
4. Regarding the scaling experiments, no systematic and significant influence of different sample sizes on these parameters could be found neither for the tensile nor for the fracture tests.
5. The fracture surfaces following tensile tests show river patterns at all test temperatures.

6. The fracture surfaces following fracture toughness tests show river and dimple patterns at all test temperatures, but more clearly at low temperatures.

In conclusion it should be pointed out that SINTIMID shows excellent mechanical properties as derived from the tensile test at all temperatures down to 4.2 K. The low fracture toughness of SINTIMID is similar to other materials of this class and increases slightly at cryogenic temperatures. The present results on the scaling experiments confirm the feasibility of mechanical tests leading to intrinsic material parameters on very small test samples needed for irradiation experiments.

ACKNOWLEDGEMENTS

The authors are greatly indebted to SINTIMID GesmbH., Reutte/Tyrol, Austria, for providing us with the test samples of SINTIMID. Technical support from Mr.H.Niedermaier, Mr.E.Tischler, and Mr.W.Hametner is acknowledged. This work is supported in part by the Federal Ministry of Science and Research, Vienna, Austria, under contract # 77.011/2-25/89.

REFERENCES

1. P. C. Michael, D. Aized, E. Rabinowicz and I. Iwasa, Mechanical properties and static friction behaviour of epoxy mixes at room temperature and at 77 K, Cryogenics 30:775-786 (1990).
2. G. Hartwig, B. Kneifel and K. Pöhlmann, Fracture properties of polymers and composites at cryogenic temperatures, Adv. Cryog. Eng. 32:169-177 (1986).
3. G. Hartwig and K. Pöhlmann, Low temperature fracture strain at high deformation rates, Adv. Cryog. Eng. 30:83-88 (1984).
4. R. R. Coltman, jr. and C. E. Klabunde, Mechanical strength of low-temperature-irradiated polyimides: A five-to-tenfold improvement in dose-resistance over epoxies, J. Nucl. Mat. 103&104:717-721 (1981).
5. S. Nishijima, S. Ueta and T. Okada, The effects of low temperature irradiation effect on the cryogenic fatigue resistance of epoxy resin used in superconducting magnets, Cryogenics 21:312-313 (1981).
6. H. W. Weber and E. K. Tschegg, Test program for mechanical strength measurements on fiber reinforced plastics exposed to radiation environments, Adv. Cryog. Eng. 36:863-869 (1990).
7. M. B. Kasen, Cryogenic properties of filamentary-reinforced composites: an update, Cryogenics 21:323-340 (1981).
8. E. K. Tschegg, K. Humer and H. W. Weber, Influence of test geometry on tensile strength of fibre reinforced plastics at cryogenic temperatures, Cryogenics 31:312-318 (1991).
9. E. K. Tschegg, E. Kubasta and W. Steiner, A top loading 200kN tensile testing cryostat, Cryogenics 19:269-271 (1979).
10. E. K. Tschegg, K. Humer and H. W. Weber, Mechanical properties and fracture behavior of polyimide (SINTIMID) at cryogenic temperatures, Cryogenics (in press).
11. Y. Murakami, Soc.Mat.Sci.Jap., Stress intensity factors handbook, Vol.2:643 (1987).
12. J. A. Hinkley and S. L. Mings, Fracture toughness of polyimide films, Polymer 31:75-77 (1990).
13. M. J. Zhang and F. X. Zhi, Crack growth mechanism in some polyamides, Polymer 29:2152-2158 (1988).
14. K. L. De Vries and L. E. Hornberger, Macroscopic, microscopic and molecular aspects of fracture in polymers, Polymer Degradation and Stability 24:213-240 (1989).

SEVERAL PROPERTIES OF IMPREGNATING EPOXY RESINS

USED FOR SUPERCONDUCTING COILS

>Hideshige Moriyama, Yoshiyuki Inoue, Hisayasu Mitsui,
>Yoshinao Sanada
>
>Toshiba Corporation
>Yokohama, Japan
>
>Yoshio Kobayashi
>
>Super-GM
>Osaka, Japan

ABSTRACT

In order to find the most appropriate vacuum-pressure impregnating epoxy resin for superconducting coils, three types of epoxy resin were experimentally investigated from various points of view. Those epoxy resins were amine curing type A, anhydride curing type B with tertiary amine accelerator, and anhydride curing type C without accelerator.

From the investigations, resin A was found to be superior in thermal shock resistance than resins B and C because of its large tensile fracture strain at cryogenic temperature and its small cure shrinkage. However, the test coil impregnated with resin A experienced more repeated premature quenches than the coil impregnated with resin C. This result might be related to the crackings which started from microscopic defects, and which occurred gradually in resin A compared with resins B and C. Moreover, the viscosity of resin A increased more rapidly during the impregnating process than that of resins B and C, so that resins B and C appear to be more appropriate for large superconducting coils than resin A.

INTRODUCTION

As a type of vacuum-pressure impregnating epoxy resin usable for superconducting coils, amine curing type epoxy resin is well-known[1]. The curing reaction of this resin progresses and the resin viscosity increases, during the impregnation process. Therefore, other resins that possess a long life time for low viscosity impregnation are desirable for large superconducting coils which require a long time for complete impregnation[2]. However, such resins and the resin properties have not been reported sufficiently so far.

It is known that cracking in cured resin facilitates the quenching of superconducting coils, and that the cracking is related to the training characteristics of superconducting coils[3,4]. Nevertheless, the causes of cracking have not been thoroughly investigated.

We selected epoxy resins belonging to the following three curing types, and studied the characteristics of these resins.
  (1) Amine curing type
  (2) Acid anhydride curing type with tertiary amine accelerator
      (hereinafter called "accelerated curing type")
  (3) Acid anhydride curing type

Table 1. Sample resins

Resin	Type of curing reaction	Impregnation temperature	Cure condition
A	Amine curing	R.T.	50°C × 10h + 120°C × 15h
B_1	Accelerated curing	50°C	120°C × 10h
B_2	Accelerated curing	50°C	80°C × 15h + 120°C × 10h
C	Acid anhydride curing	60°C	120°C × 48h

The life time for low viscosity impregnation, cracking resistance and other mechanical properties, curing shrinkage, thermal contraction, and the training characteristics of impregnated and cured superconducting coils were measured. The investigation on measured properties yielded a finding in the cause of cracking, and substantiated the suitability of accelerated curing resin and acid anhydride curing resin as superconducting coil impregnation resins. In this paper, the results of the investigation of such properties are summarized.

SPECIMENS

Bisphenol A epoxy resins divided into three types by curing reaction, as shown in Table 1, were used. The amine curing resin A was selected as a representative impregnating resin commonly used for superconducting coils, the accelerated curing resin B and the acid anhydride curing resin C were selected for the superior cracking resistance as described later. Resin B_1 and B_2 were of the same compositions, but were cured under different conditions.

TEST METHODS AND RESULTS

Life time for low viscosity impregnation

Increases in the resin viscosity were measured while maintaining the resin temperature at the impregnatable level as shown in Table 1. Since the viscosities of resins B and C were too high for impregnating at room temperature, the two resins were maintained at 50°C and 60°C respectively. Fig.1 shows the result of resin viscosity increases. The resin viscosities increase in order of resin A > B > C. This indicates that a longer time is available for impregnating with resin C, B than resin A.

Tensile, compressive and bending properties

The specimen resins A, B_1, and C were tested for tension, compression and bending at room temperature (R.T.) and at 77K, according to JIS K 6911, and the strengths and the elastic moduli were obtained. Especially, the resins were found to have a significant difference in tensile strength compared to other properties. At R.T. and 77K, the tensile stress - strain relationships were nearly linear up to the fracture. Fig.2 shows the result of tensile fracture strains. Resin A at 77K and resin B_1 at R.T. possess large values of tensile fracture strain.

Adhesive strength

The tensile adhesive strength and the tensile shear adhesive strength of the specimen resins A, B_1 and C were measured according to JIS K 6849 and JIS K 6850, at R.T. and 77K. SUS304 pieces were used as the adherend. Fig.3

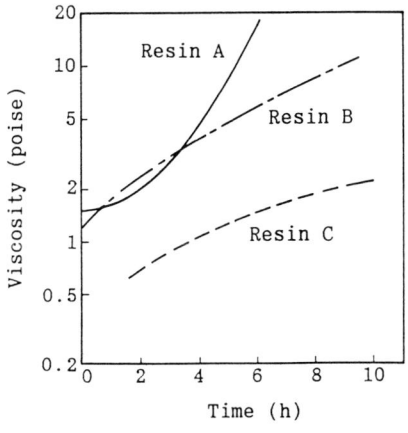

Fig.1 Increases of the viscosities

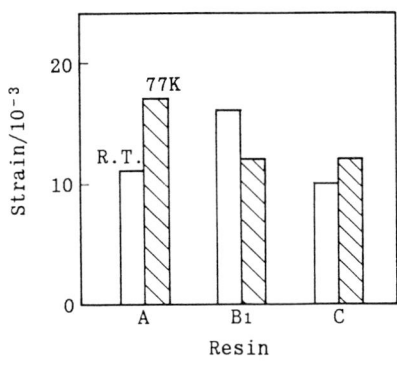

Fig.2 Tensile fracture strain

shows the mean values of adhesive strength of each resin. The value of tensile adhesive strength is noted to become greater in order of resin A < B1 ≃ C, while the value of tensile shear adhesive strength is relatively closer.

Dynamic visco-elasticity

Resins with low tanδ peak temperature at α relaxation of dynamic visco-elasticity and with large β relaxation, are generally tough[5]. Therefore, these resins are assumed to possess high cracking resistance. The dynamic visco-elasticities of resin A, B1, B2 and C measured at 11 Hz, are shown in Fig.4. The tanδ peak temperature at α relaxation is lower and β relaxation is slightly larger in resin A than in resins B1, B2 and C. Consequently, the cracking resistance of resin A is assumed to be higher than that of resin B and C.

Cracking resistance

Since a cracking resistance test method using thermal shocks between normal temperature and cryogenic temperature has not been fully established[1], an original method was designed and used. Fig.5 shows the test sample configuration. Rods with five diameters 0 (no rod), 5, 10, 15, and 20 mm, were

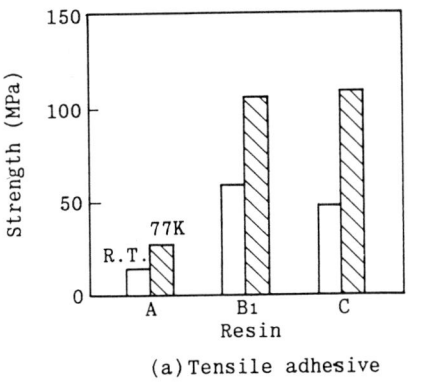

(a) Tensile adhesive

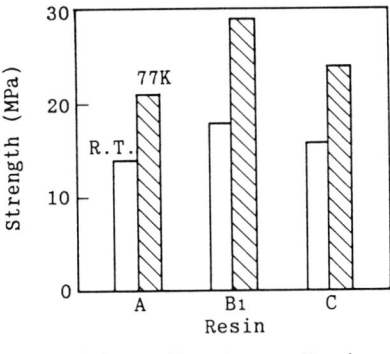

(b) Tensile shear adhesive

Fig.3 Adhesive strengths

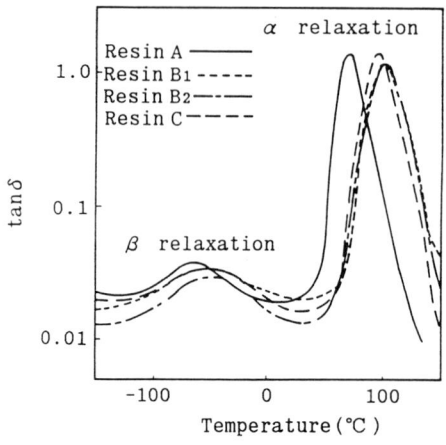

Fig.4   Dynamic visco-elasticities

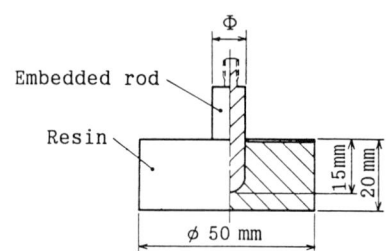

Fig.5   Shape of the sample for testing resistance

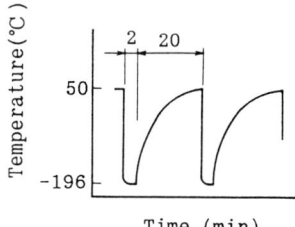

Fig.6   Condition of the thermal shock cycle

embedded. The thermal shock cycles shown in Fig.6 were applied to those test samples up to 20 cycles, and the number of cycles up to the crack occurrence was measured as the cracking resistance life, and plotted in Fig.7. With resin A, the longest life for all rod diameters is over 20 cycles, and the plots of short lives less than several cycles are not many. In contrast, the longest life of resin C is shortened by larger rod diameters, and the plots of shorter lives increase in number. The comparison between the frequency of long life cases and of short life cases with less than 5 cycles, indicates that the cracking resistance is superior in order of resin A > B1 > C. This high cracking resistance of resin A coincides with our previously mentioned assumption based on the dynamic visco-elasticity.

Cure shrinkage

As shown in Fig.8, the resins were cured at various temperatures in cylindrical SUS304 containers which were previously coated by a releasing agent, inside diameter D of the container and outside diameter D' of the cured resin were measured at room temperature, and the cure shrinkage rate was calculated by the following equation:

$$\text{cure shrinkage rate} = \frac{D - D'}{D} \times 100 \ (\%)$$

Since resin B and resin C did not cure at the curing temperature of resin A, the two resins were cured above 80°C and 120°C respectively. To study the relationship between the cure shrinkage and the cracking resistance described before, the same containers were used for those tests. Fig.9 shows the measured values of cure shrinkage. The cure shrinkage decreases with lowering the curing temperature, and the cure shrinkage is greater in order of resin A < resin B ≦ resin C. As large cure shrinkage is known to produce large residual stress, the residual stress of the cracking resistance test sample was estimated to be higher in order of resin A < B1 ≦ C.

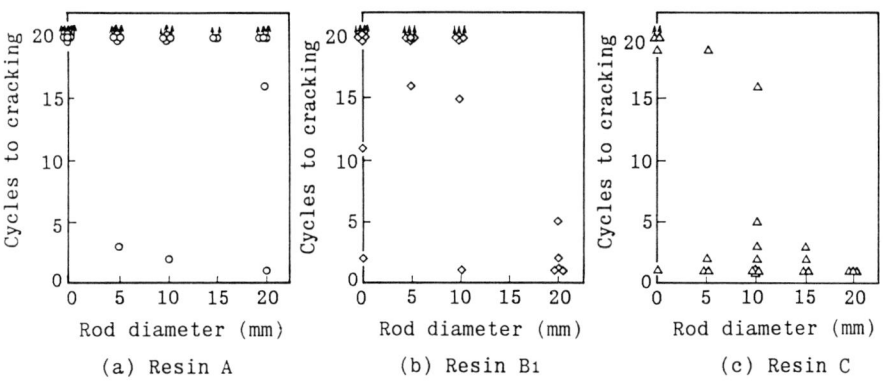

Fig.7 Cracking resistance lives

(a) Resin A   (b) Resin B₁   (c) Resin C

Thermal contraction

　　The thermal contractions of resin A and C between room temperature and dry ice temperature, liquid nitrogen temperature, liquid helium temperature were measured with a method using strain gauges[6]. Fig.10 shows the result of thermal contraction. Because of no significant difference between the value of resin A and of resin C, the substantial difference in cracking resistance between the two resins was considered not due to thermal contraction.

Training characteristics of test coils

　　Test coils wound with polyvinyl formal insulated NbTi superconducting wire with 1 mm diameter, were impregnated by vacuum-pressure method with resin A and C, each resin having clearly different cracking resistance, and then the resins were cured. The training characteristics of the test coils were measured. Fig.11 shows the dimension of the coils and the test result. The current of the ordinate is the ratio of quench current at each excitation to the critical current of short superconducting conductor. The maximum current is nearly 100% with both resins. This result indicates that both

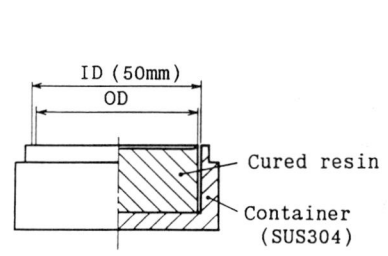

Fig.8 Measurement method of the cure shrinkage

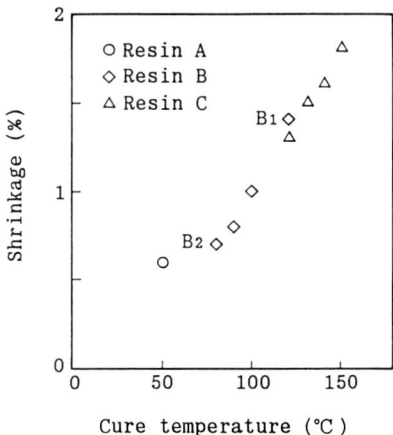

Fig.9 Relationship between the cure shrinkage and the cure temperature

343

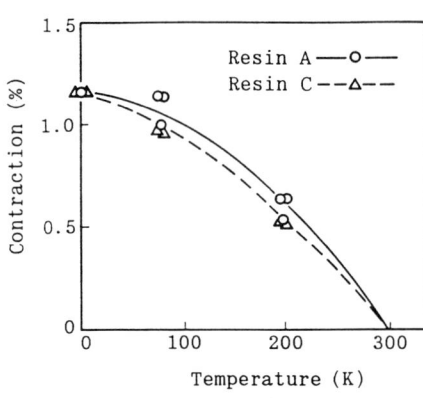
Fig.10 Thermal contractions

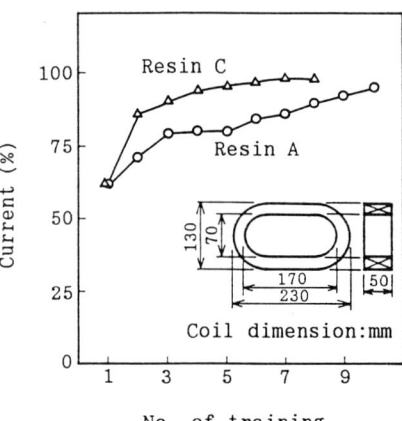
Fig.11 Characteristics of the trainings

resins are usable for superconducting coil impregnation. It is noted that the premature quenches of the coil impregnated with resin A, possessing higher cracking resistance, are needed to repeat more than the coil impregnated with resin C.

ANALYSIS AND OBSERVATION

Thermal stress analysis of cracking resistance test sample

The thermal stress of the test sample cooled to 77K was analyzed with the finite element method. Fig.12 shows an example of the analysis results. The analysis showed the presence of maximum tensile stress and maximum shear stress near the point P in the diagram. The circumferential component of the tensile stress increased with the diameter of imbedded rod.

Fracture surfaces of cracking resistance test samples

The SEM observation of the fracture surfaces of test samples which exhibited short cracking resistance life compared with those of the long life, revealed some peeling between the resin and the imbedded rod in the crack starting areas which was near the point P in Fig.12. Fig.13a shows the peeling surface of resin. According to the wrinkle shape in the fracture

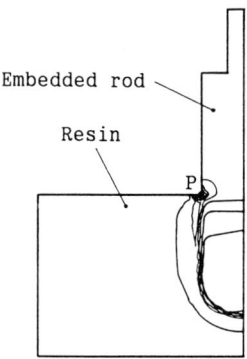
Fig.12 Contour of the maximum principal stress

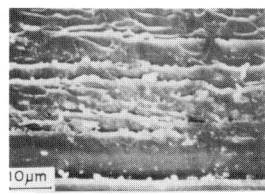 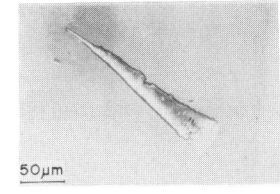 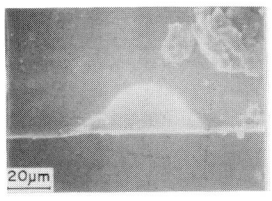

a) Peeling b) Solid impurity c) Void

Fig.13 Microscopic defects of cracking resistance samples

surface, the peeling can be ascribed to the resin shrinkage during the curing process. Also, there were some samples which included a solid impurity (Fig.13b) or a void (Fig.13c) in the crack starting area.

DISCUSSION

Causes of cracking

When determining the cause of cryogenic cracking, considering both cases of the long life and the short life to cracking is important, because of the irregularity in distribution of the cracking resistance life shown in Fig.7.

The cause of long life to cracking could be explained from the relationship between resin properties and maximum stress. The thermal stress analysis of cracking resistance test sample at 77K, indicates the presence of maximum tensile stress and maximum shear stress near the point P in Fig.12. These stresses can correspond to the tensile fracture strain and the tensile shear adhesive strength. Since the circumferential component of the tensile stress near the point P increases with the diameter of the embedded rod, the long life is expected to decrease with increasing the rod diameter. When the resin has been cured at the lower temperature, the smaller residual stress, and the higher cracking resistance are expected. All of these considerations suggest the higher cracking resistance of the amine curing resin A.

From the fracture surface observation of the cracking resistance test samples, the short cracking resistance life is caused by the presence of microscopic defects. Because of the brittleness of resin at cryogenic temperature, cracks tend to develop at microscopic defects such as peeling, solid impurity and void.

Relationship between crackings and training effect

The training characteristics in Fig.11 might be related to the cracking which starts from a microscopic defect. That is, premature quenching is induced by the cracking which generates heat energy above a certain level, and the number of the microscopic defects is finite. With resins of high cracking resistance such like resin A, many training might be repeated until a final crack starts from the finite defects, while with resins of reasonable cracking resistance such like resin C, less trainings might be repeated. However, with very high cracking resistance resin that has no significant defects, the quench current may attain to the critical level at the first excitation.

CONCLUSION

1. The cracking resistance tests with thermal shock cycles between normal and cryogenic temperature gave two types of cracking resistance lives, the long life and short life. The long life was related to the tensile

fracture strain at cryogenic temperature, the tensile shear adhesive strength and the cure shrinkage, while the short life was caused by microscopic defects.
2. The superconducting coils impregnated with resins of high cracking resistance and of reasonable cracking resistance, differed in training characteristics. Coils impregnated with reasonable cracking resistance resin, increased the quench current close to the critical current level by fewer trainings. This phenomenon was related to the short cracking resistance life which was caused by microscopic defects.
3. From the viewpoint of cracking resistance, accelerated curing resin and acid anhydride curing resin were thought to be suitable as superconducting coil impregnants. Moreover, because of their longer impregnatable time, the accelerated curing and acid anhydride curing resins were expected to be advantageous for impregnating large superconducting coils.

ACKNOWLEDEMENT

The study of several mechanical properties was performed as a part of "R&D on Superconducting Technology for Electric Power Apparatuses" as a subject of Super-GM under the Moonlight Project of the Agency of Industrial Science and Technology, MITI, being consigned by the New Energy and Industrial Technology Development Organization(NEDO).

REFERENCES

1. D. Evans and J. T. Morgen, Epoxide Resins for Use at Low Temperatures, Nonmetallic Materials and Composites at Low Temperatures 2, p. 73 (1982).
2. M. A. Green et al., Vacuum Impregnation with Epoxy of Large Superconducting Magnet Structure, Nonmetallic Materials and Composites at Low Temperatures, p. 409 (1982).
3. H.Fujita et al., Experimental and Theoretical Investigation of Mechanical Disturbances in Epoxy-impregnated Superconducting Coils 4, Cryogenics, Vol.25, p. 323(1985).
4. J. W. Ekin et al., Effect of Strain on Epoxy-impregnated Superconducting Composites, Nonmetallic Materials and Composites at Temperatures, p.301 (1979).
5. U. T. Kreibrich et al., Polymers in Low Temperature Technology, Nonmetallic Materials and Composites at Low Temperatures, p. 1 (1979).
6. KYOWA Engineering News, No. 283 (1981).

SHEAR BEHAVIOR OF GLASS-REINFORCED SYSTEMS AT LOW TEMPERATURES

P. E. Fabian, C. S. Hazelton, and R. P. Reed

Composite Technology Development, Inc.
2400 Central Avenue, Suite H
Boulder, Colorado, U.S.A.

ABSTRACT

Uniaxially reinforced composites exhibit nonlinear load displacements when loaded in shear, even at very low temperatures. The nonlinearity can be attributed to both elastic and plastic deformation. Information on nonlinear and plastic strain can be obtained by closely analyzing data from low-temperature shear tests. This paper describes the nonlinear elastic and plastic deformation resulting from solid-rod torsion tests at low temperatures. Nonlinear elastic and plastic strain was measured as a function of shear stress to detect the onset of irreversible deformation. At both 77 and 4 K, nonlinear strain measurements also showed finite strain differences among various composite materials, including unique strain variations of different epoxy resin systems. A third-order stress-strain relationship for predicting yield strength was investigated.

INTRODUCTION

Low-temperature studies of composite materials have demonstrated that it is essential to understand their nonlinear response to shear in order to predict their behavior under operating conditions. Solid-rod torsion tests were used to analyze S-2 Glass®-reinforced composites* in shear. The solid-rod torsion test is an effective method for characterizing the shear performance of this type of material,[1] and the only shear test to yield "true" or "pure" shear.[2]

After torsion testing of various resin systems, including epoxies, bismaleimides, and polyimides, we have observed that all the systems displayed some nonlinear characteristics, even at cryogenic temperatures as low as 4 K. That is, in a typical torsion test plot of stress vs. strain (shown in Fig. 1), the response does not remain linear through failure, and it remains in the elastic, or reversible, region during part of this nonlinear curve.

By recognizing that reversible nonlinear response does exist and that permanent deformation does not necessarily begin at the apex of linearity of

---

*S-2 Glass® is a registered trademark of Owens-Corning Fiberglas Corporation.

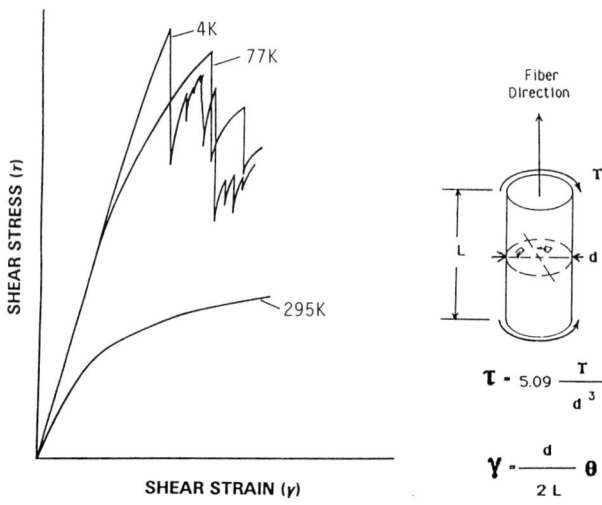

Fig. 1. Material response of a glass-reinforced epoxy during the solid-rod torsion test.

the stress-strain curve, designs can be optimized by identifying the point at which permanent deformation begins. This may lead to a better understanding of the fatigue or creep characteristics of composites.

EXPERIMENTAL PROCEDURES

The specimens were uniaxially reinforced rods, 3.2 mm in diameter. These specimens were tested with the Torsion Testing Facility (TTF) developed by CTD. The TTF employs a probe assembly that can be inserted into a cryogenic Dewar for testing directly in the cryogenic liquid. A torsional load cell measures the load, and a rotary variable differential transformer determines the angle of rotation or displacement of the specimen. An x-y plotter produces a curve of torsional load vs. angle of rotation; this curve can be easily analyzed for shear strength, shear modulus, and shear strain.

After extensive torsion testing of CTD's materials to find their ultimate shear properties at 295, 77, and 4 K, two distinct types of strain were apparent at lower temperatures: nonlinear elastic strain and plastic strain. Nonlinear elastic strain is strain that does not permanently deform the specimen nor permanently alter its properties; thus, after unloading, the specimen returns to its original condition. Plastic strain begins when the material sustains some permanent deformation. This is similar to the elastic limit of a mild steel, usually near the yield point, where Hooke's law can no longer be applied.[3] These two types of strain were experimentally identified after loading a specimen in torsion to 80 to 90% of its ultimate strength, unloading it, and analyzing the resulting curve, such as the example shown in Fig. 2. By extending the parallel, unloading leg of the curve down to the x-axis (the dotted line), the total nonlinear strain can be measured. The plastic strain can likewise be measured since it is simply the residual strain at zero load. The nonlinear elastic strain is defined experimentally as the difference between the total nonlinear strain and the plastic strain.

To follow the progression of the nonlinear elastic strain region into the plastic strain region of a material, the torsion tests were run at 77 K with specimens initially loaded to 353.1 N·mm (50 in·oz) and then unloaded.

The same procedure was followed with the specimen loaded in increasing increments of 70.6 N·mm (10 in·oz) and then immediately unloading the specimen. This pattern was followed until the specimen failed. From the resulting curves, the nonlinear elastic and plastic strains can be obtained as they change with increasing torsional load.

Three materials were tested in this manner: CTD-102, a glass-reinforced epoxy resin system, CTD-310, a glass-reinforced polyimide resin system, and 304 stainless steel. The stainless steel was tested at the same stress levels as the composite materials. At these levels, the steel was assumed to be in its linear, elastic stress-strain region. Therefore, any measurable nonlinear strain or behavior observed in the stainless steel tests was subtracted from the nonlinear strains measured in the composite materials. In this way, any test equipment slippage or compliance introduced from the machinery or grips was eliminated from the test results.

RESULTS

Figure 3 shows the nonlinear strain of CTD-310 and CTD-102 as a function of shear stress. As the shear stress increased, the nonlinear strain increased for both materials, although at different rates. The polyimide reached its nonlinear range at a lower shear stress than the epoxy, but the epoxy, in turn, exhibited a greater degree of nonlinearity than the polyimide. Nevertheless, both materials displayed the same type of nonlinear behavior at 77 K, as evidenced by the similarity of the curves shown in Fig. 3.

The plastic strain was also measured (Fig. 4), and, like the nonlinear strain, plastic strain increased with increasing stress. The polyimide reached the plastic range at a lower stress level than the epoxy system, although the plastic strain was larger in the epoxy.

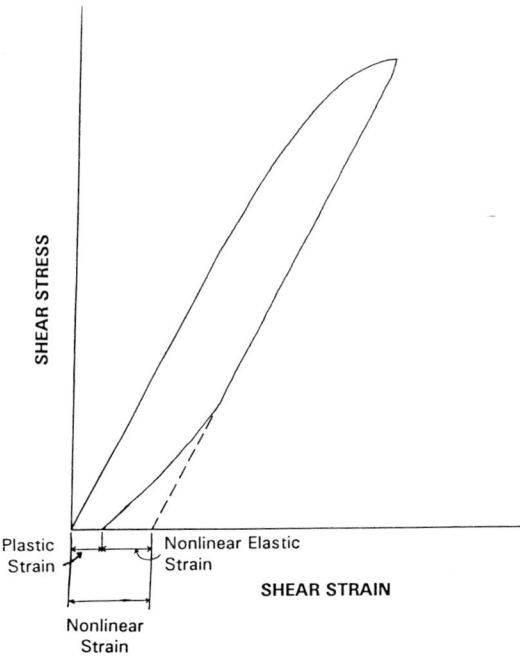

Fig. 2. Graphical definitions of the two types of nonlinear strain encountered in the torsion test.

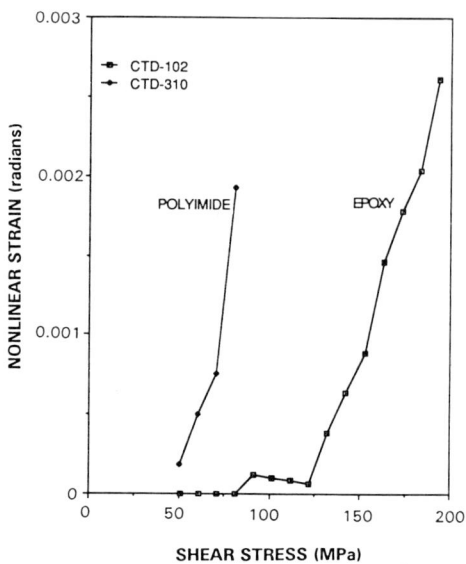

Fig. 3. Nonlinear strain behavior of unidirectional S-2-Glass®-reinforced polyimide and epoxy systems at 77 K.

One important observation from these two plots (nonlinear strain vs. shear stress and plastic strain vs. shear stress) is that, for both materials, the plastic strain is less than the nonlinear strain. In other words, most of the nonlinear section of the stress-strain curve is nonlinear elastic, and only a small portion of the curve represents permanent deformation or plastic strain.

In addition to the two different strains associated with the nonlinear response in composite materials, there is also a material dependence corresponding to the nonlinear strains. To show this dependence, the total non-

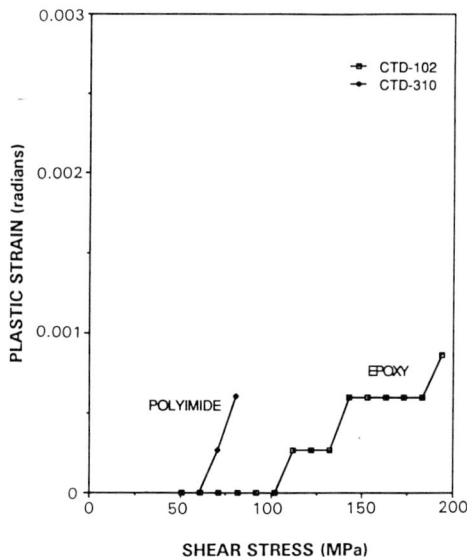

Fig. 4. Plastic strain behavior of unidirectional S-2-Glass®-reinforced polyimide and epoxy systems at 77 K.

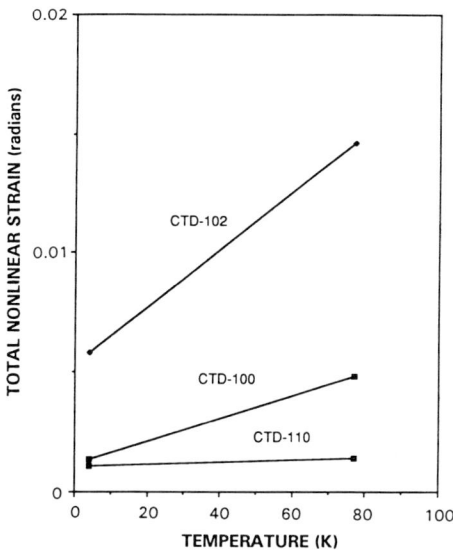

Fig. 5. Temperature dependence of nonlinear response for unidirectional, S-2 Glass®-reinforced epoxy systems.

linear strain of three epoxy resin systems was measured at 77 and 4 K. The resin systems were CTD-100, a DGEBA epoxy/Aromatic-aliphatic amine blend; CTD-102, a DGEBA epoxy/anhydride; and CTD-110, a multifunctional epoxy. As seen in Fig. 5, each epoxy system had its own finite nonlinear strain at each temperature, showing that each epoxy performs differently at cryogenic temperatures.

ANALYTICAL PROCEDURE

An analytical technique was used for numerical determination of the point at which irreversible damage began to accumulate in the material. By assuming that the nonlinear reversible response of the composite can be described by a cubic stress-strain relationship, data taken from experimental test curves can be dissected to find the point where damage to the specimen becomes irreversible and, thus, find the "yield point" of the composite.

The approach was based on the standard linear relationship for shear stress and shear strain:

$$\tau_{12} = G_{12}(\gamma_{12}),$$

where τ_{12} is shear stress, G_{12} is shear modulus, and γ_{12} is shear strain. The equation was modified to take into account the nonlinearity of the test curve. Hahn[4] derived such a relationship from a fourth-order strain-energy function. Adapting both for our application, the resulting nonlinear stress-strain relationship is

$$\tau_{12} = G_{12}(\gamma_{12}) + Q_{6666}(\gamma_{12})^3,$$

where $Q_{6666} < 0$ is the nonlinear modulus. A least squares fit of the nonlinear stress-strain equation (Fig. 6) can then be performed from the original test data. As the magnitude of Q_{6666} increases, the nonlinear region of the reversible stress-strain curve increases. The point at which the fitted curve deviates from the test data represents the onset of permanent deformation. Table 1 gives the experimental results for a glass-reinforced epoxy system.

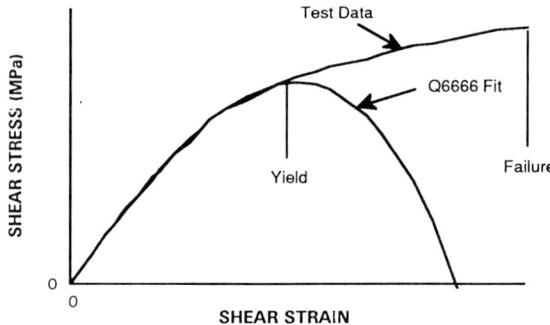

Fig. 6. Example of a least squares fit of the nonlinear stress-strain relationship to an actual test curve.

Table 1. Yield Strength Calculated from the Experimental Test Curve

Temperature (K)	Shear Strength (MPa)	Yield Strength (MPa)	Initial Modulus (GPa)	Curve Modulus (GPa)	Nonlinear Modulus (GPa)
295	84.8	77.9	2.32	2.29	-311.0
77	156.5	135.8	7.93	7.58	-142.7
4	184.8	155.1	10.27	9.79	-57.2

The process requires analysis of many data points from each test and becomes quite tedious. However, by automating the calculation of Q_{6666}, this technique may become an effective method for predicting the onset of the plastic strain region, enabling a better understanding of observed shear behavior at low temperatures.

CONCLUSIONS

From this study we conclude

1. Nonlinear strain was observed, to some extent, in all composites tested at cryogenic temperatures, even at 4 K.
2. The two components of shear strain observed at low temperatures can be classified as (1) nonlinear elastic strain prior to permanent deformation or (2) plastic strain, which is associated with permanent deformation prior to failure.
3. Nonlinear strain is material dependent, as well as temperature dependent, at cryogenic temperatures.
4. A "yield strength" can be calculated from composite shear stress-strain data at low temperatures by using a third-order strain energy function.

REFERENCES

1. M. B. Kasen, Strain-controlled torsional test method for screening the performance of composite materials at cryogenic temperatures, J. Mater. Sci. 23:830-834 (1988).

2. D. W. Wilson, An overview of test methods used for shear characterization of advanced composite materials, in: *Advances in Cryogenic Engineering—Materials*, vol. 36, R. P. Reed and F. R. Fickett, eds., Plenum, New York (1989), p. 793.
3. E. P. Popov, *Mechanics of Materials*, Prentice-Hall, New Jersey (1976).
4. H. T. Hahn and S. W. Tsai, Nonlinear elastic behavior of unidirectional composite laminae, *J. Compos. Mater.* 7:102 (1973).

SHEAR FRACTURE TESTS (MODE II) ON FIBER REINFORCED PLASTICS AT ROOM AND
CRYOGENIC TEMPERATURES

Elmar K. Tschegg[+], Karl Humer[++] and Harald W. Weber[++]

[+]Institut für Angewandte und Technische Physik
Technische Universität, A-1040 Wien, Austria
[++]Atominstitut der österreichischen Universitäten
A-1020 Wien, Austria

ABSTRACT

Because of their excellent mechanical properties at low temperatures glass-fiber reinforced plastics have been used for an increasing number of applications. Hence, the assessment of their mechanical properties at low temperatures under various loading conditions (tension, compression and in particular shear) have become of special interest. In the present contribution, we report on a new (experimentally simple) method designed to measure the shearing properties and the fracture behavior in mode II on notched rectangular specimens, which is especially well suited for measurements on small samples, e.g. for irradiation experiments. Results on a two dimensional reinforced epoxy (ISOVAL 10), which were obtained at room temperature and at 77 K, will be presented. The fracture-mechanical characterization of the material in mode II (shear mode) will be treated as a limiting case of the new testing method. Results of acoustic-emission investigations as well as data pertaining to variations of the shear area will be discussed. Advantages and disadvantages of this new technique will be assessed and compared to other shear test methods.

INTRODUCTION

A full assessment of the mechanical properties of fiber reinforced plastics comprises investigations of their mechanical behavior under shear. In the past, many different specimen geometries have been proposed to determine the inter- and intralaminar shear properties, such as: the single-notched-tension (SNT) and the cracked-lap-shear (CLS) specimen[1], the double-notched-tension (DNT) specimen[2], the three-point-bending (TPB) specimen[3], the end-notched-flexure (ENF) and the end-loaded-split (ELS) specimen[4], the double-notched Iosipescu specimen[5,6], the thin-tubular specimen[7], and the double-notched-compression (DNC) specimen[8] as an improvement of the Iosipescu specimen.

Because of the specimen geometry and/or the specimen size some of these shear test methods cannot be employed at low testing temperatures and in combination with gamma- and particle irradiation programs. Furthermore, an evaluation of fracture-mechanical tests on fiber reinforced plastics within the linear elastic fracture-mechanical concept

(LEFM-concept) requires that the crack tip and the crack length be determined with sufficient accuracy. When testing a composite, the matrix always fails before fracture, and hence, "bridging" of the fibers occurs around the crack tip, which masks the exact location of the real crack tip. Hence, the application of fracture-mechanical concepts, which require the exact crack length for the calculation of the fracture-mechanical properties, is not feasible.

The simple and small specimen geometry used in the present contribution permits an uncomplicated measurement of the shear behavior in the temperature range from room temperature to cryogenic temperatures and is very well suited for irradiation experiments. The load-displacement curve contains all the information needed for an application of the *fracture-energy-concept*[11,13], in which a knowledge of the crack length is not required. We will report on measurements on a two dimensional glass-fiber reinforced epoxy laminate (ISOVAL 10) carried out at room temperature and at 77 K and deduce fracture-mechanical material parameters which can be used for design purposes.

EXPERIMENTAL

Material and test geometry

The geometry and the dimensions of the rectangular shear specimen which is double-notched on both sides, are shown in Fig.1a. Cubic specimens of a similar geometry made of concrete have been investigated by Davies[9,10] previously. In order to select the specimen geometry (size and shear heights) for the present experiments (Fig.1a), finite element (FE) calculations (for isotropic materials) have been employed. The corresponding results show that the ratio of the stress-intensity-factors in mode II and mode I (K_{II}/K_{I}) is ≥ 9 in all cases, and hence, mostly shear conditions prevail when loading the specimen. The specimens were prepared from a 2 mm thick plate of a two dimensional E-glass fiber reinforced epoxy laminate (ISOVAL 10, ISOVOLTA AG, Wiener Neudorf, Austria). The fiber orientations of the two dimensional woven glass-fabric (warp x woof: 17 x 8) are parallel to the sides of the rectangular specimens. Before testing a sample, the starter notch was always sharpened with a razor blade.

Testing machine and test procedures

The loading arrangement and the specimen in the test position are shown in Fig.1b. The specimen is placed on to a compressive plate. The

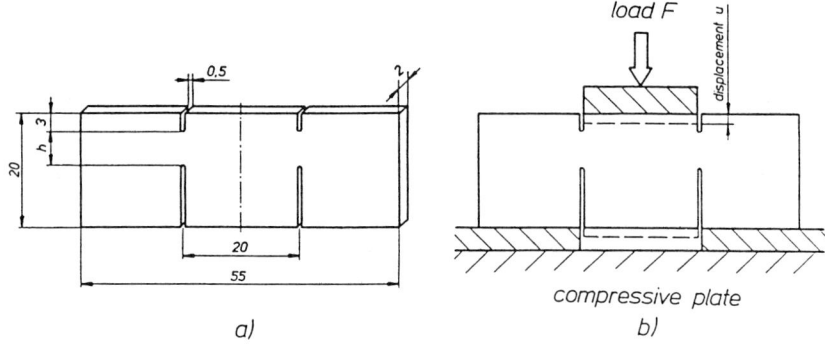

Fig. 1. Geometry and dimensions of the shear test specimens (a) and schematic view of the test arrangement (b).

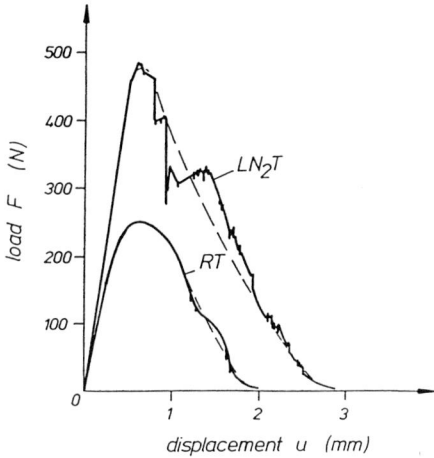

Fig. 2. Typical load displacement curves for samples with a shear height of 6 mm measured at room temperature and 77 K.

load is applied vertically over a specially shaped stamp using a stiff tensile testing machine in combination with a force-reversal arrangement. The crosshead speed was kept constant at 0.5 mm min^{-1} throughout the experiments. During the measurements both the vertical displacement of the compressive plate and the applied load were recorded on an XY-recorder. Because of the stiff testing machine and the strong and stiff design of the testing device, the crack propagation occurs with a stable crack growth mechanism.

The evaluation of this kind of results is described Ref.[11]. From the load-displacement curve the specific fracture energy G_F, which is a measure of the resistance against crack propagation, can be directly calculated using the following equation:

$$G_F(N/mm) = W(Nmm)/A(mm^2)$$

where W is the total energy (area under the load-displacement curve) and A is the shear area. Since our sample contains two shear areas, A is given by 2 × thickness × shear height. Furthermore, the maximum applied load F_{max} as well as a "shear stress" (F_{max}/A) can be determined directly from the load-displacement curve.

In addition, acoustic-emission (AE) investigations have been made during all the tests, in order to obtain more detailed qualitative information about the fracture process. Furthermore, in order to investigate the influence of the shear height in the specimen on the specific fracture energy, measurements on samples with shear heights of 6, 8 and 10 mm, respectively, were carried out at room temperature. Four measurements were made on each sample geometry (shear height h) and at all test temperatures.

RESULTS

Typical load-displacement curves (F versus u, F/u), which are representative of all measurements on samples with a shear height of 6 mm, are shown in Fig. 2 for room temperature and 77 K, respectively. The results are summarized in Table I. As mentioned above, all data presented in the table refer to average values calculated from four measurements, the deviations being always below 5%. As a general result, we note that the specific fracture energy G amounts to about 130 N/mm at

Table I. Test temperatures, average results for the maximum load, the "shear stress" and the specific fracture energy for samples with a shear height of 6 mm.

Temperature (K)	maximum load F_{max} (N)	"shear stress" F_{max}/A (MPa)	specific fracture energy G_F (N/mm)
293	2530	105	131
77	4610	192	280

room temperature and increases by about a factor of two (280 N/mm) at 77 K. Both the maximum load (~2500 N at room temperature) and the calculated "shear stress" (~105 MPa at room temperature) increase by ~80% (~4600 N and ~190 MPa, respectively) at 77 K.

Concerning the AE investigations, a comparison of the AE-count rates (AE-events per second) and the total number of AE-events at room temperature and 77 K is shown in Fig.3. The total amount of the AE-events is ~30% higher and the maximum AE-count rate ~100% higher at room temperature compared to 77 K.

Test results pertaining to the influence of the shear height of the specimen on the material parameters are presented in Table II. With increasing shear height from 6 to 10 mm the maximum load increases continuously by ~50% (from ~2500 N to ~3800 N), whereas the "shear stress" decreases in a similar way by ~10% (from 105MPa to 94MPa). With regard to the specific fracture energy no systematic and significant influence of the shear height is noticed. The total decrease or increase of the specific fracture energy is less than ~10%. Fotographs of two test samples following shear fracture at room temperature and 77 K, re-

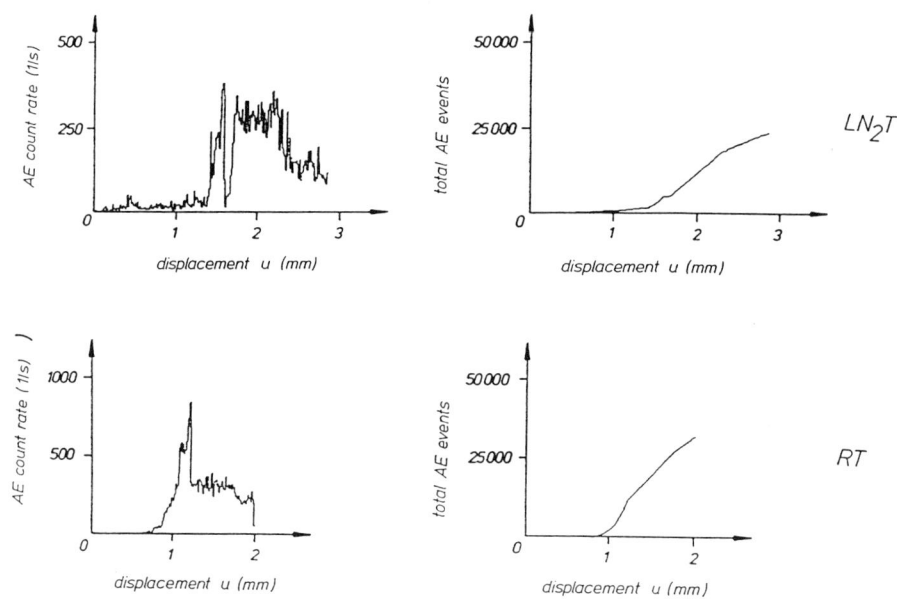

Fig. 3. AE-count rates and total number of AE-events versus displacement u at room temperature and 77 K.

Table II. Shear heights and test temperature, average results for the maximum load, the "shear stress" and the specific fracture energy.

shear height h (mm)	Temp. (K)	maximum load F_{max} (N)	"shear stress" F_{max}/A (MPa)	specific fracture energy G_F (N/mm)
6	293	2530	105	131
8	293	3240	101	145
10	293	3770	94	143

spectively, are shown in Fig.4. Clearly, more "plastic" deformations can be observed following fracture at 77 K due to matrix crazing and fiber deformations.

DISCUSSION

The recorded load-deformation curves should be discussed in detail. Two characteristic parts of this curve (Fig.2) can be distinguished at both test temperatures (room temperature, 77 K). Before reaching the maximum load, a linear elastic material response can be observed. In addition, slight "ductile" deformations occur shortly before the maximum load is reached at room temperature. At the maximum load the crack initiation takes place and the crack propagates under stable crack growth conditions. This second part of the load-displacement curve (F/u), the so called "strain-softening-area", is more clearly serrated at 77 K. We believe that these serrations reflect the successive pull-out of fibers and fiber fractures during the crack propagation. Compared to the ductile matrix behavior at room temperature, the matrix becomes clearly more brittle at low temperatures[12], and hence, the serrations can be observed more clearly at 77 K.

As pointed out in the introduction, a problem of fracture-mechanical testing on fiber reinforced plastics is the determination of the crack

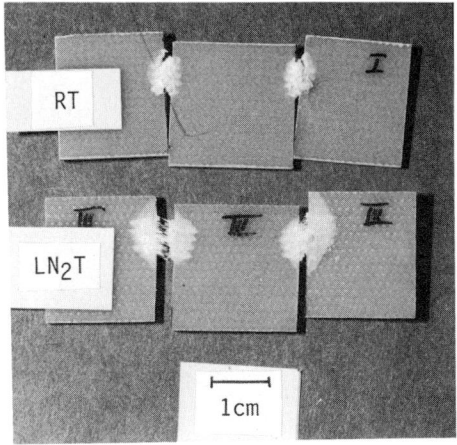

Fig. 4. Photographs of the test samples following shear fracture at room temperature and 77 k.

tip and of the crack length with sufficient accuracy. Fracture-mechanical concepts, such as the stress-intensity-concept (K-concept) as well as the J-integral-concept (J-concept) and the crack-resistance-curve concept (R-concept) require explicitly the exact crack length for fracture-mechanical calculations. In comparison to these concepts, no accurate determination of the crack length and of the crack elongation is needed for fracture-mechanical calculations using the fracture-energy-concept[11,13]. As a further advantage of this concept it should be noted, that the F/u-curve is sufficient to characterize the material behavior. All information needed for the application of the fracture-energy-concept is contained in this curve: The specific fracture energy G_F as well as the maximum load F_{max} and the "shear stress" F_{max}/A can be determined directly from the F/u-curve. If the G_F-values are independent of specimen size, they are characteristic material data for design purposes. Accordingly, the G_F-values can also be applied to study the qualitative influence of sample irradiation on the mechanical shear properties of the materials.

Based on the F/u-curves, the shear-stress/shear-strain curves (τ/ε-curves) can be calculated with finite element methods (FEM). These curves characterize the mechanical behavior of a material completely and are the basis for FE calculations of the fracture behavior of components. More details concerning the application of FEM to the calculation of the τ/ε-curves are currently being worked out[14]. The fact that the τ/ε-curves can be obtained from the measured F/u-curves only with the application of FEM is considered to represent a certain disadvantage of the fracture-energy-concept.

Another disadvantage is related to the fact, that both G_F and the F/u-curve could depend on the sample geometry. Hence, "size effects" have to be investigated carefully, if the fracture-energy-concept is to be used for material characterizations. With regard to this problem, no systematic and significant influences of the shear height of the samples on the specific fracture energy G_F have been found (Table II). With increasing shear height the maximum applied load increased continuously by ~50%, whereas the "shear stress" decreased in a similar way by ~10%. Investigations of the influence of sample size and thickness on the load-displacement curves and the G_F-values, respectively, are in progress[14].

As can be seen in Fig.4, clearly more "plastic" deformation occurs following shear fracture at 77 K. Because the matrix is brittle at low temperatures, the matrix crazes enable the fibers to yield and bend, and hence, the process zone is larger at 77 K. Since more energy is needed for the formation of a larger process zone, the specific fracture energy G_F (being a measure of the resistance against crack propagation) is higher at 77 K (i.e. by about a factor of two, Fig.2). Shear fracture tests carried out at 4.2 K are currently under way[14].

Concerning the AE investigations, the plots showing count rates versus displacement (Fig.3) display a maximum AE-count rate and further peaks, which correspond accurately to the first serration (crack initiation) and further serrations (crack propagation) of the F/u-curves (Fig.2). The brittleness of the matrix at low temperatures leads to matrix crazing and a small AE-count rate also before the crack initiation takes place, whereas the ductile matrix fails shortly before the crack is initiated at room temperature. Hence, AE-count rates are recorded also shortly before the crack was initiated. A more continuous distribution of the AE activity can be observed at 77 K as compared to room temperature.

Finally, some advantages and disadvantages of the shear fracture test method presented in this contribution, should be emphasized. The test method and the device are certainly experimentally simple and the test can be done in all top-loading cryostats suitable for tensile test machines with a force-reversal arrangement. Furthermore, the sample mounting is very simple (especially important for irradiated samples!) and the sample preparation quick and cheap. Two disadvantages should be noted. Firstly, a minimum sample thickness is necessary to avoid buckling, and secondly the evaluation of the τ/ε-behavior requires FEM as mentioned above.

CONCLUSION

As pointed out in the introduction, the assessment of the mechanical properties of fiber reinforced plastics, especially under shear loading conditions, is of considerable interest for many applications. Several different sample geometries have been proposed to determine the shear response of fiber reinforced plastics, but no shear test method has received general acceptance or is suitable for all requirements. In this paper a new specimen geometry has been proposed for the characterization of the shear fracture properties based on the fracture-energy-concept.

The main results of this test method and of the tests obtained on a fiber reinforced plastic (ISOVAL 10), may be summarized as follows:
1. The test method and the device are experimentally simple and the test can be done at all temperatures down to cryogenic temperatures.
2. The fracture-energy-concept does not require the knowledge of the exact crack length. The recorded load-displacement curves contain all the information needed for a full material characterization.
3. With decreasing test temperature from room temperature to 77 K, the specific fracture energy increases by about a factor of two; both the maximum applied load and the "shear stress" increase in nearly the same way (~80%) for the material investigated in the present study.
4. The load-displacement curves show linear-elastic material behavior at the beginning, and serrations (more clearly at 77 K) in the second part of the curves (strain-softening area).
5. The peaks of the acoustic-emission count rates are accurately correlated with the serrations of the load-displacement curve. More uniform acoustic emission activity was observed at 77 K.
6. No significant and systematic influence of the shear height on the specific fracture energy was observed in the tested range.

ACKNOWLEDGEMENTS

The authors are greatly indebted to ISOVOLTA AG, Wiener Neudorf, Austria, for providing us with the test samples of ISOVAL 10. This work is supported in part by the Federal Ministry of Science and Research, Vienna , Austria, under contract # 77.011/2-25/89.

REFERENCES

1. A. C. Garg, Intralaminar and interlaminar fracture in graphite/epoxy laminates, Eng. Fract. Mech. 23:719-733 (1986).
2. S. Nishijima, T. Okada, T. Hirokawa, J. Yasuda and Y. Iwasaki, Radiation damage of organic composite material for fusion magnet, Cryogenics 31:273-276 (1991).
3. S. Egusa, Anisotropy of radiation-induced degredation in mechanical

properties of fabric-reinforced polymer-matrix composites, J. Mat. Sci. 25:1863-1871 (1990).
4. S. Hashemi, A. J. Kinloch and J. G. Williams, The analysis of interlaminar fracture in uniaxial fibre-polymer composites, Proc. R. Soc. Lond. A 427:173-199 (1990).
5. D. F. Adams and D. E. Walrath, Iosipescu shear properties of SMC Composite materials, Composite materials: Testing and Design (Sixth Conference). ASTM STP 787, American Soc. for Testing and Materials, 19-33 (1982).
6. J. A. Barnes, M. Kumosa and D. Hull, Theoretical and experimental evaluation of the Iosipescu shear test, Comp. Sci. Tech. 28:251-268 (1987).
7. G. S. Giare, A. Herold, V. Edwards and R. R. Newcomb, Fracture toughness of unidirectional graphite fibre reinforced/epoxy composite in mode II (forward shear), using a thin tubular specimen, Eng. Fract. Mech. 30:531-545 (1988).
8. P. Ifju and D. Post, A compact double notched specimen for in-plane shear testing, preprint.
9. J. Davies, C. W. A. Yim and T. G. Morgan, Determination of fracture parameters of a punch-through shear specimen, Int. J. of Cement Comp. and Leightweight Concrete 9:33-41 (1987).
10. J. Davies, Numerical study of punch-through shear specimen in mode II testing for cementitious materials, Int. J. of Cement Comp. and Leightweight Concrete 10:3-14 (1988).
11. E. K. Tschegg, Strain softening behavior of fiber reinforced materials, Mat. Sci. Eng. submitted (1991).
12. E. K. Tschegg, K. Humer and H. W. Weber, Influence of test geometry on tensile strength of fibre reinforced plastics at cryogenic temperatures, Cryogenics 31:312-318 (1991).
13. A. Hillerborg, Analysis of one single crack, in Proc. "Fracture Mechanics of Concrete, Developments in Civil Engineering", Ed. by F. Wittmann, Elsevier Amsterdam, 7:223-249 (1983).
14. E. K. Tschegg, K. Humer and H. W. Weber, to be published.

COMPRESSION AND SHEAR TESTS OF VACUUM-IMPREGNATED COMPOSITES AT CRYOGENIC TEMPERATURES

N. J. Simon, R. P. Reed, and R. P. Walsh

National Institute of Standards and Technology
Boulder, CO, 80303, USA

ABSTRACT

To evaluate filament-reinforced, vacuum-impregnated, epoxy-resin systems at cryogenic temperatures, new compression-shear tests have been developed for stainless-composite-stainless sandwiches. Results at 76 K are reported and analyzed for a series of weaves and surface finishes.

INTRODUCTION

Superconducting magnets for fusion energy systems require insulation that can be applied in the field to components that may experience further heat treatment at high temperatures. Because geometries may be complex, some of this insulation will consist of fiberglass layers that are subsequently vacuum-impregnated with a thermosetting resin. Only an epoxy resin system will fulfill the requirements of low viscosity and a long, ~24-h, set-up time. Current conceptual designs of fusion reactors, such as the ITER (International Thermonuclear Experimental Reactor), project a radiation dose of about 5×10^7 Gy at 4 K for the fiberglass-epoxy insulation system.

Owing to these requirements for insulation and the lack of a relevant design data base, screening tests must be devised that can use specimens small enough to be irradiated and tested under cryogenic conditions. The method of specimen preparation must simulate practical methods of insulation fabrication. Under the expected operating conditions, magnet insulation must also withstand simultaneous shear and compression stresses; maxima projected from the ITER conceptual design are 30 MPa and 400 MPa, respectively. Both Okada and Nishijima[1,2] and McManamy et al.[3,4] have reported that the apparent shear strength varies with the applied compressive stress. The McManamy apparatus is not designed for use at 4 K. The Okada and Nishijima fixture design does not permit the application of an unrestrained shear stress, and it loads two specimens in parallel, so the failure stress could be indeterminate, by a factor of 2. Thus, a new test is required to measure the interrelation of compression and shear stresess on composites bonded to metals.

Advances in Cryogenic Engineering (Materials), Vol. 38
Edited by F.R. Fickett and R.P. Reed, Plenum Press, New York, 1992

EXPERIMENTAL TEST FIXTURE AND SPECIMEN PREPARATION

The test fixture shown in Fig. 1 is used in a screw-driven test machine with a maximum load of 10 kN. Compressive loads are applied to the test fixture through an inverted structure. Since the center piece of the test fixture is free to move, there is no shear constraint and indeterminate shear strain does not accumulate in the specimen before failure. Because the two specimens are loaded in series, there is no uncertainty in the failure stress. Fixtures have been constructed for test angles, θ, of 0°, 15°, 30°, 45°, 60°, 75°, and 90°. Details of the fixture are slightly different at 15 and 30°; at 0°, a one-specimen, pure lap-shear design is used. Usually only one specimen fails during a test; the other specimen is reused in successive tests, but specimen history is tracked, in case accumulated damage in successive tests should prove to affect shear strength.

Sandwich-style specimens designed to simulate magnet fabrication conditions are shown in Fig. 2a. These specimens are vacuum-impregnated with epoxy in a mold, as indicated schematically in Fig. 2b. The recesses in the test fixture are square, with a depth slightly less than the thickness of the chip in the sandwich specimen. AISI 316LN was used for the chip to simulate a potential magnet construction alloy and to provide high yield strength. The surface treatment of the chip can be varied, as indicated below. The mold design allows for the simultaneous fabrication of 62 specimens; this permits preparation of enough specimens with identical resin cure conditions that the effect of parameters such as chip surface finish and type, as well as fiberglass surface finish, volume fraction, and layup, can be examined. A fiberglass cutter is used to prepare 13 x 13 mm plies for the specimens. Plies with raveled or misaligned edges are discarded. In the preliminary studies reported here, sandblasting followed by vapor degreasing with trichloroethane was the chief method of surface preparation, although a few specimens were prepared with a very thin surface layer of silane. The epoxy resin used was a diglycidylether of bisphenol-A with an anhydride curing agent. Cure schedule was 5 h at 110°C followed by 16 h at 125°C. This epoxy resin has been previously tested[5] and is relatively resistant to a dose of 5 x 10^7 Gy: about 56% of the torsion shear strength of 201 MPa at 76 K was retained after irradiation with a fission spectrum at 4 K, warmup, and subsequent testing at 76 K. Two E-glass fabrics and one S-glass reinforcing fabric have been used in the preliminary tests reported here. All plies had the same warp-fill alignment and were aligned in the same orientation in the test fixture (shear parallel to warp). All fabrics had a similar epoxy-compatible silane finish. The glass volume fractions used, 0.2 to 0.3, were approximately the minimums that the mold could accommodate, without significant variation in the laminate thickness, since the mold design depends upon the presence of sufficient fiberglass filler to fix the position of the upper chip.

OBJECTIVES

Evaluation of the test fixture, methods of specimen preparation, and preliminary screening tests were carried out in parallel. The purpose of test matrix 1 (Table 1) was to assess the effect of surface roughness and a silane surface primer, and to compare the performance of two types of E-glass plain weaves: a coarse weave with a high fraction of areal open space, and a tight weave with a very low fraction of open space. The tight weave also had an areal density that was one-half of the areal density of the first, and each ply was about one-half the thickness of the coarse weave. The effect of a test temperature of 295 versus 76 K was also examined, but only a few specimens were allocated to 295-K tests since results were expected to be inferior to 76-K results and the insulation must perform at cryogenic temperatures. All tests were run in the 45° configuration.

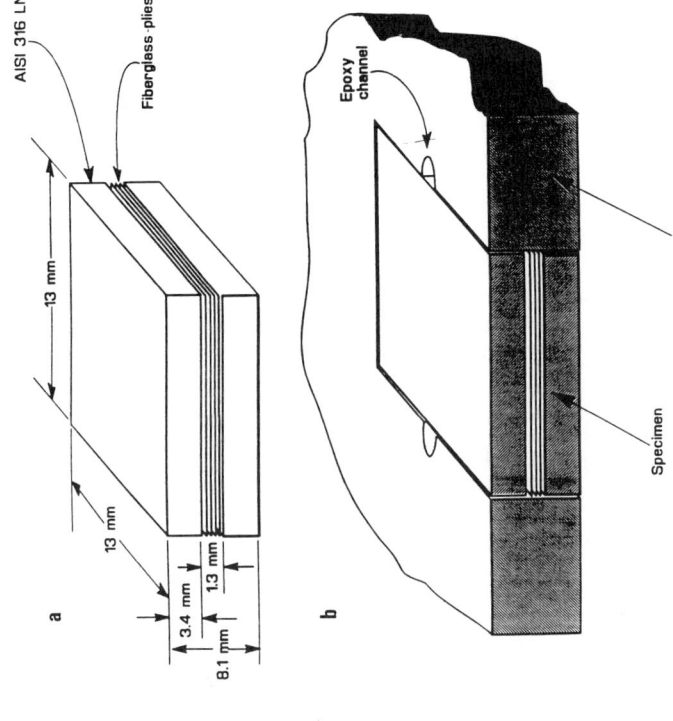

Figure 2 a. Test specimen (schematic).
b. Specimen in mold used for vacuum epoxy impregnation and first stage cure. Upper and lower flat plates with channels for epoxy infiltration are not shown.

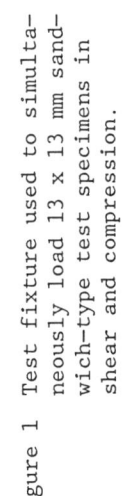

Figure 1 Test fixture used to simultaneously load 13 x 13 mm sandwich-type test specimens in shear and compression.

365

Table 1. Test Matrix 1, 76 K

Aim No. of Specimens	Chip Surface Preparation	Glass Type	Weave Type	Areal Density, mg/cm²	Areal Open Fraction	Volume Fraction	Glass Finish	Fixture Angle
4	SB-100; unprimed	E	Plain - 2532	25	0.176	0.22	Silane	45°
2	SB-100; silane	E	Plain - 2532	25	0.176	0.22	Silane	45°
4	SB-100; unprimed	E	Plain - 1165	12.5	0.033	0.26	Silane	45°
2	SB-100; silane	E	Plain - 1165	12.5	0.033	0.26	Silane	45°
4	SB-250; unprimed	E	Plain - 2532	25	0.176	0.22	Silane	45°
2	SB-250; silane	E	Plain - 2532	25	0.176	0.22	Silane	45°
4	SB-250; unprimed	E	Plain - 1165	12.5	0.033	0.26	Silane	45°
2	SB-250; silane	E	Plain - 1165	12.5	0.033	0.26	Silane	45°

SB - 100 = sand blasted, 100 grit   SB - 250 = sand blasted, 250 grit

Table 2. Test Matrix 2, 76 K

Aim No. of Specimens	Chip Surface Preparation	Glass Type	Weave Type	Areal Density, mg/cm²	Areal Open Fraction	Volume Fraction	Glass Finish	Fixture Angle
8	SB-100; unprimed	E	Plain - 2532	25	0.176	0.22	Silane	45°
2	SB-100; unprimed	E	Plain - 2532	25	0.176	0.22	Silane	0°
8	SB-100; unprimed	E	Plain - 2532	25	0.176	0.30	Silane	45°
2	SB-100; unprimed	E	Plain - 2532	25	0.176	0.30	Silane	0°
9	SB-100; unprimed	S	Plain - 4533	19	0.186	0.23	Silane	45°
2	SB-100; unprimed	S	Plain - 4533	19	0.186	0.23	Silane	0°
9	SB-100; unprimed	S	Plain - 4533	19	0.186	0.29	Silane	45°
2	SB-100; unprimed	S	Plain - 4533	19	0.186	0.29	Silane	0°
8	SB-100; unprimed	E	Plain - 1165	12.5	0.033	0.22	Silane	45°
2	SB-100; unprimed	E	Plain - 1165	12.5	0.033	0.22	Silane	0°
8	SB-100; unprimed	E	Plain - 1165	12.5	0.033	0.30	Silane	45°
2	SB-100; unprimed	E	Plain - 1165	12.5	0.033	0.30	Silane	0°

SB - 100 = sand blasted, 100 grit

Test matrix 2 (Table 2) was constructed to use the best surface preparation determined in test matrix 1, but to vary the fiber volume fraction between approximately 0.2 and 0.3, and to test an S-glass (required for radiation resistance) with characteristics comparable to the best-performing weave determined from test matrix 1. All test matrix 2 tests were performed at 76 K, where most of the total differential thermal contraction present at 4 K has already occurred. Specimens were tested in both the 45° and 0° configuration. Only a few specimens could be allocated to 0° testing, since most were needed to evaluate the 6 other variables (3 weaves x 2 volume fractions) at 45°. The specimens used for the 0° tests had survived several rounds of tests at 45°, so results are high. The number of test specimens listed in Table 2 is the production aim number, and not the actual number tested, due to some failures in the specimen production.

RESULTS OF SHEAR-COMPRESSION TESTS

Figure 3 and Table 3 show the results from test matrix 1 at 76 K, without primer. The open weave reinforcement is superior to the tight weave, in mean 45° shear strength, when combined with the rougher surface finish of the chip. For the finer surface finish, results with both weave styles are comparable, but the performance is below that obtained with the rough surface. Shear strengths obtained with the silane primer on the chips are highly variable, perhaps because the thickness of the primer coat was not adequately controlled. Therefore, these results are not listed in Table 3.

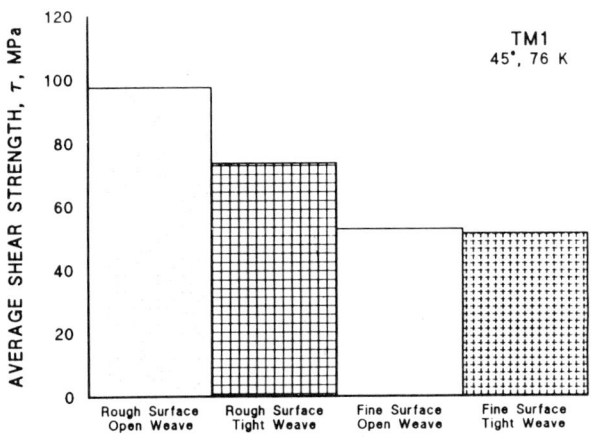

Figure 3 Mean specimen shear strengths at 45° from rough surface, sandblasted with 100 grit; fine surface, sandblasted with 250 grit; open weave (E-2532) and tight weave (E-1165).

Results from four specimens in each condition tested at 295 K are as follows: (1) 52.6 ± 3.8 MPa, rough grit, open weave; (2) 43.2 ± 16.3 MPa, rough grit, tight weave; (3) 43.5 ± 4.8 MPa, fine grit, open weave; (4) 40.0 ± 6.6 MPa, fine grit, tight weave. Thus, at 295 K, within the standard deviations for the small amounts of data obtained, there is no significant difference between the two types of fiberglass weave; a modest superiority of the rougher surface is evident, with the open-weave reinforcement only. As expected, the mean 295-K shear strengths were lower than the 76-K shear strengths for the same condition.

Figures 4 and 5 and Table 4 present the results of test matrix 2. The S-4533 weave is comparable in 45° shear strength to the coarse E-2532 weave. Both weaves have a similar areal fraction of open space (Table 2) and a similar number of filaments per bundle or weave thread. Thickness is 0.28 mm for the coarse E-glass and 0.25 mm for the S-glass. The mass areal density is lower for the S-glass weave, but this does not appear to strongly affect the shear strength. Volume fraction also does not appear to exert a consistent effect, at least over the range from about 0.2 to 0.3.

Two specimens of each of the 6 types were used to test shear strength at $\theta = 0°$ ("pure shear"), except that only 1 specimen of E-1165 at the lower volume fraction was available. Results were lower than at 45°, by a factor of ~2 to 3, as expected from previous results,[1,2] since compression stress has a considerable strengthening effect upon shear strength. Not enough data are available to discriminate between reinforcement types on the basis of this test.

Table 3. Results from Test Matrix 1, 45°

No. of Specimens	Chip Surface Preparation	Fiberglass Type	Mean Shear Strength, MPa	Standard Deviation, MPa	Mean Compressive Stress, MPa
5	SB-100	E - 2532	97.5	±21.4	97.5
4	SB-100	E - 1165	73.6	±16.8	73.6
4	SB-250	E - 2532	52.8	±21.1	52.8
4	SB-250	E - 1165	51.4	±18.6	51.4

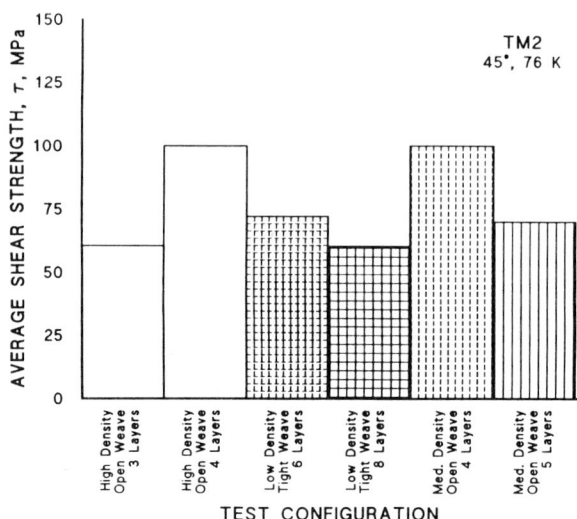

Figure 4 Mean specimen shear strengths at 45° (all with 100-grit sandblasted surface) from different volume fractions of high density, open weave (E-2532); low density, tight weave (E-1165); and medium density, open weave (S-4533).

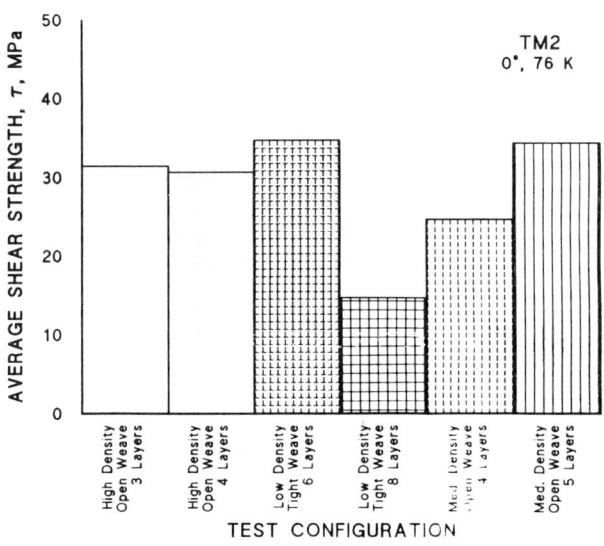

Figure 5 Mean specimen shear strengths at 0° (all with 100-grit sandblasted surface); designations as for Fig. 4.

Table 4. Results from Test Matrix 2, 45° and 0°

No. of Specimens	Chip Surface Preparation	Fiberglass Type	Volume Fraction	Mean Shear Strength, MPa	Standard Deviation, MPa	Mean Compressive Stress, MPa
4	SB-100	E - 2532	0.22	60.6	± 19.0	60.6
2	SB-100	E - 2532	0.22	31.5	± 7.9	0
5	SB-100	E - 2532	0.30	100.1	± 7.5	100.1
2	SB-100	E - 2532	0.30	30.7	± 0.7	0
6	SB-100	S - 4533	0.23	100.1	± 19.3	100.1
2	SB-100	S - 4533	0.23	24.7	± 7.3	0
8	SB-100	S - 4533	0.29	70.0	± 30.1	70.0
2	SB-100	S - 4533	0.29	34.3	± 1.0	0
5	SB-100	E - 1165	0.22	72.2	± 28.1	72.2
1	SB-100	E - 1165	0.22	34.7	—	0
4	SB-100	E - 1165	0.30	60.2	± 18.3	60.2
2	SB-100	E - 1165	0.30	14.7	± 10.0	0

TEST FIXTURE EVALUATION: FAILURE MODE

The histograms presented in Fig. 6 depict the location of the failures for both test matrixes. Most failures were adhesive (at the chip-composite surface) rather than cohesive or interlaminar (within the composite). From the amount of data available from test matrix 1 (Fig. 6b), no bias for any of the 4 adhesive failure positions is clearly evident. However, Fig. 6c shows a low failure rate at the upper chip positions (1) and (3) for test matrix 2. Further analysis showed that the failures in test matrix 2 occurred predominantly at the chip that was uppermost during epoxy infiltration; we believe this may reflect a tendency for the upper chip to "float"; that is, the volume fraction of fiberglass was so low that a weak resin-rich interface may have occurred at the upper chip surface. Since the chips were placed in the test apparatus with the top side (during infiltration) down, a predominance of failures at positions (2) and (4) would be expected in view of the bias related to infiltration position. This problem was not present in test matrix 1: the sandwich specimens were loaded from the opposite side of the middle mold plate for that run, and the mold plate proved to be asymmetrical with regard to width of the slot for the specimen chips. Therefore, in test metrix 1, friction may have prevented the upper chip from floating.

CONCLUSIONS

In equally-loaded shear-compression tests of fiberglass reinforced epoxy specimens, a rougher, 100-grit surface finish resulted in higher average shear strength than a finer, 250-grit finish. Use of a coarse, more open plain weave of E-glass improves shear strength in failures that are primarily adhesive. Shear strength at 76 K was higher than at room temperature, and failures again were primarily adhesive. Shear strength of the S-glass plain weave, intermediate in areal density between the two E-glass weaves tested, is comparable to the best results with high areal density E-glass. A limited amount of data indicated that "pure" shear strengths at $\theta = 0°$ were lower by a factor of 2 to 3 than the shear strengths measured with an equal compressive load applied in a 45° test fixture.

ACKNOWLEDGMENTS

E.S. Drexler was responsible for specimen preparation, testing, and data reduction. The excellent design and machining services of J. A. Shepic were essential to the success of the novel test fixture. Permission to borrow the mold for specimen preparation was kindly granted by C. Bushnell of Princeton Plasma Physics Laboratory and T. J. McManamy of Oak Ridge National

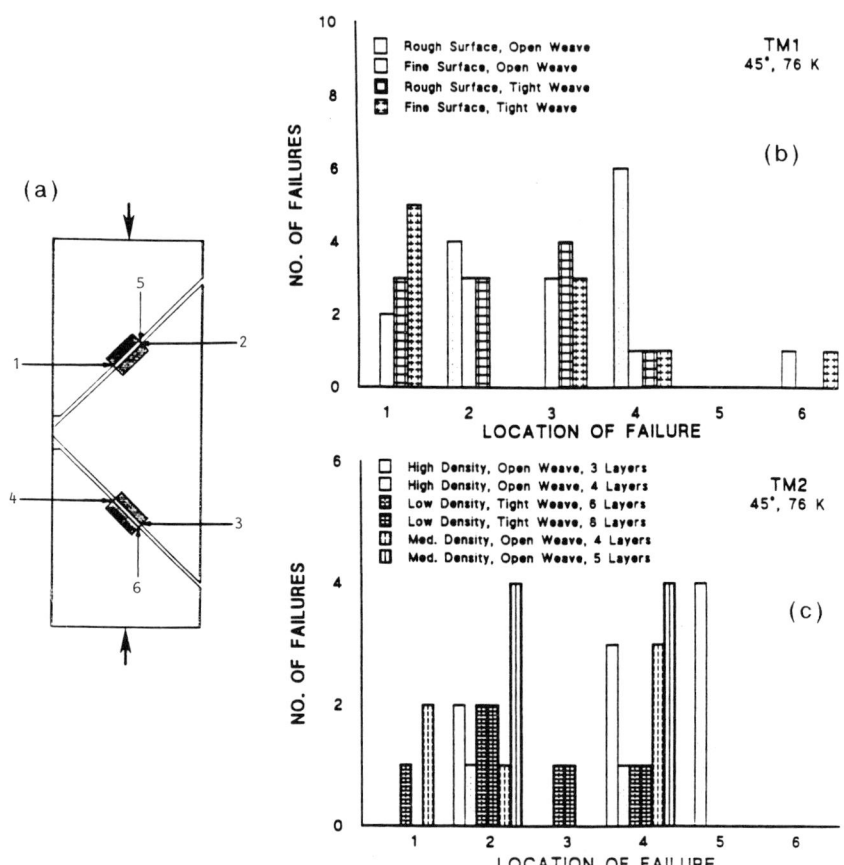

Figure 6 a. Position of failure in test fixture: 1–4 are adhesive failures; 5 and 6 are cohesive failures.
b. Number of 45° adhesive and cohesive failures, test matrix 1; designations as in Fig. 3.
c. Number of 45° adhesive and cohesive failures, test matrix 2; designations as in Fig. 4.

Laboratory. Indispensable cooperation, technical advice, and epoxy infiltration and curing were provided by N. A. Munshi and C. S. Hazelton of Composite Technology Development, Inc. Useful discussions on the design of test fixtures and other matters with T. Okada and S. Nishijima are also acknowledged.

REFERENCES

1. T. Okada and S. Nishijima, Adv. Cryogenic Eng. 36:811 (1990).
2. J. Yasuda et al., Adv. Cryogenic Eng. 36:985 (1990).
3. T. J. McManamy, G. Kanemoto, and P. Snook, "Insulation Irradiation Test Programme for the Compact Ignition Tokamak," Cryogenics 31:277 (1991).
4. T.J. McManamy, J.E. Brasier, and P. Snook, "Insulation Interlaminar Shear Strength Testing with Compression and Irradiation", in Proc. IEEE 13th Symposium on Fusion Engineering, IEEE, New York, Vol. 1: 342 (1990).
5. N. A. Munshi, "Radiation-Resistant Epoxy Resin System for TF and Other Superconducting Coil Fabrication," in Adv. Cryogenic Eng. 38: xxx (1992).

CRYOGENIC FATIGUE TESTING OF GLASS
REINFORCED EPOXY TUBES

M. K. Abdelsalam

Applied Superconductivity Center
University of Wisconsin
Madison, Wisconsin 53706

ABSTRACT

In superconductive magnetic energy storage (SMES) units, cold to warm struts are required to transmit the magnetic radial load from the 1.8 K helium dewar surface to the 300 K vacuum vessel surface with a minimum heat leak. Glass reinforced epoxy tubes are proposed for such struts. Subscale samples of strut material and configuration are tested. Ultimate compressive strength testing at both ambient and liquid nitrogen temperatures in addition to the compression-compression fatigue life at 77 K are reported. Scanning electron micrographs of the failed specimens are discussed. The average ultimate compressive strength at 77 K is 128 ksi, about 50% higher than room temperature strength. The low temperature fatigue life at a stress level of 60% of the ultimate strength exceeds one half million cycles.

INTRODUCTION

Cold to warm struts are used in the design of the SMES engineering test model to transfer the radial magnetic forces from the magnet surface to the outer wall of the vacuum vessel. These load-bearing struts function also as thermal insulators between the cold magnet at liquid helium temperature and the warm outer vacuum wall at ambient temperature. The temperature gradient across the strut can be as high as 300 K. Optimization of the refrigeration power requirements for different strut materials showed that E-glass reinforced epoxy is superior to all structural metals and to most of the non-metallic composite materials. The strut tubular configuration increases its resistance to global structural instability. Mechanical and thermal testing have been performed on these tubes to qualify their use in the SMES design. We report here on the ultimate compressive strength at ambient and LN_2 temperatures and on the compression-compression fatigue life at LN_2 temperature.

EXPERIMENTAL PROCEDURE

Two different test equipments were used in the testing, one for the ultimate strength and the other for the fatigue life. In the ultimate strength testing, a million pound capacity test machine with a large load frame was used. This hydraulically driven machine is equipped with an MTS controller for load control and system shut off after sample failure. The cyclic fatigue testing was performed using the much smaller load frame shown in Figure 1. It consists of three thick steel plates connected with four tension rods. The hydraulic actuator and the load cell are located between the bottom and the intermediate plates. A thick piece of G-10 insulator is placed at the bottom of the test dewar located on top of the intermediate plate. The sample is centered on top of a spherical head sitting on a stainless steel base plate inside the test dewar. A stainless steel thick rod transmits the load from the top of the sample to the top plate. The top plate rests horizontally on four half nuts with a clearance between it and the steel rod. Four nuts at each of the top and bottom plates tie the four tension rods and adjust the alignment of the top and bottom plates so that the load is applied axially on the sample. A 3000 psi, 15 gal/min hydraulic power supply charges the pressure cylinder. An MTS closed loop control system (Fig. 2) with a digital function generator controls and generates the sinusoidal load signal. The load is read via the calibrated load cell.

Grooved end caps on both ends of the composite tube prevent edge brooming. These caps restrain the ends on both the inside and the outside surfaces of the tube. In all

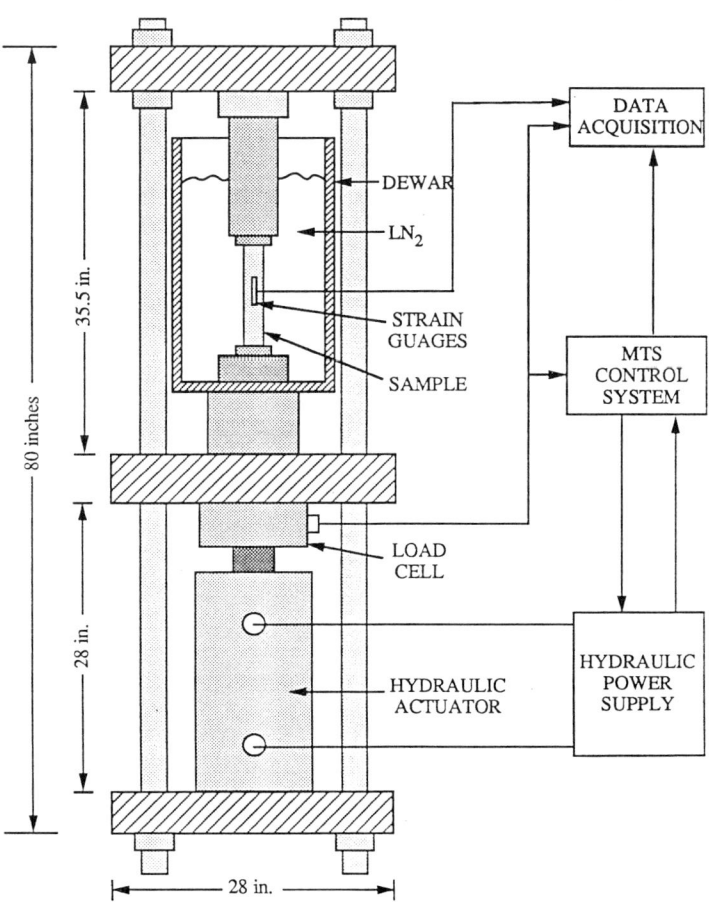

Figure 1 Fatigue Test Machine

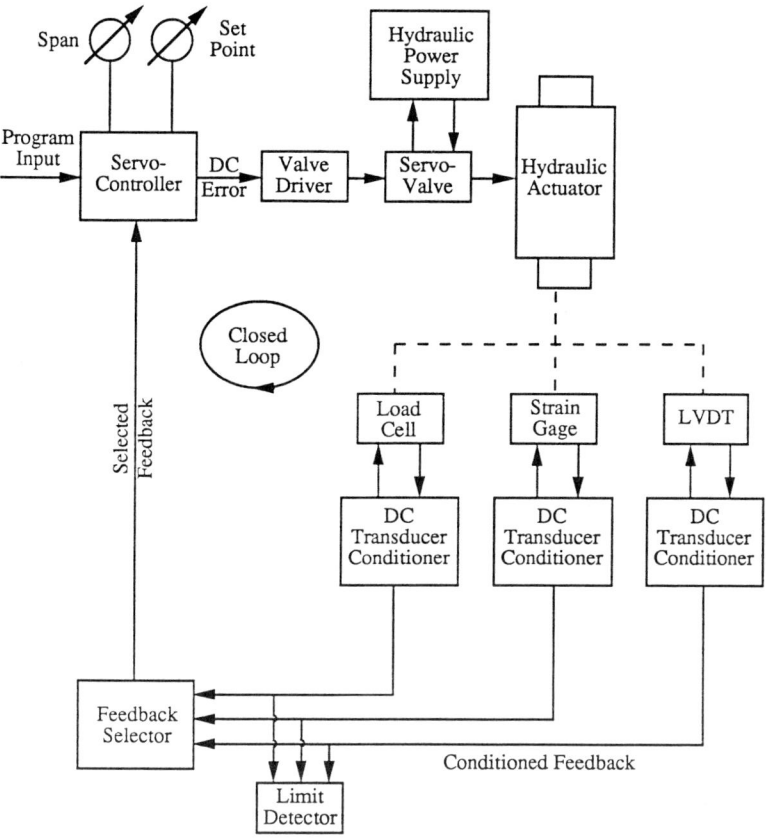

Figure 2 Feedback control circuit for the fatigue test

tests, strain gauges are attached to the middle section of the tubes and are used to align the load axially. Bending strains are kept within 8% of the average compressive strain. In addition to the axial gauges, two strain gauges are also attached around the circumference of the tube to measure the transverse strain. Dummy strain gauges are used to compensate for the temperature effects. These gauges are mounted on a dummy sample which is kept during testing at the same environment as the test sample.

A liquid nitrogen test dewar is specially designed and built for these tests. The dewar consists of a non metallic composite inner wall and a steel outer wall with a thick layer of foam insulation in between. The dewar base is strengthened with a steel plate on each side. A thick plate of G-10 glass reinforced epoxy composite is used to insulate the dewar base. The dewar is successfully tested both at ambient temperature and filled with nitrogen. The LN_2 level inside the test dewar is maintained well above the top of the sample using an automatic refill controller. The cyclic fatigue testing required about 480 liters of liquid nitrogen daily.

Sinusoidal cyclic loading at three peak loads, 0.60, 0.70 and 0.80, of the LN_2 ultimate strength is used. The minimum load is kept at 10% of the ultimate load. Cyclic loading frequency ranges between 0.5 and 1 Hz.

TEST SAMPLES

The composite tubular samples are shown in Fig. 3. The 9" long sample has a 2.125" ID and 2.583" OD in the 5" long middle section graded to 2.738" OD at both ends. The samples are manufactured by Spaulding Fiber Company, Inc. using a filament winding technique. The fiber fraction in the matrix is 55% as estimated by the manufacturer. The winding pattern is $[\pm 10_3, \pm 81.5]_3$ at the middle straight section.

RESULTS AND DISCUSSION

Table 1 lists the values of the ultimate compressive strength, Young's modulus and Poisson's ratio at both ambient and LN_2 temperatures. Stress strain relationship is plotted in Figures 4 and 5 for the temperatures. It is seen that cooling the tubes down to 77 K increases the ultimate compressive strength by more than 50%. On the other hand Young's modulus increases by only 15% and Poisson's ratio is almost independent of temperature in that range. The fatigue life at three different peak load levels is listed in Table 2 and plotted in Fig. 6. The 129 Kips peak load test (60% of σ_u) is stopped after 526,000 cycles

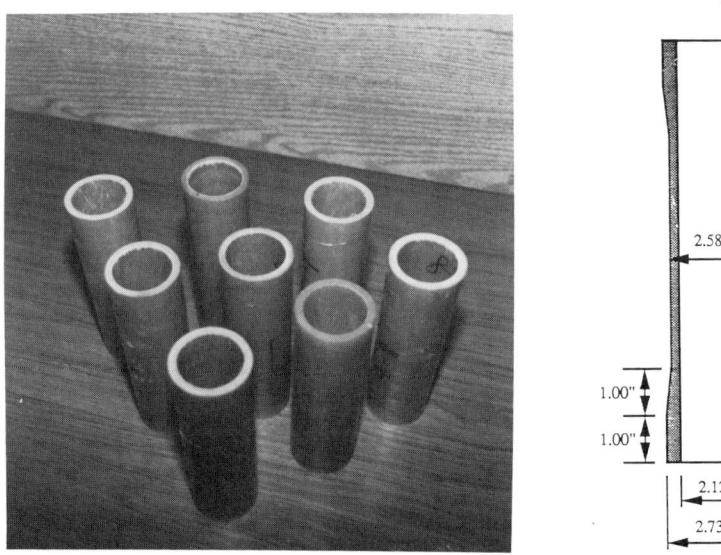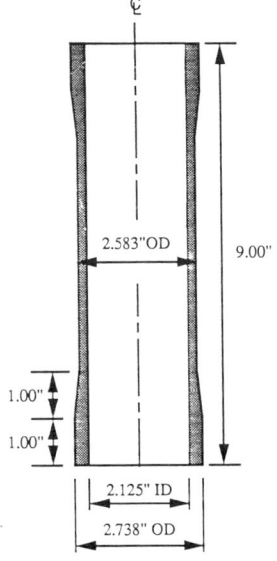

Figure 3 Cross section and photos of the strut tubes

Table 1. Ultimate Compressive Strength of E-Glass Reinforced Tubes

Property	300 K	77 K
Ultimate compressive strength (ksi)	80.7 ± 12.7	126.9 ± 9.7

Young's modulus (Mpsi) :
  Tangential
  Secant (at failure)
	4.77
	4.29
		5.60
		5.00

Poisson's ratio	0.24 - 0.28	0.26 - 0.29

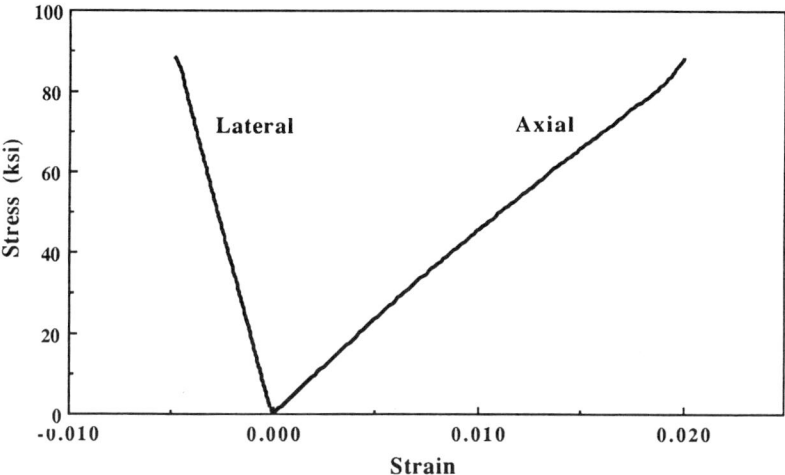

Figure 4 Ambient stress strain relationship for the E-glass reinforced epoxy tubes.

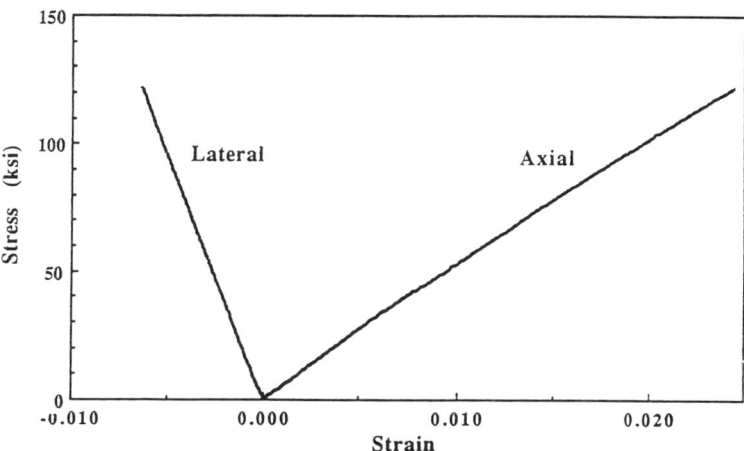

Figure 5 LN_2 stress strain relationship of the E-glass reinforced epoxy tubes.

Table 2. Fatigue Life of E-Glass Filament Composite Tubes in LN_2

Sample	Peak Load (% of σ_u)	Number of Cycles
4-6	80%	152
2-8	80%	23,731
2-6	80%	43,809
3-8	70%	59,293
4-10	70%	72,096
3-6	70%	113,012
3-5	70%	247,780
3-9	60%	> 526,000

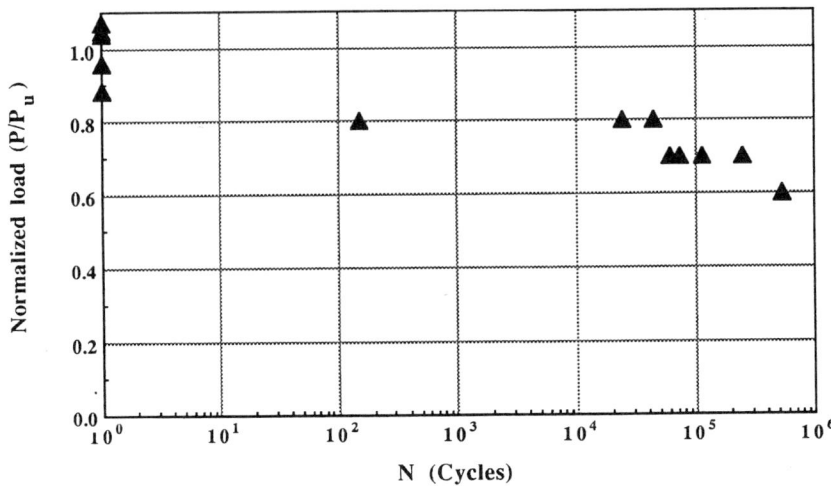

Figure 6 Fatigue life of the E-glass filament wound composite tubes at LN_2 temperature.

with no apparent failure detected in the sample. Scanning electron micrographs (SEM) for the low temperature failed specimens are shown in Figures 7 and 8.

In both ultimate and cyclic load testing the failed samples show delamination, buckling of the ±10° laminae and tensile failure of the ±80° laminae. However, the first sign of failure in the ultimate load testing seems to be a delamination of the inner laminates.

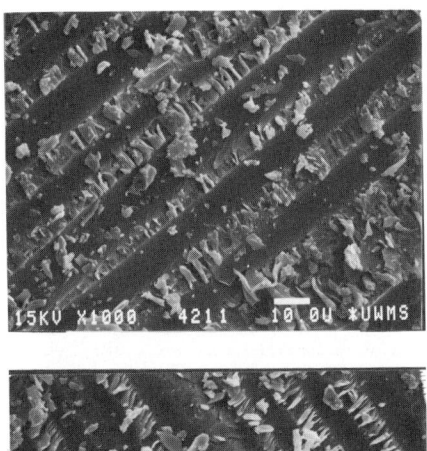

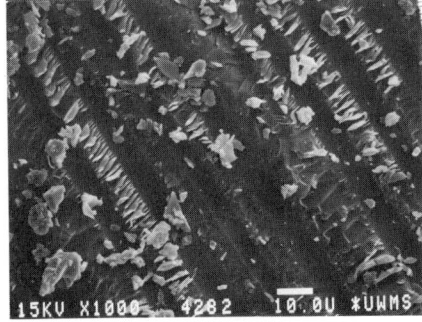

Figure 7 Scanning electron micrographs showing the fiber matrix interface of the low temperature ultimate load tested samples. Fiber matrix debonding is clearly seen. (10 u = 10 micrometers)

Figure 8 Micrographs of the fragmented matrix after fatigue cycling at LN_2 temperature.

The ±10° laminae on the inside of the tube separate and buckle inward before a complete failure takes place. This has been verified by loading a sample close to its ultimate load and then unloading it. Examining this sample clearly shows that in addition to the delamination and buckling of the ±10° laminae, each lamina is sheared into strips 3 to 8 mm wide along the fiber lines. The micrographs of the completely failed samples show also a debonding between the fiber and matrix. The two micrographs of Fig. 7 show the matrix-fiber interface for both the 10° and 80° laminae. The smooth surface of the interface after failure illustrates the debonding between the fiber and the matrix.

In the fatigue failed samples, delamination at the interface between the inner set of the ±10° and the ±80° laminates causes some of the samples to separate into two tubes held together at the sample holder. Delamination and buckling of the inner ±10° layers are similar to the case of the ultimate load. Examining the SEM graphs (Fig. 8) shows that the fibers are still bonded to the matrix. However, a large degree of fragmentation of the matrix itself is clearly seen.

GLASS-FILM-GLASS HYBRID ORGANIC COMPOSITES FOR FORCED-FLOW FUSION MAGNETS

S. Ueno, S. Nishijima, T. Okada and M. Maruyama*

Osaka University, ISIR, Ibaraki, Osaka, Japan
*Arisawa Mfg. Co., Ltd., Joetsu, Niigata, Japan

ABSTRACT

The interlaminar shear strength of glass-film-glass organic composites has been tested to check the integrity of the insulating materials used in forced-flow superconducting magnets.

The glass-film-glass (GFG) composites are used as the adhesive insulating material in forced-flow superconducting magnets. The interlaminar shear strength (ILSS) is an important characteristic in maintaining a highly rigid magnet.

Compressive shear tests were performed on GFG materials with different process conditions. The ILSS values of GFG material were found to be low compared to those of conventional composites. The radiation damage of ILSS in GFG material was also examined. It was found that a change of failure mode was brought about by reactor irradiation.

INTRODUCTION

Coil rigidity in fusion magnets ought to be high enough to withstand the electromagnetic forces. The pool-boiling magnets, however, are not highly rigid because of the high volume fraction of the cooling channels. Insulating materials with low Young's moduli and high thermal contractions in the thickness direction also decrease the rigidity of fusion magnets. To increase the rigidity of the magnets, forced-flow superconducting magnets should be used instead of pool-boiling magnets.

A schematic illustration of a forced-flow superconducting magnet and insulator is shown in Fig. 1. In the forced-flow superconducting magnet, the superconducting wires are packed in a stainless steel conduit. Since the conduits are to be spirally wrapped with insulating materials, prepreg materials are used for the insulator. After the coil winding, the prepreg is cured to make the superconducting windings a unit body. Consequently, the insulating material has important roles, not only as the electric insulator but also as the adhesive to increase the coil rigidity in the forced-flow magnet.[1,2]

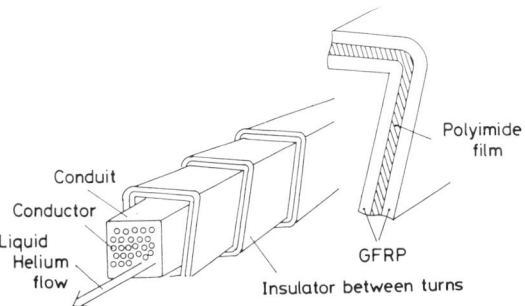

Fig. 1. Schematical illustration of forced-flow superconductor and insulator.

The volume of the insulator, even in the forced-flow fusion magnet, should be minimized to increase the coil rigidity, although the dielectric strength becomes small. To meet the opposite demands, polyimide film is inserted between the prepregs, as shown in Fig. 1. (Hereafter, the glass-film-glass insulating material is called GFG in this work). The polyimide film increases the dielectric strength of the insulator.

Another problem we encounter in the GFG materials is the weak interface strength between the polyimide film and the glass-fiber-reinforced plastic (GFRP). The interlaminar failure at the boundary that occurs at low stress levels results in the degradation of the coil rigidity. To increase the interlaminar shear strength, we tried mat treatment of the polyimide film; the effect of this surface treatment was evaluated by contact angle measurement. The interlaminar shear strength (ILSS) of GFG material was measured at cryogenic temperatures. The effect of nuclear irradiation on the ILSS of GFG insulating material was investigated to determine the suitability of the material for fusion superconducting magnets.

EXPERIMENTAL PROCEDURES

Specimens

Mat treatment was performed on the 25-μm-thick polyimide film. The film was sandwiched between prepreg tapes, which were made of an epoxy matrix and E-glass reinforcement. Untreated film was also used for comparison.

Measurement of Contact Angle

To evaluate the surface tension of treated polyimide film, the contact angle was measured. Figure 2 presents a schematic illustration of contact angle measurement. The contact angle was measured by taking photographs of drops of water and glycerol. A small contact angle means high wettability

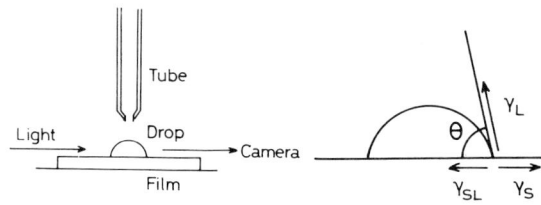

Fig. 2. Measurement of contact angles.

Table 1. Test Conditions of Contact Angle Measurements.

Temperature	293 K
Humidity	lower than 30%
Volume of Drop	0.1 ml

Table 2. Surface Tension of Liquid.

Liquid	Surface Tension ($\times 10^{-4}$ N/m)	Component of Surface Tension ($\times 10^{-4}$ N/m)	
		Dispersion	Polarity
Water	72.8	21.8	51.0
Glycerol	63.4	37.0	26.4

and, hence, a high adhesive strength. The effect of aging on the surface tension was also evaluated after the treated films were stored in a desiccator for three days. The test conditions and the characteristics of the drops are shown in Tables 1 and 2, respectively.[3] From the contact angles, the surface tensions of the films were calculated, which are independent of the characteristics of the drops. The relationship between the contact angle, Θ, and the surface tension, γ, can be represented by the following formulas;[4]

$$\gamma_s = \gamma_{sl} + \gamma_l \cos\Theta$$
$$\gamma_{sl} = \gamma_s + \gamma_l - 2(\gamma_s^d \gamma_l^d)^{\frac{1}{2}} - 2(\gamma_s^p \gamma_l^p)^{\frac{1}{2}}$$
$$(1 + \cos\Theta)\gamma_l = 2(\gamma_s^d \gamma_l^d)^{\frac{1}{2}} + 2(\gamma_s^p \gamma_l^p)^{\frac{1}{2}}$$

where s and l mean solid and liquid, and d and p mean dispersion and polarity, respectively. γ_{sl} is the interface tension between solid and liquid.

Measurement of Interlaminar Shear Strength

Three layers of polyimide film were sandwiched between the prepreg and was cured. The cured specimen was cut into the ILSS specimens, whose shape and size are shown in Fig. 3. This shape was chosen in order to induce interface failure at the boundary between the film and the GRFP. By the compressing the sample, the load at the failure was measured and the ILSS was calculated. The compressive test speed was set at 1 mm/min. Five specimens made under the same conditions were tested. The tests were performed at both room (RT) and liquid nitrogen temperature (LNT). SEM observations of the fracture surfaces were made after the tests.

Reactor Irradiation

Reactor irradiation was performed at the Low Temperature Loop of Kyoto University Reactor. The irradiation temperature was below 20 K.   The

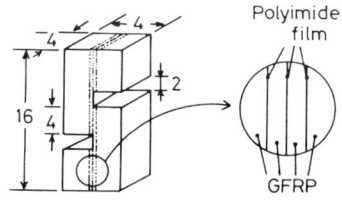

Fig. 3. Shape and size of ILSS test specimen.

Table 3. Surface Tension and Its Components of Mat treated (MT) and untreated (UT) Polyimide Film

Specimen	Contact Angle		Surface Tension		
	Water	Glycerol	γ_s (X10^{-4} N/m)	γ_s^d	γ_s^p
MT	65	49	44.2	31.3	12.9
MT after 3days	69	55	39.8	27.6	12.2
UT	71	55	40.4	30.1	10.3

interlaminar shear test was performed after irradiation at LNT without increasing the temperature to RT. The highest dose was approximately 8 MGy.

RESULTS AND DISCUSSION

Contact Angle and Surface Tension

In Table 3, the measured contact angles and the calculated surface tensions are presented. The surface tension of solid is given by the dispersion, γ^d, and polar component, γ^p;[5]

$$\gamma_s = \gamma_s^d + \gamma_s^p$$

The contact angle of the mat treated polyimide film was smaller than that of untreated film. It means that the surface tension of the polyimide film was increased by the mat treatment. The increase in surface tension is caused by the increase in the polar component. This suggests that polar compounds be added to the surface. After three-day interval in a dessicator, the effect of film treatment vanished.

ILSS and SEM Observation

In Fig. 4, SEM observations of fracture surfaces are shown. The untreated and mat treated surfaces are shown in (a) and (b), respectively. At both RT and LNT, the epoxy resin adhered to the mat treated film surface. On the other hand, the epoxy resin did not adhere to the untreated film surface. This demonstrates that mat treatment increases adhesive strength.

In Table 4, the ILSS values obtained in this experiment are shown. In contrast to the SEM observation, ILSS values of mat treated and untreated film are not much different. The surface treatment of the film for cryogenic use will be established for practical applications.

Reactor Irradiation

The mat treatment was thought to increase the adhesive strength between polyimide film and GFRP. Thus, the mat treated GFG materials were chosen and irradiated in the reactor.

The change of ILSS in the GFG materials induced by reactor irradiation is shown in Fig. 5. The degradation of GFRP is also presented in Fig. 6. Before irradiation, the ILSS values of GFG materials are less than half those of the GFRP.

The ILSS of the GFG material irradiated to 7.6 MGy is only 25% of those of unirradiated specimens. The degradation of ILSS in GFG materials, therefore, would be a serious problem in superconducting fusion magnets.

Figure 6 shows SEM observations of irradiated specimens of GFG materials. Before irradiation (a), the fracture occurred at the boundary between

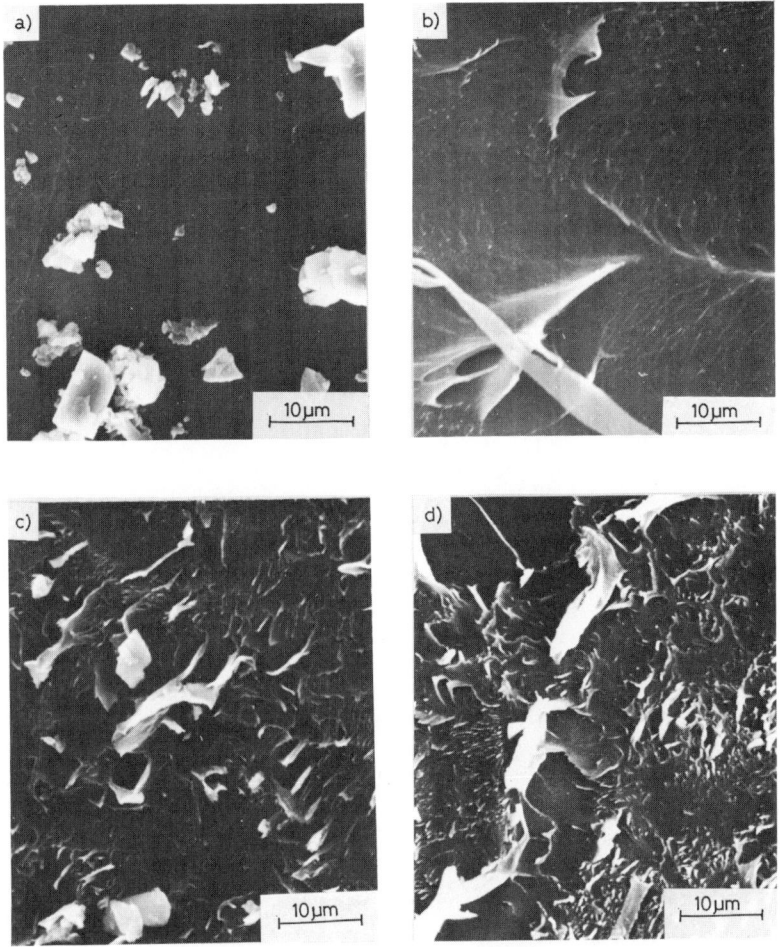

Fig. 4. SEM observations of fracture surfaces ;
a) untreated, tested at RT
b) untreated, tested at LNT
c) mat treated, tested at RT
d) mat treated, tested at LNT

Table 4. Interlaminar Shear Strength (ILSS) of Mat Treated (MT) and Untreated (UT) Polyimide Film

	ILSS (MPa) at	
	Room Temperature	Liquid Nitrogen Temperature
MT	38.3 ± 1.52	58.1 ± 13.3
UT	39.3 ± 2.32	57.7 ± 9.40

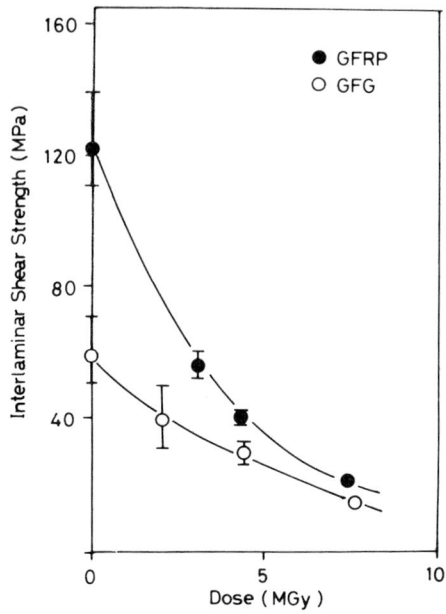

Fig. 5. Degradation of ILSS induced by nuclear irradiation.

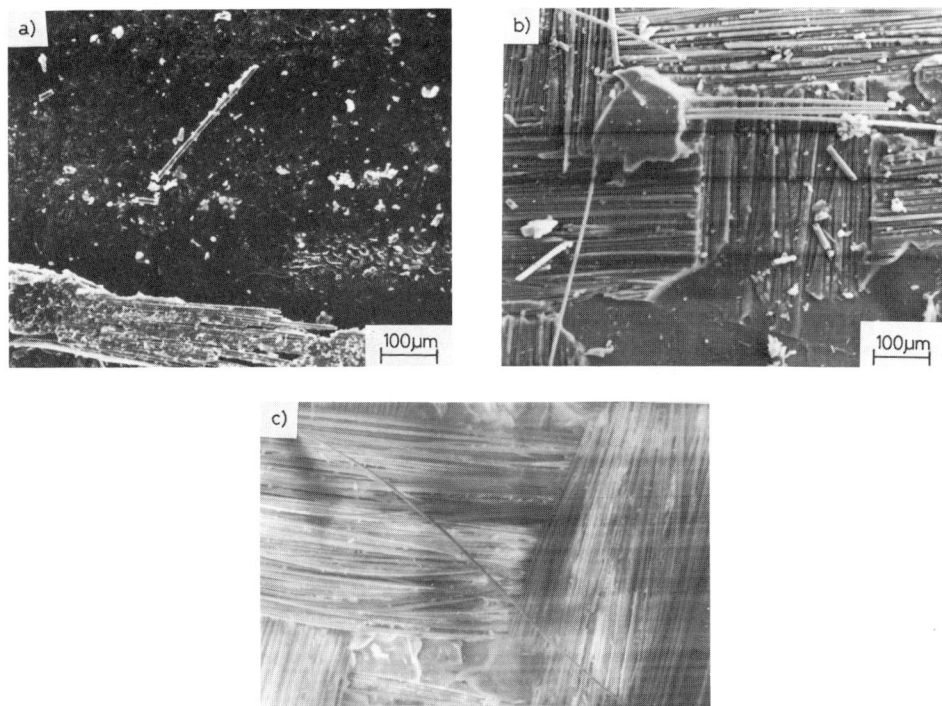

Fig. 6. SEM observations of GFG fracture surfaces ;
a) before irradiation
b) after 4.5 MGy irradiation
c) after 7.6 MGy irradiation

the film and the GFRP. At 4.5 MGy irradiation (b), the fracture area moved toward the GFRP. At 7.6 MGy irradiation (c), only the boundary failure between the glass fiber and matrix was observed. In this case, the glass fiber used was E-glass, and hence, boron oxide was present in the glass fibers. Therefore, a nuclear reaction occurred between the thermal neutrons and the boron; the alpha particles, which degrade the interface strength between the fiber and matrix, were emitted. This is why the failure area changed with the dose.

CONCLUSIONS

Polyimide film was treated in a mat treatment, aiming to increase the interface strength in GFG materials. The GFG materials were irradiated to investigate the radiation damage to their ILSS, because GFG materials are being considered for applications in fusion magnets. The following conclusions were drawn:

1. Mat treatment increased the wettability and surface energy of polyimide film. Consequently, the epoxy matrix adhered firmly to the treated polyimide film.

2. The ILSS of the GFG material was degraded markedly by reactor irradiation. This is partly due to the nuclear reaction $B(n,\alpha)$. Because of this, the failure area changed from the film surface to the glass fiber surface with the dose. Boron-free glass fiber should be used as the reinforcement of GFG material, along with radiation-resistant matrix. The interface behavior between the film and the GFRP is a potential problem, even in the GFG materials in which boron-free glass fibers are used.

3. Studies of film-surface treatments, such as plasma treatment, are needed to achieve a suitable GFG insulator for fusion magnets.

REFERENCES

1. H. Nakajima, K. Okuno, H. Tsuji, K. Yoshida, K. Koizumi, T. Isono, E. Yaguchi, H. Shimane, and S. Shimamoto, Mechanical characteristics in the experiment of the Nb-Ti demo poloidal coils, Proc. MT-11, 824 (1989)
2. K. Okuno, H. Tsuji, Y. Takahashi, H. Nakajima, K. Kawano, T. Ando, T. Hiyama, M. Nishi, E. Tada, K. Yoshida, K. Koizumi, T. Kato, T. Isono, H. Yamamura, M. Satoh, J. Yoshida, N. Itoh, M. Oshikiri, H. Nisugi, M. Konno, E. Kawagoe, Y. Kamiyauchi, M. Hasegawa, Y. Matsuzaki, and S. Shimamoto, The first experiment of the 30-kA Nb-Ti demo poloidal coils, Proc. MT-11, 812 (1989)
3. T. Hirotsu and S. Ohnishi, Surface modification of some fluorine polymer films by glow discharges, J. Adhesion 11:60 (1980)
4. T. Kasemura, S. Ozawa, and K. Hattori, Surface modification of flourinated polymers by microwave plasmas, J. Adhesion (in Japanese) 25:222 (1989)
5. J. R. Dann, Forces involved in the adhesive process, J. Colloid and Interface Sci. 32:2-302 (1969)

FRACTURE-MECHANICAL CHARACTERIZATION OF FIBER REINFORCED PLASTICS IN THE

CRACK-OPENING-MODE (MODE I)

Elmar K. Tschegg[+], Karl Humer[++] and Harald W. Weber[++]

[+]Institut für Angewandte und Technische Physik
Technische Universität, A-1040 Wien, Austria
[++]Atominstitut der österreichischen Universitäten
A-1020 Wien, Austria

ABSTRACT

In view of applications of fiber reinforced plastics at cryogenic temperatures, the mechanical properties of these materials have to be tested both at low temperatures and under various loading conditions (e.g. in tension, compression, shear and in particular in the crack opening mode). The main difficulty with a characterization of the crack growth (mode I) in glass-fiber reinforced materials lies in the fact that the crack length as well as the crack tip cannot be assessed with sufficient accuracy because of delamination and bridging of broken and unbroken fibers. Hence, linear elastic fracture-mechanics cannot be employed. In the present work, first attempts to characterize crack growth in mode I under quasi-static loading conditions in terms of fracture-mechanics have been made. Tests and evaluation procedures based on the fracture-energy-concept, which does not require the knowledge of the exact crack length, have been made at room temperature and at 77 K using a two dimensional glass-fiber reinforced epoxy (ISOVAL 10). Details of the technique as well as results of acoustic-emission investigations and of the influence of sample shape on the fracture-mechanical quantities will be presented. Advantages and disadvantages of the new technique will be discussed.

INTRODUCTION

In the past, many different specimen geometries have been employed to characterize the inter- and the intralaminar properties in the crack-opening-mode (mode I), such as the double-cantilever-beam (DCB) specimen[1,2,3], the compact-tension (CT) and the three-point-bending (TPB) specimen[3] as well as the center-notched-tension (CNT) specimen[3,4]. In order to deduce fracture-mechanical quantities, the stress-intensity-concept (K-concept) as well as the energy-release-concept (G-concept) and the crack-resistance-curve-concept (R-concept) have been used. For fracture-mechanical calculations based on these concepts, among other parameters an accurate knowledge of the crack length and the crack elongation is required.

As can be seen in Fig.1., this is in fact a problem for fiber reinforced materials. Firstly, when testing these materials, the matrix

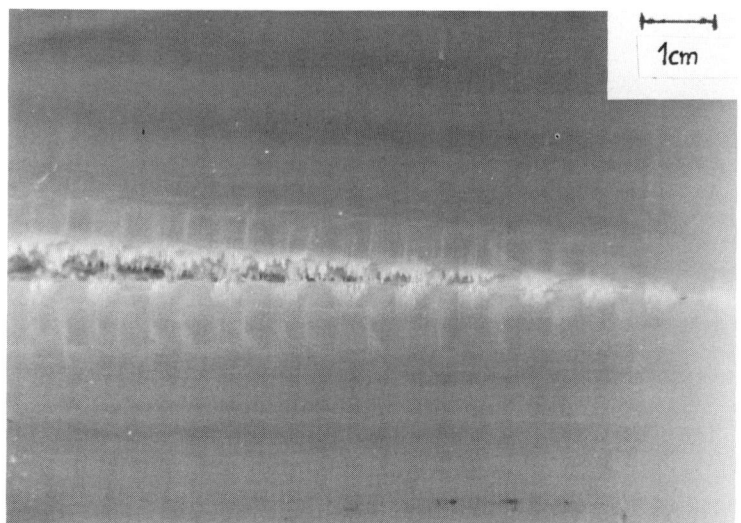

Fig. 1. "Bridging" of the fibers around the crack tip in a fiber reinforced material.

always fails before the complete fracture takes place, and hence, "bridging" of the fibers occurs around the crack tip and masks the exact location of the real crack tip. Hence, the application of fracture-mechanical concepts leads to large inaccuracies. Secondly, the "fiber-bridging" of fiber reinforced materials leads to a stress distribution around the crack tip, which is different compared to a linear elastic material and hence cannot be described by the linear elastic fracture-mechanical method (LEFM). A schematic view of these differences in material behavior is shown in Fig.2. Hence, concepts based on the LEFM are not suitable for fiber reinforced materials.

The *fracture-energy-concept*[5], which does not require the knowledge of the exact crack length for fracture-mechanical calculations, can be used for fiber reinforced plastics as discussed in detail elsewhere[6]. The simple and small specimen geometry described in the present contribution permits an experimentally simple measurement of the fracture behavior of composites in the crack-opening-mode (mode I) at all temperatures down to cryogenic temperatures. The main results of such tests are load-displacement curves. These curves contain all the information needed for the description of the fracture behavior of fiber reinforced plastics. The data measured on a two dimensional glass-fiber reinforced epoxy laminate (ISOVAL 10) at room temperature and at cryogenic temperatures are fracture-mechanical material parameters and can be used to solve technical fracture problems with FEM.

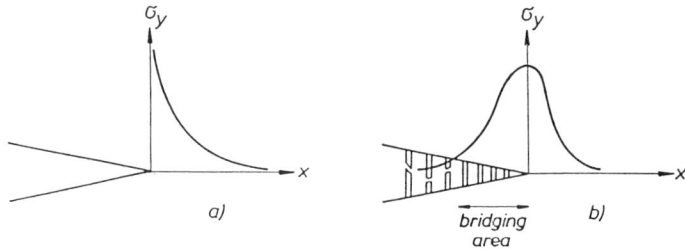

Fig. 2. Schematic view of the stress distribution of a linear elastic material (a) and of a reinforced material (b).

EXPERIMENTAL

The main goal of the present contribution is the determination of the fracture-mechanical properties of fiber reinforced plastics in mode I during stable crack growth using a wedge loading device. In the past, a wedge splitting test has been employed successfully by Hillemayr[7] to measure the fracture toughness in mode I on CT-specimens of concrete materials[7]. In the present study, a modification of this loading device in combination with a new simple specimen geometry[8,9] is described.

The geometry and the dimensions of the rectangular specimen are shown in Fig.3a. The sample has a rectangular groove with a starter notch at the bottom of this groove. The specimens were prepared from a 2 mm thick plate of a two dimensional E-glass-fiber reinforced epoxy laminate (ISOVAL 10, ISOVOLTA AG, Wiener Neudorf, Austria). The fiber orientations of the woven glas fabric (warp x woof: 17 x 8) are parallel to the sides of the rectangular specimens. Before testing a sample, the starter notch was always sharpened with a razor blade.

The wedge loading device and the specimen in the test position are shown in Fig.3b and c. Two load transmission pieces with bearings are placed in the groove (Fig.3b). After inserting a wedge between the bearings, the force F_M can be transmitted from the testing machine to the specimen. The slender wedge (angle ~8°) produce a large horizontal force component F_H and a small vertical force component F_v on the specimen. The horizontal force F_H splits the specimen in a similar way as in the notched bending test. If the wedge angle is small enough, the vertical force F_v does not influence the test results. This has been verified by measurements on concrete[10].

The load is applied vertically using a stiff tensile testing machine in combination with a force-reversal arrangement. The crosshead speed was kept constant at 0.5 mm min^{-1} throughout the experiments. If the test machine as well as the loading device are sufficiently stiff, stable crack growth conditions can be achieved during splitting the specimen. During the measurements both the vertical displacement v of the compressive plate and the applied load from the testing machine F_M were recorded on an XY-recorder (load-displacement curve F_M versus v, F_M/v).

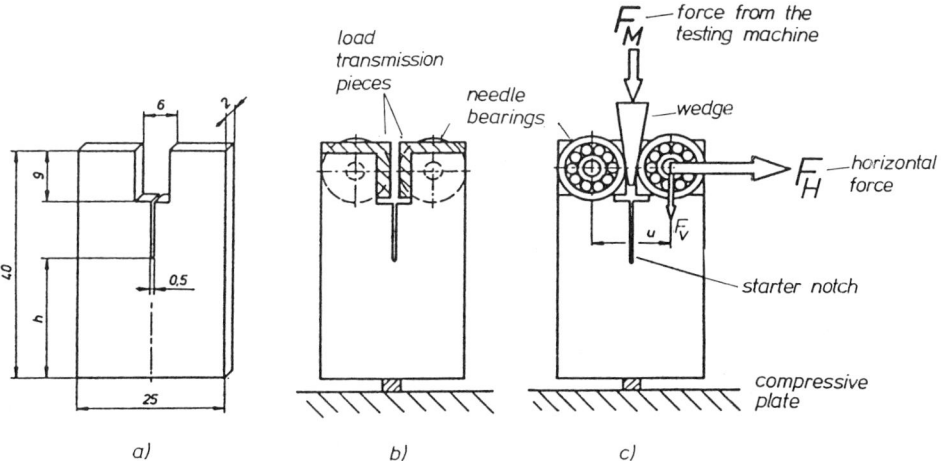

Fig. 3. Geometry and dimensions of the splitting test specimens (a) and schematic view of the test arrangement (b) and (c).

The evaluation of results is described in Ref.[6]. The horizontal force F_H and the crack-opening-displacement (COD) u (Fig.3c) are calculated from the following equations:

$$F_H = F_V/2 \times \tan\alpha \qquad (1)$$

and

$$u = 2 \times v \times \tan\alpha \qquad (2)$$

where α is the wedge angle and v is the vertical displacement. The area under the F_H/u-curve characterizes the work which is needed to split the specimen and is denoted by the term "fracture energy". Dividing this energy by the fracture surface (e.g. in our sample geometry: sample thickness x ligament length h), the specific fracture energy G_F (being a measure of the resistance against crack propagation) is obtained. Furthermore, the maximum applied load F_{Hmax} can be determined directly from the F_M/v-curve using equation (1). The measurements were carried out at room temperature and at 77 K in order to assess the specific fracture energy G_F in mode I and the maximum applied load F_{Hmax}. For the loading device, needle bearings were used at room temperature, and PTFE bearings for testing at 77 K. The friction coefficient of the bearings was small and friction effects on the test results could be neglected.

In addition, acoustic-emission (AE) investigations have been employed throughout the experiments to obtain detailed qualitative information about the fracture process. Furthermore, the influence of the ligament length h of the sample on the fracture-mechanical quantities was investigated. The experiments were carried out at room temperature with ligament lengths of 14, 17, 20 and 23 mm, respectively. In order to determine average data, 4 measurements were made on each sample geometry (ligament length h) and at all test temperatures.

RESULTS

Load-displacement curves (F_M/v), which are representative of all measurements on samples with a ligament length of 20 mm, are shown in Fig.4 for room temperature and 77 K, respectively. All test results presented in Table I refer to average values from four measurements; the deviations are always below 5%. According to Table I the data show an increase of the maximum horizontal load by a factor of nearly four (from ~800 N at room temperature to ~2700 N at 77 K). With regard to the specific fracture energy (~13 N/mm at room temperature) we find an increase by about a factor of seven to ~95 N/mm at 77 K.

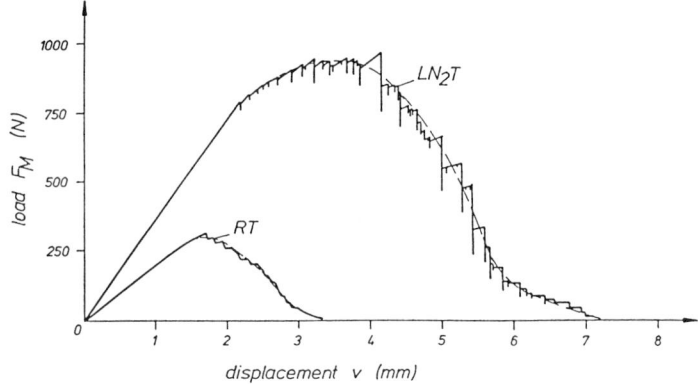

Fig. 4. Typical load-displacement curves for samples with a ligament length of 20 mm measured at room temperature and 77 K.

Temperature	maximum horizontal load F_{Hmax}	specific fracture energy G_F
(K)	(N)	(N/mm)
293	822	13,3
77	2723	94,6

Table I. Test temperatures, average results for the maximum horizontal load and the specific fracture energy for samples with a ligament length of 20 mm.

Fig.5 shows the results of AE experiments carried out at room temperature and at 77 K. The total number of AE-events is about the same at both temperatures, whereas the maximum AE-count rates are about three times higher at room temperature.

Test results pertaining to the influence of the ligament length of the specimen on the parameters at room temperature are presented in Table II. With increasing ligament length from 14 to 23 mm the maximum

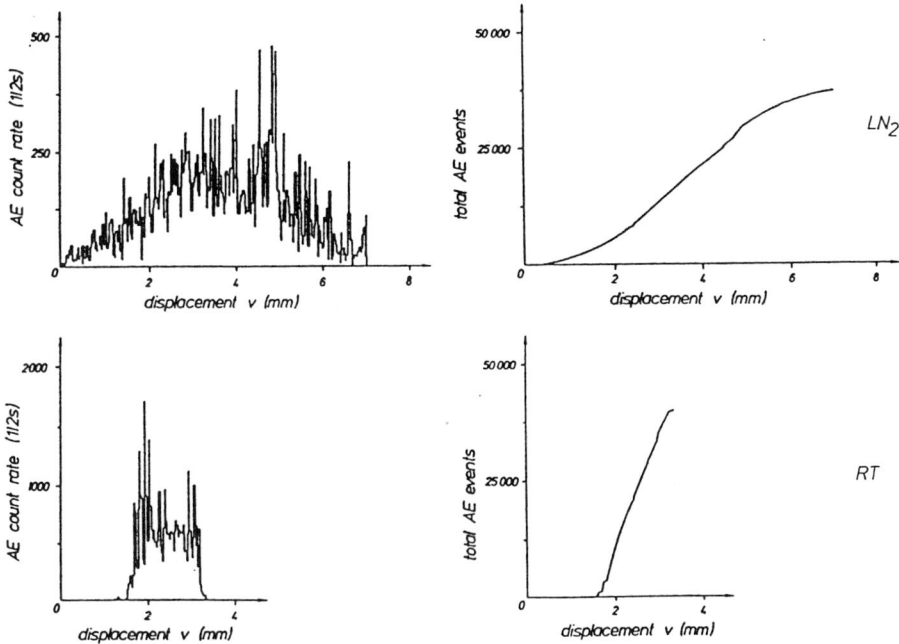

Fig. 5. AE-count rates and total number of AE-events versus displacement v at room temperature and 77 K.

ligament length h	Temp.	maximum horizontal load F_{Hmax}	specific fracture energy G_F
(mm)	(K)	(N)	(N/mm)
14	293	562	12,1
17	293	672	12,2
20	293	822	13.3
23	293	978	13.5

Table II. Ligament lengths and test temperature, average results for the maximum horizontal load and the specific fracture energy.

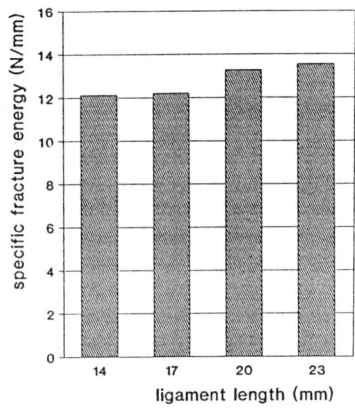

Fig. 6. Average results for the specific fracture energy versus ligament length at room temperature.

load increases by a factor of two (from ~500 N to ~1000 N). The specific fracture energy amounts to about 12 N/mm (for ligament lengths of 14 and 17 mm) and increases slightly (~10%) to about 13 N/mm (for ligament lengths of 20 and 23 mm). The dependence of the specific fracture energy on the ligament length is plotted in Fig.6. A fotograph of two test samples following fracture in the crack-opening-mode (mode I) at room temperature and 77 K, respectively, is shown in Fig.7. Compared to room temperature, more "plastic" deformations (larger process zone) can be observed following fracture at 77 K.

DISCUSSION

With regard to the load-displacement curves (F_M/v), two characteristic parts of these curves (Fig.4) can be distinguished at both test temperatures (room temperature, 77 K). Before reaching the maximum load, linear elastic material response occurs at both test temperatures. Slight "plasticity" can be observed shortly before the maximum load is reached at 77 K. This area of the F_M/v-curve shows also small serrations. We believe that they are caused by the fact, that the matrix becomes more

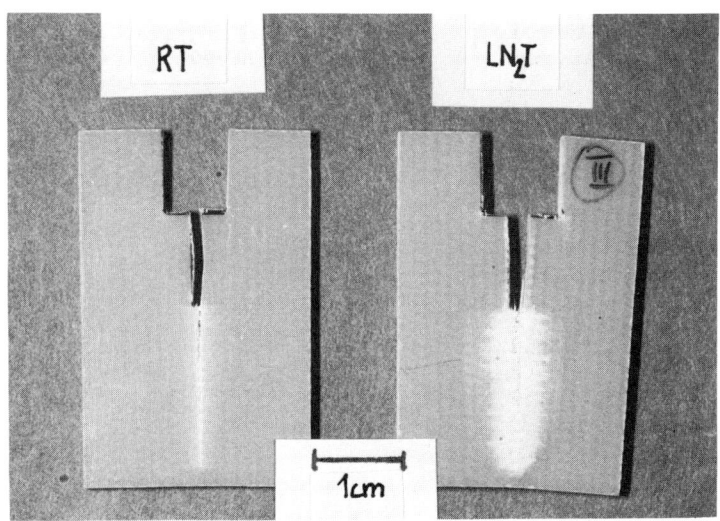

Fig. 7. Photographs of the test samples following fracture in the crack-opening-mode (mode I) at room temperature and 77 K.

brittle at low temperatures[11] as compared to room temperature, and therefore crazes. At room temperature, the crack initiation takes place when the maximum load is reached. Then the crack propagates under stable crack growth conditions. Because of the serrations and the broad shape of the maximum of the F_M/v-curve (upper curve in Fig.4), the crack initiation at 77 K cannot be located exactly. Again the crack propagates under stable crack growth conditions at 77 K. This second part of the F_M/v-curve, the so called "strain-softening-area", is more clearly serrated at 77 K. We believe that these serrations reflect the successive pull-out of fibers and fiber fractures during crack propagation.

As pointed out in the introduction, a problem of fracture-mechanical testing on fiber reinforced plastics is the determination of the exact crack tip and of the crack length with sufficient accuracy. In order to avoid this problem, the fracture-energy-concept[5] was used, which does not require the knowledge of the exact crack length for fracture-mechanical calculations. Advantages and disadvantages compared to other fracture-mechanical concepts (e.g. the K-, G- or R-concept) as well as the application of FEM to fracture-mechanical material parameters are discussed in detail elsewhere[6,12].

One of the disadvantages of the fracture-energy-concept is related to the fact, that both G_F and the F_M/v-curve could depend on the sample geometry, and hence, "size effects" have to be investigated carefully. With regard to this problem, the influence of the ligament length of the specimen on G_F has been investigated at room temperature. With increasing ligament length, the test results show only a slight increase of G_F by ~10% (Fig.6). Since G_F may depend on the sample geometry, further investigations of "size effects" are currently under way[12].

According to Fig.7, clearly a larger process zone (more "plastic" deformation) occurs following fracture at 77 K. As mentioned above, the matrix becomes more brittle at low temperatures, and hence, matrix crazes enable the fibers to yield and bend, which results in a larger process zone at 77 K. Since more energy is needed for the formation of a larger process zone, G_F (being a measure of the resistance against crack propagation) is higher at 77 K (i.e. by about a factor of seven, Fig.3). Splitting tests in the crack opening mode (mode I) carried out at 4.2 K are currently under way[12].

The results of the AE investigations (Fig.5) are different at both test temperatures. At room temperature, no AE-events occur before the crack initiation takes place. Beginning from this moment peaks in the AE-count rates are recorded, which correspond accurately to the serrations of the F_M/v-curve (Fig.4). At 77 K, the brittleness of the matrix leads to matrix crazing and to small AE-count rates also before the crack initiation takes place, whereas the further peaks in the AE-count rates correspond again to the serrations of the F_M/v-curve. Hence, a more continuous distribution of the AE activity can be observed at 77 K as compared to room temperature.

Finally, advantages and disadvantages of the test method should be discussed. The test method and the device are experimentally simple and the test can be done in all top-loading cryostats suitable for tensile test machines with a force-reversal arrangement. Furthermore, the sample mounting is simple (especially important for irradiated samples!) and the sample preparation quick and cheap. Certain disadvantages are related to the fact that a minimum sample thickness is necessary to avoid buckling, and secondly, that the evaluation requires FEM for the determination of real material parameters (stress-strain curve) from the load-displacement curve.

CONCLUSION

The assessment of fracture-mechanical properties of fiber reinforced plastics in the crack-opening-mode (mode I) is of considerable interest for many applications. Several different sample geometries have been proposed for mode I tests on fiber reinforced plastics, but no test method has received general acceptance or is suitable for all measurements. In this paper an new splitting test method has been proposed for the characterization of fracture-mechanical material properties in the crack opening mode (mode I) based on the fracture-energy-concept. This allows us to determine complete τ/ε-curves.

The main results of this test method and of the test results, obtained on a fiber reinforced plastic (ISOVAL 10), may be summarized as follows:

1. The test method and the device are experimentally simple and the test can be done at a wide range of temperatures including cryogenic temperatures.
2. The fracture-energy-concept does not require the knowledge of the exact crack length. The recorded load-displacement curves contain all the information needed for a full material characterization.
3. With decreasing test temperature from room temperature down to 77 K, both the specific fracture energy and the maximum load increase (by factors of seven and four, respectively) for the material investigated in the present study.
4. The load-displacement curves show linear elastic material behavior at the beginning and serrations (more clearly at 77 K) in the second part of the curves (strain-softening-area).
5. The peaks in the acoustic-emission count rates are accurately correlated with the serrations of the load-displacement curve. More uniform acoustic-emission activity was observed at 77 K.
6. With increasing ligament length only slight increases (~10%) of the specific fracture energy are observed.

ACKNOWLEDGEMENTS

The authors are greatly indebted to ISOVOLTA AG, Wiener Neudorf, Austria, for providing us with the test samples of ISOLAL 10. This work is supported in part by the Federal Ministry of Science and Research, Vienna, Austria, under contract # 77.011/2-25/89.

REFERENCES

1. S. Hashemi, A. J. Kinloch and J. G. Williams, Mechanics and mechanisms of delamination in a poly(ether sulphone)-fibre composite, Comp. Sci. Tech. 37:429-462 (1990).
2. G. M. Newaz and J. Ahmad, A simple technique for measuring mode I delamination energy release rate in polymeric composites, Eng. Fract. Mech. 33:541-552 (1989).
3. A. C. Garg, Intralaminar and interlaminar fracture in graphite/epoxy laminates, Eng. Fract. Mech. 23:719-733 (1986).
4. P. K. Sarkarand S. K. Maiti, Prediction of mode I fracture toughness of a laminated fibre composite from matrix fracture toughness of the basic layer, Eng. Fract. Mech. 38:71-82 (1991).
5. A. Hillerborg, Analysis of a single crack, in Proc. "Fracture Mechanics of Concrete, Developments in Civil Engineering", Ed. by F. Wittmann, Elsevier Amsterdam, 7:223-249 (1983).

6. E. K. Tschegg, Strain softening behavior of fiber-reinforced materials, <u>Mat. Sci. Eng.</u> submitted (1991).
7. B. Hillemeier, Bruchmechanische Untersuchungen des Rißfortschritts in zementgebundenen Werkstoffen; thesis, Universität Karlsruhe (1976).
8. E. K. Tschegg, "Prüfeinrichtung zur Ermittlung von bruchmechanischen Kennwerten sowie hiefür geeignete Prüfkörper", Patent No. 390328, 1986, Patent application 31.1.1986.
9. E. K. Tschegg, New equipment for fracture tests on concrete, <u>J. Mat. Struct.</u> submitted (1990).
10. H. N. Linsbauer and E. K. Tschegg, Determination of fracture energy of cementitious materials subjected to environmental influences, Final report A1, COST 502, No. HNL-01-89 (1989).
11. E. K. Tschegg, K. Humer and H. W. Weber, Influence of test geometry on tensile strength of fibre reinforced plastics at cryogenic temperatures, <u>Cryogenics</u> 31:312-318 (1991).
12. E. K. Tschegg, K. Humer, H. W. Weber, to be published.

FRACTO-EMISSION FROM CRYSTALLINE AND NON-CRYSTALLINE MATERIALS

AT CRYOGENIC TEMPERATURES

>Shigehiro Owaki and Toichi Okada
>ISIR, Osaka University
>8-1 Mihogaoka, Ibaraki, Osaka 567, Japan
>
>Sumio Nakahara and Kiyoshi Sugihara
>Dept. Mechanical Engineering, Kansai University
>3-3-35 Yamatecho, Suita, Osaka 564, Japan

ABSTRACT

Electron emissions from metals, FRP and ceramics during the fracture process (fracto-emission) at cryogenic temperatures have been reported in previous conferences. In this paper, the FEs from some crystalline (single and poly-crystal) and non-crystalline (glassy state) insulating materials are compared and discussed. The FEs from sodium silicate glass and pure fused silica are very similar to those from poly-crystalline ceramics, but are different from those from synthetic quartz (single crystal) especially for fracture at room temperature. From experimental results, the fracture modes, generation of free electrons and its compensation in these materials are discussed.

INTRODUCTION

Emissions of low energy electrons from material surfaces have been reportedly observed during and after various kinds of sample treatments such as deformation, abrasion and phase transformation.[1-3] Among them, the emission during and after the fracture of solid state materials, called fracto-emission (FE), in which emissions of ions and photons are involved, is supposed to depend on the fracture mode and the material characteristics.[4-8] Generation mechanism of emitted electrons and relaxation process of the charge irregularity on the freshly created fracture surface are interesting subjects in not only fracture mechanism research but also surface physics.

The three-point bending test provides electron emission during the fracture process and maintains the freshly created fracture surface if the test was performed in a high vacuum. The sample temperatures during the test affect the fracture mode and surface relaxation, especially at cryogenic temperatures. They must give fracto-emission different aspects. The test is performed in a cryostat with a clean vacuum better than 10^{-4} Pa around the sample holder.

The authors reported experimental results about fracto-emission from fiber reinforced plastics (FRP), brittle metals and advanced ceramics at room and cryogenic temperatures in previous conferences [9-12] and the other. [13-15] Each material showed unique characteristics of fracto-emission and some of the mechanism were explained, while some of them remained unsolved. That is, we obtained the following conclusion;

1) Time distribution of FE is available for monitoring the occurrence of local micro-fractures and the fracture modes (fracture of matrix, interface between fiber and matrix, and the fiber itself) induced by the deformation of FRP. [13,15]
2) The origins of FE from pure iron deformed at 30 K are due to some micro-cracks at the boundary of twins, and to many cleaved surfaces of the crystal grains formed with low temperature deformation. Increase of the frequency and intensity of FE from high-carbon steel are due to many kinds of surface changes such as cracks in ferrite, cracks within cementite and separation of the cementite-ferrite interface with the brittleness increase. [11,14]
3) Increase of the peak intensity and the decay time of FE from poly-crystalline alumina depends on the magnitude of stiffness and fracture strength, and on the extent of the transgranular and micro-cracking at low temperatures. [12]
4) The decay time of FE from insulating materials is considerably longer than that of metals because of low electronic conductivity on fracture surface. [11]

Generally in insulating materials, electronic behavior and fracture mode are expected to change sensitively with temperature decrease. However, the fracture mechanism at cryogenic temperatures has not yet been so much investigated, though employment of ceramics is developing at those temperatures. According to our research program of fracto-emission from insulating materials, two kinds of glasses and single crystal plates of synthetic quartz are examined and compared with experimental results of poly-crystalline alumina plate, and the results are presented in this paper.

EXPERIMENT

Fracto-emission (only electrons) from thin plates of glasses and synthetic quartz during deformation by three point bending were observed basically in the same manner and apparatus as the previous experiments. [9-12] A cryostat was evacuated by a turbo-molecular pump to a pressure of 10^{-4} Pa at room temperature (RT). As the pressure decrease with temperature, a considerably clean vacuum was secured around fracture surface of the sample. The specimen temperature decreased to 30 K with liquid He and to 90 K with liquid N_2 cooling.

Two kinds of glass were prepared as specimens of glassy state materials. One was sodium silicate glass, commercially available as a deck glass plate for optical microscopes, and the other was pure fused quartz, in which no optical absorption band was observed in the UV and visible region of wavelength longer than 200 nm. The dimensions (mm) of the bending test pieces were 50 x 10 x (1.0 or 0.5).

As single crystal specimens, blocks were quarried from X- and Z-region of quartz crystals made by hydro-thermal synthesis. They were cut to plates of which planes are perpendicular to X or Z axis and of the same size as above (we call them as X and Z specimens respectively). Generally Z-region crystals are known to be purer than those of the X-region. This was confirmed with the fact that an X-region plate was densely colored dark violet with high energy electron irradiation, while only small colored dots appeared in that of the Z-region.

Some of the specimens of each kind of materials were irradiated by electrons of an energy of 21 MeV from a linear accelerator to dope lattice defects like color centers which may affect the features of fracto-emission. Among the specimens, only pure fused quartz were not visually colored but the optical absorption edge slightly shifted to longer wavelength. Sodium glass was colored densely brown but discolored gradually during storage at room temperature. After one month storage for radiation cooling down, the flex test was performed using the specimens of slightly colored sodium glass.

Electrons emitted from the specimen surface opposite to the press rod controlled from outside the cryostat were detected by a ceramic electron multiplier and were counted as pulses after amplification. The counted pulses were stored as a time distribution before and after fracture in a personal computer. The branched signals were integrated and displayed on an analog recorder simultaneously with those of strain gage and an Au-Fe-chromel thermo-couple attached to the specimen.

A push rod for three point flex test equipped with a micro-gage to monitor its stroke was operated slowly watching the electron emission every 0.25 mm. Different from FRP and brittle metals, no electron emissions from glass and crystal specimens were observed before fracture. At the instant of fracture, electron emission showed a peak intensity and then decayed either quickly or slowly.

As the uniformity and the reproducibility of fracto-emission data are known to be not so good, experiments in same condition were performed on about 3 to 5 specimens.

EXPERIMENTAL RESULTS AND DISCUSSION

Time distributions of fracto-emission from sodium glass plates are shown in Fig. 1. The emission in Fig. 1 (a) was observed during fracture process at room temperature. The emission does not have so high a peak intensity and ends in a short time. With decrease of specimen temperature, however, the peak intensity increase and the decay time become longer as shown in Fig. 1 (b). This trend is more clear in the case of fused quartz as shown in Fig. 2, in which (a) shows the emission during fracture at room temperature, and (b) at 90 K and (c) at 30 K.

These experimental results are very similar to those of poly-crystalline alumina ceramics, of which stiffness and fracture strength increased with decrease in temperature. The features of fracto-emission at low temperatures were explained to be related with the increase of these mechanical properties, which induces much energy released at fracture, and with the change of fracture mode, in which the amount of micro-cracks increases and transgranular facets dominate with decrease in temperature.[12] Moreover, it is supposed that decrease of charge mobility prevents to compensate an uneven distribution of charge on the fresh fracture surface and causes FE decay longer. The mechanical strength of glasses increases with decrease in temperature but the change of fracture mode has not yet been observed. There may be the other factors affecting these features of fracto-emission.

Radiation effects on fracto-emission were examined for both glasses and quartz crystals. The effects are scarcely found in sodium glass as Fig. 3 (a) shows, but in fused quartz, Fig 3. (b), the effects provide a higher peak intensity and longer decay time of fracto-emission same as in polycrystalline ceramics. Although the types of lattice defects like color centers induced by high energy radiation and fracture modes will be essen-

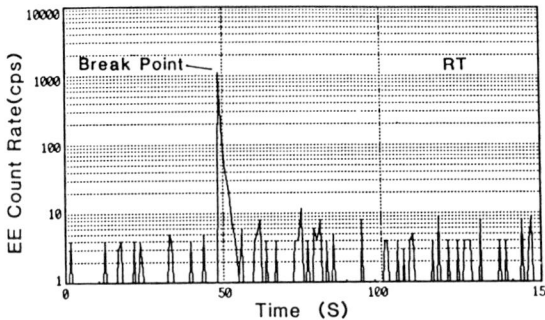

Fig. 1

Time distribution of fracto-emission from sodium silicate glass plate.

(a) FE was observed during fracture process at room temperature.

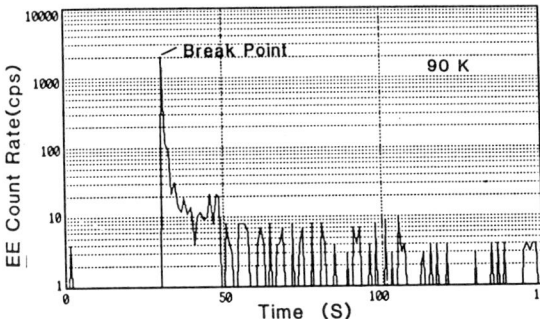

(b) FE at 90 K.

tially different between fused quartz and poly-crystalline ceramics, the features of their fracto-emission are very similar. More experimental data are necessary for detailed discussion.

In the case of quartz crystals, the FE features are considerably different from those of poly crystalline ceramics and glasses. That is, the emission from X specimens even at room temperature features high peak intensity and long decay time as shown in Figs. 4 and 5 (a). Moreover, the decay time during fracture at 90 K became slightly shorter than that at room temperature as shown in Fig. 5 (b). The irradiation, however, induces longer decay times similar to those in poly-crystalline ceramics and fused quartz. The emission distributions from irradiated X specimens are shown in Fig. 5 (c). Figure 5 is displayed in log/log scale to clearly show the emission peak at the instant of fracture and the decay time depending on the relaxation process of the fracture surface.

The high intensity and the long decay time of the emission from quartz crystal during fracture at room temperature suggest that the charge separation are densely generated from fairly perfect crystalline lattice, and that there are a few defects related with impurity filling role of the charge recombination on the fracture surface in the pure quartz crystal. Though irradiation induces some defects, they may prevent rather than enhance the recombination.

Although Dickinson et al. reported so much about the fracto-emission from single crystals like Si, MgO and LiF,[16,7,8] much more data are necessary for explanation of their phenomena. As for the emission from X specimens at 90 K, we have little information to explain the shorter decay time which is quite different from those of ceramics and glasses. Of course, the stiffness and the fracture strength of quartz crystals are to increase with decrease in temperature.

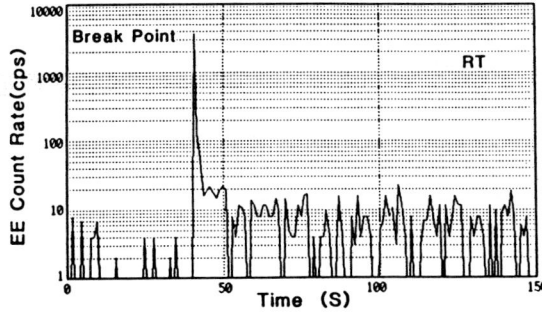

Fig. 2

Time distribution of fracto-emission from pure fused quartz plate.

(a) FE was observed during fracture process at room temperature.

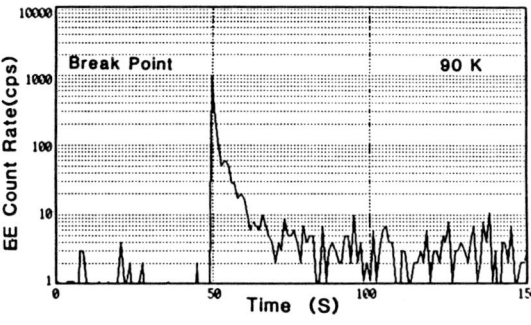

(b) FE at 90 K.

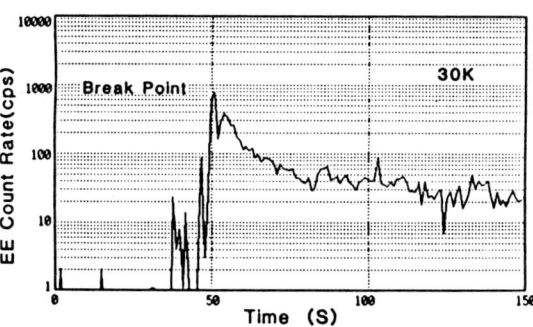

(c) FE at 30 K.

The electron emission from the quartz crystal Z specimen is basically not different from that of the X specimen, but often showed the anomalous behavior that the emission did not decay quasi-exponentially but continued at random for more than ten minutes after fracture. The purity of the Z specimen is higher than that of the X specimen, and there is no cleavage facet in quartz single crystal. Crystal channeling may be related with these phenomena.

CONCLUSION

The electron emissions from glasses during fracture at room temperature have low peak intensity and short decay time, and the intensity and the decay time increase with decrease of the sample temperature. These features are similar to those of pure alumina (99.9 %) poly-crystalline ceramics. Radiation effects on the emission bring high intensity and long decay time in pure fused quartz same as in the ceram-

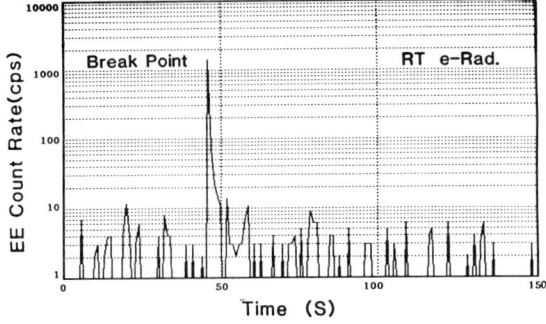

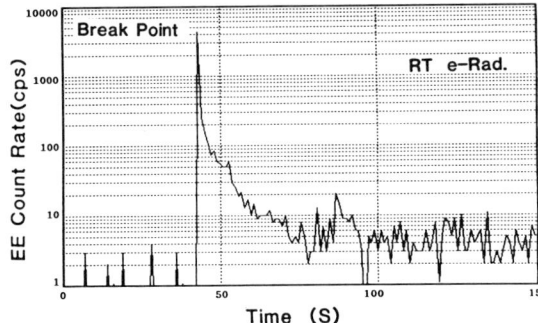

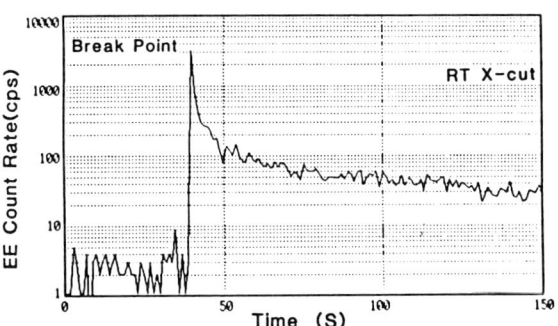

Fig. 3

Radiation effects on fracto-emission observed at room temperature.

(a) FE from sodium silicate glass plate.

(b) FE from pure fused quartz plate.

Fig. 4

Time distribution of fracto-emission from quartz single crystal plate of X region. FE was observed during fracture process at room temperature.

ics. These effects are not remarkable in sodium silicate glass because radiation induced defects may be annealed for period of the storage even at room temperature.

As for the emission from quartz single crystals, it is clear that the peak intensity is high and the decay time is long even in fracture at room temperature and the decay time scarcely changes at 90 K. Further experiments about this material have to be done, because there remain many problems to be solved.

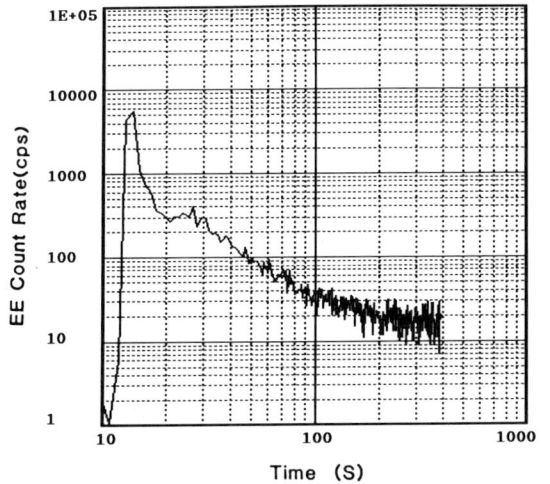

Fig. 5

Time distribution of fractoemission from quartz single crystal plate of X region.

(a) FE was observed during fracture process at room temperature.

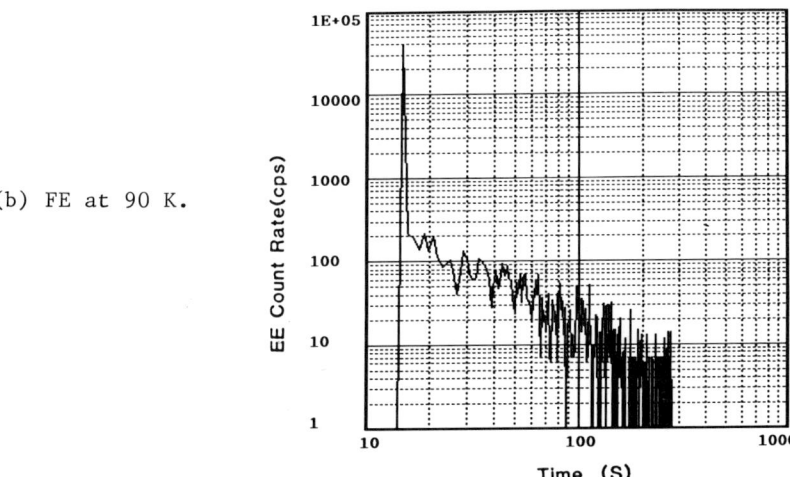

(b) FE at 90 K.

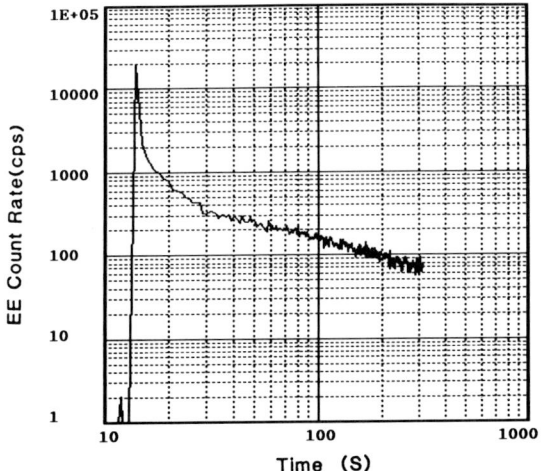

(c) FE from irradiated specimen was observed at room temperature.

ACKNOWLEDGMENT

The irradiation on specimens was performed by an electron linear accelerator in the Radiation Laboratory, ISIR, Osaka University. The authors wish to thank the persons of the machine group there for their help. Also we wish to thank the graduate students in our group at ISIR for their assistance in the completion of this paper.

REFERENCES

1) P. Braunlich and J. T. Dickinson ; "Proceedings of 6th International Symposium on Exoelectron Emission and Application" Rostock, East Germany (1979) 9
2) H. Glaefeke ; Exoemission in "Thermally Stimulated Relaxation in Solids" ed. P. Braunlich, Springer Verlag, Berlin, (1983)
3) J. T. Dickinson, M. K. Park, E. E. Donaldson and L. C. Jensen ; J. Vac. Sci. Technol. 20:436 (1982)
4) J. T. Dickinson, L.C. Jensen and A. Jahan-Latibari ; J. Vac. Technol. A 2:1112 (1984)
5) A. V. Poletaev and S. Z. Shmurak ; Sov. Phys. Solid State ; 26:2147 (1984)
6) J. T. Dickinson, A. Jahan-Latibari and L. C. Jensen ; J. Mater. Sci. 20:229 (1985)
7) S. C. Langford, J. T. Dickinson and L.C. Jensen ; J. Appl. Phys. 62:1437 (1987)
8) J. P. Mathison, S. C. Langford and J. T. Dickinson ; J. Appl. Phys. 65:1923 (1989)
9) S. Owaki, K. Katagiri, T. Okada, S. Nakahara and K. Sugihara ; Adv. Cryog. Eng. Mater. 34:283 (1988)
10) S. Nakahara, T. Fujita, K. Sugihara, S. Owaki, K. Katagiri and T. Okada ; Adv. Cryog. Eng. Mater. 34:91 (1988)
11) S. Owaki, K. Katagiri, T. Okada, S. Nakahara and K. Sugihara ; Adv. Cryog. Eng. Mater. 36:1361 (1990)
12) S. Nakahara, T. Fujita, K. Sugihara, S. Owaki, K. Katagiri and T. Okada ; Adv. Cryog. Eng. Mater. 36:1201 (1990)
13) S. Nakahara, T. Fujita, K. Sugihara, S. Owaki, K. Katagiri and T. Okada ; Jpn. J. Appl. Phys. 24-4:198 (1985)
14) S. Owaki, K. Katagiri, T. Okada, S. Nakahara and K. Sugihara ; "Proceeding of 9th International Symposium on Exoelectron Emission and Application" Wroclaw, Poland (1988) 262
15) S. Nakahara, T. Fujita, K. Sugihara, S. Owaki, K. Katagiri and T. Okada ; "Proceeding of 9th International Symposium on Exoelectron Emission and Application" Wroclaw, Poland (1988) 254
16) S. C. Langford, D. L. Doering and J. T. Dickinson ; Phys. Rev. Lett. 59:2795 (1987)

FRICTION AND WEAR OF RADIATION RESISTANT COMPOSITES, COATINGS AND CERAMICS IN VACUUM AND LOW TEMPERATURE ENVIRONMENT

A. Lipski and M. Ruschman

Fermi National Accelerator Laboratory*
Batavia, Illinois

ABSTRACT

Superconducting magnets used in accelerators such as the Superconducting Super Collider (SSC) are exposed to fast neutron and gamma irradiation. Some of the components used in the SSC cryostat are fabricated from organic materials with low radiation resistance. The slide bearing material presently supporting the magnet assembly contains Teflon and must be replaced with a material of improved radiation resistance. A group of sliding materials, most of which have suitable radiation resistance, were tested under conditions of pressure, temperature, velocity, and vacuum typically encountered in normal cryostat operation. As this was a preliminary screening test, the samples were only cooled to liquid nitrogen temperature. The group of materials tested consists of composites, coated base metals, and ceramics. The criteria was to maintain a low coefficient of friction throughout the experiment in spite of changes in temperature and vacuum. Subsequent tests will expose finalist materials to fast neutron irradiation at liquid helium temperatures. This paper describes the experimental setup and presents data of the friction coefficient measurements taken for the various samples.

INTRODUCTION

The effect of radiation on materials has been of great concern to the designers of high-energy accelerators. The subject of high-energy radiation damage to polymers used in accelerators has been the subject of several studies and publications. It has been pointed out[1] that the Teflon(trade name used by DuPont) resin, used in the present cold mass slide material, could be affected by the high levels of radiation found in the SSC dipole magnets.

The cold mass assembly is supported and laterally restrained by cradle assemblies at five locations along its length. Aside from the center cradle which is fixed, the other four cradles allow for thermal expansion or contraction of the cold mass along its axis, as shown in figure 1. The cold mass assembly slides on four bearing pads mounted in the sliding cradle as illustrated in figure 2. The present bearing pads are DU (trade name used by Garlock Bearing) bearings which contain Teflon (PTFE), impregnated in the porous bronze as well as in the form of an overlay.

This paper describes the work performed to select one or more materials with comparably low coefficients of friction, yet better radiation resistance properties, with which to replace the DU bearing material.

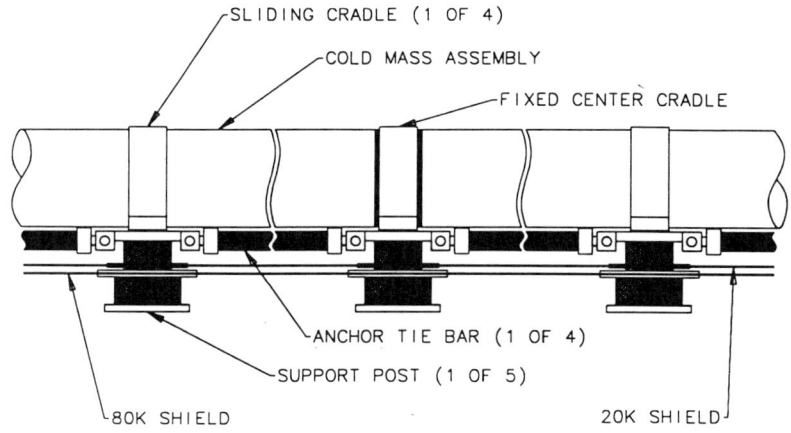

Fig. 1. SSC 50mm Dipole Cryostat Suspension System

MATERIAL SELECTION

The primary requirements for selecting the slide material were:

- Maintain friction coefficient of 0.20 to 0.22 at operating conditions.
- Maintain performance capabilities in radiation levels exceeding 3×10^3 gray over a 20 year period.

In addition the following operating conditions will have to be met:

- Temperature: Operating temperature of approximately 4 K; storage temperature of approximately 340 K

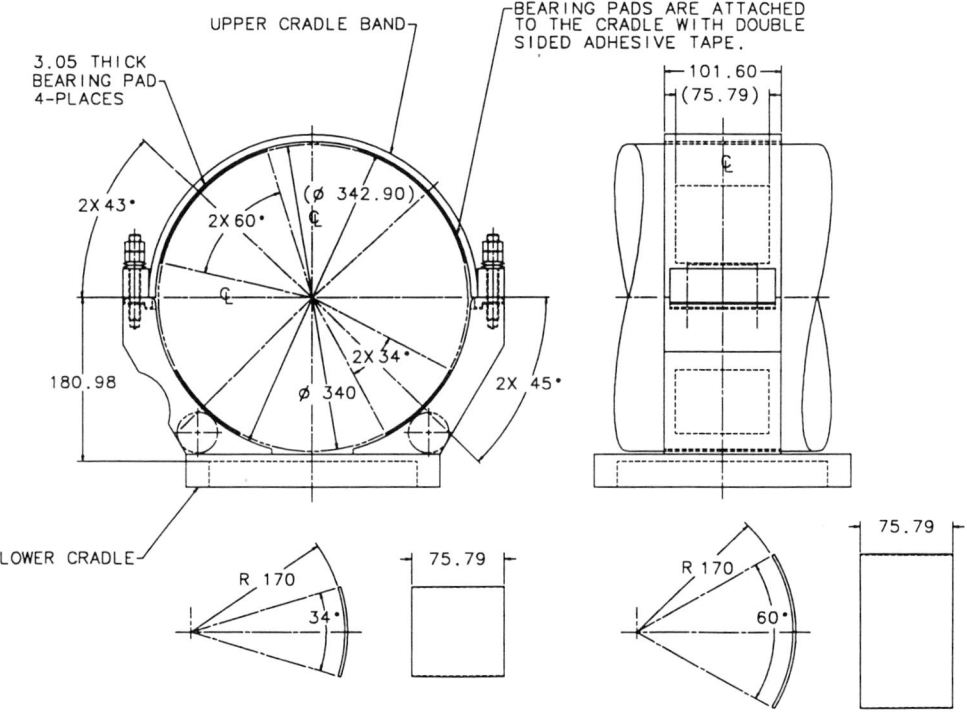

Fig. 2. Cradle Assembly and Bearing Pad Detail

Table 1. Sliding Test Samples

Coating/Metal	Composite	Ceramic
CoNiInMoS$_2$/Inconel 718	Vespel SP-1	AlPO$_4$+Al$_2$O$_3$ (SiC - continuous fiber)
Graphite (embedded) / bronze - copper	Vespel SP-3	Glass bonded Mica
	APC-2/AS4 (PEEK) (0° fiber orientation)	
Everlube 860/316 SS		
Ti-TiN/316 SS	Torlon 4301	
WS$_2$/316 SS	PEEK/carbon (short fiber)	

- Vacuum: 10^{-7} Torr at the operating temperature
- Pressure: 4.14 MPa - dynamic at the operating temperature.
  8.28 MPa - static at 320 K
- Humidity: 0% at the operating temperature.

Other performance criteria which have been taken into consideration are aging and wear.

The slide samples are tested for approximately 900 cycles of duty. A cycle denotes expansion and contraction of a magnet (±2.5 cm). Over a 24 hour period the magnet will contract 2.5 cm when cooled from room to operating temperatures. It is also assumed that the slide material will be exposed to air after being irradiated several times during the course of its lifetime.

The selected materials can be divided into three groups: coatings (of metals), composites and ceramics, as shown in Table 1.

EXPERIMENTAL SETUP

The test setup was made to simulate the actual conditions of the slides in the magnet. However, since this is a preliminary test, several variations from the actual situation were adopted. (See material selection)

- The low temperature is approximately 80 K and the vacuum is 10^{-5} Torr.
- The support posts are made of solid stainless steel to minimize post deflection
- The loading pressure is the equivalent to that of a dynamic loading during operating conditions. (See material selection)

The samples were secured on the stationary portion of the sliding apparatus. All samples were tested sliding against 316 stainless steel plates with a surface finish of 0.50 µm to simulate the cold mass skin. Each set of samples was fitted with a new set of stainless steel plates. The stainless steel plates were mounted on to the moving portion of the sliding apparatus. The complete sliding apparatus was placed inside a vacuum vessel as shown in figure 3.

In order to maintain close simulation between the test and the actual magnet performance the following system requirements were imposed:

- Stroke = ± 2.54 cm
- Stroke velocity = .0025 m/sec
- Total contact area = 12.9 cm^2
- Dead load weight = 485 kg

The friction force was measured by a load cell in line with the actuation system. The load cell readings and corresponding temperature readings were recorded by a computerized data

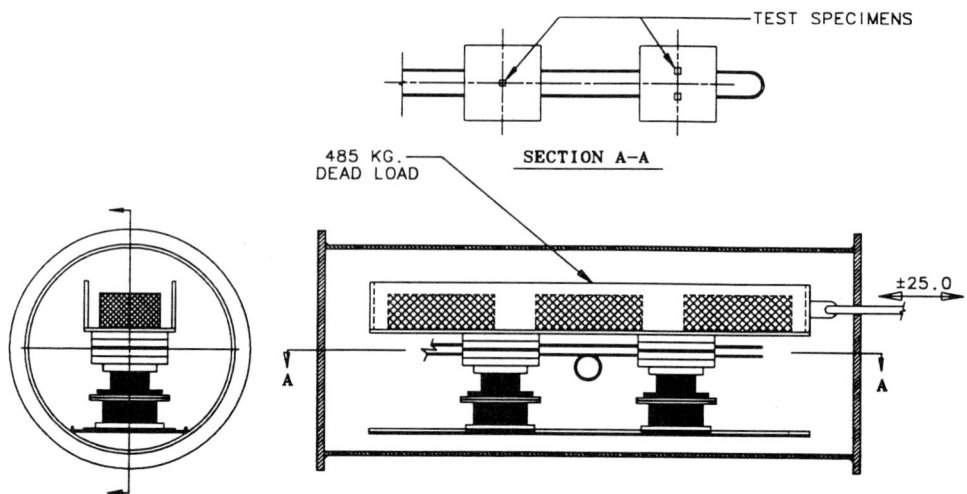

Fig. 3. Sliding Test Apparatus

acquisition system. The coefficient of friction (COF) was calculated by dividing the friction force by the normal force (dead weight).

The actuation system included a closed loop controller which produced a constant speed reciprocating motion, thus eliminating inertia from affecting the load cell measurements.

Data for dynamic COF is obtained from 900 cycles as shown in Table 2. The force data for the dynamic COF (coefficient of friction) was generated by taking an average of six readings per cycle. In addition, measurements of static coefficient of friction (force to slip) as well are taken at six separate instances as described in Table 2. The static COF was measured for a single cycle over a length of 1.5mm traveling at a velocity of .025mm/sec.

RESULTS AND DISCUSSION

As can be seen from Table 3 as well as from figures 4 and 5, three coatings sucessfully completed the test. However the COF of the graphite (embedded) / bronze-copper was the lowest within the required range. Its COF never exceeded 0.20 throughout the test. The embedded solid lubricant bearing is composed of a solid lubricant bearing (in a plug form) and base metal. The sliding medium consists of a pattern of embedment and a surface lubrication which is used for startup.(See figure 8) The other two coatings which met the test, requirements

Table 2. Sliding Test Sequence

Cycle #	Pressure (Torr)	Temperature (K)
Dynamic COF		
1-300	10^{-5}	80
300-600	10^{-5}-10^{-3}	warm-up (80 - 285)
600-900	10^{-3}	285
Static COF		
1	760	285
2	10^{-3}	285
3	10^{-5}	80
300	10^{-5}	80
600	10^{-3}	285
900	10^{-3}	285
900	10^{-3}	285

Table 3. Summary of Test Results

Material Tested	Friction Coefficient			Remarks
	0-300 cycles	300-600 cycles	600-900 cycles	
DU (Bronze + Teflon)	.16	.18	.12	
Compopsites				
Vespel SP-1	.37	-	-	Test terminated after 50 cycles @ 80K
Vespel SP-3	.18	.18 - .40	-	Test terminated after 400 cycles @ 200K
APC-2/AS4 (0°fiber orientation))	.13	.33	.14	
Torlon 4301	.34	.36	.23	
PEEK/carbon	.28	.37	.38	
Coatings				
$CuN_iI_nM_oS_2$ / Inconel 718	.09	.22	.16	
Graphite (embedded) /Bronze-copper	.08	.15	.19	
Everlube 860 / 316 SS	.08	.21	.21	
T_i-T_iN / 316 SS	.42	-	-	Test terminated after 25 cycles @ 80K
WS_2 / 316 SS	.40	-	-	Test terminated after 25 cycles @ 80K

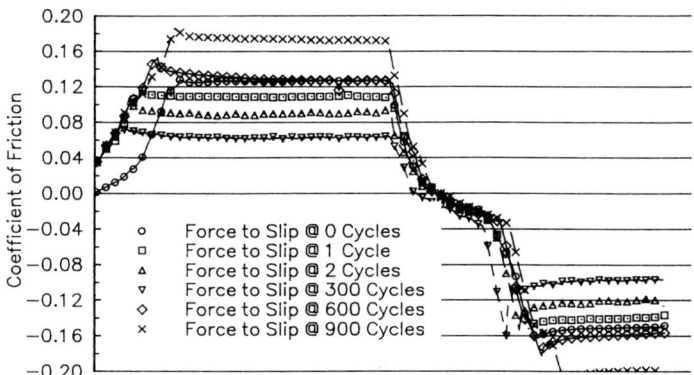

Fig. 4. Force to Slip Plots - Graphite (embedded)/Bronze-Copper

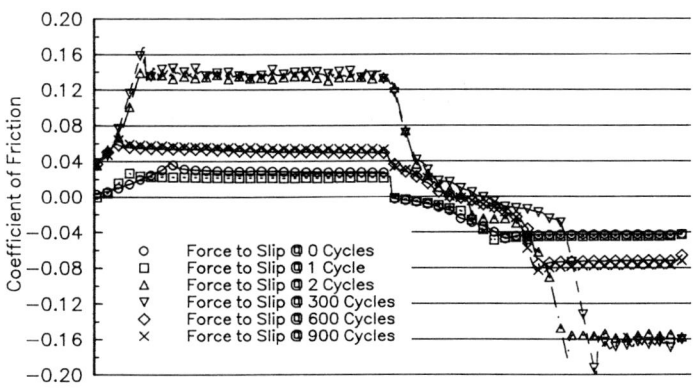

Fig. 5. Force to Slip Plots - DU (Bronze + Teflon)

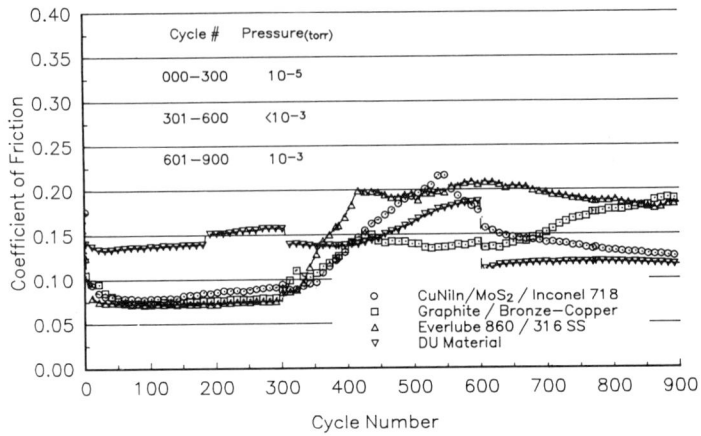

Fig. 6. Comparison Plots of COF for Coated Metals

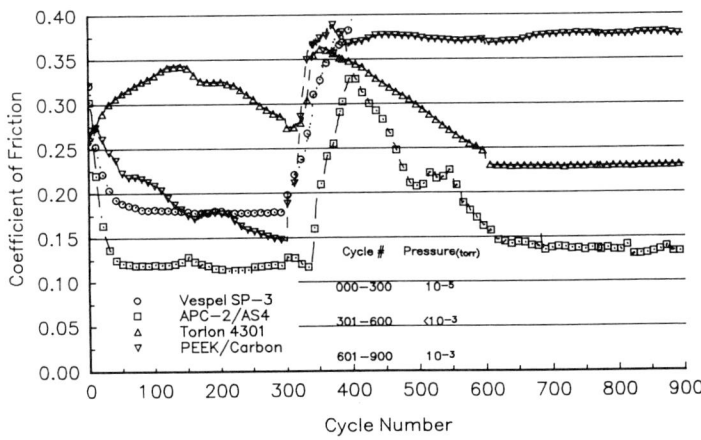

Fig. 7. Comparison Plots of COF for Composites

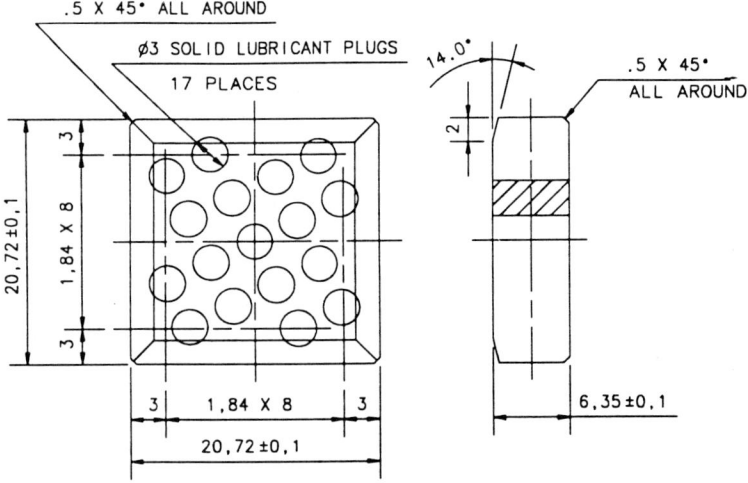

Fig. 8. Graphite (embedded)/Bronze-Copper (Courtesy - Oiles America Corp.)

were the Everlube 860 / 316 SS and the CuNiInMoS$_2$ / Inconel 718. Over all, the test results show that the COF of the coating group is lower than that of the composite group. The APC-2 / AS4 (0° fiber orientation) had the lowest COF among the composites group. As evident in the comparison plots (figures 6 and 7) the COF increased when temperatures increased from 80K to room temperature for all samples. However there is no uniform pattern evidenced in the behavior of the various materials at room temperatures. While the COF for composites changed erratically over a wide range of valves, that of the coated metals maintained a more settled nature with respect to temperature changes and number of cycles.

The materials in the tests which were terminated failed by exceeding the upper limit of COF. (Upper limit was set at 1890 N) In the composite group, the materials which failed exhibit stick - slip behavior and the high COF seems to result predominantly due to adhesive frictional interaction between the stainless steel and the composite material.

Graphite (embedded)/bronze-copper is the only sample of which the COF continued to increase during the last portion of the test. In comparison the COF of all other samples dropped or stabilized.

Two samples of the composite group contained PEEK (Polyetheretherketone) resin. The APC-2/AS4 (0° fiber orientation) showed substantially better performance over the Peek/carbon which contained short stock carbon fibers.

All the coating samples which completed the test contained MoS$_2$ as part of their coating.

CONCLUSION

This work concentrated on the selection of materials which have known good resistance to radiation yet unknown COF at vacuum and low temperatures. The test attempted to simulate the actual conditions occurring inside a dipole cryostat during cool down and warm up. Over all it was found that coatings of metals performed better than bearings made of solid composites. The dynamic COF increased in direct proportion to the rise in temperature and then either decreased or stabilized (in most cases) during the last portion of the test at room temperature. In most of the cases the number of cycles did not seem to have an adverse effect on the COF. The COF of the coatings appear to be more uniform than that of the various composite materials. Although three materials met the required COF of 0.20 - 0.22, the investigation for a replacement material for the DU bearings is far from complete. The materials which took part in this test so far represent only a fraction of a more comprehensive list of coatings and composites which need to be evaluated. Ceramic materials have the potential to be good candidates, thus further evaluation and testing are recommended.

In addition, further testing and evaluation is required with finalist material such as graphite (embedded)/bronze-copper or the Everlube 860/316 SS. Prior to selecting a replacement material for the DU the following subjects should be investigated.

- Long term effects of high vacuum (10^{-7} Torr) on the material and its compatibility in the cryostat environment.
- Bearing ability to perform adequately after being irradiated as well as exposed to air.
- Wear mechanism.
- Behavior under multiple thermal cycles.

ACKNOWLEDGEMENTS

The work as presented was performed at Fermi National Accelerator Laboratory, which is operated by Universities Research Association, Inc. under contract with the U.S. Department of Energy.

REFERENCES

1. DuPont, "Radiation Tolerance of Teflon Resins," January - February 1969 issue of The Journal of Teflon.

AN INNOVATIVE PROCESS FOR THE IMPREGNATION

OF MAGNET COILS AND OTHER STRUCTURES

D Evans and J T Morgan
Rutherford Appleton Laboratory
Chilton, Oxon. OX11 0QX. UK

ABSTRACT

The need to bond and encapsulate the conductor in many types of superconductive magnet is well understood. The choice of material for bonding may influence the performance of the coil and the technique used for its application. This paper considers the merits of three types of 'bonding' procedure and presents details of an innovative method for the vacuum impregnation of coils. The process has been developed and evaluated in an attempt to remove some of the uncertainties of the vacuum impregnation process that traditionally may be eliminated only with the use of sealed mould tools and high quality vacuum chambers. For large magnets or for mass production, this process may lead to a reduction in tooling and plant costs, together with reduced resin consumption and improved health and safety factors. The process has been developed in conjunction with a resin system that has been designed to exhibit excellent thermal shock characteristics and to minimise preparation time and post impregnation cleaning requirements.

INTRODUCTION

There are, essentially, three methods of bonding coils and these may be defined as follows:

(i)   Vacuum impregnation.

(ii) Winding with liquid resin ('wet' winding).

(iii) Winding with pre-impregnated ('pre-preg') cloth or braid.

The salient points of each method are presented in Table 1. Each technique has its merits, advocates and adversaries and there are situations where any one technique may be the obvious choice. The authors reported [1] their experiences winding and bonding a large (approximately six metre diameter) solenoid using (iii) above, and a recent report [2] compared the mechanical performance of coils bonded using different techniques and concluded that those manufactured by vacuum impregnation offered the best mechanical characteristics.

Table 1 Comparison of 3 Coil Bonding Procedures

	Vacuum Impregnation	Wet Winding	Pre-impregnation
Tooling	Vacuum tank plus coil enclosure (mould tool). Generally high cost.	Simple tooling generally. Low cost.	Simple tooling generally. Low cost.
Quality	Good (best) quality if impregnation successful. Disastrous if air leaks during impregnation.	Resin drainage and void formation likely.	Concern about quality of bonding because of uncertain resin distribution.
Cleanliness of operations	Clean but extensive cleaning of equipment after use.	Often extremely messy.	Reasonably clean.
Quality control during winding	Good.	Good up to the point at which resin gels. Messy to unwrap and re-wind.	Good with 'tacky' materials, poor with dry pre-pregs. Re-winding difficult.
Abrasion resistance of glass wrapping	Dry glass cloth has poor abrasion resistance.	Dry glass cloth has poor abrasion resistance	Pre-preg tapes have good abrasion resistance
Effect of delays in the winding process	No problems.	Some problems may arise if using RT cure.	No problems unless shelf life' of pre-preg is exceeded.
Storage and handling of materials	Resin mixing and vacuum degassing equipment necessary.	Mixing equipment needed but not degassing.	Refrigeration may be required
Resin wastage	High	Very high because of pot life limitations.	High if pre-preg not used within its shelf life.
Cure temperature	Moderately elevated.	Room temperature or elevated.	Approx 120C.
Demoulding time	Extensive time for de-moulding and cleaning equipment.	Extensive for cleaning up and minor corrections.	Generally little cleaning.

A new process of vacuum impregnation has been developed to overcome many of the difficulties and disadvantages of the methods traditionally used, yet retaining many of their advantages. The moulding technique is complemented with a resin system that has been developed specifically for low temperature use and for processing by the vacuum impregnation route.

THE VACUUM IMPREGNATION PROCESS

There are a number of variations in general use with this technique and some of these are shown diagrammatically in Figure 1. The so called 'flood filling' method (Figure 1a) is widely used since it demands only relatively inexpensive tooling in addition to the vacuum chamber. However, relative to the material required for the impregnation of a coil, it does use large quantities of resin. It also has the attendant problem of removing the coil from the excess cured resin at the end of

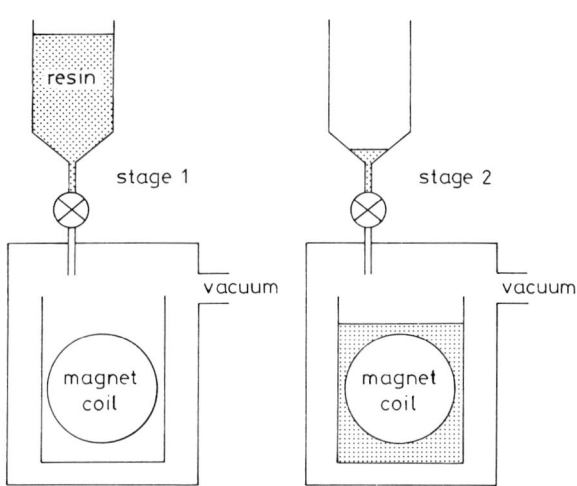

Vacuum Impregnation: Flood Filling
Figure 1a   Figure 1b

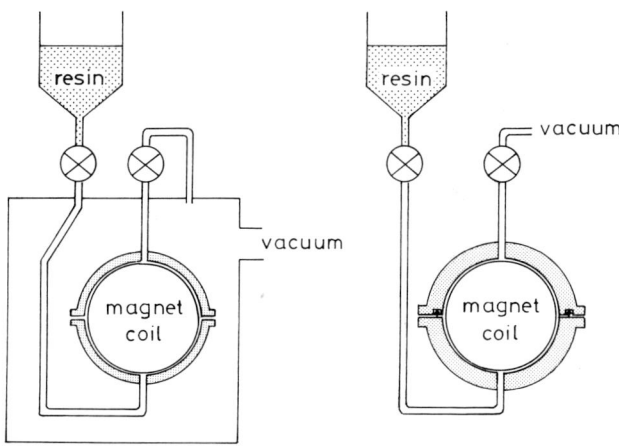

Figure 1c. Alternative Techniques for Vacuum Impregnation.

the process. The preparation of large quantities of resin is difficult, time consuming and increases quality assurance problems together with health and safety risks.

The 'sealed mould route' requires a resin tight mould (note - not necessarily vacuum sealed) and a vacuum chamber. This technique is illustrated in Figure 1b and, provided the mould is resin tight, it may be possible to apply atmospheric pressure to the resin before releasing the vacuum in the chamber. The needs are clearly for high quality tooling and a vacuum chamber to accommodate the mould. This technique minimises the amount of resin used in processing but demands the use of a vacuum chamber for long periods of time. A mould tool that is reliably vacuum tight allows one to dispense with a vacuum chamber and proceed with the impregnation process without the need for further tooling (Figure 1c). While this technique minimises resin use it is inefficient in the use of expensive tooling.

The technique that is now presented combines the best features of each method of vacuum impregnation, viz:

(i) Minimum quantities of resin are required.

(ii) Only simplified tooling is necessary.

(iii) A vacuum chamber is not required.

The method uses the "vacuum membrane" technique that has, for some time, been used for processing resin based fibrous composite materials for the aerospace industry. Since coil shapes and sizes vary significantly, the general principles only are depicted in Figure 2. In the UK, typical suppliers of these materials would be Fothergill Tygaflor Ltd of Aerovac Systems (Keighley) Ltd. In the USA, Zip-Vac Inc.

To minimise the processing problems associated with any of the techniques of vacuum impregnation described above, a pre-catalysed resin system has been developed that is suitable for low temperature applications. Using this resin system, the vacuum impregnation process may be described as follows.

Using sheets of rigid plastics materials or metal, prepare ports for vacuum pumping and for resin inlet, which should be on opposite faces

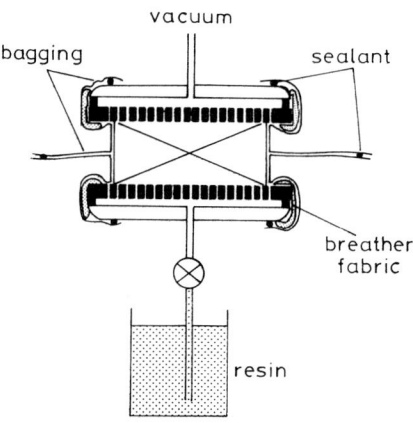

Figure 2. Vacuum Impregnation using 'Bag' Technique

of the tool, with the resin inlet(s) at the lowest point and the vacuum pumping port(s) at the highest. Ensure that all edges of tooling etc. are smooth and free from defects that may puncture the membrane. Cover edges with breather fabric and additionally, position some 'breather' material at pumping and resin inlet ports if necessary. Allow sufficient 'membrane' to enter recesses that may be associated with the mould tool, seal the film using the sealant strip and apply a vacuum. Note that if insufficient film is allowed it may be drawn into recesses, stretched and broken. Mould tools that are of a significant size should include a number of resin inlet ports at strategic positions to ensure adequate resin flow. (See Resin Flow & Distribution). Using this process, pressures of one-twentieth of a millibar have been achieved with coils at $40^{\circ}C$. Note that after impregnation the pressure on the resin in the coil rises to atmospheric without venting the sealed system.

PRE-CATALYSED RESIN SYSTEM

A pre-catalysed resin system is one that has been formulated to contain all resin components and a latent catalyst, resulting in a system that has a long usable life at room and slightly elevated temperatures but requires no further mixing or preparation prior to use. Such systems minimise preparation, contact and cleaning time and offer opportunities for batch quality assurance testing.

There are a number of problems inherent in the processing of liquid epoxide resins and these may be briefly described as follows:-

(a) Multi-part resin systems need to be proportioned accurately and mixed thoroughly. Quality assurance requirements may be compromised if great care is not taken at these stages. 'Thorough mixing' is difficult to define and control without specialised equipment.

(b) Physical handling and control of contamination is difficult when significant quantities of resin are to proportioned and mixed.

(c) The preparation and mixing of resins requires 'adequate' ventilation, a condition which is not readily defined further.

(d) Following impregnation, transfer lines, tubes and valves etc must be cleaned before the resin is allowed to gel.

Additionally, mixed resin have a relatively short 'usable life'. Quality assurance testing of each batch of material is time consuming, difficult and in most cases, not undertaken. At the Rutherford Appleton Laboratory, a 'one part' resin system has been formulated for low temperature use and, more specifically, for the vacuum impregnation of structures using the 'vacuum membrane' technique. (Resin RAL 1PF).

The viscosity/time characteristics for this system are shown in Figure 3 and the mechanical and thermal properties are shown in Table 2, where data is also presented, for comparison, on a thermally shock resistant system, previously reported [3]. In non-standard tests [3], the system described has exhibited thermal shock resistance comparable with the best systems described by the authors. The essential characteristics that have been defined as promoting thermal shock resistance, ie toughness at room temperature, a low second order glass transition temperature and low contraction on cooling, have been retained.

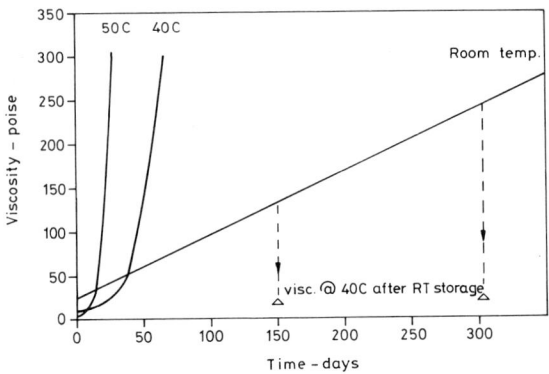

Figure 3. Viscosity v Time for Resin 1PF

Table 2 Properties of One Part Resin System - RAL 1PF

GEL TIME

Temp C	Time - Minutes
80	4600
100	800
120	160

INITIAL VISCOSITY

Temp C	Viscosity - Poise
25	40
40	10
50	5
60	3
80	1.5

MECHANICAL PROPERTIES
In Flexure (mean of 5 replicates)

Temp K	Modulus GPa	Stress MPa at break	Strain %
295	2.5	did not break	
77	7.1	204	2.9

(for comparison mix 222)

295	1.8	did not break	
77	7.1	177	2.4

INTEGRATED THERMAL CONTRACTION

RT to 77K 0.009

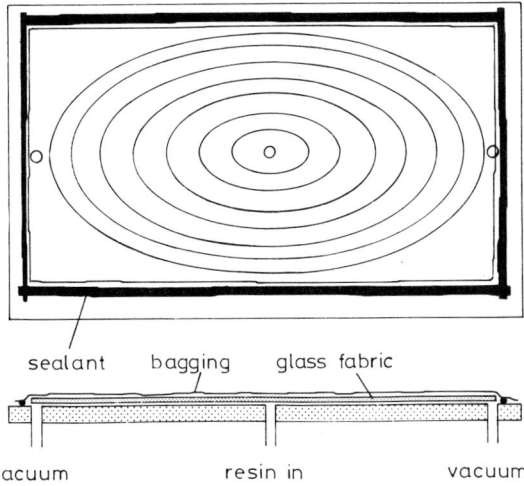

Figure 4. Resin Flow Pattern in Experimental Impregnation of Glass Fabric.

RESIN DISTRIBUTION AND FLOW

There is much folklore surrounding the vacuum impregnation process but little in the way of scientific rules. It is clear that a significant time is required for the 'vacuum degassing' of any structure[4] and that premature resin impregnation will result in reduced resin flow and the entrapment of gas within the structure. Work is currently underway, aimed at quantifying some of the variables and unknowns in coil impregnation technology, in an attempt to establish a set of general principles. An experiment in this series is shown diagrammatically in Figure 4 and typical results in Figure 5. The 'structure' consisted of twelve layers of plain weave E-glass fabric, on an aluminium alloy backing sheet which also carried the two pumping ports and the central resin inlet. The structure was closed using a nylon membrane with a sealant strip. During impregnation the system was continuously pumped up to the point where resin closed the ports. The resin flow in the 'X' and 'Y' direction was in direct proportion to the length and width of the panel, resulting in the 'oval' pattern shown in Figure 4. Without continuous pumping, poor results were obtained for the flow of resin as

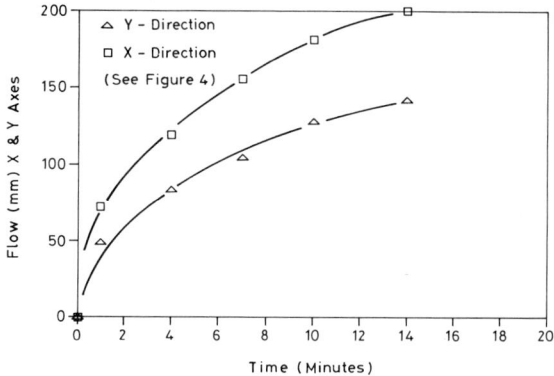

Figure 5. Resin Flow and Distribution.

measured in these experiments, where only a limited time had been allowed for outgassing. Where 'complete' outgassing had been achieved, discontinuing vacuum pumping at the start of resin impregnation did not materially influence the results. This work is continuing.

CONCLUSIONS

The vacuum membrane technique, widely used for the processing of composite materials has been applied to the vacuum processing of superconductive coils and other structures. It is proved to be an economical and technically efficient method for the vacuum impregnation of these structures and has much to commend it. In addition, a pre-catalysed resin system has been developed for low temperature use that offers improvements to coil performance, quality assurance and health and safety matters, relative to 'two part' systems. Preliminary results are presented that demonstrate the importance of fully understanding the principles involved in the vacuum impregnation of high impedance structures.

REFERENCES

1. D Evans and J T Morgan "Aspects of the bonding and insulation of a large superconducting solenoid". Cryogenics 1988, Vol 28, April.

2. G W Knight and D Evans "Evaluation of the structural integrity of superconducting magnet coils". Cryogenics 1991, Vol 31, April.

3. D Evans and J T Morgan "Epoxide Resin for use at low temperature" - Non Metallic Materials and Composites at Low Temperature 2. Plenum Press 1982, pp 73-87.

4. D Evans, S J Robertson, S Walmsley and J Wilson, "Measurement of the permeability of carbon fibre/PEEK composites". Cryogenic Materials 1988, Vol 2, Structural Materials, ICMC, pp 755-763.

LARGE SCALE TESTS OF COMPOSITE SUPPORT STRUTS FOR

SUPERCONDUCTING MAGNETIC ENERGY STORAGE RINGS[†]

R.P. Walsh, R.P. Reed, and J.D. McColskey

Materials Reliability Division
National Institute of Standards and Technology
Boulder, CO 80303

M. Tupper and E. Johnson

General Atomics
San Diego, CA 92121

ABSTRACT

Cold-to-warm struts will be used in superconducting magnet energy storage plants to support the superconducting magnet ring. A prototype design of the structural support struts was tested by simulating the in-service, thermal and mechanical, conditions. The maximum strength measured, with the simulated in-service conditions, was 519 MPa (75 ksi). Problems with the design of the end fittings for the struts prohibited determination of the material's ultimate strength.

INTRODUCTION

The design for the struts uses fiber-reinforced polymer (FRP) composite tubes. These FRP tubes must withstand cyclic compressive loads, in a vacuum environment, at temperatures ranging from cryogenic (1.8 K) to ambient (295 K).

As a precursor to the full scale tests, a 4.4 MN capacity servo-hydraulic test machine was modified for testing at cryogenic temperatures. The details of the cryostat design are presented in the first part of this report along with details pertaining to instrumentation and strut specifications.

The test program divides into three distinct areas:

  (1) Ultimate compressive strength tests of short specimens.
  (2) Ultimate compressive strength tests of full scale specimens.
  (3) Compressive fatigue tests of full scale specimens.

The results of each of these tests are discussed in the final section of the report. For more detail of any part of this test program see Reference 1.

---

† Contribution of NIST, not subject to copyright.

Figure 1. 4.45-MN (1000-kip) Test Machine and Cryostat

MECHANICAL TESTING: FACILITIES AND INSTRUMENTATION

Facility Development

Two types of full scale qualification tests were desired: ultimate compressive strength tests and compressive cyclic fatigue tests. The tests required mechanical loading (to loads in excess of 3000 kN) of the composite tube and an imposed thermal gradient (4 to 295 K) over the length of the tube. To meet these specifications we developed a mechanical test cryostat for a 4.4 MN capacity servo-hydraulic test machine (Figure 1).

The specimen was tested between two compression platens; one platen was maintained at room temperature while the other was cooled using liquid helium. The cold platen represents the superconducting coil operating at 1.8 K, and the warm platen represents the ambient temperature support trench that the struts are anchored to. In service, each coil support strut thermally loads the 1.8 K coil. To minimize this thermal load, the struts are equipped with liquid nitrogen thermal intercepts at a strategic position along the length of the strut. To best simulate the service environment, the test specimen and test fixture were also equipped with a liquid nitrogen intercept at the appropriate location. Since the specimen is long (2 m), the cryostat was designed to have the cold end isolated deep in the bottom of a closed bottom super-insulated dewar. This allowed the warm end to be at the top, thermally anchored to the test machine's heavy framework.

The system was designed with the specimen in a vacuum chamber, cooled mainly by conduction. Liquid helium (4 K) and liquid nitrogen (76 K) flowed through tubing that kept them isolated from the test chamber vacuum space. For the lower platen (maraging steel) single-walled copper tubing was soldered directly to the bottom and the perimeter of the steel disc. The liquid nitrogen thermal intercept was a copper ring with the copper tubing soldered to it. This ring was thermally anchored to both the specimen and fixture at the appropriate location.

A large (760 L/min) mechanical vacuum pump was used to evacuate the chamber. Lower vacuum pressure than the pump could attain was possible due to cryo-pumping from the liquid helium cooled components inside of the test chamber.

Instrumentation

Load, displacement, and temperature data were recorded with a computer data acquisition system and an analog x-y chart recorder. Three alternative methods of displacement measurement were used: localized specimen strain measurement using strain gages bonded directly on the specimen, overall specimen displacement measurement using LVDT's (linear variable differential transformer) from specimen end to end, and the test machine's stroke displacement on short specimen tests. Load was measured using the machine's compression load cell. Temperature data were obtained for all the thermal gradient tests using type E (chromel/ constantan) thermocouples. Each thermocouple bead was securely attached to the specimen or fixture at strategic locations along the length of the apparatus.

SUPPORT STRUT SPECIFICATIONS

A tubular geometry was chosen for the compressive support columns. The tubes, shown schematically in Figure 2, were fabricated from E-glass fiber reinforced epoxy resin. They were manufactured using pre-preg tape and a hand lay-up process. The majority of the fibers were oriented 0° to the loading axis. Inner and outer wraps of ± 45°-oriented fibers were used to react hoop stresses. The global fiber volume of each tube is approximately 60%.

Another detail of the design is that the cross-sectional area of the tube is monotonically decreasing for the upper one-third of length. The reduction in diameter was done to use the cold temperature strengthening and in turn reduce the area of the thermal conductive path.

Most of the test specimens were equipped with end fittings designed to prevent brooming of the tube ends. There were two types of end fittings used which are shown schematically in Figures 3A and 3B. The self-aligning end fitting is a three-piece part that is designed to improve the axiality of compressive load on the tubes.

EXPERIMENTAL RESULTS AND DISCUSSION

Ultimate Compressive Strength: Short Specimens

Four short specimens of the full scale diameter tube were tested at 295 K. Two were 0.3 m long, and two were 0.6 m long. The two shorter

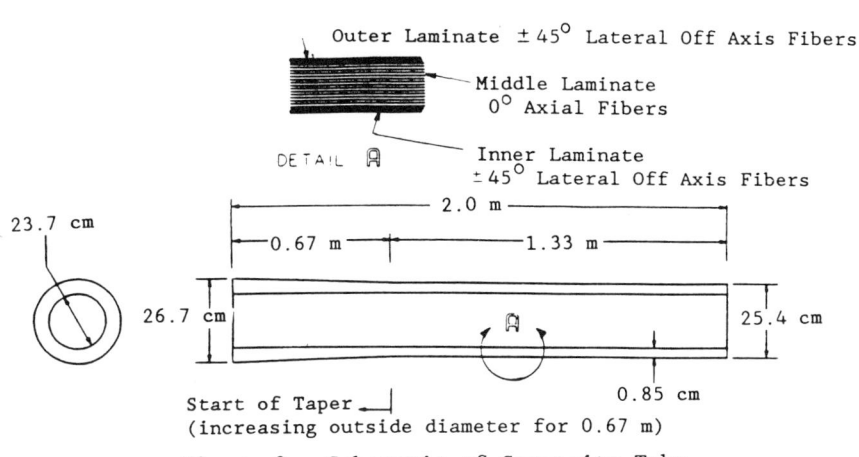

Figure 2. Schematic of Composite Tube

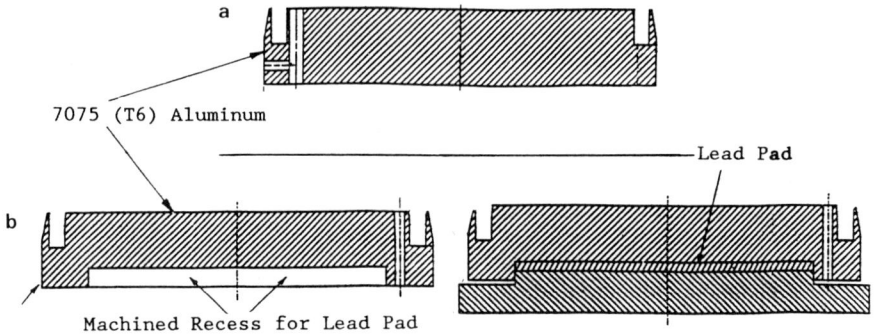

Figure 3. Rigid End Fitting (top) and Self-Aligning End Fitting (bottom)

tubes (numbers 005-2 and 006-2) were tested without any special end fittings. The ends of the composite tubes were set directly on the compression platens. The 0.6 m long tubes (numbers 005-1 and 006-1) were equipped with aluminum (7075-T6) rigid end fittings (Figure 3A).

Specimen 005-1 failed at a maximum compressive load of 1.6 MN (369 kips). Specimen 006-1 failed at a maximum compressive load of 1.7 MN (384 kips). The failure mode for both was shell buckling which was preceded by a large longitudinal crack running the length of the specimen. For each test the end fittings exhibited signs of failure, radial cracking on the exterior surface.

Specimen 005-2 with no end fittings failed at a compressive load of 2.4 MN (536 kips). Specimen 006-2 failed at a compressive load of 2.4 MN (538 kips). The mode of failure for both was brooming on one end of the specimen.

Summary of Results of Short Specimen Tests

The results of all tests are shown in Table 1. The short specimens without end fittings failed at higher stresses than the longer specimens with end fittings. It cannot be concluded that end fittings were

Table 1. Test Result Summary Sheet

Specimen No.	Length (m)	Thermal Conditions	End Fitting*	Ult. Load (kN) (kips)	Ult. Stress (MPa) (ksi)	Failure Mode†
005-1	0.66	295 K, isothermal	2	1641 369	254 36.9	1,4
006-1	0.58	295 K, isothermal	2	1708 384	265 38.4	1,4
006-2	0.35	295 K, isothermal	1	2393 538	371 53.8	2
005-2	0.30	295 K, isothermal	1	2384 536	370 53.6	2
009	2.00	120 to 280 K gradient	2	3345 752	519 75.2	1,3,4
007	2.00	25 to 273 K gradient	3	1284 282	194 28.2	5
008	2.00	35 to 273 K gradient	3	1156 260	179 26.0	5

_____*End Fitting Notes_____

1) no end fittings
2) ultimate load (rigid) end fittings
3) self-aligning end fittings

_____†Failure Mode Notes_____

1) longitudinal crack initiated at end
2) brooming at end
3) column buckling
4) shell buckling
5) end fitting failure

detrimental to the tube strength since there was also the variable of specimen length. The initial signs of failure of the end fittings indicate a problem area near a geometric stress concentration (the root radius at the base of the machined groove).

Ultimate Compressive Strength: Long Specimen

One full scale cryogenic test (specimen 009) is discussed in this section. The full length (~2 m) specimen was equipped with the rigid aluminum end fittings described previously. Liquid nitrogen (76 K) was used to cool the lower two-thirds of the test fixture and specimen. The goal was to test a full length tube with an imposed cryogenic thermal gradient, and also to verify the cryostat thermal efficiency before liquid helium (4 K) tests.

The desired conditions would be isothermal for the lower two-thirds of the specimen at 76 K, then a gradient from 76 K to ambient (295 K) over the upper one-third of specimen length. The actual gradient attained on the specimen ranged from 115 K to 278 K and is shown schematically in Figure 4 along with the temperature gradients of two other specimens that will be discussed later. The specimen failed with a loud noise at 3.34 MN (752 kips).

Summary of Results of Long Specimen Test

Specimen 009 failed in the mid-region, where the liquid nitrogen thermal intercept was attached (Figure 5). At this location, several factors could influence the stress field in the material. The mechanical attachment of the thermal intercept could increase the localized stress on the specimen since it confines diametral strain when the tube is compressed. This is also the location along the length of the specimen where the specimen changes from a constant cross-sectional area (see Figure 2).

The ultimate compressive strength of specimen 009 was 518 MPa (75 ksi). The temperature in the region of fracture was 140 K. The elastic modulus calculated from the load-displacement curve was 45 GPa (6.5 Msi). Total strain to failure was 0.0135. The specimen failed catastrophically

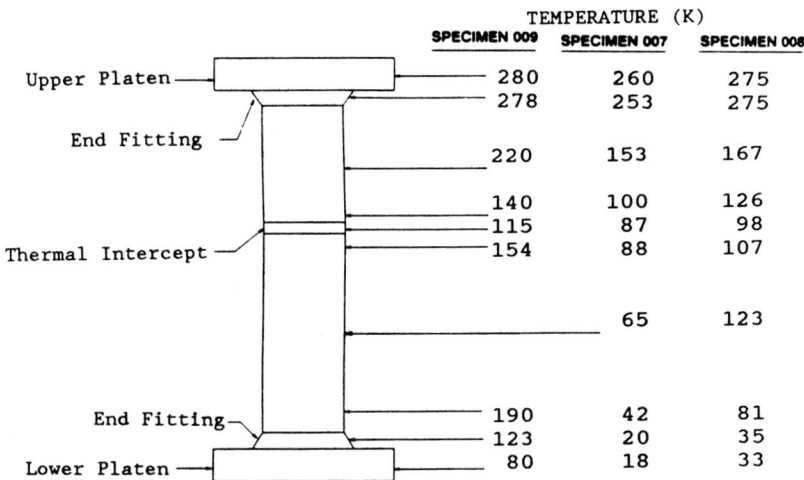

Figure 4. Thermal Gradient for Specimen Numbers 009, 007 and 008

Figure 5. Photograph of Failed Specimen Number 009

by a combination of failure modes: shell buckling, column buckling, and longitudinal cracking. Both end fittings exhibited the initial signs of failure described for the previous short specimen tests: circumferential cracking on the exterior surface.

The location of the failure is a potential problem that needs further design evaluation. It is also important that the specimen failed in a cold (140 K) region. Further down the specimen (Figure 4) the temperature was higher (190 K) and the stress was equivalent (518 MPa). The end fittings again exhibited cracking which possibly induced premature failure of the FRP tubes.

Fatigue Tests

Two extreme thermal gradient tests, using liquid helium cooling, are discussed in this section. The first (specimen 007) was a cyclic fatigue test that had multiple failure events before total catastrophic failure at 385 cycles. The second (specimen 008) was a single-cycle test performed with approximately the same environmental and mechanical conditions to help analyze the results of the first test. The two fatigue specimens were fitted with self-aligning end fittings. The difference in design between the two types of end fittings is shown in Figure 3.

The tube was to be cyclically loaded (22 000 cycles) at a frequency of 0.25 Hz between compressive stresses of 276 MPa (40 ksi) and 27.6 MPa (4 ksi). The 276 MPa (40 ksi) stress is based on 1.33 times the operating design stress of 207 MPa (30 ksi). The thermal gradient for both specimens just before testing are shown in Figure 4.

The first load cycle on specimen 007 was controlled manually with the machine in displacement control. The load-displacement curve, shown in Figure 6, was linear to 1.2 MN (282 kips), when a loud noise occurred.

The specimen load dropped to 1.0 MN (230 kips), and the load-displacement curve shifted an amount that indicated 3.5 mm (0.14 in) of platen-to-platen closing. The specimen was then unloaded and was observed to have the same linear elastic slope as before the noise. The specimen was then reloaded to the desired maximum fatigue load of 1.8 MN (400 kips) without incident. The specimen was unloaded and the test was continued with the computer controlled fatigue waveform. The fatigue cycle continued for 385 cycles until another loud noise occurred. The fatigue cycle was stopped and again the specimen was still able to sustain a load. The specimen was reloaded and the load-displacement curve had shifted again. This reloading cycle was curiously normal until another loud noise at 1.7 MN (380 kips) signaled total catastrophic failure of the specimen: no load could be sustained with further displacement of the actuator.

The specimen failed at the bottom (or the cold end). The failure mode appeared to be shell buckling with numerous longitudinal cracks emanating from the bottom end of the specimen.

Both of the end fittings on the specimen failed. The lower (cold) end fitting completely fractured into multiple pieces. This may be the reason for the initiation of longitudinal cracks in the specimen. The upper (warm) end fitting had radial cracks visible on the outside surface but had not completely fractured as in the lower end fitting.

Due to uncertainty of dominating failure mode on the test of specimen 007, we decided to perform a test on specimen 008 with as close to the same conditions as possible. The thermal gradient measured just before testing is shown schematically in Figure 4.

The specimen was loaded manually in displacement control. The load vs. displacement curve, shown in Figure 6B, was linear up to point A where a loud noise was heard. The load at this point was 1.16 MN (260 kips) and it dropped instantly to 0.93 MN (210 kips). This load drop, curve shift, and loud noise indicated some type of failure so we stopped the test to inspect the specimen.

The test specimen remained intact but the lower end fitting was completely fractured into multiple pieces. The failure of the lower end fitting was essentially identical to the failure of the lower end fitting on specimen 007. The specimen exhibited the start of numerous longitudinal cracks emanating from the bottom of the specimen.

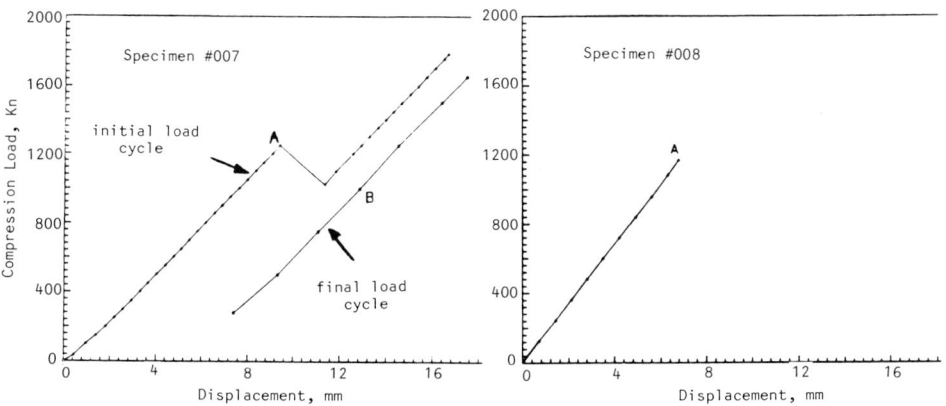

Figure 6. Load-Displacement Curves for Specimen Numbers 007 and 008

The load at which this end fitting failed was within 8% of the load at the first failure event in the previous test. This coincidence of load and the similarity of the two failed end fittings lead us to conclude that the first failure event in the previous test also occurred in the lower (cold) end fitting.

Summary of Results of Fatigue Tests

Examination of specimen 007 indicated that the specimen and both end fittings failed. The specimen failed at the cold end. The end fitting at the cold end had fractured into multiple pieces while the warm end fitting fractured into two pieces.

The single cycle compressive test of specimen 008 caused the cold end fitting to fail. The fracture of the end fitting into multiple pieces was essentially identical to the previous failure of the cold end fitting on specimen 007.

The most important conclusion from these tests is that the compressive self-aligning end fittings must be redesigned to develop the full strength of the composite tube. Previous tests performed using the rigid aluminum end fittings suggested a strength problem. The self-aligning end fittings contain more geometric stress concentrations than the rigid end fitting. This combined with the more brittle behavior of the 7075-T6 aluminum alloy at cold temperatures was detrimental to end fitting performance. Both tests failed on the first load cycle at almost identical stresses (186 ± 8.5 MPa). For specimen 007, it was not possible to verify the cold end fitting failure as the first failure event but that conclusion was supported by the test of specimen 008.

Since the failure of the cold end fittings prohibited further determination of the composite tube properties, the only other conclusion from the test on specimen 007 is that the fatigue life of the tube for the documented thermal and mechanical conditions is greater than 385 cycles for the loading conditions described above.

SUMMARY AND CONCLUSIONS

The results of all tests of the FRP support tubes are summarized in Table 1. For each test, the thermal conditions, ultimate load, and failure mode are indicated. We conclude the following:

(1) The failure stress on the full scale, ultimate compressive strength specimen (specimen 009) was nearly 20% lower than expected.

(2) The end fittings for the tubes present the greatest obstacle to loading the tubes to their ultimate strength.

(3) The stress concentration of the end fittings is a problem that is magnified with the present design of the self-aligning end fittings.

REFERENCES

1. NISTIR 3966, "A Cryogenic 4.4 MN Mechanical Test System for Full Scale Tests of Composite Support Struts," R.P. Walsh, R.P. Reed, J.D. McColskey, W. Fehringer, and J.R. Berger.

FRICTION AND WEAR OF A THREE-DIMENSIONAL FABRIC-REINFORCED PLASTIC AT ROOM TEMPERATURE AND LIQUID NITROGEN TEMPERATURE

S.Nishijima*, T.Okada, P.C.Michael[†] and Y. Iwasa[†]

ISIR Osaka University Ibaraki Osaka Japan
[†]FBNML MIT Cambridge MA USA

ABSTRACT

The friction and wear characteristics for copper slid against ZI-005, a newly-developed, three-dimensional-fabric-reinforced plastic, were examined at room temperature (300 K) and at liquid nitrogen temperature (77 K) to determine its suitability for use in cryogenically-cooled electromagnets. For comparison, the friction and wear characteristics for copper slid against G-10, a conventional two-dimensional-fabric-reinforced plastic, were also measured. The friction coefficient for the copper/G-10 pair increased markedly with sliding distance at 300K, achieving a steady-state value of 0.75; at 77K the friction coefficient is much lower, ~0.3, and relatively independent of distance. By contrast, the friction coefficient for copper/ZI-005 is equal to ~0.5, independent of both sliding distance and temperature. The magnitude of the nondimensional wear coefficients for copper/ZI-005 are equal to ~10^{-4}, independent of temperature, and are approximately 100 times greater than those for copper/G-10 (~10^{-6}). The magnitude of these wear coefficients, together with an examination of the composites' surface morphologies, suggest that the surface interactions for copper/ZI-005 are predominantly abrasive, whereas the interactions for copper/G-10 are generally adhesive. Because of its tendency towards abrasive friction, ZI-005 should provide a much stabler friction behavior when used as an insulator between highly-stressed copper windings.

INTRODUCTION

Three-dimensional-fabric-reinforced plastics (hereafter named 3DFRPs) have been developed for use as insulating and/or structural materials in superconducting magnets.[1,2] In these materials, the reinforcing fibers are oriented not only in the plane of the composite but also in the thickness (z) direction, hence their z-direction thermal-dimensional stabilities and mechanical rigidities are high. Several applications for 3DFRPs have been previously studied and include both conventional and superconducting magnets, including those for Tokamak devices.[3] Earlier investigations have also confirmed the 3DFRPs high interlaminar shear strengths and that their

---

*Visiting scientist to the Francis Bitter National Magnet Laboratory, September and October, 1989.

resistance to radiation damage is higher than conventional composites.[4,5] Fabric-reinforced composites are used in superconducting magnets not only as electrical insulators but also to form liquid helium cooling channels and to transmit mechanical stresses between the windings and the magnet's support structure. Consequently, the structural properties of these insulators at cryogenic temperatures have been widely studied. The interfacial behavior between the insulator and the superconducting materials/structural materials is also of considerable importance, but has not been studied as extensively. For instance, relative motion between the insulator and superconductor can reduce a magnet's structural rigidity,[6,7] and frictional heating produced during a conductor "slip" is often of sufficient intensity to initiate a quench.[8,9] Although 3DFRPs exist in many configurations, thin materials are usually used in superconducting magnets. In the 3DFRP examined in this study, the fibers used for z-directional reinforcement are oriented along the surface on the materials in the warp direction, hence its surface is rougher than conventional composites. Because of this difference in its surface geometry, the interfacial behavior between the 3DFRP and a superconducting wires is expected to be different than that of a more conventional composite. Because of the importance of the insulator's interfacial properties to successful magnet operation, this study was conducted to investigate the friction and wear properties of a simulated, copper-stabilized superconductor slid against 1) a conventional composite material, G-10 and 2) the 3DFRP ZI-005. The stabilized superconductor was simulated during the study by the use of commercial purity copper.

EXPERIMENTAL

Samples

The specifications of the G-10 and ZI-005 samples are shown in Table 1. The ZI-005 composite has T-glass (similar to S-glass) reinforcement in a BT-resin matrix while the G-10 uses E-glass reinforcement in an epoxy matrix.

Friction and Wear Tests

The rotational pin-on-disk test apparatus developed previously[10] was used for this study. A 59mm x 59mm square test sample was bolted to the bottom of a copper flange mounted at the end of the motor driven shaft. The three copper balls which served as the counterface during an experiment were mounted in a brass holder, equally spaced on a 51mm bolt circle diameter, and attached to the outer sliding friction assembly. Before the samples were installed in the apparatus, their surfaces were wiped clean with methyl alcohol. By locating the composite specimen between two copper surfaces, we simulated the insulator geometry typically

Table 1. Specifications of the test materials.

	G-10	ZI-005
Resin Material	Epoxy	Bismalemide triazin
Fiber Material	E glass	T glass
Resin, vol%	54	50
Glass, vol%	46	50
Warp direction	57	57
Fill direction	43	34
Thickness direction	0	9
Thickness, mm	0.89	1.1

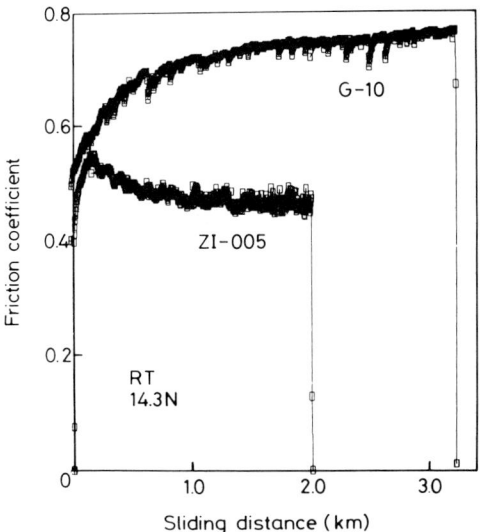

Fig. 1. Friction coefficient vs. sliding distance plots for G-10/copper and ZI-005/copper at 300K.

used in superconducting magnets. Two load levels, 7.6N and 14.3N, were used. With an assumed Hertzian stress distribution, these load levels correspond to interfacial stress levels of 150-190MPa at 300K and 300-390MPa at 77K; these stress levels are estimated to be similar to those found in large superconducting magnets. During the tests a 20mm/s sliding speed was to simulate micro-slip speed previously measured for wires in a superconducting winding model.[9]

The friction force output was monitored on an oscilloscope and recorded by a computerbased data acquisition system. Friction coefficients were calculated by dividing the friction forced by the normal force. At the conclusion of each experiment, the wear surfaces on the copper balls and the FRP were examined under an optical microscope. The wear of the copper specimens was also determined by measuring the diameter of the wear scars and calculating the wear volume.

RESULTS AND DISCUSSION

Figure 1 shows friction coefficient vs sliding distance graphs for copper slid against G-10 and ZI-005 at 300K while Figure 2 shows similar graphs at 77K. The data in these graphs were obtained at a 14.3N normal force. At 300K the friction coefficient against the G-10 increases with sliding distance before approaching a stable value of about 0.75. The ZI-005 shows small peak near the beginning of sliding before setting to value about 0.5. At 77K the friction coefficient against G-10 demonstrates a small peak during the early sliding stages and settles gradually to a value of around 0.3. The results against the ZI-005 at 300K and at 77K are similar to those at 300K; Hence, the newly developed ZI-005 show a different frictional behavior compared with that of G-10.

Figure 3 presents oscillograms of the friction force output vs time traces during tests on G-10 and ZI-005 samples, respectively. The traces were obtained after 20 hours of sliding (1440m) at 300K with a 7.6N normal force. The friction coefficients during these tests closely reproduced the sliding distance dependence presented in Figures 1 and 2,

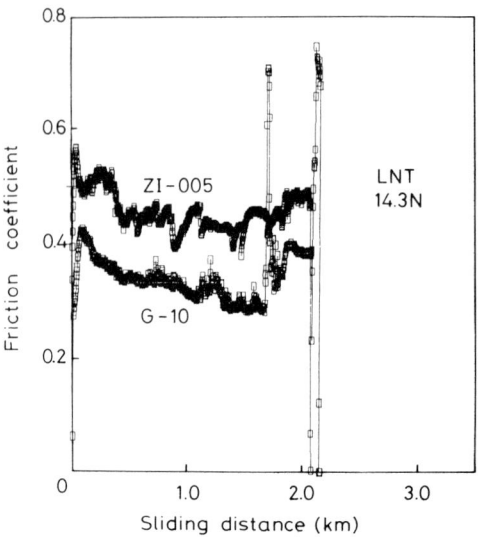

Fig. 2. Friction coefficient vs. sliding distance plots for G-10/copper and ZI-005/copper at 77K.

hence it was confirmed that the normal force has little effect on frictional behavior. In each of these traces, the regular variation of the friction force about its normal value is attributed to the mesh spacing of the composite's reinforcing fibers. For the G-10 composite both the normal value of the friction force and the amplitude of the friction force variation increase with sliding distance. On the other hand, both the average friction force and the friction force variation for the ZI-005 remain small and relatively constant.

Figure 4 presents micrographs of the copper wear surfaces produced after 20-hour sliding at 300K under a 7.6N normal force. Based on the wear scar diameters in these micrographs, the wear volume of the copper pins slid against G-10 (Fig. 4a) was approximately $0.0013mm^3$, while the wear volume against ZI-005 (Fig. 5b) was $0.15mm^3$. Nondimensional, Archard-type wear coefficients were then calculated to help identify the principal wear mechanisms for the copper specimens using the following equation:

V=kLS/3P

where V is the wear volume, L the normal force, S the total distance slid, P the penetration hardness of the worn material, and k the wear

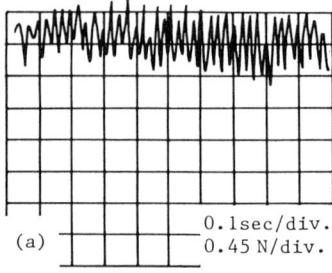
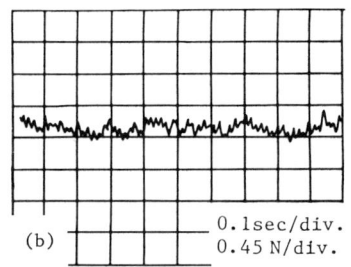

Fig. 3. Oscillograms of friction force vs. time trace at 300K after a 1400 m sliding distance for (a) copper/G-10 and (b) copper/ZI-005. Vertical : 0.45N/div. Holisontal : 0.1sec/div.

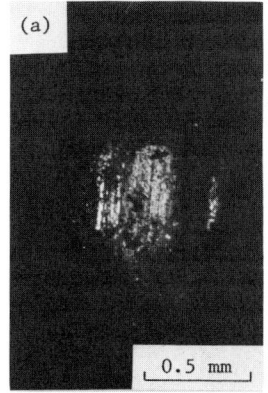 

Fig. 4. Microphotographs of copper surface after wear test for 20 hours at 300K under 7.6N load (a) against G-10 and (b) against ZI-005

coefficient. Table 2 presents the results of these calculations, based on an estimated copper hardness of 800MPa at 300K and 1000MPa at 77K.

The most notably thing to note about the results in Table 2 is that the wear coefficient values for copper against ZI-005 are about 100 times greater than those for copper against G-10. Secondly, since laboratory wear coefficients do not generally reproduce any better than to within a factor of two or three, it also appears that the copper wear mechanism against each composite is unaffected by changes in either the normal force or the test temperature. The wear coefficient value of $\sim 10^{-6}$ for the copper/G-10 pairs is generally indicative of adhesive wear for a metal slid against a non-metal, while the wear coefficient values of $\sim 10^{-4}$ for the copper/ZI-005 pairs is often indicative of mild abrasive wear.[11]

The identification of the dominant wear mechanism for the copper against each composite is consistent with their surface morphologies. As Figure 5 indicates, the z-direction reinforcing fibers in ZI-005 project rigidly above the composite's surface and hence they are in a favorable orientation for abrading the copper counterface. By contrast, the reinforcing fibers in the G-10 lie predominantly along the plane of the composite.

 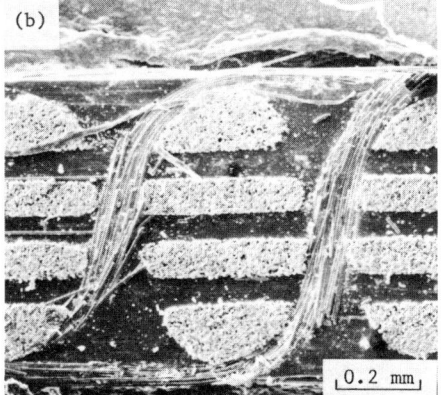

Fig. 5. Cross sectional photomicrographs of the test materials. (a)G-10 and (b)ZI-005.

Table 2. Copper wear coefficients obtained under several conditions.

Material	Temperature(K)	Load(N)	Wear Coefficient
G-10	300	14.3	9.1×10^{-7}
	300	7.6	8.0×10^{-7}
	77	14.3	1.5×10^{-6}
	77	7.6	2.1×10^{-6}
ZI-005	300	14.3	1.6×10^{-4}
	300	7.6	9.3×10^{-5}
	77	14.3	5.5×10^{-5}
	77	7.6	1.5×10^{-4}

Figure 6 presents optical micrographs of (a)G-10 and (b)ZI-005 wear surfaces obtained at 300K under a 7.6-N normal load. The appearance of several large, strongly-adherent copper particles in the G-10 wear track provides further evidence in support of the adhesive wear mechanism. Alternately, the considerably smaller size of the copper fragments found on the ZI-005 surface likewise helps to support the contention of abrasive wear against the 3DFRP.

Based on a review of the experimental evidence, the frictional behavior of each composite is interpreted as follows. As sliding commences between the copper/G-10 pairs at 300K the adsorbed surface films and oxides initially present on the copper surface are gradually removed, resulting in increased adhesion between the sliding materials and a corresponding increase in the friction coefficient.[12] Further increases in the friction coefficient also occur due to the copper-to-copper friction that results during slidings against the adherent copper particles. Although we do not have a clear explanation for the large reduction that occurs in the G-10's friction coefficient when cooled from 300K to 77K earlier investigations suggest two possibilities. A recent study attributes the much smaller 77-K friction coefficients for unfilled epoxy resins to the large reduction in their viscoelastic damping properties that occur upon cooling to below their glass transition temperatures,[13] while an earlier study reports that the friction coefficients for copper/copper pairs are much lower at 77K because the copper's smaller strain-harding rate reduces the tendency towards extensive junction growth.[14] For copper/ZI-005 pairs, the friction is primarily abrasive because of the z-fibers projecting above the 3DFRP surface, hence the friction coefficient is relatively independent of test temperature.

Fig. 6. Microphotographs of surface morphology of (a)G-10 and (b) ZI-005, each tested for 20 hours at 300K under 7.6N load.

CONCLUSIONS

The friction and wear characteristics of copper slid against 1) G-10, a conventional 2DFRP, and 2) ZI-005, a newly-developed 3DFRP, were drawn regarding the relative merits of using each composite as an insulating material for superconducting magnets.

The frictional interaction between copper and G-10 is predominantly adhesive, whereas the interaction against ZI-005 is largely abrasive. Consequently, the friction coefficients against ZI-005 are much more stable with respect to both sliding distance (or sliding time) and temperature. The temporal stability of ZI-005's friction coefficient should help to minimize the tendency towards stick-slip behavior, while its frictional temperature stability would enable a much more reliable estimate of the force transmission between windings, particularly in pulsed, liquid nitrogen cooled, normal magnets where thermal cycling is an important consideration. Thus, the newly developed 3DFRP, ZI-005, offers some advantages over conventional composites as an insulating material in superconducting magnets.

REFERENCES

1. S.Nishijima, Y.A.Wang, T.Okada, T.Uemura, T.Hirokawa, and J.Yasuda, Adv. Cryog. Eng, 34:59 (1988)
2. Y.A.Wang, S.Nishijima, T.Okada, T.Uemura, T.Hirokawa, and J.Yasuda, Proc. ICMC, Shenyang (1988) 765
3. T.J.McNanamy, J.E.Brasier and P.Snook to be published in IEEE 13th Symposium on Fusion Engineering, Knoxville (1989)
4. S.Nishijima, T.Okada, T.Hirokawa, J.Yasuda, and T.Uemura, Proc. ICMC, Shenyang (1988) 817
5. S.Nishijima, T.Nishiura, T.Okada, T.Hirokawa, J.Yasuda, and Y.Iwasaki, Adv. Cryog. Eng, 36:877 (1990)
6. Y.Hattori, Y.Yoshida, and H.Nakajima, Proc MT-9, Zurich (1984) 371
7. K.Yoshida, M.F.Nishi, Y.Takahashi, H.Tsuji, K.Koizumi, K.okuno, and T.Ando, IEEE Trans MAG 25:1488 (1989)
8. R.S.Kensley, H.Maeda and Y.Iwasa, Cryogenics 21:479 (1981)
9. H.Maeda, O.Tsukamoto, and Y.Iwasa, Cryogenics 22:287 (1982)
10. R.S.Kensley and Y.Iwasa, Cryogenics 20:25 (1980)
11. E.Rabinowicz, Tribology I, in "Center for Advanced Engineering Study", MIT (1974)
12. N.P.Suh and H.C.Sin Wear, 69:91 (1981)
13. P.C.Micheal et al. Cryogenics 30:775 (1990)
14. F.P.Bowden and T.H.C. Childs, Proc Roy Soc London (1969) 312A 451

MODIFICATION OF THE ASTM D 3039 TENSILE SPECIMEN
FOR CRYOGENIC APPLICATIONS

T.J. Eisenreich and D.S. Cox

General Dynamics Space Systems Division

ABSTRACT

The test procedure ASTM D 3039 was written to generate Resin Matrix Composite (RMC) tensile properties under ambient temperature conditions. However, when this procedure was applied to unidirectional graphite/epoxy tape laminates at cryogenic temperatures problems arose. General Dynamics Space Systems Division (GDSS) has developed an alternate specimen design to be used at cryogenic temperatures. This paper explains the methodology, analysis and results to proof the new specimen concept.

INTRODUCTION

Mechanical testing of resin matrix composite (RMC) materials is a subject that is receiving increased attention, as RMCs are becoming more prevalent in structural applications. As a result, existing test methods and procedures are receiving greater scrutiny for numerous reasons, not the least of which is the realization throughout industry that many of the common test methods employed are, for various reasons, inadequate. Much of the attention has focused on the wide data scatter and inconsistent failure modes commonly encountered in composites testing, and the need for standard test methods that can alleviate these inconsistencies.

For the most part, nearly all of the standard test methods that are currently accepted and used are designed for ambient (i.e., room temperature) testing. At GDSS, we are faced with the challenge of testing RMC materials over a wide temperature range, and often in non-accessible and/or hazardous environments. In particular, cryogenic testing in liquid nitrogen (77K) and in liquid hydrogen (20K) present the greatest challenges and generate the most concerns due to the deleterious affects of cryogenic exposure on composite material behavior and the hazards associated with liquid cryogens.

The most difficult mechanical test to run for composite materials has been the tension test. The difficulty of testing composite specimens in tension is due to the high strength, stiffness, and anisotrophy of the RMC materials being tested. Typically, RMC cryogenic tension tests have resulted in noticeable inconsistencies with regard to strength values, property scatter, and failure modes/locations. Much of this has been attributed to inadequate specimen design and non-uniformity of load transfer to the specimen during testing. To address the problems encountered with cryogenic tension testing of composite materials, investigations were initiated to identify and eliminate the source(s) of these problems.

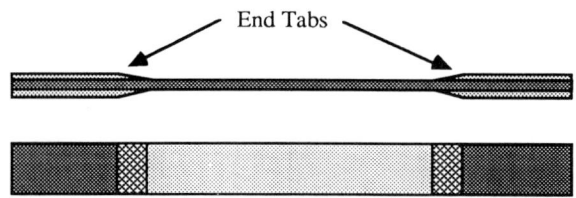

Figure 1 Tension Specimen Configuration per ASTM Standard D 3039 Utilized at GDSS.

SPECIMEN BACKGROUND

Traditionally, the most widely used and accepted tension test method in the USA has been the American Society for Testing and Materials (ASTM) D 3039 "Test Method for Tensile Properties of Fiber-Resin Composites". A diagram of the standard ASTM D 3039 specimen is shown in Figure 1.

The ASTM D 3039 test specimen features parallel edges with reinforcing tabs bonded to each end of the specimen. The use of reinforcing tabs on composite tension specimens serve the dual purpose of transferring load from the test fixture to the specimen and ensuring that load is transferred as uniformly as possible with minimal stress concentrations. The ASTM D 3039 test specimen is designed to be tested using either friction-type wedge grips or hydraulic/pneumatic actuated grips. Testing in cryogenic liquids precludes the use of the latter; thus, frictional wedge grips are employed for all composite tension testing performed at GDSS. There are inherent problems, however, with utilizing these frictional wedge grips at cryogenic temperatures. The primary problem is the interaction between the reinforcing tabs of the specimen and the faying surfaces of the serrated wedge grips. The mechanism of gripping the specimen relies on the ability of the grip surfaces to "bite" into the glass/epoxy tabs to grasp the specimen until failure occurs. Both frictional and compressive forces are acting on the tabs to achieve gripping of the specimen. The nature of the epoxy resin matrix of the tabs is such that at cryogenic temperatures, the resin becomes very hard, and the serrated wedge grip surfaces are unable to penetrate ("bite") into the tab surfaces. The result is that the specimen slips out of the test grips during testing, and the test must be repeated, often with the same results.

A secondary problem that also exists with testing composite specimens in cryogenic environments is the large coefficient of thermal expansion (CTE) mismatch between the composite specimen and the metal test fixture. For typical epoxy resins, the CTE is generally in the range of 30-50 $\mu in/in-°F$, while for metal alloys generally utilized for test fixturing, the CTE is approximately 7-8 $\mu in/in-°F$. Consequently, the composite test specimen contracts five to six times more in magnitude than the metal test fixture. The result of this differential CTE is that the test specimen contracts so much more than the metal tool that the contact between the grip surfaces and the specimen tabs is minimized and, again, specimen slipping occurs.

In developing RMC materials for cryogenic applications, it was recognized that the problems associated with cryogenic tensile testing required resolution to ensure data generated was accurate and consistent. Furthermore, the need to lower the cost of cryogenic testing by achieving better reliability and repeatability was also a strong driver in initiating these studies. Toward this end, efforts were begun to identify and evaluate the factors affecting cryogenic tension testing of RMC materials.

Six (6) specimen/gripping concepts were proposed during these studies. Each concept was evaluated and weighted against the following list of requirements :

1)   Concept must be able to handle any ply orientation (cross-ply and unidirectional).
2)   Concept must meet ASTM D 3039 recommended gauge length.
3)   Concept must work in liquid hydrogen.

4) Concept must avoid stress concentrations at specimen-tab interface which are partly due to the CTE mismatch.
5) Concept must allow for reasonable tolerances and misalignment.
6) Concept must compensate/prevent peeling of the reinforcing tabs.
7) Concept must allow for variable laminate thickness (.030" - .120").
8) Concept must protect the fibers from being damaged by the grips.

The results of this study (Reference 1) concluded a wedge-shaped tab bonded to a straight sided specimen solved the primary problem of specimen slippage from the test fixture. The reason for this, as shown in Figure 2, is due to the wedge design of the tabs, which mechanically wedge into the test grips when load is applied, thereby insuring no specimen slippage occurs during testing. In addition, the differential CTE problems encountered with the ASTM D 3039 specimen are not as pronounced with the wedge tab configuration, as more load can be applied to the specimen, resulting in a tighter grip, thereby offsetting the effect of the CTE mismatch.

FINAL CRYOGENIC SPECIMEN DEVELOPMENT

Although specimen slipping and CTE mismatch problems were essentially solved with the wedge tab geometry, several issues regarding the new specimen design required further attention prior to settling on a final specimen configuration. It was clear that by changing the geometry of the reinforcing tabs, we would be significantly altering the way in which load would be introduced into the test specimen. There were many unknown factors associated with making such drastic changes in test specimen geometry, particularly in light of the limited experimental data that had been compiled utilizing the new design. Specific concerns focused on minimizing stress concentrations at the tab/specimen interface and eliminating peel loads at the feather edge of the tab. In addition, the tapered geometry of the tabs indicated that much higher shear loads would be transferred into the specimen via the tab adhesive.

DESIGN APPROACH

Much of the data reported in the literature has concluded that tab design and geometry have a large influence on both the failure mode/location and ultimate strength values of composite tension specimens (References 2 and 3). One important factor of tab design is the taper angle that is used, as the magnitude of this angle has a large affect on the stress

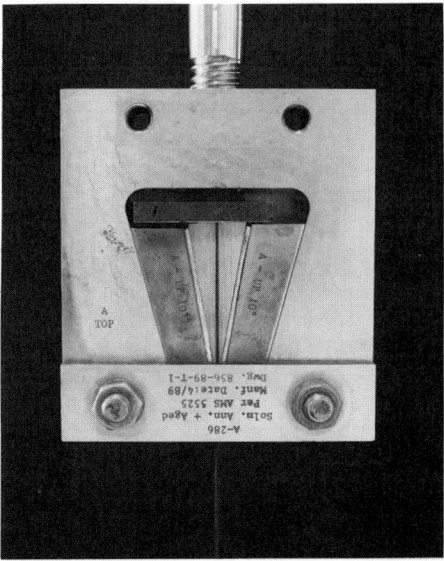

Figure 2 Wedge Tab Composite Specimen Design Insures Positive Gripping Between Test Fixture and Specimen.

peak at the tab/specimen interface. Studies indicate that as the taper angle decreases, the stress peak also decreases proportionately. In addition, factors such as tab material, tab layup (ply orientation), surface preparation for bonding (Reference 4), and adhesive used to bond tabs to panels play a large role in the load transfer characteristics between the tab and the composite specimen, particularly in cryogenic environments.

Because of the large number of factors influencing the geometric configuration for the wedge tab design, Taguchi methods were employed in determining final verification/analysis of the wedge tab specimen. The value of the Taguchi approach is that it provides the means to evaluate several experimental parameters simultaneously, thereby eliminating the need to perform more conventional full-factorial experiments. This results in appreciable cost savings, and the orthogonality of the Taguchi method compensates for the inherent interactive affects that occur between separate and distinct parameters. Thus, an L^8 orthogonal array was established, using the experimental parameters (factors) that were determined to have the greatest affect on specimen behavior.

Six control factors, each with two levels, were selected for the study. The logic behind the selection of each of the factors is indicated in Table 1.

Tab materials selected included 3M SP 1003 glass/epoxy prepreg, and Fiberite HyE 1034 (T300/934) carbon/epoxy prepreg. The wide difference in modulus between glass/epoxy (2-5Msi) and carbon/epoxy ($\approx$25Msi) was desired to determine the affect modulus has on load transfer and uniformity of stress distribution at cryogenic temperatures.

Previous empirical data and studies reported in the literature indicated that a taper angle of 10° was considered optimum for ASTM D 3039 tab configurations. Because of the change in tab geometry and load vectors with the new specimen design, indications were that an angle smaller than 10° would further alleviate stress concentrations at the tab/specimen interface. Thus, a 5° taper angle was included as a second-level control factor.

The thermal contraction mismatch between the low CTE carbon/epoxy unidirectional test specimen and the tab has been suspected as a key deficiency in past cryogenic tension specimen designs. To evaluate this affect, the ply orientations of the tab material candidates were altered from 100% 0° orientation to 50%-0°, 50%-90° [0/90]s orientation. (Tab orientation referenced with respect to the specimen orientation). It was expected that 0° carbon/epoxy tabs would show greater compatibility with the 0° carbon/epoxy test specimens than would the 0° glass/epoxy tabs because of a better CTE match. Effects of

Table 1. Justification for Selection of Experimental Control Factors.

CONTROL FACTOR	REASON SELECTED
1. Tab material	Modulus of tab material influences stress distribution into test specimen
2. Tab Angle	Tab angle has large affect on magnitude of stress/peak at tab/specimen interface
3. Tab Ply Orientation	Ply orientation govern the CTE of the tabs, particularly important during cryogenic exposure
4. Tab Length	Length of tabs determines amount of bond surface area for load transfer through the adhesive between tab and specimen.
5. Adhesive	Selection of proper adhesive critical in ensuring load is transferred from the tab to specimen. Shear strength, peel strength, strain and modulus are all important adhesive properties at cryogenic temperatures.
6. Surface Preparation	Surface preparation of adherends has significant affect on bond joint integrity

Table 2. Configuration of Test Panels for Tension Specimen Development

Panel	Tab Material	Tab Length	Tab Angle	Tab Layup	Adhesive	Surface Preparation
-1	Gl/Ep	2.00 in	10°	0°	EA 9394	Peel Ply
-2	Gl/Ep	3.00 in	10°	0°/90°	Metlbond 1133	Hand Abrade
-3	Gl/Ep	2.00 in	5°	0°	Metlbond 1133	Hand Abrade
-4	Gl/Ep	3.00 in	5°	0°/90°	EA 9394	Peel Ply
-5	Gr/Ep	2.00 in	10°	0°	EA 9394	Hand Abrade
-6	Gr/Ep	3.00 in	10°	0°/90°	Metlbond 1133	Peel Ply
-7	Gr/Ep	2.00 in	5°	0°	Metlbond 1133	Peel Ply
-8	Gr/Ep	3.00 in	5°	0°/90°	EA 9394	Hand Abrade

Gl/Ep = Glass/Epoxy
Gr/Ep = Graphite/Epoxy

the 0/90° oriented tabs were unknown, thus this was included as a second level experimental control factor.

By increasing the tab length 50% from 2.0 inches to 3.0 inches, the faying surface area between the specimen and the tab was also increased by 50%. The implication of this was the expectation that the applied shear load from the tab, through the adhesive, to the specimen could increase by 50% for a given adhesive. Thus, it was anticipated that the longer tab would provide greater load-carrying capacity.

The adhesives utilized in the study were selected for two primary reasons. First, based on comparative data of many adhesive candidates, Hysols EA 9394 and BASF Metlbond 1133 exhibited the best cryogenic properties (shear strength and peel strength). Second, we wanted to evaluate the affects that a paste adhesive (EA 9394) versus a film adhesive (Metlbond 1133) would have on load transfer and resistance to tab failure (tab peeling from the specimen) during cryogenic testing. Of secondary interest was the affect controlling bondline thickness would also have on load transfer/tab survival. Supported film adhesives inherently have bondline thickness control, while pastes, unless filled with microspheres, will generally result in arbitrary bondline thicknesses. For our experiments, the EA 9394 adhesive was applied without microspheres.

Intimately related to adhesive selection/behavior is the surface preparation techniques utilized on the adherends. In the case of the tab materials, all faying surfaces were prepared using a Scotchbrite™ pad, followed by solvent wiping (hand abrading). For the carbon/epoxy test specimen surfaces, both peel ply and hand abrading techniques were utilized prior to bonding the tabs. Although it was expected that the peel ply surfaces would be cleaner, the surfaces prepared using the Scotchbrite™ pad were expected to contain more mechanical bonding sites as a result of the abrading, thereby increasing the propensity for cohesive (as opposed to adhesive) failures.

A commonly used carbon/epoxy prepreg material - T300/934 - was selected as the test panel material because of the large database of properties that existed for this material. Twenty-ply unidirectional test panels were chose because of previous difficulties encountered in testing thick panels at cryogenic temperatures. Also, the unidirectional orientation, although necessary for design evaluations, is the most difficult orientation to test because of the high strength levels required for failure and also because of the random failure modes/locations typically encountered due to inherent material anisotrophy. Thus, the test panels represented a "worst-case" scenario for cryogenic testing of RMC materials.

EXPERIMENTAL PROCEDURE

Eight 20 ply unidirectional panels were laid up, cured, and prepared per the Taguchi experimental parameters, as shown in Table 2. Prior to preparing the different test panel configurations, it was necessary to design and fabricate ancillary tooling to support both the preparation and testing of the specimens. Because of the wide disparity in test specimen geometry/dimensions, a new test fixture was designed and built to accommodate all of the specimen configurations.

In preparing the wedge tabs, a milling procedure was employed to produce the taper angles. To facilitate this operation, a vacuum chuck was designed and built to hold the tab panels during fabrication. Once the tab machining operation was completed, the faying surfaces of the specimen panels and tabs were prepared per the specified procedure. The tabs were then assembled with the specimen panels. Upon competing the assembly, the bonding jig was placed in to a heated platen press for cure. After curing, the specimen panels were then removed from the bonding fixture and cut into 1.00 inch wide specimens for testing.

Tension testing of the specimens was performed in a cryostat at -320°F. Test specimens were loaded into the test fixture as shown in Figure 3. Five replicates were tested for each specimen configuration, bringing the total number of specimens tested to 40.

RESULTS OF SPECIMEN DEVELOPMENT TESTS

Test results were compiled and compared graphically in Figure 4. Note from Figure 4 that from strictly a statistical comparison, panel number 1 or 7 appears to produce the highest mean strength, lowest standard deviation, and highest percentage of valid failures than any of the other configuration tested. (Note : a valid failure was defined as any failure 1 specimen width away from the tabs in the test section). However, utilizing Taguchi methodology required a much more detailed analysis in selecting the optimum specimen configuration.

The results of the Taguchi analysis are shown in Table 3. Note only three values were used from each panels evaluated (highest and lowest were eliminated). The response tables for the mean and signal-to-noise values were used to analyze which of the control

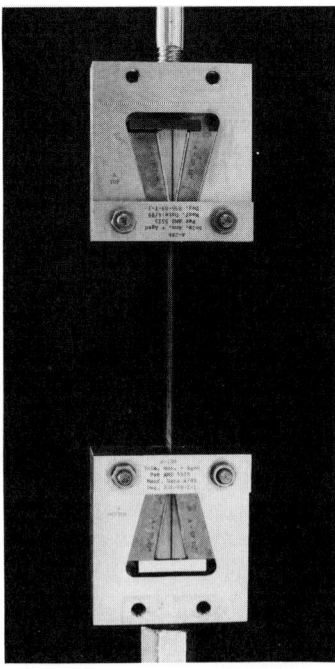

Figure 3   Wedge Tab Specimen Loaded into the Test Fixture, Ready for Tension Testing.

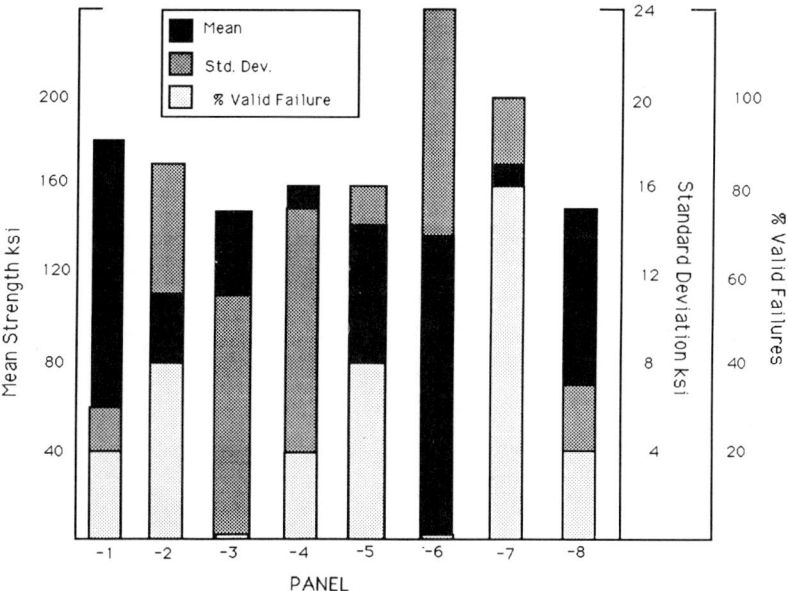

Figure 4. Graphical Results of Tension Testing.

factors used in the panel configuration has the greatest effect on tensile strength values. Note that in the mean response table, the tab angle, tab layup, and the surface preparation techniques produced the largest percent-differences between the two levels evaluated.

These results are consistent with those presented in the signal-to-noise ratio response table. For the other control factors evaluated, there was not enough of a difference between levels 1 and 2 to draw any strong conclusions. Thus, the selection of these factors were based on cost and/or ease of fabrication. Consequently, utilizing the Taguchi analyses, the following optimum selections were made:

Tab Material:	SP1003 glass/epoxy (based on cost)
Tab Angle:	5° taper
Tab Layup:	0° (100%) orientation

Table 3 Resulting Analysis of Tension Test Data Utilizing Taguchi Methodology.

RESPONSE TABLE FOR MEAN

Mean Strength	Tab Matl	Tab Angle	Interaction	Tab Layup	Tab Length	Adhesive	Surface Prep
Level 1	148015	141107	149990	157493	150724	154757	159117
Level 2	148808	155716	146833	139329	146099	142065	137705
Difference	0.53 %	9.83 %	2.13 %	12.22 %	3.12 %	8.54 %	14.39 %

RESPONSE TABLE FOR SIGNAL/NOISE

Mean Strength	Tab Matl	Tab Angle	Interaction	Tab Layup	Tab Length	Adhesive	Surface Prep
Level 1	103.21	102.69	103.82	103.41	103.71	103.71	103.84
Level 2	103.23	103.75	103.16	102.62	103.03	102.73	102.60
Difference	0.03	1.05	0.12	1.19	0.37	0.98	1.24

Uopt = 184,162   S/N = 105.45
Uexist = 150,993   S/N exist = 103.38

Tab Length: 2.00 inches
Adhesive: EA9394
Surface Preparation: Peel Ply

CONCLUSION

The wedge tab geometry has eliminated many of the problems associated with cryogenic tensile testing of resin matrix composite laminates. Vaild failures, one specimen width away from the tabs, are commonly being achieved. Material properties (strength, modulus, and failure strain) are being generated with less scatter using the wedge tab geometry, than was being generated using the ASTM D 3039 specimen configuration. The wedge tab specimen has been successfully used to generate material properties in liquid hydrogen and in liquid nitrogen on cross-ply and unidirectional high strength and intermediate modulus fiber/resin laminates.

ACKNOWLEDGEMENTS

The authors wish to thank M.J. Yokota and E.M. Gilchrist for their technical assistance, E.L. Christian for tooling design, J.A. Haile and N.P. Dyson for tooling fabrication, R.E. Headding for specimen fabrication and tabbing and J.L. Lloyd, J.M. Elion, G.L. Hill and T. Burke for conducting the tests.

REFERENCES

Ref. 1 Elion, J. M., et. al. Low Cost Resin Matrix Composites Materials and Processes, GDSS -ERR-088-602, Dec. 1988.

Ref. 2 Cunningham, M. E., Schoultz, S. V., and Toth, J. M., Jr., "Effect of End-Tab Design Tension Specimen Stress Concentration," Recent Advance in Composites in the United States and Japan. ASTM STP 864, J. R. Vinson and M. Taya, Eds., American Society for Testing and Materials, Philadelphia, 1985, pp., 253-262.

Ref. 3 Tarnopol' skii, Y. M., Static Test Methods for Composites, Van Nostrand Reinhold, Co., New York, 1985, pp. 32-74./

Ref. 4 Pocius, A., and Wenz, R. P., "Mechanical Surface Preparation of Graphite-Epoxy Composite for Adhesive Bonding," SAMPE Journal, September/October 1985, pp., 50-58.

THERMAL INSULATING SUPPORT SYSTEMS FOR RADIATION ENVIRONMENTS

Y. Ohtani, S. Nishijima, and T. Okada

ISIR Osaka University
Ibaraki, Osaka, Japan

K. Asano

Arisawa Manufacturing Company, Ltd.
Joetsu, Niigata, Japan

ABSTRACT

Thermal insulation for use in support systems for superconducting fusion magnets in radiation environments has been investigated. To develop a high-performance support system, we determined the cryogenic characteristics of advanced composites materials, such as those reinforced with alumina or carbon fibers, and designed model systems based on their characteristics. Thermal performance was influenced mainly by fiber selection. Mechanical performance, strength as well as the rigidity, were evaluated in terms of support-system requirements. The effects of radiation on the support systems were also assessed on the basis of damage resulting from irradiation of the materials.

INTRODUCTION

The superconducting magnets for fusion reactors will operate in a radiation environment. The insulating and supporting materials for the superconducting magnets will also be exposed to radiation. Organic composite materials are used for insulation and support, not only for their insulating properties, but also for their machinability and mechanical properties, which are superior to those of inorganic materials. On the other hand, organic composites are sensitive to radiation; their degradation is expected to cause difficulties in magnet operation. This study focused on support systems constructed with organic materials, taking into consideration the operating conditions of fusion magnets. The support system was designed to meet the requirements for support materials, including radiation resistance. The effects of material degradation on the support system were clarified. The results suggest that orientation may improve the performance of materials for fusion magnets.

DESIGN OUTLINE

In the design of the support system, two conditions were established: (1) filament-winding (hereinafter called FW) pipe would be used, and (2) the thermal contraction should be minimized. In FW pipe, the characteristics

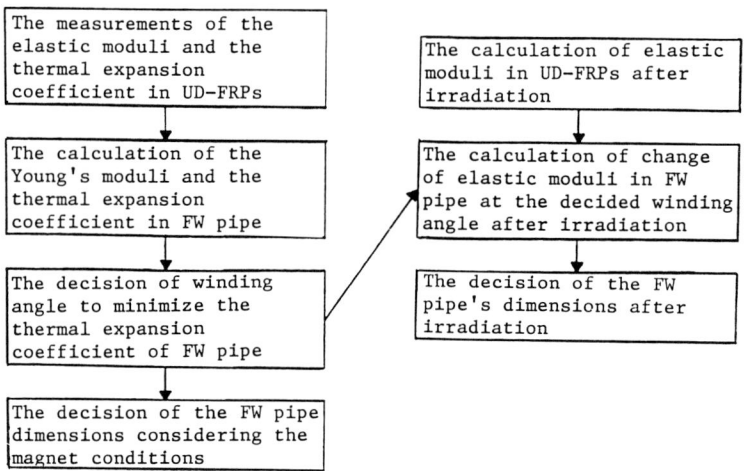

Fig. 1. Flow chart for the design of FW pipe.

varied with the winding angle, and hence, the winding angle is very significant in this design.

The flow chart of this design is presented in Fig. 1. The design procedure was (1) measure the elastic moduli and thermal expansion coefficients in unidirectional fiber-reinforced plastics (UD-FRPs), (2) calculate Young's moduli and thermal expansion coefficients of the FW pipe, (3) determine the winding angle that would minimize the thermal expansion coefficient, and (4) determine the dimensions of the FW pipe by considering the magnet conditions. In the same manner, the procedure for taking into account the effects of irradiation was (1) calculate the elastic moduli of UD-FRPs after irradiation, (2) calculate the change of elastic moduli in FW pipe at the winding angle selected after irradiation, and (3) determine the dimensions of the FW pipe after irradiation.

ELASTIC MODULI AND THERMAL EXPANSION COEFFICIENT

Unidirectional Fiber-Reinforced Plastics

The elastic moduli and thermal expansion coefficients of UD-FRPs were measured. Advanced composites, such as alumina-fiber-reinforced plastic (ALFRP), carbon-fiber-reinforced plastic (CFRP), and glass-fiber-reinforced plastic (GFRP), were selected for the support system.

Figures 2 and 3 present the off-axis-angle dependences of the elastic moduli and those of the thermal expansion coefficients, respectively. The points plotted in the figures were measured values. The off-axis-angle dependence of Young's modulus and the thermal expansion coefficient can be calculated from the material parameters and Equations (1) and (2):[1,2]

$$1/E = \cos^4 \vartheta / E_L + \sin^4 \vartheta / E_T + \sin^2 \vartheta \cos^2 \vartheta \, (1/G_{LT} - 2\nu_L/E_L) \quad (1)$$

$$\alpha = \alpha_L \cos^2 \vartheta + \alpha_T \sin^2 \vartheta \quad (2)$$

where E_L is Young's modulus in the fiber direction in UD-FRP; E_T, that in the perpendicular direction; G_{LT}, shear modulus; ν_L, Poisson's ratio, α_L, the thermal expansion coefficient in the fiber direction; α_T, that in the transverse direction; ϑ, the off-axis angle; E, Young's modulus in the ϑ direction; and α, the thermal expansion coefficient in the ϑ direction. The

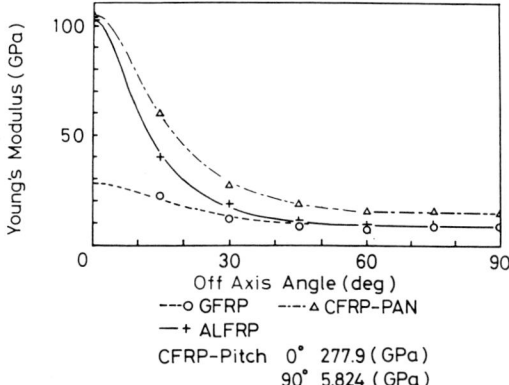

Fig. 2. Off-axis-angle dependence of Young's modulus in UD-FRPs.

curves in Figs. 2 and 3 show the results of the calculations. Good agreement between the calculations and measurements was confirmed. The results of the calculations were used in the following steps.

Filament-Winding Pipe

The Young's moduli and the thermal expansion coefficients of FW pipes (made from the materials tested, as discussed in the previous section) were calculated. The measured parameters of these materials used in the calculation were E_L, E_T, ν_L, G_{LT}, α_L, and α_T. The derivations of Young's modulus and thermal expansion coefficient follow:[1,3]

Derivation of Young's modulus:

$$1/E_x = 1/\underline{E}_x - \chi^2 \underline{G}_{xy}$$
$$1/E_y = 1/\underline{E}_x - \psi^2 \underline{G}_{xy}$$
$$\nu_x/E_x = \nu_y/E_y = (\underline{\nu}_x/\underline{E}_x) + \chi\psi\underline{G}_{xy}$$
$$1/G_{xy} = 1/\underline{G}_{xy} - \chi(\chi + \psi\underline{\nu}_y)\underline{E}_x/(1 - \underline{\nu}_x\underline{\nu}_y) - \psi(\psi + \chi\underline{\nu}_x)\underline{E}_y/(1 - \underline{\nu}_x\underline{\nu}_y)$$

$$1/\underline{E}_x = \ell^4/E_L + m^4/E_T + (1/G_{LT} - 2\nu_L/E_L)\ell^2 m^2$$
$$1/\underline{E}_y = m^4/E_L + \ell^4/E_T + (1/G_{LT} - 2\nu_L/E_L)\ell^2 m^2$$
$$1/\underline{G}_{xy} = 4[(1 + \nu_L)/E_L + (1 + \nu_T)/E_T]\ell^2 m^2 + (\ell^2 - m^2)^2/G_{LT}$$

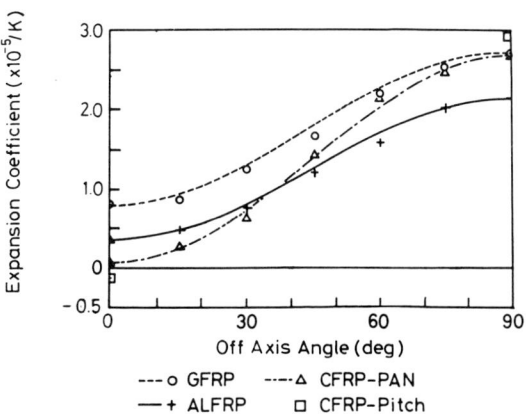

Fig. 3. Off-axis-angle dependence of thermal expansion coefficient in UD-FRPs.

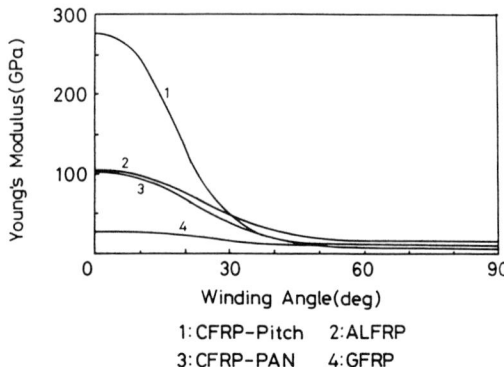

Fig. 4. Winding-angle dependence of axial Young's modulus in FW pipes.

$$\nu_x/E_x = \nu_y/E_y = \nu_L(\ell^4 + m^4)/E_L + (1/G_{LT} - 1/E_L - 1/E_T)\ell^2 m^2$$
$$\chi = 2[m^2/E_T - \ell^2/E_L + (1/G_{LT} - 2\nu_L/E_L)(\ell^2 - m^2)/2]\ell m$$
$$\psi = 2[\ell^2/E_T - m^2/E_L + (1/G_{LT} - 2\nu_L/E_L)(\ell^2 - m^2)/2]\ell m$$

Derivation of thermal expansion coefficient:

$$\alpha_{hx} = d_{11}\alpha_L + d_{12}\alpha_T$$
$$\alpha_{hy} = d_{21}\alpha_L + d_{22}\alpha_T$$

$$d_{11} = [\ell_2(\ell^2 - m^2)(1 - \nu_L\nu_T)E_L'E_T' + 4\ell^2 m^2 G_{LT}(E_L' + \nu_L E_T')]/\Delta$$
$$d_{12} = [-m_2(\ell^2 - m^2)(1 - \nu_L\nu_T)E_L'E_T' + 4\ell^2 m^2 G_{LT}(E_T' + \nu_T E_L')]/\Delta$$
$$d_{21} = [-m_2(\ell^2 - m^2)(1 - \nu_L\nu_T)E_L'E_T' + 4\ell^2 m^2 G_{LT}(E_L' + \mu_L E_T')]/\Delta$$
$$d_{22} = [\ell_2(\ell^2 - m^2)(1 - \nu_L\nu_T)E_L'E_T' + 4\ell^2 m^2 G_{LT}(E_T' + \nu_T E_L')]/\Delta$$
$$\Delta = (\ell^2 - m^2)^2(1 - \nu_L\nu_T)E_L'E_T' + 4\ell^2 m^2 G_{LT}(E_L' + E_T' + 2\nu_T E_L')$$

$$\ell = \cos \xi, \quad m = \sin \xi, \quad E_L' = E_L/(1 - \nu_L\nu_T), \quad E_T' = E_T/(1 - \nu_L\nu_T)$$

Subscripts x and y indicate characteristics in the longitudinal and circumferential directions; ξ is the winding angle; and α_h is the thermal expansion coefficient of FW pipe. Figures 4 and 5 present the calculated winding-angle dependence of Young's modulus and thermal expansion coefficient in the axial direction of the FW pipe.

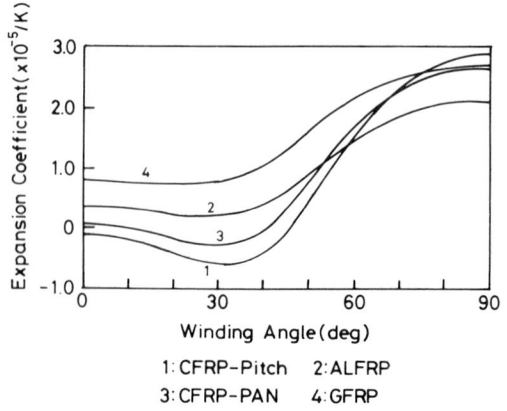

Fig. 5. Winding-angle dependence of the axial thermal coefficient in FW pipe.

Determination of the Winding Angle

The winding angle for the minimum thermal expansion coefficient in the axial direction was determined from Fig. 5. The GFRP had the largest thermal expansion coefficient. The advanced composites could be acceptable. From this point on, on the basis of the thermal properties, only ALFRP material was considered.

DETERMINATION OF THE PIPE'S DIMENSIONS

Although operating conditions as a support system were taken into consideration, the pipe was basically designed for room-temperature (RT) operation because mechanical properties at RT are usually inferior to those at cryogenic temperatures. The following factors were considered: (1) geometry, (2) buckling, (3) wall crippling, and (4) compressive strength.

Geometry. The cross section of the support system must be smaller than that of the superconducting magnet:[4]

$$t + r \leq 285,$$

where t is the thickness of the pipe, and r is the inner radius in mm.

Buckling. Euler buckling occurs in a long bar under a longitudinal compressive load, that is,

$$P \geq \pi^2 EI/\ell^2,$$

where P is compressive load; E, longitudinal Young's modulus; I, second moment of area; and ℓ, length of column.

Wall crippling. When pipe walls are thin, even pipes with large inner radii show mechanical instability. This phenomenon, called "wall crippling," occurs when the axial stress is larger than a certain value, which is given by

$$\sigma_{cr.s} = (E_x E_y/3)/[(1 - \nu_x \nu_y)^{\frac{1}{2}} \, t/r]$$

$$\sigma_{cr.v} = \{[(E_x E_y)^{\frac{1}{2}} + 2G_{xy}(1 - \nu_x \nu_y) + E_x \nu_y)]/[(E_x E_y)^{\frac{1}{2}} + E_x E_y/(2G_{xy} - E_x \nu_y)]^{\frac{1}{2}}\} \sigma_{cr.s},$$

where $\sigma_{cr.s}$ is the symmetrical crippling stress, and $\sigma_{cr.v}$ is the asymmetrical crippling stress. When the axial stress reaches the smaller value defined by the above equations, wall crippling takes place.

Compressive strength. Under compressive load, there are three compressive failure modes that vary with the winding angle of the FW pipe:[1] (1) compressive failure in the fiber direction, including fiber buckling, (2) tensile failure in the direction transverse to the fibers, and (3) shear failure along the fibers. The failure stresses of the three failure modes (F_L, F_T, and $F_{LT\phi}$) were calculated for the selected winding angle; they are illustrated in Fig. 6. The lowest strength of the three at a given winding

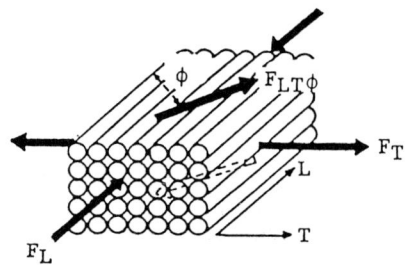

Fig. 6. Three failure modes of FW pipe under compressive load.

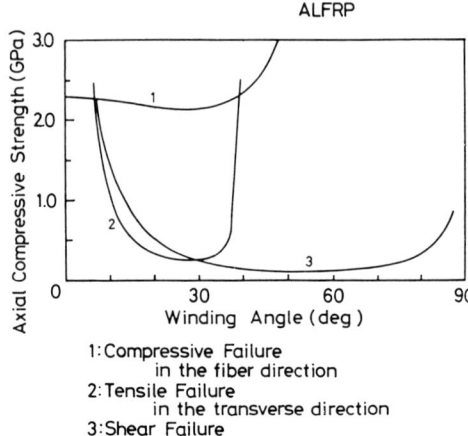

Fig. 7. Winding-angle dependence of axial compressive strength in FW pipe.

angle provides the compressive strength. The three compressive strengths are defined, with reference to Fig. 7, as follows:

$$\sigma_L = (\cos^2 \xi + \chi G_{xy} \sin 2\xi)\sigma_c$$
$$\sigma_T = -(\sin^2 \xi + \chi G_{xy} \sin 2\xi)\sigma_c$$
$$\tau_{LT\phi} = (\sin 2\xi/2 - \chi G_{xy} \cos 2\xi)\cos \phi \ \sigma_c$$

where σ_c is the stress to the pipe; σ_L, the stress in the fiber direction; σ_T, stress in the transverse direction; and $\tau_{LT\phi}$, shear stress. The lowest shear strength, $F_{LT\phi}$, varies with ξ and ϕ and is given (in GPa) by

$$F_{LT\phi} = (5.5 - \phi/90) \times 9.8 \times 10^{-3} - \sigma_T/4$$

From the above formulas, the compressive strength of the pipe was calculated. The winding angle had already been determined in the previous step. In the design, the pipe's dimensions were selected so that it could withstand the axial compressive load of about a hundred tons, that is,

$$\sigma_c \geq P/[\pi t(2r + t)]$$

The result is shown in Fig. 8. The hatched area shows the acceptable dimensions of the pipe.

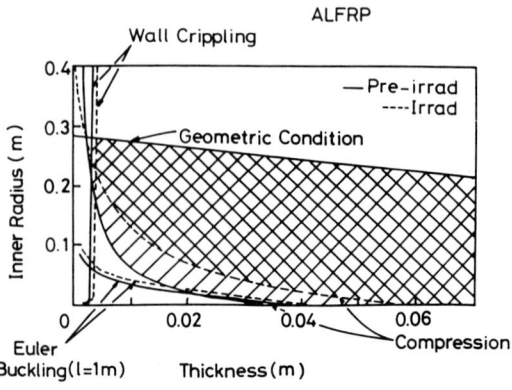

Fig. 8. Acceptable dimensions of FW pipe used for support systems.

DESIGN AFTER IRRADIATION

The radiation-induced changes in elastic moduli were thought to be caused by the degradation of Young's moduli in the matrices. The changes in the elastic moduli of the matrices was estimated from experimental data.[6] The law of mixtures was used to estimate the elastic moduli of FRP after irradiation. For compressive strength, experimental data were used.

The acceptable dimensions of the support system after 10-MGy irradiation are shown in Fig. 8. The acceptable area is narrower after irradiation than before irradiation. Especially affected was the compressive strength of the pipes: interlaminar failure and tensile failure occurred in the transverse direction in the FRP. It was concluded that the radiation damage of FRP for support systems must be carefully considered in order to achieve stable operation of superconducting fusion magnets.

CONCLUSIONS

A support system for fusion magnets was designed for minimum thermal contraction on the basis of measured properties of the material; radiation effects on the support system were also taken into account. The following conclusions were drawn:

1. For this design, the effects of irradiation that reduce the performance of this support system were identified, and thus the problems of UD-FRP material in fusion applications were specified: tensile strength in the direction transverse to the fibers and interlaminar shear strength.

2. The material performance of this design ought to be evaluated to confirm the significance of this design.

REFERENCES

1. M. Uemura, H. Iyama, and Y. Noguchi, Compressive fracture strength of helically wound composite cylinder, *J. Jap. Soc. Aeron. Space Sci.* 24:496 (1976).
2. T. Ishikawa and S. Kobayashi, Thermal expansion coefficients of unidirectional fiber-reinforced composites, Part 1, Analysis. *J. Jap. Soc. Aeron. Space Sci.* 25:394 (1977).
3. M. Uemura, H. Iyama, and Y. Yamaguchi, Thermal expansion coefficient and residual stresses in filament-wound CFRP materials, *J. Jap. Soc. Aeron. Space Sci.* 26:471 (1978).
4. Design group report for the new large helical device, from Design Proposal of ISIR Osaka University, Osaka, Japan (March 1989).
5. T. Hayashi, On the elastic instability of orthogonal anisotropic cylindrical shell, *J. Soc. Nav. Architects Jap.* 81:85 (1949).
6. D. Evans and J. T. Morgan, A review of the effects of ionizing radiation on plastic materials at low temperatures, in *Advances in Cryogenic Engineering—Materials*, vol. 28, Plenum, New York (1981), p. 147.

FATIGUE BEHAVIOR OF UD-CARBON-FIBRE COMPOSITES AT CRYOGENIC TEMPERATURES

G. Hartwig and K. Pannkoke

Kernforschungszentrum Karlsruhe, Inst. für Materialforschung II
7500 Karlsruhe 1, Federal Republic of Germany

INTRODUCTION

The present paper describes the static and dynamic properties of unidirectional (UD) fibre composites with various types of carbon-fibres and different matrices under tensile threshold loading at 77 K.
It is the objective to make a general statement as to how the characteristics of the individual components of fibre and matrix are affecting the static and dynamic properties of a UD-composite. It turns out that the static properties are controlled by the fibre properties while the fatigue limit is significantly influenced by the matrix.

MATERIALS AND SPECIMENS

UD-composites have been investigated with three types of carbon-fibres embedded in duroplastic or thermoplastic matrices.

Carbon-fibres:	T 300, AS 4 (high tensile fibres)	
	M 40 JB (high modulus fibre)	
Matrices:	epoxies (semiflexible E162/E113)	Shell
	(brittle V913)	Ciba Geigy
	thermoplastic matrix: PEEK	ICI

The quality of the composites has been tested by ultrasonic-scanning. Only materials of similar quality were used for cutting specimens. The gauge length of the specimens had a reduced thickness. The dimensions are sketched in Fig. 1. This type of specimen has been selected out of several types in a previous investigation /1/. This specimen yielded the lowest data scatter and a good reliability.

MEASUREMENTS

Tensile threshold fatigue cycling has been performed in a cryogenic testing machine at a frequency of 75 to 80 Hz /2/. No significant heating of the sample interior occurs at these rather high testing frequencies since the loss-factor is small and the LN_2 environment effects good cooling conditions /3/.

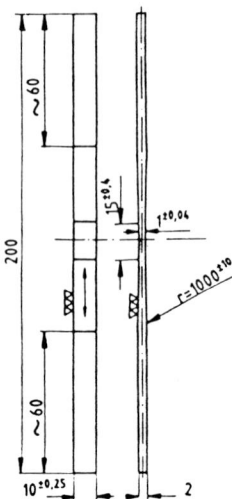

Figure 1.   Dimensions of UD-composite specimens (mm)

RESULTS AND DISCUSSION

Static Properties

As seen from Table 1, ultimate tensile stress and strain of UD-composites are determined by the carbon-fibre type and nearly independent of the matrix.

Fatigue properties

Very different, however, is the fatigue behavior shown in Figs. 2 to 5, where stress and strain life diagrams are plotted. As seen from Fig. 1, the degradation of strength is largest for composites with AS4-fibres which, however, have the highest static strength. The highest fatigue stress limit is exhibited by the composite with high modulus fibres (M 40 JB).

The curves of Fig. 2 can be converted to strain life diagrams by making use of the fact that the modulus are essentially unchanged during fatigue cycling /3/.

Both Figures 3 and 4 show an asymptotic behavior at large load cycles. The fatigue strain limit is nearly independent of the fibre type if the matrix is the same. This is true for a semiflexible epoxy matrix and a tough thermoplastic matrix (PEEK).

Table 1.   mechanical properties at 77 K

Fibre	Matrix	$\sigma_{UT\,\parallel}$ [MPa]	$\varepsilon_{UT\,\parallel}$ [%]	$E_\parallel$ [GPa]	ν_{21}	σ_D [MPa]	σ_D / σ_{UT}
AS 4	EP (semiflexible)	2522 ± 97	1,48 ± 0,05	167 ± 6	0,36	1100	0,44
AS 4	PEEK	2489 ± 97	1,56 ± 0,07	154 ± 8	0,41	840	0,34
T 300	EP (semiflexible)	1792 ± 37	1,14 ± 0,05	159 ± 7	0,35	1100	0,61
T 300	EP (brittle)	1664 ± 99	1,06 ± 0,08	141 ± 11	0,36	1400	0,84
T 300	PEEK	1641 ± 72	1,1 ± 0,05	151 ± 7	0,37	900	0,55
T 300	PC 3200	1698 ± 155	1,08 ± 0,04	149 ± 2	0,37	750	0,44
M 40JB	EP (semiflexible)	2104 ± 105	0,91 ± 0,03	229 ± 12	0,31	1550	0,74

$\sigma_{UT\,\parallel}$...ultimate tensile stress, $E_\parallel$...Young's modulus, σ_D...fatigue stress limit
$\varepsilon_{UT\,\parallel}$...ultimate tensile strain, ν_{21}...Poisson's ratio

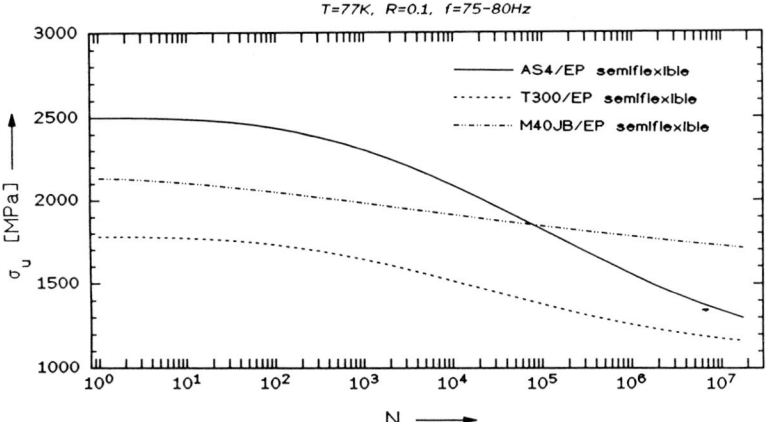

Figure 2. Stress-life diagrams of UD composites with different fibres but the same matrix (semiflexible epoxy) versus load cycles N.

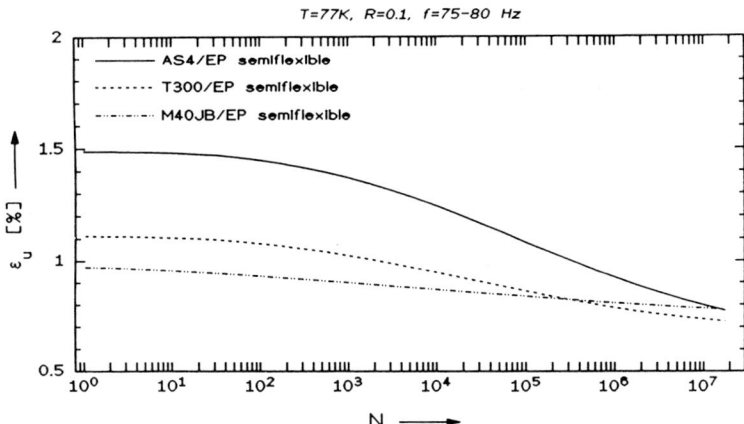

Figure 3. Strain-life-diagrams of UD-composites with different fibres but the same matrix (semiflexible epoxy).

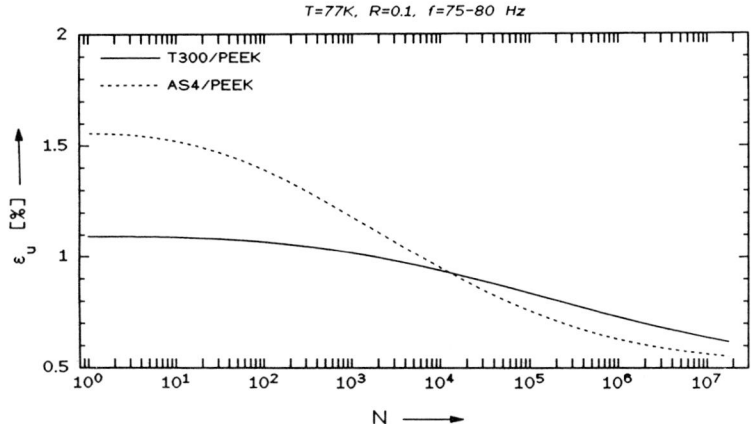

Figure 4. Strain-life-diagrams of UD-composites with different fibres but the same matrix (thermoplastic PEEK)

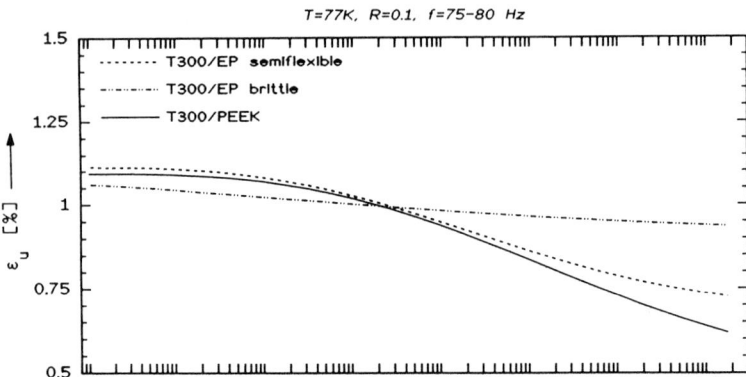

Figure 5. Strain-life-diagrams for T300 - composites with three different matrices.

A different situation has been found for the cases when one fibre type is combined with different matrices. This is shown in Fig. 5 where strain-life-diagrams are plotted for T300-composites with three different matrices.

The fatigue strain limit depends significantly on the matrix type. A similar situation holds for AS4 composites with different matrices as shown in Figure 6.

It is astonishing that the highest fatigue strain limit is achieved with a brittle epoxy matrix while composites with tougher matrices, such as PEEK or a semiflexible epoxy degrade much more. A similar situation has been found for UD-composites with a thermoplastic PC-matrix /3/.

The results of these investigations support the following conclusions for UD-carbon-fibre composites.
1) The static ultimate tensile stress and strain are determined by the fibre type and are nearly independent of the matrix.
2) The fatigue stress limit depends on both the fibre and the matrix.
3) The fatigue strain is nearly independent of the fibre but depends on the matrix. The strain amplitude at the fatigue limit seems to be the decisive parameter.
4) UD-composites with brittle epoxy matrices show higher fatigue limits than those with tough matrices, especially with thermoplastic matrices.

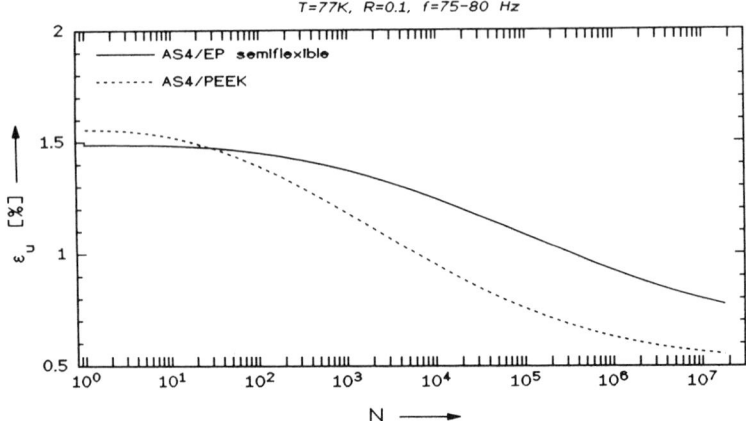

Figure 6. Strain-life-diagrams for AS 4 - composites with two different matrices.

An explanation of this unexpected behavior is difficult. Micrographs of cycled specimens may help. It has been found that composites with a brittle epoxy matrix show a much higher density of transverse cracks than those with a thermoplastic matrix, e.g. PEEK /4/. It might be the case that more strain energy is consumed at a high crack density. In addition, each crack tip of a multicrack system forms a lower stress concentration than those of a system with a poor crack density. A high stress concentration at transverse crack tips in the vicinity of a fibre surface might be a failure mode of the transversely weak carbon-fibre. No crack propagation occurs in the matrix at strain amplitudes below the fatigue strain limit.

References

(1) A. Mayer and K. Pannkoke, KFK report No. 4887 to be published.
(2) G. Hartwig, Encyclopedia of Polymer Science and Engineering, Vol 4. Wiley and Sons, New York, USA (1986)
(3) K. Ahlborn, Mechanical properties of CFRTP for cryogenic applications, PhD Thesis Technical University of Karlsruhe, FRG (1988).
(4) K. Pannkoke and H.J. Wagner, Fatigue properties of unidirectional composites at cryogenic temperatures, Cryogenics (1991) 31 248